Lecture Notes in Mathematics 1932

FONDAZIONE CIME
ROBERTO CONTI
CENTRO INTERNAZIONALE MATEMATICO ESTIVO
INTERNATIONAL MATHEMATICAL SUMMER CENTER

C.I.M.E. means Centro Internazionale Matematico Estivo, that is, International Mathematical Summer Center. Conceived in the early fifties, it was born in 1954 and made welcome by the world mathematical community where it remains in good health and spirit. Many mathematicians from all over the world have been involved in a way or another in C.I.M.E.'s activities during the past years.

So they already know what the C.I.M.E. is all about. For the benefit of future potential users and co-operators the main purposes and the functioning of the Centre may be summarized as follows: every year, during the summer, Sessions (three or four as a rule) on different themes from pure and applied mathematics are offered by application to mathematicians from all countries. Each session is generally based on three or four main courses (24–30 hours over a period of 6-8 working days) held from specialists of international renown, plus a certain number of seminars.

A C.I.M.E. Session, therefore, is neither a Symposium, nor just a School, but maybe a blend of both. The aim is that of bringing to the attention of younger researchers the origins, later developments, and perspectives of some branch of live mathematics.

The topics of the courses are generally of international resonance and the participation of the courses cover the expertise of different countries and continents. Such combination, gave an excellent opportunity to young participants to be acquainted with the most advance research in the topics of the courses and the possibility of an interchange with the world famous specialists. The full immersion atmosphere of the courses and the daily exchange among participants are a first building brick in the edifice of international collaboration in mathematical research.

C.I.M.E. Director
Pietro ZECCA
Dipartimento di Energetica "S. Stecco"
Università di Firenze
Via S. Marta, 3
50139 Florence
Italy
e-mail: zecca@unifi.it

C.I.M.E. Secretary
Elvira MASCOLO
Dipartimento di Matematica
Università di Firenze
viale G.B. Morgagni 67/A
50134 Florence
Italy
e-mail: mascolo@math.unifi.it

For more information see CIME's homepage: http://www.cime.unifi.it

CIME's activity is supported by:

- Istituto Nationale di Alta Mathematica "F. Severi"
- Ministero dell'Istruzione, dell'Università e delle Ricerca
- Ministero degli Affari Esteri, Direzione Generale per la Promozione e la Cooperazione, Ufficio V
- E.U. under the Training and Mobility of Researchers Programme UNESCO ROSTE
- This course was also supported by the research project PRIN 2004 "Control, Optimization and Stability of Nonlinear Systems: Geometric and Analytic Methods"

Andrei A. Agrachev · A. Stephen Morse
Eduardo D. Sontag · Héctor J. Sussmann
Vadim I. Utkin

Nonlinear and Optimal Control Theory

Lectures given at the
C.I.M.E. Summer School
held in Cetraro, Italy
June 19–29, 2004

Editors:
Paolo Nistri
Gianna Stefani

Springer

Fondazione
CIME
Roberto Conti

Andrei A. Agrachev
SISSA-ISAS
International School for Advanced Studies
via Beirut 4
34014 Trieste, Italy
agrachev@sissa.it

Eduardo D. Sontag
Héctor J. Sussmann
Department of Mathematics, Hill Center
Rutgers University
110 Frelinghuysen Rd
Piscataway, NJ 08854-8019, USA
sontag@math.rutgers.edu
sussmann@math.rutgers.edu

A. Stephen Morse
Department of Electrical Engineering
Yale University
PO Box 208267
New Haven CT 06520-8284, USA
morse@sysc.eng.yale.edu

Vadim I. Utkin
Department of Electrical Engineering
205 Dreese Laboratory
The Ohio State University
2015 Neil Avenue
Columbus, OH 43210, USA
utkin@ee.eng.ohio-state.edu

Paolo Nistri
Dipartimento di Ingegneria
dell'Informazione
Facoltà di Ingegneria
Università di Siena
via Roma 56
53100 Siena, Italia
pnistri@dii.unisi.it
http://www.dii.unisi.it/~pnistri/

Gianna Stefani
Dipartimento di Matematica Applicata
"G. Sansone"
Facoltà di Ingegneria
Università di Firenze
via di S. Marta 3
50139 Firenze, Italia
gianna.stefani@unifi.it
http://poincare.dma.unifi.it/~stefani/

ISBN 978-3-540-77644-4
DOI 10.1007/978-3-540-77653-6

ISBN 978-3-540-77653-6 (eBook)

Lecture Notes in Mathematics ISSN print edition: 0075-8434
ISSN electronic edition: 1617-9692

Library of Congress Control Number: 2007943246

Mathematics Subject Classification (2000): 93B50, 93B12, 93D25, 49J15, 49J24

Cover design: *design & production* GmbH, Heidelberg

Printed on acid-free paper

9 8 7 6 5 4 3 2 1

springer.com

Preface

Mathematical Control Theory is a branch of Mathematics having as one of its main aims the establishment of a sound mathematical foundation for the control techniques employed in several different fields of applications, including engineering, economy, biology and so forth. The systems arising from these applied Sciences are modeled using different types of mathematical formalism, primarily involving Ordinary Differential Equations, or Partial Differential Equations or Functional Differential Equations. These equations depend on one or more parameters that can be varied, and thus constitute the control aspect of the problem. The parameters are to be chosen so as to obtain a desired behavior for the system. From the many different problems arising in Control Theory, the C.I.M.E. school focused on some aspects of the control and optimization of nonlinear, not necessarily smooth, dynamical systems. Two points of view were presented: Geometric Control Theory and Nonlinear Control Theory. The C.I.M.E. session was arranged in five six-hours courses delivered by Professors A.A. Agrachev (SISSA-ISAS, Trieste and Steklov Mathematical Institute, Moscow), A.S. Morse (Yale University, USA), E.D. Sontag (Rutgers University, NJ, USA), H.J. Sussmann (Rutgers University, NJ, USA) and V.I. Utkin (Ohio State University Columbus, OH, USA).

We now briefly describe the presentations.

Agrachev's contribution began with the investigation of second order information in smooth optimal control problems as a means of explaining the variational and dynamical nature of powerful concepts and results such as Jacobi fields, Morse's index formula, Levi-Civita connection, Riemannian curvature. These are primarily known only within the framework of Riemannian Geometry. The theory presented is part of a beautiful project aimed at investigating the connections between Differential Geometry, Dynamical Systems and Optimal Control Theory.

The main objective of Morse's lectures was to give an overview of a variety of methods for synthesizing and analyzing logic-based switching control systems. The term "logic-based switching controller" is used to denote a controller whose subsystems include not only familiar dynamical components (integrators, summers, gains, etc.) but logic-driven elements as well. An important category of such control systems are those consisting of a process to be controlled, a family of fixed-gain or variable-gain candidate controllers, and an "event-drive switching logic" called a supervisor whose job is to determine in real time which controller should be applied to the process. Examples of supervisory control systems include re-configurable systems, and certain types of parameter-adaptive systems.

Sontag's contribution was devoted to the input to state stability (ISS) paradigm which provides a way of formulating questions of stability with respect to disturbances, as well as a method to conceptually unify detectability, input/output stability, minimum-phase behavior, and other systems properties. The lectures discussed the main theoretical results concerning ISS and related notions. The proofs of the results showed in particular connections to relaxations for differential inclusions, converse Lyapunov theorems, and nonsmooth analysis.

Sussmann's presentation involved the technical background material for a version of the Pontryagin Maximum Principle with state space constraints and very weak technical hypotheses. It was based primarily on an approach that used generalized differentials and packets of needle variations. In particular, a detailed account of two theories of generalized differentials, the "generalized differential quotients" (GDQs) and the "approximate generalized differential quotients" (AGDQs), was presented. Then the resulting version of the Maximum Principle was stated.

Finally, Utkin's contribution concerned the Sliding Mode Control concept that for many years has been recognized as one of the key approaches for the systematic design of robust controllers for complex nonlinear dynamic systems operating under uncertainty conditions. The design of feedback control in systems with sliding modes implies design of manifolds in the state space where control components undergo discontinuities, and control functions enforcing motions along the manifolds. The design methodology was illustrated by sliding mode control to achieve different objectives: eigenvalue placement, optimization, disturbance rejection, identification.

The C.I.M.E. course was attended by fifty five participants from several countries. Both graduate students and senior mathematicians intensively followed the lectures, seminars and discussions in a friendly and co-operative atmosphere.

As Editors of these Lectures Notes we would like to thank the persons and institutions that contributed to the success of the course. It is our pleasure to thank the Scientific Committee of C.I.M.E. for supporting our project: the Director, Prof. Pietro Zecca and the Secretary, Prof. Elvira Mascolo for their support during the organization. We would like also to thank Carla Dionisi for her valuable and efficient work in preparing the final manuscript for this volume.

Our special thanks go to the lecturers for their early preparation of the material to be distributed to the participants, for their excellent performance in teaching the courses and their stimulating scientific contributions.

We dedicate this volume to our teacher Prof. Roberto Conti, one of the pioneers of Mathematical Control Theory, who contributed in a decisive way to the development and to the international success of Fondazione C.I.M.E.

Siena and Firenze, May 2006

Paolo Nistri
Gianna Stefani

Contents

Geometry of Optimal Control Problems and Hamiltonian Systems

A.A. Agrachev

SISSA-ISAS, International School for Advanced Studies, via Beirut 4, 34014 Trieste, Italy
Steklov Institute of Mathematics, Moscow, Russia
agrachev@ma.sissa.it

Preface

These notes are based on the mini-course given in June 2004 in Cetraro, Italy, in the frame of a C.I.M.E. school. Of course, they contain much more material that I could present in the 6 h course. The idea was to explain a general variational and dynamical nature of nice and powerful concepts and results mainly known in the narrow framework of Riemannian Geometry. This concerns Jacobi fields, Morse's index formula, Levi-Civita connection, Riemannian curvature and related topics.

I tried to make the presentation as light as possible: gave more details in smooth regular situations and referred to the literature in more complicated cases. There is an evidence that the results described in the notes and treated in technical papers we refer to are just parts of a united beautiful subject to be discovered on the crossroads of Differential Geometry, Dynamical Systems, and Optimal Control Theory. I will be happy if the course and the notes encourage some young ambitious researchers to take part in the discovery and exploration of this subject.

Acknowledgments. I would like to express my gratitude to Professor Gamkrelidze for his permanent interest to this topic and many inspiring discussions and to thank participants of the school for their surprising and encouraging will to work in the relaxing atmosphere of the Mediterranean resort.

1 Lagrange Multipliers' Geometry

1.1 Smooth Optimal Control Problems

In these lectures we discuss some geometric constructions and results emerged from the investigation of smooth optimal control problems. We will consider problems with integral costs and fixed endpoints. A standard formulation of such a problem is as follows: Minimize a functional

$$J_{t_0}^{t_1}(u(\cdot)) = \int_{t_0}^{t_1} \varphi(q(t), u(t))\, dt, \tag{1}$$

where

$$\dot{q}(t) = f(q(t), u(t)), \quad u(t) \in U, \quad \forall t \in [t_0, t_1], \tag{2}$$

$q(t_0) = q_0$, $q(t_1) = q_1$. Here $q(t) \in \mathbb{R}^n$, $U \subset \mathbb{R}^k$, a *control function* $u(\cdot)$ is supposed to be measurable bounded while $q(\cdot)$ is Lipschitzian; scalar function φ and vector function f are smooth. A pair $(u(\cdot), q(\cdot))$ is called an *admissible pair* if it satisfies differential (2) but may violate the boundary conditions.

We usually assume that Optimal Control Theory generalizes classical Calculus of Variations. Unfortunately, even the most classical geometric variational problem, the length minimization on a Riemannian manifold, cannot be presented in the just described way. First of all, even simplest manifolds, like spheres, are not domains in $\mathbb{R}^n$. This does not look as a serious difficulty: we slightly generalize original formulation of the optimal control problem assuming that $q(t)$ belongs to a smooth manifold M instead of $\mathbb{R}^n$. Then $\dot{q}(t)$ is a tangent vector to M, i.e., $\dot{q}(t) \in T_{q(t)}M$ and we assume that $f(q, u) \in T_qM$, $\forall q, u$. Manifold M is called the *state space* of the optimal control problem.

Now we will try to give a natural formulation of the length minimization problem as an optimal control problem on a Riemannian manifold M. Riemannian structure on M is (by definition) a family of Euclidean scalar products $\langle \cdot, \cdot \rangle_q$ on T_qM, $q \in M$, smoothly depending on q. Let $f_1(q), \ldots, f_n(q)$ be an orthonormal basis of T_qM for the Euclidean structure $\langle \cdot, \cdot \rangle_q$ selected in such a way that $f_i(q)$ are smooth with respect to q. Then any Lipschitzian curve on M satisfies a differential equation of the form:

$$\dot{q} = \sum_{i=1}^{n} u_i(t) f_i(q), \tag{3}$$

where $u_i(\cdot)$ are measurable bounded scalar functions. In other words, any Lipschitzian curve on M is an admissible trajectory of the control system (3). The Riemannian length of the tangent vector $\sum\limits_{i=1}^{n} u_i f_i(q)$ is $\left(\sum\limits_{i=1}^{n} u_i^2\right)^{1/2}$. Hence the length of a trajectory of system (3) defined on the segment $[t_0, t_1]$ is $\ell(u(\cdot)) = \int_{t_0}^{t_1} \left(\sum\limits_{i=1}^{n} u_i^2(t)\right)^{1/2} dt$. Moreover, it is easy to derive from the Cauchy–Schwarz inequality that the length minimization is equivalent to the minimization of the functional $J_{t_0}^{t_1}(u(\cdot)) = \int_{t_0}^{t_1} \sum\limits_{i=1}^{n} u_i^2(t)\, dt$. The length minimization problem is thus reduced to a specific optimal control problem on the manifold of the form (1), (2).

Unfortunately, what I have just written was wrong. It would be correct if we could select a smooth orthonormal frame $f_i(q)$, $q \in M$, $i = 1, \dots, n$. Of course, we can always do it locally, in a coordinate neighborhood of M but, in general, we cannot do it globally. We cannot do it even on the two-dimensional sphere: you know very well that any continuous vector field on the two-dimensional sphere vanishes somewhere. We thus need another more flexible formulation of a smooth optimal control problem.

Recall that a *smooth locally trivial bundle* over M is a submersion $\pi : V \to M$, where all *fibers* $V_q = \pi^{-1}(q)$ are diffeomorphic to each other and, moreover, any $q \in M$ possesses a neighborhood O_q and a diffeomorphism $\Phi_q : O_q \times V_q \to \pi^{-1}(O_q)$ such that $\Phi_q(q', V_q) = V_{q'}$, $\forall q' \in O_q$. In a less formal language one can say that a smooth locally trivial bundle is a smooth family of diffeomorphic manifolds V_q (the fibers) parameterized by the points of the manifold M (the base). Typical example is the tangent bundle $TM = \bigcup\limits_{q \in M} T_qM$ with the canonical projection π sending T_qM into q.

Definition. A smooth control system with the state space M is a smooth mapping $f : V \to TM$, where V is a locally trivial bundle over M and $f(V_q) \subset T_qM$ for any fiber V_q, $q \in M$. An admissible pair is a bounded[1] measurable mapping $v(\cdot) : [t_0, t_1] \to V$ such that $t \mapsto \pi(v(t)) = q(t)$ is a Lipschitzian curve in M and $\dot{q}(t) = f(v(t))$ for almost all $t \in [t_0, t_1]$. Integral cost is a functional $J_{t_0}^{t_1}(v(\cdot)) = \int\limits_{t_0}^{t_1} \varphi(v(t))\, dt$, where φ is a smooth scalar function on V.

Remark. The above more narrow definition of an optimal control problem on M was related to the case of a *trivial bundle* $V = M \times U$, $V_q = \{q\} \times U$. For the length minimization problem we have $V = TM$, $f = \mathrm{Id}$, $\varphi(v) = \langle v, v \rangle_q$, $\forall v \in T_qM$, $q \in M$.

Of course, any general smooth control system on the manifold M is locally equivalent to a standard control system on $\mathbb{R}^n$. Indeed, any point $q \in M$ possesses a coordinate neighborhood O_q diffeomorphic to $\mathbb{R}^n$ and a mapping $\Phi_q : O_q \times V_q \to \pi^{-1}(O_q)$ trivializing the restriction of the bundle V to O_q; moreover, the fiber V_q can be embedded in $\mathbb{R}^k$ and thus serve as a set of control parameters U.

Yes, working locally we do not obtain new systems with respect to those in $\mathbb{R}^n$. Nevertheless, general intrinsic definition is very useful and instructive even for a purely local geometric analysis. Indeed, we do not need to fix specific coordinates on M and a trivialization of V when we study a control system defined in the intrinsic way. A change of coordinates in M is actually a smooth transformation of the state space while a change of the trivialization results in the feedback transformation of the control system. This means that an intrinsically defined control system represents actually the whole class of

[1] The term "bounded" means that the closure of the image of the mapping is compact.

systems that are equivalent with respect to smooth state and feedback transformations. All information on the system obtained in the intrinsic language is automatically invariant with respect to smooth state and feedback transformations. And this is what any geometric analysis intends to do: to study properties of the object under consideration preserved by the natural transformation group.

We denote by $L_\infty([t_0,t_1];V)$ the space of measurable bounded mappings from $[t_0,t_1]$ to V equipped with the L_∞-topology of the uniform convergence on a full measure subset of $[t_0,t_1]$. If V were an Euclidean space, then $L_\infty([t_0,t_1];V)$ would have a structure of a Banach space. Since V is only a smooth manifold, then $L_\infty([t_0,t_1];V)$ possesses a natural structure of a smooth Banach manifold modeled on the Banach space $L_\infty([t_0,t_1];\mathbb{R}^{\dim V})$.

Assume that $V \to M$ is a locally trivial bundle with the n-dimensional base and m-dimensional fibers; then V is an $(n+m)$-dimensional manifold.

Proposition 1.1. *Let $f : V \to TM$ be a smooth control system; then the space $\mathcal{V}$ of admissible pairs of this system is a smooth Banach submanifold of $L_\infty([t_0,t_1];V)$ modeled on $\mathbb{R}^n \times L_\infty([t_0,t_1];\mathbb{R}^m)$.*

Proof. Let $v(\cdot)$ be an admissible pair and $q(t) = \pi(v(t))$, $t \in [t_0,t_1]$. There exists a Lipschitzian with respect to t family of local trivializations $R_t : O_{q(t)} \times U \to \pi^{-1}(O_{q(t)})$, where U is diffeomorphic to the fibers V_q. The construction of such a family is a boring exercise which we omit.

Consider the system

$$\dot{q} = f \circ R_t(q,u), \quad u \in U. \tag{4}$$

Let $v(t) = R_t(q(t),u(t))$; then R_t, $t_0 \le t \le t_1$, induces a diffeomorphism of an L_∞-neighborhood of $(q(\cdot),u(\cdot))$ in the space of admissible pairs for (4) on a neighborhood of $v(\cdot)$ in $\mathcal{V}$. Now fix $\bar{t} \in [t_0,t_1]$. For any $\hat{q}$ close enough to $q(\bar{t})$ and any $u'(\cdot)$ sufficiently close to $u(\cdot)$ in the L_∞-topology there exists a unique Lipschitzian path $q'(\cdot)$ such that $\dot{q}'(t) = f \circ R_t(q'(t),u'(t))$, $t_0 \le t \le t_1$, $q'(\bar{t}) = \hat{q}$; moreover the mapping $(\hat{q},u'(\cdot)) \mapsto q'(\cdot)$ is smooth. In other words, the Cartesian product of a neighborhood of $q(\bar{t})$ in M and a neighborhood of $u(\cdot)$ in $L_\infty([t_0,t_1],U)$ serves as a coordinate chart for a neighborhood of $v(\cdot)$ in $\mathcal{V}$. This finishes the proof since M is an n-dimensional manifold and $L_\infty([t_0,t_1],U)$ is a Banach manifold modeled on $L_\infty([t_0,t_1],\mathbb{R}^m)$. □

An important role in our study will be played by the "evaluation mappings" $F_t : v(\cdot) \mapsto q(t) = \pi(v(t))$. It is easy to show that F_t is a smooth mapping from $\mathcal{V}$ to M. Moreover, it follows from the proof of Proposition 1.1 that F_t is a submersion. Indeed, $q(t) = F_t(v(\cdot))$ is, in fact a part of the coordinates of $v(\cdot)$ built in the proof (the remaining part of the coordinates is the control $u(\cdot)$).

1.2 Lagrange Multipliers

Smooth optimal control problem is a special case of the general smooth conditional minimum problem on a Banach manifold $\mathcal{W}$. The general problem consists of the minimization of a smooth functional $J : \mathcal{W} \to \mathbb{R}$ on the level sets $\Phi^{-1}(z)$ of a smooth mapping $\Phi : \mathcal{W} \to N$, where N is a finite-dimensional manifold. In the optimal control problem we have $\mathcal{W} = \mathcal{V}$, $N = M \times M$, $\Phi = (F_{t_0}, F_{t_1})$.

An efficient classical way to study the conditional minimum problem is the Lagrange multipliers rule. Let us give a coordinate free description of this rule. Consider the mapping

$$\bar{\Phi} = (J, \Phi) : \mathcal{W} \to \mathbb{R} \times N, \quad \bar{\Phi}(w) = (J(w), \Phi(w)), \ w \in \mathcal{W}.$$

It is easy to see that any point of the local conditional minimum or maximum (i.e., local minimum or maximum of J on a level set of Φ) is a critical point of $\bar{\Phi}$. I recall that w is a critical point of $\bar{\Phi}$ if the differential $D_w\bar{\Phi} : T_w\mathcal{W} \to T_{\bar{\Phi}(w)}(\mathbb{R} \times N)$ is *not* a surjective mapping. Indeed, if $D_w\bar{\Phi}$ were surjective then, according to the implicit function theorem, the image $\bar{\Phi}(O_w)$ of an arbitrary neighborhood O_w of w would contain a neighborhood of $\bar{\Phi}(w) = (J(w), \Phi(w))$; in particular, this image would contain an interval $((J(w) - \varepsilon, J(w) + \varepsilon), \Phi(w))$ that contradicts the local conditional minimality or maximality of $J(w)$.

The linear mapping $D_w\bar{\Phi}$ is not surjective if and only if there exists a nonzero linear form $\bar{\ell}$ on $T_{\bar{\Phi}(w)}(\mathbb{R} \times N)$ which annihilates the image of $D_w\bar{\Phi}$. In other words, $\bar{\ell}D_w\bar{\Phi} = 0$, where $\bar{\ell}D_w\bar{\Phi} : T_w\mathcal{W} \to \mathbb{R}$ is the composition of $D_w\bar{\Phi}$ and the linear form $\bar{\ell} : T_{\bar{\Phi}(w)}(\mathbb{R} \times N) \to \mathbb{R}$.

We have $T_{\bar{\Phi}(w)}(\mathbb{R} \times N) = \mathbb{R} \times T_{\Phi(w)}N$. Linear forms on $(\mathbb{R} \times N)$ constitute the adjoint space $(\mathbb{R} \times N)^* = \mathbb{R} \oplus T^*_{\Phi(w)}N$, where $T^*_{\Phi(w)}N$ is the adjoint space of $T_{\Phi(w)}M$ (the *cotangent space* to M at the point $\Phi(w)$). Hence $\bar{\ell} = \nu \oplus \ell$, where $\nu \in \mathbb{R}$, $\ell \in T^*_{\Phi(w)}N$ and

$$\bar{\ell}D_w\bar{\Phi} = (\nu \oplus \ell)\,(d_wJ, D_w\Phi) = \nu d_wJ + \ell D_w\Phi.$$

We obtain the equation

$$\nu d_wJ + \ell D_w\Phi = 0. \tag{5}$$

This is the Lagrange multipliers rule: if w is a local conditional extremum, then there exists a nontrivial pair (ν, ℓ) such that (5) is satisfied. The pair (ν, ℓ) is never unique: indeed, if α is a nonzero real number, then the pair $(\alpha\nu, \alpha\ell)$ is also nontrivial and satisfies (5). So the pair is actually defined up to scalar multiplier; it is natural to treat this pair as an element of the projective space $\mathbb{P}\left(\mathbb{R} \oplus T^*_{\Phi(w)}N\right)$ rather than an element of the linear space.

The pair (ν, ℓ) which satisfies (5) is called the *Lagrange multiplier* associated to the critical point w. The Lagrange multiplier is called *normal* if $\nu \neq 0$ and abnormal if $\nu = 0$. In these lectures we consider only normal Lagrange multipliers, they belong to a distinguished coordinate chart of the projective space $\mathbb{P}\left(\mathbb{R} \oplus T^*_{\Phi(w)}N\right)$.

Any normal Lagrange multiplier has a unique representative of the form $(-1, \ell)$; then (5) is reduced to the equation

$$\ell D_w \Phi = d_w J. \tag{6}$$

The vector $\ell \in T^*_{\Phi(w)} N$ from (6) is also called a normal Lagrange multiplier (along with $(-1, \ell)$).

1.3 Extremals

Now we apply the Lagrange multipliers rule to the optimal control problem. We have $\Phi = (F_{t_0}, F_{t_1}) : \mathcal{V} \to M \times M$. Let an admissible pair $v \in \mathcal{V}$ be a critical point of the mapping $(J_{t_0}^{t_1}, \Phi)$, the curve $q(t) = \pi(v(t))$, $t_0 \le t \le t_1$ be the corresponding trajectory, and $\ell \in T^*_{(q(t_0), q(t_1))}(M \times M)$ be a normal Lagrange multiplier associated to $v(\cdot)$. Then

$$\ell D_v (F_{t_0}, F_{t_1}) = d_v J_{t_0}^{t_1}. \tag{7}$$

We have $T^*_{(q(t_0), q(t_1))}(M \times M) = T^*_{q(t_0)} M \times T^*_{q(t_1)} M$, hence ℓ can be presented in the form $\ell = (-\lambda_{t_0}, \lambda_{t_1})$, where $\lambda_{t_i} \in T^*_{q(t_i)} M$, $i = 0, 1$. Equation (7) takes the form

$$\lambda_{t_1} D_v F_{t_1} - \lambda_{t_0} D_v F_{t_0} = d_v J_{t_0}^{t_1}. \tag{8}$$

Note that λ_{t_1} in (8) is uniquely defined by λ_{t_0} and v. Indeed, assume that $\lambda'_{t_1} D_v F_{t_1} - \lambda_{t_0} D_v F_{t_0} = d_v J_{t_0}^{t_1}$ for some $\lambda'_{t_1} \in T^*_{q(t_1)} M$. Then $(\lambda'_{t_1} - \lambda_{t_1}) D_v F_{t_1} = 0$. Recall that F_{t_1} is a submersion, hence $D_v F_{t_1}$ is a surjective linear map and $\lambda'_{t_1} - \lambda_{t_1} = 0$.

Proposition 1.2. *Equality (8) implies that for any $t \in [t_0, t_1]$ there exists a unique $\lambda_t \in T^*_{q(t)} M$ such that*

$$\lambda_t D_v F_t - \lambda_{t_0} D_v F_{t_0} = d_v J_{t_0}^{t} \tag{9}$$

and λ_t is Lipschitzian with respect to t.

Proof. The uniqueness of λ_t follows from the fact that F_t is a submersion as it was explained few lines above. Let us proof the existence. To do that we use the coordinatization of $\mathcal{V}$ introduced in the proof of Proposition 1.1, in particular, the family of local trivializations $R_t : O_{q(t)} \times U \to \pi^{-1}(O_{q(t)})$. Assume that $v(t) = R_t(q(t), u(t))$, $t_0 \le t \le t_1$, where $v(\cdot)$ is the referenced admissible pair from (8).

Given $\tau \in [t_0, t_1]$, $\hat{q} \in O_{q(\tau)}$ let $t \mapsto Q_\tau^t(\hat{q})$ be the solution of the differential equation $\dot{q} = R_t(q, u(t))$ which satisfies the condition $Q_\tau^\tau(\hat{q}) = \hat{q}$. In particular, $Q_\tau^t(q(\tau)) = q(t)$. Then Q_τ^t is a diffeomorphism of a neighborhood of $q(\tau)$ on a neighborhood of $q(t)$. We define a Banach submanifold $\mathcal{V}_\tau$ of the Banach manifold $\mathcal{V}$ in the following way:

$$\mathcal{V}_\tau = \{v' \in \mathcal{V} : \pi(v'(t)) = Q_\tau^t(\pi(v'(\tau))),\ \tau \le t \le t_1\}.$$

It is easy to see that $F_{t_1}\Big|_{\mathcal{V}_\tau} = Q_\tau^{t_1} \circ F_\tau\Big|_{\mathcal{V}_\tau}$ and $J_\tau^{t_1}\Big|_{\mathcal{V}_\tau} = a_\tau \circ F_\tau$, where $a_\tau(\hat{q}) = \int\limits_\tau^t \varphi(\Phi_t(Q_\tau^t(\hat{q}), u(t)))\, dt$. On the other hand, the set $\{v' \in \mathcal{V} : v'|_{[t_0,\tau]} \in \mathcal{V}_\tau\big|_{[t_0,\tau]}\}$ is a neighborhood of v in $\mathcal{V}$. The restriction of (8) to $\mathcal{V}_\tau$ gives:

$$\lambda_{t_1} D_v \left(Q_\tau^{t_1} \circ F_\tau\right) - \lambda_{t_0} D_v F_{t_0} = d_v J_{t_0}^\tau + d_v \left(a_\tau \circ F_\tau\right).$$

Now we apply the chain rule for the differentiation and obtain:

$$\lambda_\tau D_v F_\tau - \lambda_{t_0} D_v F_{t_0} = d_v J_{t_0}^\tau,$$

where $\lambda_\tau = \lambda_{t_1} D_{q(\tau)} Q_\tau^{t_1} - d_{q(\tau)} a_\tau$. $\square$

Definition. A Lipschitzian curve $t \mapsto \lambda_t$, $t_0 \leq t \leq t_1$, is called a *normal extremal* of the given optimal control problem if there exists an admissible pair $v \in \mathcal{V}$ such that equality (9) holds. The projection $q(t) = \pi(\lambda_t)$ of a normal extremal is called a (normal) *extremal path* or a (normal) *extremal trajectory*.

According to Proposition 1.2, normal Lagrange multipliers are just points of normal extremals. A good thing about normal extremals is that they satisfy a nice differential equation which links optimal control theory with a beautiful and powerful mathematics and, in many cases, allows to explicitly characterize all extremal paths.

1.4 Hamiltonian System

Here we derive equations which characterize normal extremals; we start from coordinate calculations. Given $\tau \in [t_0, t_1]$, fix a coordinate neighborhood $\mathcal{O}$ in M centered at $q(\tau)$, and focus on the piece of the extremal path $q(\cdot)$ which contains $q(\tau)$ and is completely contained in $\mathcal{O}$. Identity (9) can be rewritten in the form

$$\lambda_t D_v F_t - \lambda_\tau D_v F_\tau = d_v J_\tau^t, \tag{10}$$

where $q(t)$ belongs to the piece of $q(\cdot)$ under consideration. Fixing coordinates and a local trivialization of V we (locally) identify our optimal control problem with a problem (1), (2) in $\mathbb{R}^n$. We have $T^*\mathbb{R}^n \cong \mathbb{R}^n \times \mathbb{R}^n = \{(p,q) : p, q \in \mathbb{R}^n\}$, where $T_q^*\mathbb{R}^n = \mathbb{R}^n \times \{q\}$. Then $\lambda_t = \{p(t), q(t)\}$ and $\lambda_t D_v F_t \cdot = \langle p(t), D_v F_t \cdot \rangle = D_v \langle p(t), F_t \rangle$.

Admissible pairs of (2) are parameterized by $\hat{q} = F_\tau(v')$, $v' \in \mathcal{V}$, and control functions $u'(\cdot)$; the pairs have the form: $v' = (u'(\cdot), q'(\cdot; \hat{q}, u'(\cdot)))$, where $\frac{\partial}{\partial t} q'(t; \hat{q}, u'(\cdot)) = f\left(q'(t; \hat{q}, u'(\cdot)), u'(t)\right)$ for all available t and $q'(\tau; \hat{q}, u(\cdot)) = \hat{q}$. Then $F_t(v') = q'(t; \hat{q}, u'(\cdot))$.

Now we differentiate identity (10) with respect to t: $\frac{\partial}{\partial t}D_v\langle p(t),F_t\rangle = \frac{\partial}{\partial t}d_v J_\tau^t$ and change the order of the differentiation $D_v\frac{\partial}{\partial t}\langle p(t),F_t\rangle = d_v\frac{\partial}{\partial t}J_\tau^t$. We compute the derivatives with respect to t at $t=\tau$:

$$\frac{\partial}{\partial t}\langle p(t),F_t\rangle\big|_{t=\tau} = \langle \dot p(\tau),\hat q\rangle + \langle p(\tau), f(\hat q, u'(\tau))\rangle, \quad \frac{\partial}{\partial t}J_\tau^t\big|_{t=\tau} = \varphi(\hat q, u'(\tau)).$$

Now we have to differentiate with respect to $v'(\cdot) = (u'(\cdot), q'(\cdot))$. We however see that the quantities to differentiate depend only on the values of $u'(\cdot)$ and $q'(\cdot)$ at τ, i.e., on the finite-dimensional vector $(u'(\tau),\hat q)$. We derive:

$$\dot p(\tau) + \frac{\partial}{\partial q}\langle p(\tau), f(q(\tau),u(\tau))\rangle = \frac{\partial \varphi}{\partial q}(q(t),u(t)),$$

$$\frac{\partial}{\partial u}\langle p(\tau), f(q(\tau),u(\tau))\rangle = \frac{\partial \varphi}{\partial u}(q(\tau),u(\tau)),$$

where $v(\cdot) = (q(\cdot),u(\cdot))$.

Of course, we can change τ and perform the differentiation at any available moment t. Finally, we obtain that (10) is equivalent to the identities

$$\dot p(t) + \frac{\partial}{\partial q}\left(\langle p(t), f(q(t),u(t))\rangle - \varphi(q(t),u(t))\right) = 0,$$

$$\frac{\partial}{\partial u}\left(\langle p(t), f(q(t),u(t))\rangle - \varphi(q(t),u(t))\right) = 0,$$

which can be completed by the equation $\dot q = f(q(t),u(t))$. We introduce a function $h(p,q,u) = \langle p, f(q,u)\rangle - \varphi(q,u)$ which is called the *Hamiltonian* of the optimal control problem (1), (2). This function permits us to present the obtained relations in a nice Hamiltonian form:

$$\begin{cases} \dot p = -\dfrac{\partial h}{\partial q}(p,q,u) \\ \dot q = \dfrac{\partial h}{\partial p}(p,q,u) \end{cases}, \quad \frac{\partial h}{\partial u}(p,q,u) = 0. \tag{11}$$

A more important fact is that system (11) has an intrinsic coordinate free interpretation. Recall that in the triple (p,q,u) neither p nor u has an intrinsic meaning; the pair (p,q) represents $\lambda \in T^*M$ while the pair (q,u) represents $v \in V$. First we consider an intermediate case $V = M\times U$ (when u is separated from q but coordinates in M are not fixed) and then turn to the completely intrinsic setting.

If $V = M\times U$, then $f : M\times U \to TM$ and $f(q,u) \in T_qM$. The Hamiltonian of the optimal control problem is a function $h : T^*M \times U \to \mathbb{R}$ defined by the formula $h(\lambda,u) = \lambda(f(q,u)) - \varphi(q,u)$, $\forall \lambda \in T_q^*M,\ q\in M,\ u \in U$. For any $u \in U$ we obtain a function $h_u \stackrel{def}{=} h(\cdot,u)$ on T^*M. The cotangent bundle T^*M possesses a canonical symplectic structure which provides a standard way to associate a *Hamiltonian vector field* to any smooth function on T^*M. We will recall this procedure.

Let $\pi : T^*M \to M$ be the projection, $\pi(T_q^*M) = \{q\}$. The *Liouville* (or *tautological*) differential 1-form ς on T^*M is defined as follows. Let $\varsigma_\lambda : T_\lambda(T^*M) \to \mathbb{R}$ be the value of ς at $\lambda \in T^*M$, then $\varsigma_\lambda = \lambda \circ \pi_*$, the composition of $\pi_* : T_\lambda(T^*M) \to T_{\pi(\lambda)}M$ and the cotangent vector $\lambda : T_{\pi(\lambda)}M \to \mathbb{R}$. The coordinate presentation of the Liouville form is: $\varsigma_{(p,q)} = \langle p, dq\rangle = \sum\limits_{i=1}^{n} p^i dq^i$, where $p = (p^1, \dots, p^n)$, $q = (q^1, \dots, q^n)$. The *canonical symplectic structure* on T^*M is the differential 2-form $\sigma = d\varsigma$; its coordinate representation is: $\sigma = \sum\limits_{i=1}^{n} dp^i \wedge dq^i$. The Hamiltonian vector field associated to a smooth function $a : T^*M \to \mathbb{R}$ is a unique vector field $\mathbf{a}$ on T^*M which satisfies the equation $\sigma(\cdot, \mathbf{a}) = da$. The coordinate representation of this field is: $\mathbf{a} = \sum\limits_{i=1}^{n} \left(\frac{\partial a}{\partial p_i} \frac{\partial}{\partial q_i} - \frac{\partial a}{\partial q_i} \frac{\partial}{\partial p_i} \right)$. Equations (11) can be rewritten in the form:

$$\dot{\lambda} = \mathbf{h}_u(\lambda), \quad \frac{\partial h}{\partial u}(\lambda, u) = 0. \tag{12}$$

Now let V be an arbitrary locally trivial bundle over M. Consider the Cartesian product of two bundles:

$$T^*M \times_M V = \{(\lambda, v) : v \in V_q, \lambda \in T_q^*M, q \in M\}$$

that is a bundle over M whose fibers are Cartesian products of the correspondent fibers of V and T^*M. Hamiltonian of the optimal control problem takes the form $h(\lambda, v) = \lambda(f(v)) - \varphi(v)$; this is a well-defined smooth function on $T^*M \times_M U$. Let $\mathfrak{p} : T^*M \times_M V \to T^*M$ be the projection on the first factor, $\mathfrak{p} : (\lambda, v) \mapsto \lambda$. Equations (11) (or (12)) can be rewritten in the completely intrinsic form as follows: $(\mathfrak{p}^*\sigma)_v(\cdot, \dot{\lambda}) = dh$. One may check this fact in any coordinates; we leave this simple calculation to the reader.

Of course, by fixing a local trivialization of V, we turn the last relation back into a more convenient to study (12). A domain $\mathcal{D}$ in T^*M is called regular for the Hamiltonian h if for any $\lambda \in \mathcal{D}$ there exists a unique solution $u = \bar{u}(\lambda)$ of the equation $\frac{\partial h}{\partial u}(\lambda, u) = 0$, where $\bar{u}(\lambda)$ is smooth with respect to λ. In particular, if U is an affine space and the functions $u \mapsto h(\lambda, u)$ are strongly concave (convex) and possess minima (maxima) for $\lambda \in \mathcal{D}$, then $\mathcal{D}$ is regular and $\bar{u}(\lambda)$ is defined by the relation

$$h(\lambda, \bar{u}(\lambda)) = \max_{u \in U} h(\lambda, u) \quad \left(h(\lambda, \bar{u}(\lambda)) = \min_{u \in U} h(\lambda, u) \right).$$

In the regular domain, we set $H(\lambda) = h(\lambda, \bar{u}(\lambda))$, where $\frac{\partial h}{\partial u}(\lambda, \bar{u}(\lambda)) = 0$. It is easy to see that (12) are equivalent to one Hamiltonian system $\dot{\lambda} = \mathbf{H}(\lambda)$. Indeed, the equality $d_{(\lambda, \bar{u}(\lambda))}h = d_\lambda h_{\bar{u}(\lambda)} + \frac{\partial h_{\bar{u}(\lambda)}}{\partial u} du = d_\lambda h_{\bar{u}(\lambda)}$ immediately implies that $\mathbf{H}(\lambda) = \mathbf{h}_{\bar{u}(\lambda)}(\lambda)$.

1.5 Second Order Information

We come back to the general setting of Sect. 2 and try to go beyond the Lagrange multipliers rule. Take a pair (ℓ, w) which satisfies (6). We call such pairs (normal) *Lagrangian points.* Let $\Phi(w) = z$. If w is a regular point of Φ, then $\Phi^{-1}(z) \cap O_w$ is a smooth codimension $\dim N$ submanifold of $\mathcal{W}$, for some neighborhood O_w of w. In this case w is a critical point of $J\big|_{\Phi^{-1}(z)\cap O_w}$. We are going to compute the Hessian of $J\big|_{\Phi^{-1}(z)}$ at w without resolving the constraints $\Phi(w) = z$. The formula we obtain makes sense without the regularity assumptions as well.

Let $s \mapsto \gamma(s)$ be a smooth curve in $\Phi^{-1}(z)$ such that $\gamma(0) = w$. Differentiation of the identity $\Phi(\gamma(s)) = z$ gives:

$$D_w\Phi\dot{\gamma} = 0, \quad D_w^2\Phi(\dot{\gamma},\dot{\gamma}) + D_w\Phi\ddot{\gamma} = 0,$$

where $\dot{\gamma}$ and $\ddot{\gamma}$ are the first and the second derivatives of γ at $s = 0$. We also have:

$$\frac{d^2}{ds^2}J(\gamma(s))|_{s=0} = D_w^2 J(\dot{\gamma},\dot{\gamma}) + D_w J\ddot{\gamma} \overset{\text{eq.}(6)}{=},$$

$$D_w^2 J(\dot{\gamma},\dot{\gamma}) + \ell D_w\Phi\ddot{\gamma} = D_w^2 J(\dot{\gamma},\dot{\gamma}) - \ell D_w^2\Phi(\dot{\gamma},\dot{\gamma}).$$

Finally,

$$\operatorname{Hess}_w(J\big|_{\Phi^{-1}(z)}) = (D_w^2 J - \ell D_w^2\Phi)\big|_{\ker D_w\Phi}. \tag{13}$$

Proposition 1.3. *If quadratic form (13) is positive (negative) definite, then w is a strict local minimizer (maximizer) of $J\big|_{\Phi^{-1}(z)}$.*

If w is a regular point of Φ, then the proposition is obvious but one can check that it remains valid without the regularity assumption. On the other hand, without the regularity assumption, local minimality does not imply nonnegativity of form (13). What local minimality (maximality) certainly implies is nonnegativity (nonpositivity) of form (13) on a finite codimension subspace of $\ker D_w\Phi$ (see [7, Ch. 20] and references there).

Definition. A Lagrangian point (ℓ, w) is called *sharp* if quadratic form (13) is nonnegative or nonpositive on a finite codimension subspace of $\ker D_w\Phi$.

Only sharp Lagrangian points are counted in the conditional extremal problems under consideration. Let Q be a real quadratic form defined on a linear space E. Recall that the *negative inertia index* (or the *Morse index*) $\operatorname{ind} Q$ is the maximal possible dimension of a subspace in E such that the restriction of Q to the subspace is a negative form. The *positive inertia index* of Q is the Morse index of $-Q$. Each of these indices is a nonnegative integer or $+\infty$. A Lagrangian point (ℓ, w) is sharp if the negative or positive inertia index of form (13) is finite.

In the optimal control problems, $\mathcal{W}$ is a huge infinite dimensional manifold while N usually has a modest dimension. It is much simpler to characterize

Lagrange multipliers in T^*N (see the previous section) than to work directly with $J\big|_{\Phi^{-1}(z)}$. Fortunately, the information on the sign and, more generally, on the inertia indices of the infinite dimensional quadratic form (13) can also be extracted from the Lagrange multipliers or, more precisely, from the so called $\mathcal{L}$-*derivative* that can be treated as a dual to the form (13) object.

$\mathcal{L}$-derivative concerns the linearization of (6) at a given Lagrangian point. In order to linearize the equation we have to present its left- and right-hand sides as smooth mappings of some manifolds. No problem with the right-hand side: $w \mapsto d_w J$ is a smooth mapping from $\mathcal{W}$ to $T^*\mathcal{W}$. The variables (ℓ, w) of the left-hand side live in the manifold

$$\Phi^* T^* N = \{(\ell, w) : \ell \in T^*_{\Phi(w)},\ w \in \mathcal{W}\} \subset T^*N \times \mathcal{W}.$$

Note that Φ^*T^*N is a locally trivial bundle over $\mathcal{W}$ with the projector $\pi : (\ell, w) \mapsto w$; this is nothing else but the *induced bundle* from T^*N by the mapping Φ. We treat (6) as the equality of values of two mappings from Φ^*T^*N to $T^*\mathcal{W}$. Let us rewrite this equation in local coordinates.

So let $N = \mathbb{R}^m$ and $\mathcal{W}$ be a Banach space. Then $T^*N = \mathbb{R}^{m*} \times \mathbb{R}^m$ (where $T_z N = \mathbb{R}^{m*} \times \{z\}$), $T^*\mathcal{W} = \mathcal{W}^* \times \mathcal{W}$, $\Phi^*T^*N = \mathbb{R}^{m*} \times \mathbb{R}^m \times \mathcal{W}$. Surely, $\mathbb{R}^{m*} \cong \mathbb{R}^m$ but in the forthcoming calculations it is convenient to treat the first factor in the product $\mathbb{R}^{m*} \times \mathbb{R}^m$ as the space of linear forms on the second factor. We have: $\ell = (\zeta, z) \in \mathbb{R}^{m*} \times \mathbb{R}^m$ and (6) takes the form

$$\zeta \frac{d\Phi}{dw} = \frac{dJ}{dw}, \quad \Phi(w) = z. \tag{14}$$

Linearization of system (14) at the point (ζ, z, w) reads:

$$\zeta' \frac{d\Phi}{dw} + \zeta \frac{d^2\Phi}{dw^2}(w', \cdot) = \frac{d^2 J}{dw^2}(w', \cdot), \quad \frac{d\Phi}{dw} w' = z'. \tag{15}$$

We set

$$\mathcal{L}^0_{(\ell,w)}(\bar{\Phi}) = \{\ell' = (\zeta', z') \in T_\ell(T^*N) : \exists w' \in \mathcal{W} \text{ s.t. } (\zeta', z', w') \text{ satisfies (15)}\}.$$

Note that subspace $\mathcal{L}^0_{(\ell,w)}(\bar{\Phi}) \subset T_\ell(T^*N)$ does not depend on the choice of local coordinates. Indeed, to construct this subspace we take all $(\ell', w') \in T_{(\ell,w)}(\Phi^*T^*N)$ which satisfy the linearized (6) and then apply the projection $(\ell', w') \mapsto \ell'$.

Recall that $T_\ell(T^*N)$ is a symplectic space endowed with the canonical symplectic form σ_ℓ (cf. Sect. 1.4). A subspace $S \subset T_\ell(T^*N)$ is *isotropic* if $\sigma_\ell|_S = 0$. Isotropic subspaces of maximal possible dimension $m = \frac{1}{2} \dim T_\ell(T^*N)$ are called *Lagrangian subspaces.*

Proposition 1.4. *$\mathcal{L}^0_{(\ell,w)}(\bar{\Phi})$ is an isotropic subspace of $T_\ell(T^*N)$. If* dim *$\mathcal{W} < \infty$, then $\mathcal{L}^0_{(\ell,w)}(\bar{\Phi})$ is a Lagrangian subspace.*

Proof. First we will prove the isotropy of $\mathcal{L}^0_{(\ell,w)}(\bar{\Phi})$. Let $(\zeta',z'),(\zeta'',z'') \in T_\ell(T^*N)$. We have $\sigma_\ell((\zeta',z'),(\zeta'',z'')) = \zeta'z'' - \zeta''z'$; here the symbol ζz denotes the result of the application of the linear form $\zeta \in \mathbb{R}^{m*}$ to the vector $z \in \mathbb{R}^n$ or, in the matrix terminology, the product of the row ζ and the column z. Assume that (ζ',z',w') and (ζ'',z'',w'') satisfy (15); then

$$\zeta'z'' = \zeta'\frac{d\Phi}{dw}w'' = \frac{d^2J}{dw^2}(w',w'') - \zeta\frac{d^2\Phi}{dw^2}(w',w''). \tag{16}$$

The right-hand side of (16) is symmetric with respect to w' and w'' due to the symmetry of second derivatives. Hence $\zeta'z'' = \zeta''z'$. In other words, $\sigma_\ell((\zeta',z'),(\zeta'',z'')) = 0$. So $\mathcal{L}^0_{(\ell,w)}(\bar{\Phi})$ is isotropic and, in particular, $\dim\left(\mathcal{L}^0_{(\ell,w)}(\bar{\Phi})\right) \leq m$.

Now show that the last inequality becomes the equality as soon as $\mathcal{W}$ is finite dimensional. Set $Q = \frac{d^2J}{dw^2} - \zeta\frac{d^2\Phi}{dw^2}$ and consider the diagram:

$$\zeta'\frac{d\Phi}{dw} - Q(w',\cdot) \overset{left}{\longleftarrow} (\zeta',w') \overset{right}{\longrightarrow} \left(\zeta', \frac{d\Phi}{dw}w'\right).$$

Then $\mathcal{L}^0_{(\ell,w)}(\bar{\Phi}) = right(\ker(left))$. Passing to a factor space if necessary we may assume that $\ker(left) \cap \ker(right) = 0$; this means that:

$$\frac{d\Phi}{dw}w' \quad \& \quad Q(w',\cdot) = 0 \quad \Rightarrow \quad w' = 0. \tag{17}$$

Under this assumption, $\dim \mathcal{L}^0_{(\ell,w)}(\bar{\Phi}) = \dim\ker(left)$. On the other hand, relations (17) imply that the mapping $left : \mathbb{R}^{m*} \times \mathcal{W} \to \mathcal{W}^*$ is surjective. Indeed, if, on the contrary, the map $left$ is not surjective then there exists a nonzero vector $v \in (\mathcal{W}^*)^* = \mathcal{W}$ which annihilates the image of $left$; in other words, $\zeta'\frac{d\Phi}{dw}v - Q(w',v) = 0,\ \forall \zeta', w'$. Hence $\frac{d\Phi}{dw}v = 0$ & $Q(v,\cdot) = 0$ that contradicts (17). It follows that $\dim \mathcal{L}^0_{(\ell,w)}(\bar{\Phi}) = \dim(\mathbb{R}^{m*} \times \mathcal{W}) - \dim \mathcal{W}^* = m$. □

For infinite dimensional $\mathcal{W}$, the space $\mathcal{L}^0_{(\ell,w)}(\bar{\Phi})$ may have dimension smaller than m due to an ill-posedness of (15); to guarantee dimension m one needs certain coercivity of the form $\zeta\frac{d^2\Phi}{dw^2}$. I am not going to discuss here what kind of coercivity is sufficient, it can be easily reconstructed from the proof of Proposition 1.4 (see also [5]). Anyway, independently on any coercivity one can take a finite dimensional approximation of the original problem and obtain a Lagrangian subspace $\mathcal{L}^0_{(\ell,w)}(\bar{\Phi})$ guaranteed by Proposition 1.4. What happens with these subspaces when the approximation becomes better and better, do they have a well-defined limit (which would be unavoidably Lagrangian)? A remarkable fact is that such a limit does exist for any sharp Lagrangian point. It contains $\mathcal{L}^0_{(\ell,w)}(\bar{\Phi})$ and is called the $\mathcal{L}$-derivative of $\bar{\Phi}$ at (ℓ,w). To formulate this result we need some basic terminology from set theoretic topology.

A partially ordered set $(\mathfrak{A}, \prec)$ is a *directed set* if $\forall \alpha_1, \alpha_2 \in \mathfrak{A}\ \exists \beta \in \mathfrak{A}$ such that $\alpha_1 \prec \beta$ and $\alpha_2 \prec \beta$. A family $\{x_\alpha\}_{\alpha \in \mathfrak{A}}$ of points of a topological space $\mathcal{X}$ indexed by the elements of $\mathfrak{A}$ is a generalized sequence in $\mathcal{X}$. A point $x \in \mathcal{X}$ is the limit of the generalized sequence $\{x_\alpha\}_{\alpha \in \mathfrak{A}}$ if for any neighborhood $\mathcal{O}_x$ of x in $\mathcal{X}$ $\exists \alpha \in \mathfrak{A}$ such that $x_\beta \in \mathcal{O}_x$, $\forall \beta \succ \alpha$; in this case we write $x = \lim\limits_{\mathfrak{A}} x_\alpha$.

Let $\mathfrak{w}$ be a finite dimensional submanifold of $\mathcal{W}$ and $w \in \mathfrak{w}$. If (ℓ, w) is a Lagrangian point for $\bar{\Phi} = (J, \Phi)$, then it is a Lagrangian point for $\bar{\Phi}|_{\mathfrak{w}}$. A straightforward calculation shows that the Lagrangian subspace $\mathcal{L}^0_{(\ell,w)}(\bar{\Phi}|_{\mathfrak{w}})$ depends on the tangent space $W = T_w\mathfrak{w}$ rather than on $\mathfrak{w}$, i.e., $\mathcal{L}^0_{(\ell,w)}(\bar{\Phi}|_{\mathfrak{w}}) = \mathcal{L}^0_{(\ell,w)}(\bar{\Phi}|_{\mathfrak{w}'})$ as soon as $T_w\mathfrak{w} = T_w\mathfrak{w}' = W$. We denote $\Lambda_W = \mathcal{L}^0_{(\ell,w)}(\bar{\Phi}|_{\mathfrak{w}})$. Recall that Λ_W is an m-dimensional subspace of the $2m$-dimensional space $T_\ell(T^*N)$, i.e., Λ_W is a point of the Grassmann manifold of all m-dimensional subspaces in $T_\ell(T^*N)$.

Finally, we denote by $\mathfrak{W}$ the set of all finite dimensional subspaces of $T_w\mathcal{W}$ partially ordered by the inclusion "$\subset$". Obviously, $(\mathfrak{W}, \subset)$ is a directed set and $\{\Lambda_W\}_{W \in \mathfrak{W}}$ is a generalized sequence indexed by the elements of this directed set. It is easy to check that there exists $W_0 \in \mathfrak{W}$ such that $\Lambda_W \supset \mathcal{L}^0_{(\ell,w)}(\bar{\Phi})$, $\forall W \supset W_0$. In particular, if $\mathcal{L}^0_{(\ell,w)}(\bar{\Phi})$ is m-dimensional, then $\Lambda_{W_0} = \mathcal{L}^0_{(\ell,w)}(\bar{\Phi})$, $\forall W \supset W_0$, the sequence Λ_W is stabilizing and $\mathcal{L}^0_{(\ell,w)}(\bar{\Phi}) = \lim\limits_{\mathfrak{W}} \Lambda_W$. In general, the sequence Λ_W is not stabilizing, nevertheless the following important result is valid.

Theorem 1.1. *If (ℓ, w) is a sharp Lagrangian point, then there exists $\mathcal{L}_{(\ell,w)}(\bar{\Phi}) = \lim\limits_{\mathfrak{W}} \Lambda_W$.*

We omit the proof of the theorem, you can find this proof in paper [5] with some other results which allow to efficiently compute $\lim\limits_{\mathfrak{W}} \Lambda_W$. Lagrangian subspace $\mathcal{L}_{(\ell,w)}(\bar{\Phi}) = \lim\limits_{\mathfrak{W}} \Lambda_W$ is called the *$\mathcal{L}$-derivative* of $\bar{\Phi} = (J, \Phi)$ at the Lagrangian point (ℓ, w).

Obviously, $\mathcal{L}_{(\ell,w)}(\bar{\Phi}) \supset \mathcal{L}^0_{(\ell,w)}(\bar{\Phi})$. One should think on $\mathcal{L}_{(\ell,w)}(\bar{\Phi})$ as on a completion of $\mathcal{L}^0_{(\ell,w)}(\bar{\Phi})$ by means of a kind of weak solutions to system (15) which could be missed due to the ill-posedness of the system.

Now we should explain the connection between $\mathcal{L}_{(\ell,w)}(\bar{\Phi})$ and $\operatorname{Hess}_w(J|_{\Phi^{-1}(z)})$. We start from the following simple observation:

Lemma 1.1. *Assume that $\dim \mathcal{W} < \infty$, w is a regular point of Φ and $\ker D_w\Phi \cap \ker(D^2_w J - \ell D^2_w \Phi) = 0$. Then*

$$\ker \operatorname{Hess}_w(J|_{\Phi^{-1}(z)}) = 0 \quad \Leftrightarrow \quad \mathcal{L}_{(\ell,w)}(\bar{\Phi}) \cap T_\ell(T^*_z N) = 0,$$

i.e., quadratic form $\operatorname{Hess}_w(J|_{\Phi^{-1}(z)})$ is nondegenerate if and only if the subspace $\mathcal{L}_{(\ell,w)}(\bar{\Phi})$ is transversal to the fiber $T^*_z N$.

Proof. We make computations in coordinates. First, $T_\ell(T_z^*N) = \{(\zeta', 0) : \zeta' \in \mathbb{R}^{n*}\}$; then, according to (15), $(\zeta', 0) \in \mathcal{L}_{(\ell,w)}(\bar{\Phi})$ if and only if there exists $w \in \mathcal{W}$ such that

$$\frac{d\Phi}{dw} w' = 0, \quad \frac{d^2 J}{dw^2}(w', \cdot) - \ell \frac{d^2 \Phi}{dw^2}(w', \cdot) = \zeta' \frac{d\Phi}{dw}. \tag{18}$$

Regularity of w implies that $\zeta' \frac{d\Phi}{dw} \neq 0$ and hence $w' \neq 0$ as soon as $\zeta' \neq 0$. Equalities (18) imply: $\frac{d^2 J}{dw^2}(w', v) - \ell \frac{d^2 \Phi}{dw^2}(w', v) = 0, \quad \forall v \in \ker \frac{d\Phi}{dw}$, i.e., $w' \in \ker \mathrm{Hess}_w(J\big|_{\Phi^{-1}(z)})$. Moreover, our implications are invertible: we could start from a nonzero vector $w' \in \ker \mathrm{Hess}_w(J\big|_{\Phi^{-1}(z)})$ and arrive to a nonzero vector $(\zeta', 0) \in \mathcal{L}_{(\ell,w)}(\bar{\Phi})$. □

Remark. Condition $\ker D_w\Phi \cap \ker(D_w^2 J - \ell D_w^2 \Phi) = 0$ from Lemma 1.1 is not heavy. Indeed, a pair (J, Φ) satisfies this condition at all its Lagrangian points if and only if 0 is a regular value of the mapping $(\zeta, w) \mapsto \zeta \frac{d\Phi}{dw} - \frac{dJ}{dw}$. Standard Transversality Theorem implies that this is true for generic pair (J, Φ).

1.6 Maslov Index

Lemma 1.1 is a starting point for a far going theory which allows to effectively compute the Morse index of the Hessians in terms of the $\mathcal{L}$-derivatives.

How to do it? Normally, extremal problems depend on some parameters. Actually, $z \in N$ is such a parameter and there could be other ones, which we do not explicitly add to the constraints. In the optimal control problems a natural parameter is the time interval $t_1 - t_0$. Anyway, assume that we have a continuous family of the problems and their sharp Lagrangian points: $\ell_\tau D_{w_\tau}\Phi_\tau = d_{w_\tau} J_\tau$, $\tau_0 \leq \tau \leq \tau_1$; let $\Lambda(\tau) = \mathcal{L}_{(\ell_\tau, w_\tau)}(\bar{\Phi}_\tau)$. Our goal is to compute the difference $\mathrm{ind}\,\mathrm{Hess}_{w_{\tau_1}}(J_{\tau_1}\big|_{\Phi_{\tau_1}^{-1}(z_{\tau_1})}) - \mathrm{ind}\,\mathrm{Hess}_{w_{\tau_0}}(J_{\tau_0}\big|_{\Phi_{\tau_0}^{-1}(z_{\tau_0})})$ in terms of the family of Lagrangian subspaces $\Lambda(\tau)$; that is to get a tool to follow the evolution of the Morse index under a continuous change of the parameters. This is indeed very useful since for some special values of the parameters the index could be known a priori. It concerns, in particular, optimal control problems with the parameter $\tau = t_1 - t_0$. If $t_1 - t_0$ is very small then sharpness of the Lagrangian point almost automatically implies the positivity or negativity of the Hessian.

First we discuss the finite-dimensional case: Theorem 1.1 indicates that finite-dimensional approximations may already contain all essential information. Let Q_τ be a continuous family of quadratic forms defined on a finite-dimensional vector space. If $\ker Q_\tau = 0$, $\tau_0 \leq \tau \leq \tau_1$, then $\mathrm{ind} Q_\tau$ is constant on the segment $[\tau_0, \tau_1]$. This is why Lemma 1.1 opens the way to follow evolution of the index in terms of the $\mathcal{L}$-derivative: it locates values of the parameter where the index may change. Actually, $\mathcal{L}$-derivative allows to evaluate this change as well; the increment of $\mathrm{ind} Q_\tau$ is computed via so called Maslov index of a family of Lagrangian subspaces. In order to define this index we have to recall some elementary facts about symplectic spaces.

Let Σ, σ be a symplectic space, i.e., Σ is a $2n$-dimensional vector space and σ be a nondegenerate anti-symmetric bilinear form on Σ. The skew-orthogonal complement to the subspace $\Gamma \subset \Sigma$ is the subspace $\Gamma^{\angle} = \{x \in \Sigma : \sigma(x, \Gamma) = 0\}$. The nondegeneracy of σ implies that $\dim \Gamma^{\angle} = 2n - \dim \Gamma$. A subspace Γ is isotropic if and only if $\Gamma^{\angle} \supset \Gamma$; it is Lagrangian if and only if $\Gamma^{\angle} = \Gamma$.

Let $\Pi = span\{e_1, \ldots, e_n\}$ be a lagrangian subspace of Σ. Then there exist vectors $f_1, \ldots, f_n \in \Sigma$ such that $\sigma(e_i, f_j) = \delta_{ij}$, where δ_{ij} is the Kronecker symbol. We show this using induction with respect to n. Skew-orthogonal complement to the space $span\{e_1, \ldots, e_{n-1}\}$ contains an element f which is not skew-orthogonal to e_n; we set $f_n = \frac{1}{\sigma(e_n, f)} f$. We have

$$span\{e_n, f_n\} \cap span\{e_n, f_n\}^{\angle} = 0$$

and the restriction of σ to $span\{e_n, f_n\}^{\angle}$ is a nondegenerate bilinear form. Hence $span\{e_n, f_n\}^{\angle}$ is a $2(n-1)$-dimensional symplectic space with a Lagrangian subspace $span\{e_1, \ldots, e_{n-1}\}$. According to the induction assumption, there exist $f_1, \ldots, f_{n-1}$ such that $\sigma(e_i, f_j) = \delta_{ij}$ and we are done.

Vectors $e_1, \ldots, e_n, f_1, \ldots, f_n$ form a basis of Σ; in particular, $\Delta = span\{f_1, \ldots, f_n\}$ is a transversal to Π Lagrangian subspace, $\Sigma = \Pi \oplus \Delta$. If $x_i = \sum_{j=1}^{n} (\zeta_i^j e_j + z_i^j f_j)$, $i = 1, 2$, and $\zeta_i = (\zeta_i^1, \ldots, \zeta_i^n)$, $z_i = (z_i^1, \ldots, z_i^n)^{\top}$, then $\sigma(x_1, x_2) = \zeta_1 z_2 - \zeta_2 z_1$. The coordinates ζ, z identify Σ with $\mathbb{R}^{n*} \times \mathbb{R}^n$; any transversal to Δ n-dimensional subspace $\Lambda \subset \Sigma$ has the following presentation in these coordinates:

$$\Lambda = \{z^{\top}, S_{\Lambda} z) : z \in \mathbb{R}^n\},$$

where S_{Λ} is an $n \times n$-matrix. The subspace Λ is Lagrangian if and only if $S_{\Lambda}^* = S_{\Lambda}$. We have:

$$\Lambda \cap \Pi = \{(z^{\top}, 0) : z \in \ker S_{\Lambda}\},$$

the subspace Λ is transversal to Π if and only if S_{Λ} is nondegenerate.

That is time to introduce some notations. Let $L(\Sigma)$ be the set of all Lagrangian subspaces, a closed subset of the Grassmannian $G_n(\Sigma)$ of n-dimensional subspaces in Σ. We set

$$\Delta^{\pitchfork} = \{\Lambda \in L(\Sigma) : \Lambda \cap \Delta = 0\},$$

an open subset of $L(\Sigma)$. The mapping $\Lambda \mapsto S_{\Lambda}$ gives a regular parametrization of $\Delta^{\pitchfork}$ by the $n(n+1)/2$-dimensional space of symmetric $n \times n$-matrices. Moreover, above calculations show that $L(\Sigma) = \bigcup_{\Delta \in L(\Sigma)} \Delta^{\pitchfork}$. Hence $L(\Sigma)$ is a $n(n+1)/2$-dimensional submanifold of the Grassmannian $G_n(\Sigma)$ covered by coordinate charts $\Delta^{\pitchfork}$. The manifold $L(\Sigma)$ is called Lagrange Grassmannian associated to the symplectic space Σ. It is not hard to show that any coordinate chart $\Delta^{\pitchfork}$ is everywhere dense in $L(\Sigma)$; our calculations give also a local parametrization of its complement.

Given $\Pi \in L(\Sigma)$, the subset

$$\mathcal{M}_\Pi = L(\Sigma) \setminus \Pi^\pitchfork = \{\Lambda \in L(\Sigma) : \Lambda \cap \Pi \neq 0\}$$

is called the *train* of Π. Let $\Lambda_0 \in \mathcal{M}_\Pi$, $\dim(\Lambda_0 \cap \Pi) = k$. Assume that Δ is transversal to both Λ_0 and Π (i.e., $\Delta \in \Lambda_0^\pitchfork \cap \Pi^\pitchfork$). The mapping $\Lambda \mapsto S_\Lambda$ gives a regular parametrization of the neighborhood of Λ_0 in $\mathcal{M}_\Pi$ by a neighborhood of a corank k matrix in the set of all degenerate symmetric $n \times n$-matrices. A basic perturbation theory for symmetric matrices now implies that a small enough neighborhood of Λ_0 in $\mathcal{M}_\Pi$ is diffeomorphic to the Cartesian product of a neighborhood of the origin of the cone of all degenerate symmetric $k \times k$-matrices and a $(n(n+1) - k(k+1))/2$-dimensional smooth manifold (see [1, Lemma 2.2] for details). We see that $\mathcal{M}_\Pi$ is not a smooth submanifold of $L(\Sigma)$ but a union of smooth strata, $\mathcal{M}_\Pi = \bigcup\limits_{k>0} \mathcal{M}_\Pi^{(k)}$, where $\mathcal{M}_\Pi^{(k)} = \{\Lambda \in L(\Sigma) : \dim(\Lambda \cap \Pi) = k\}$ is a smooth submanifold of $L(\Sigma)$ of codimension $k(k+1)/2$.

Let $\Lambda(\tau)$, $\tau \in [t_0, t_1]$ be a smooth family of Lagrangian subspaces (a smooth curve in $L(\Sigma)$) and $\Lambda(t_0), \Lambda(t_1) \in \Pi^\pitchfork$. We are going to define the intersection number of $\Lambda(\cdot)$ and $\mathcal{M}_\Pi$. It is called the Maslov index and is denoted $\mu_\Pi(\Lambda(\cdot))$. Crucial property of this index is its homotopy invariance: given a homotopy $\Lambda^s(\cdot)$, $s \in [t_0, t_1]$ such that $\Lambda^s(t_0), \Lambda^s(t_1) \in \Pi^\pitchfork$ $\forall s \in [0,1]$, we have $\mu_\Pi(\Lambda^0(\cdot)) = \mu_\Pi(\Lambda^1(\cdot))$.

It is actually enough to define $\mu_\Pi(\Lambda(\cdot))$ for the curves which have empty intersection with $\mathcal{M}_\Pi \setminus \mathcal{M}_\Pi^{(1)}$; the desired index would have a well-defined extension to other curves by continuity. Indeed, generic curves have empty intersection with $\mathcal{M}_\Pi \setminus \mathcal{M}_\Pi^{(1)}$ and, moreover, generic homotopy has empty intersection with $\mathcal{M}_\Pi \setminus \mathcal{M}_\Pi^{(1)}$ since any of submanifolds $\mathcal{M}_\Pi^{(k)}$, $k = 2, \ldots n$ has codimension greater or equal to 3 in $L(\Sigma)$. Putting any curve in general position by a small perturbation, we obtain the curve which bypasses $\mathcal{M}_\Pi \setminus \mathcal{M}_\Pi^{(1)}$, and the invariance with respect to generic homotopies of the Maslov index defined for generic curves would imply that the value of the index does not depend on the choice of a small perturbation.

What remains is to fix a "coorientation" of the smooth hypersurface $\mathcal{M}_\Pi^{(1)}$ in $L(\Sigma)$, i.e., to indicate the "positive and negative sides" of the hypersurface. As soon as we have a coorientation, we may compute $\mu_\Pi(\Lambda(\cdot))$ for any curve $\Lambda(\cdot)$ which is transversal to $\mathcal{M}_\Pi^{(1)}$ and has empty intersection with $\mathcal{M}_\Pi \setminus \mathcal{M}_\Pi^{(1)}$. Maslov index of $\Lambda(\cdot)$ is just the number of points where $\Lambda(\cdot)$ intersects $\mathcal{M}_\Pi^{(1)}$ in the positive direction minus the number of points where this curve intersects $\mathcal{M}_\Pi^{(1)}$ in the negative direction. Maslov index of any curve with endpoints out of $\mathcal{M}_\Pi$ is defined by putting the curve in general position. Proof of the homotopy invariance is the same as for usual intersection number of a curve with a closed cooriented hypersurface (see, for instance, the nice elementary book by J. Milnor "Topology from the differential viewpoint", 1965).

The coorientation is a byproduct of the following important structure on the tangent spaces to $L(\Sigma)$. It happens that any tangent vector to $L(\Sigma)$ at the point $\Lambda \in L(\Sigma)$ can be naturally identified with a quadratic form on Λ. Here we use the fact that Λ is not just a point in the Grassmannian but an n-dimensional linear space. To associate a quadratic form on Λ to the velocity $\dot{\Lambda}(t) \in T_{\Lambda(t)}L(\Sigma)$ of a smooth curve $\Lambda(\cdot)$ we proceed as follows: given $x \in \Lambda(t)$ we take a smooth curve $\tau \mapsto x(\tau)$ in Σ in such a way that $x(\tau) \in \Lambda(\tau)$, $\forall \tau$ and $x(t) = x$. Then we define a quadratic form $\underline{\dot{\Lambda}}(t)(x)$, $x \in \Lambda(t)$, by the formula $\underline{\dot{\Lambda}}(t)(x) = \sigma(x, \dot{x}(t))$.

The point is that $\sigma(x, \dot{x}(t))$ does not depend on the freedom in the choice of the curve $\tau \mapsto x(\tau)$, although $\dot{x}(t)$ depends on this choice. Let us check the required property in the coordinates. We have $x = (z^\top, S_{\Lambda(t)}z)$ for some $z \in \mathbb{R}^n$ and $x(\tau) = (z(\tau)^\top, S_{\Lambda(\tau)}z(\tau))$. Then

$$\sigma(x, \dot{x}(t)) = z^\top(\dot{S}_{\Lambda(t)}z + S_{\Lambda(t)}\dot{z}) - \dot{z}^\top S_{\Lambda(t)}z = z^\top \dot{S}_{\Lambda(t)}z;$$

vector $\dot{z}$ does not show up. We have obtained a coordinate presentation of $\underline{\dot{\Lambda}}(t)$:

$$\underline{\dot{\Lambda}}(t)(z^\top, S_{\Lambda(t)}z) = z^\top \dot{S}_{\Lambda(t)}z,$$

which implies that $\dot{\Lambda} \mapsto \underline{\dot{\Lambda}}$, $\dot{\Lambda} \in T_\Lambda L(\Sigma)$ is an isomorphism of $T_\Lambda L(\Sigma)$ on the linear space of quadratic forms on Λ.

We are now ready to define the coorientation of $\mathcal{M}_\Pi^{(1)}$. Assume that $\Lambda(t) \in \mathcal{M}_\Pi^{(1)}$, i.e., $\Lambda(t) \cap \Pi = \mathbb{R}x$ for some nonzero vector $x \in \Sigma$. In coordinates, $x = (z^\top, 0)$, where $\mathbb{R}x = \ker S_{\Lambda(t)}$. It is easy to see that $\dot{\Lambda}(t)$ is transversal to $\mathcal{M}_\Pi^{(1)}$ (i.e., $\dot{S}_{\Lambda(t)}$ is transversal to the cone of degenerate symmetric matrices) if and only if $\underline{\dot{\Lambda}}(t)(x) \neq 0$ (i.e., $z^\top \dot{S}_{\Lambda(t)}z \neq 0$). Vector x is defined up to a scalar multiplier and $\underline{\dot{\Lambda}}(t)(\alpha x) = \alpha^2 \underline{\dot{\Lambda}}(t)(x)$ so that the sign of $\underline{\dot{\Lambda}}(t)(x)$ does not depend on the selection of x.

Definition. We say that $\Lambda(\cdot)$ intersects $\mathcal{M}_\Pi^{(1)}$ at the point $\Lambda(t)$ in the positive (negative) direction if $\underline{\dot{\Lambda}}(t)(x) > 0$ (< 0).

This definition completes the construction of the Maslov index. A weak point of the construction is the necessity to put the curve in general position in order to compute the intersection number. This does not look as an efficient way to do things since putting the curve in general position is nothing else but a deliberate spoiling of a maybe nice and symmetric original object that makes even more involved the nontrivial problem of the localization of its intersection with $\mathcal{M}_\Pi$. Fortunately, just the fact that Maslov index is homotopy invariant leads to a very simple and effective way of its computation without putting things in general position and without looking for the intersection points with $\mathcal{M}_\Pi$.

Lemma 1.2. *Assume that* $\Pi \cap \Delta = \Lambda(\tau) \cap \Delta = 0$, $\forall \tau \in [t_0, t_1]$. *Then* $\mu_\Pi(\Lambda(\cdot)) = \operatorname{ind}S_{\Lambda(t_0)} - \operatorname{ind}S_{\Lambda(t_1)}$, *where* $\operatorname{ind}S$ *is the Morse index of the quadratic form* $z^\top Sz$, $z \in \mathbb{R}^n$.

Proof. The matrices $S_{\Lambda(t_0)}$ and $S_{\Lambda(t_0)}$ are nondegenerate since $\Lambda(t_0) \cap \Pi = \Lambda(t_1) \cap \Pi = 0$ (we define the Maslov index only for the curves whose endpoints are out of $\mathcal{M}_\Pi$). The set of nondegenerate quadratic forms with a prescribed value of the Morse index is a connected open subset of the linear space of all quadratic forms in n variables. Hence homotopy invariance of the Maslov index implies that $\mu_\Pi(\Lambda(\cdot))$ depends only on $\mathrm{ind} S_{\Lambda(t_0)}$ and $\mathrm{ind} S_{\Lambda(t_1)}$. It remains to compute μ_Π of sample curves in $\Delta^\pitchfork$, say, for segments of the curve $\Lambda(\cdot)$ such that

$$S_{\Lambda(\tau)} = \begin{pmatrix} \tau-1 & 0 & \dots & 0 \\ 0 & \tau-2 & \dots & 0 \\ \vdots & \vdots & \ddots & \vdots \\ 0 & 0 & \dots & \tau-n \end{pmatrix}. \qquad \square$$

In general, given curve is not contained in the fixed coordinate neighborhood $\Delta^\pitchfork$ but any curve can be divided into segments $\Lambda(\cdot)|_{[\tau_i,\tau_{i+1}]}$, $i = 0, \dots, l$, in such a way that $\Lambda(\tau) \in \Delta_i^\pitchfork \ \forall \tau \in [\tau_i, \tau_{i+1}]$, where $\Delta_i \in \Pi^\pitchfork$, $i = 0, \dots, l$; then $\mu_\Pi(\Lambda(\cdot)) = \sum\limits_i \mu_\Pi\left(\Lambda(\cdot)|_{[\tau_i,\tau_{i+1}]}\right)$.

Lemma 1.2 implies the following useful formula which is valid for the important class of *monotone increasing curves* in the Lagrange Grassmannian, i.e., the curves $\Lambda(\cdot)$ such that $\underline{\dot{\Lambda}}(t)$ are nonnegative quadratic forms: $\underline{\dot{\Lambda}}(t) \geq 0, \ \forall t$.

Corollary 1.1. *Assume that* $\underline{\dot{\Lambda}}(\tau) \geq 0, \ \forall \tau \in [t_0, t_1]$ *and* $\{\tau \in [t_0, t_1] : \Lambda(\tau) \cap \Pi \neq 0\}$ *is a finite subset of* (t_0, t_1). *Then*

$$\mu_\Pi(\Lambda(\cdot)) = \sum_{\tau \in (t_0,t_1)} \dim(\Lambda(\tau) \cap \Pi). \qquad \square$$

Corollary 1.1 can be also applied to the case of monotone decreasing curves defined by the inequality $\underline{\dot{\Lambda}}(t) \leq 0, \ \forall t$; the change of parameter $t \mapsto t_0 + t_1 - t$ makes the curve monotone increasing and and change sign of the Maslov index.

Let me now recall that our interest to these symplectic playthings was motivated by the conditional minimum problems. As it was mentioned at the beginning of the section, we are going to apply this stuff to the case $\Sigma = T_{\ell_\tau}(T^*M)$, $\ell_\tau \in T^*_{z_\tau}M$, $\Pi = T_{\ell_\tau}(T^*_{z_\tau}M)$, $\Lambda(\tau) = \mathcal{L}_{(\ell_\tau, w_\tau)}(\bar{\Phi}_\tau)$, where $z_\tau = \Phi_\tau(w_\tau)$. In this case, not only Λ but also Π and even symplectic space Σ depend on τ. We thus have to define Maslov index in such situation. This is easy. We consider the bundle

$$\{(\xi, \tau) : \xi \in T_{\ell_\tau}(T^*M), \ t_0 \leq \tau \leq t_1\} \tag{19}$$

over the segment $[t_0, t_1]$ induced from $T(T^*M)$ by the mapping $\tau \mapsto \ell_\tau$. Bundle (19) endowed with the symplectic structure and its subbundle

$$\{(\xi, \tau) : \xi \in T_{\ell_\tau}(T^*_{z_\tau}M)\}$$

are trivial as any bundle over a segment. More precisely, let $t \in [t_0, t_1]$, $\Sigma_t = T_{\ell_t}(T^*M)$, $\Pi_t = T_{\ell_t}(T^*_{z_t}M)$; then there exists a continuous with respect to τ family of linear symplectic mappings $\Xi_\tau : T_{\ell_\tau}(T^*M) \to \Sigma_t$ such that $\Xi_\tau(T_{\ell_\tau}(T^*_{z_\tau}M)) = \Pi_t$, $t_0 \leq \tau \leq t_1$, $\Xi_t = \mathrm{Id}$. To any continuous family of Lagrangian subspaces $\Lambda(\tau) \subset T_{\ell_\tau}(T^*M)$, where $\Lambda(t_i) \cap \Pi_{t_i} = 0$, $i = 0, 1$, we associate a curve $\Xi.\Lambda(\cdot) : \tau \mapsto \Xi_\tau \Lambda(\tau)$ in the Lagrange Grassmannian $L(\Sigma_t)$ and set $\mu(\Lambda(\cdot)) \stackrel{def}{=} \mu_{\Pi_t}(\Xi.\Lambda(\cdot))$. Homotopy invariance of the Maslov index implies that $\mu_{\Pi_t}(\Xi.\Lambda(\cdot))$ does not depend on the choice of t and Ξ_τ.

Theorem 1.2. *Assume that* $\dim \mathcal{W} < \infty$,

$$\bar{\Phi}_\tau = (J_\tau, \Phi_\tau) : \mathcal{W} \to \mathbb{R} \times M, \quad \tau \in [t_0, t_1]$$

is a continuous one-parametric family of smooth mappings and (ℓ_τ, w_τ) *is a continuous family of their Lagrangian points such that* $\ell_\tau \neq 0$, w_τ *is a regular point of* Φ_τ, *and* $\ker D_{w_\tau}\Phi_\tau \cap \ker(D^2_{w_\tau}J_\tau - \ell_\tau D^2_{w_\tau}\Phi_\tau) = 0$, $t_0 \leq \tau \leq t_1$. *Let* $z_\tau = \Phi(w_\tau)$, $\Lambda(\tau) = \mathcal{L}_{(\ell_\tau, w_\tau)}(\bar{\Phi}_\tau)$. *If* $\mathrm{Hess}_{w_{t_i}}(J_{t_i}\big|_{\Phi^{-1}_{t_i}(z_{t_i})})$, $i = 1, 2$, *are nondegenerate, then*

$$\mathrm{ind}\, \mathrm{Hess}_{w_{t_0}}(J_{t_0}\big|_{\Phi^{-1}_{t_0}(z_{t_0})}) - \mathrm{ind}\, \mathrm{Hess}_{w_{t_1}}(J_{t_1}\big|_{\Phi^{-1}_{t_1}(z_{t_1})}) = \mu(\Lambda(\cdot)).$$

Remark. If $\ell_\tau = 0$, then w_τ is a critical point of J_τ (without restriction to the level set of Φ_τ). Theorem 1.2 can be extended to this situation (with the same proof) if we additionally assume that $\ker \mathrm{Hess}_{w_\tau} J_\tau = 0$ for any τ such that $\ell_\tau = 0$.

Proof. We introduce simplified notations: $A_\tau = D_{w_\tau}\Phi_\tau$, $Q_\tau = D^2_{w_\tau}J_\tau - \ell_\tau D^2_{w_\tau}\Phi_\tau$; the $\mathcal{L}$-derivative $\mathcal{L}_{(\ell_\tau, w_\tau)}(\bar{\Phi}_\tau) = \Lambda(\tau)$ is uniquely determined by the linear map A_τ and the symmetric bilinear form Q_τ. Fix local coordinates in the neighborhoods of w_τ and z_τ and set:

$$\Lambda(A, Q) = \{(\zeta, Av) : \zeta A + Q(v, \cdot) = 0\} \in L(\mathbb{R}^{n*} \times \mathbb{R}^n);$$

then $\Lambda_\tau = \Lambda(A_\tau, Q_\tau)$.

The assumption $\ker A_\tau \cap \ker Q_\tau = 0$ implies the smoothness of the mapping $(A, Q) \mapsto \Lambda(A, Q)$ for (A, Q) close enough to (A_τ, Q_τ). Indeed, as it is shown in the proof of Proposition 1.4, this assumption implies that the mapping $left_\tau : (\zeta, v) \mapsto \zeta A_\tau + Q_\tau(v, \cdot)$ is surjective. Hence the kernel of the mapping

$$(\zeta, v) \mapsto \zeta A + Q(v, \cdot) \tag{20}$$

smoothly depends on (A, Q) for (A, Q) close to (A_τ, Q_τ). On the other hand, $\Lambda(A, Q)$ is the image of the mapping $(\zeta, v) \mapsto (\zeta, Av)$ restricted to the kernel of map (20).

Now we have to disclose a secret which the attentive reader already knows and is perhaps indignant with our lightness: Q_τ is not a well-defined bilinear form on $T_{w_\tau}\mathcal{W}$, it essentially depends on the choice of local coordinates in M.

What is well-defined is the mapping $Q_\tau\big|_{\ker A_\tau} : \ker A_\tau \to T^*_{w_\tau}\mathcal{W}$ (check this by yourself or see [3, Sect. 2.3]), the map $A_\tau : T_{w_\tau}\mathcal{W} \to T_{z_\tau}M$ and, of course, the Lagrangian subspace $\Lambda(\tau) = \mathcal{L}_{(\ell_\tau, w_\tau)}(\bar{\Phi}_\tau)$. By the way, the fact that $Q_\tau\big|_{\ker A_\tau}$ is well-defined guarantees that assumptions of Theorem 1.2 do not depend on the coordinates choice.

Recall that any local coordinates $\{z\}$ on M induce coordinates $\{(\zeta, z) : \zeta \in \mathbb{R}^{n*}, z \in \mathbb{R}^n\}$ on T^*M and $T^*_z M = \{(\zeta, 0) : \zeta \in \mathbb{R}^{n*}\}$ in the induced coordinates.

Lemma 1.3. *Given $\hat{z} \in M$, $\ell \in T^*_{\hat{z}}M \setminus \{0\}$, and a Lagrangian subspace $\Delta \in T_\ell(T^*_{\hat{z}}M)^\pitchfork \subset L(T_\ell(T^*M))$, there exist centered at $\hat{z}$ local coordinates on M such that $\Delta = \{(0, z) : z \in \mathbb{R}^n\}$ in the induced coordinates on $T_\ell(T^*M)$.*

Proof. Working in arbitrary local coordinates we have $\ell = (\zeta_0, 0)$, $\Delta = \{(Sz, z) : z \in \mathbb{R}^n\}$, where S is a symmetric matrix. In other words, Δ is the tangent space at $(\zeta_0, 0)$ to the graph of the differential of the function $a(z) = \zeta_0 z + \frac{1}{2}z^\top Sz$. Any smooth function with a nonzero differential can be locally made linear by a smooth change of variables. To prove the lemma it is enough to make a coordinates change which kills second derivative of the function a, for instance: $z \mapsto z + \frac{1}{2|\zeta_0|^2}(z^\top Sz)\zeta_0^\top$. □

We continue the proof of Theorem 1.2. Lemma 1.3 gives us the way to take advantage of the fact that Q_τ depends on the choice of local coordinates in M. Indeed, bilinear form Q_τ is degenerate if and only if $\Lambda_\tau \cap \{(0, z) : z \in \mathbb{R}^n\} \neq 0$; this immediately follows from the relation

$$\Lambda_\tau = \{(\zeta, A_\tau v) : \zeta A_\tau + Q_\tau(v, \cdot) = 0\}.$$

Given $t \in [t_0, t_1]$ take a transversal to $T_{\ell_t}(T^*_{z_t}M)$ and $\Lambda(t)$ Lagrangian subspace $\Delta_t \subset T_{\ell_t}(T^*M)$ and centered at z_t local coordinates in M such that $\Delta_t = \{(0, z) : z \in \mathbb{R}^n\}$ in these coordinates. Then $\Lambda(\tau)$ is transversal to $\{(0, z) : z \in \mathbb{R}^n\}$ for all τ from a neighborhood O_t of t in $[t_0, t_1]$. Selecting an appropriate finite subcovering from the covering O_t, $t \in [t_0, t_1]$ of $[t_0, t_1]$ we can construct a subdivision $t_0 = \tau_0 < \tau_1 < \ldots < \tau_k < \tau_{k+1} = t_1$ of $[t_0, t_1]$ with the following property: $\forall i \in \{0, 1, \ldots, k\}$ the segment $\{z_\tau : \tau \in [\tau_i, \tau_{i+1}]\}$ of the curve z_τ is contained in a coordinate neighborhood $\mathcal{O}^i$ of M such that $\Lambda_\tau \cap \{(0, z) : z \in \mathbb{R}^n\} = 0$ $\forall \tau \in [\tau_i, \tau_{i+1}]$ in the correspondent local coordinates.

We identify the form Q_τ with its symmetric matrix, i.e., $Q_\tau(v_1, v_2) = v_1^\top Q_\tau v_2$. Then Q_τ is a nondegenerate symmetric matrix and

$$\Lambda(\tau) = \{(\zeta, -A_\tau Q_\tau^{-1} A_\tau^\top \zeta^\top\}, \quad \tau_i \leq \tau \leq \tau_{i+1}. \tag{21}$$

Now focus on the subspace $\Lambda(\tau_i)$; it has a nontrivial intersection with $\{(\zeta, 0) : \zeta \in \mathbb{R}^{n*}\} = T_{\ell_{\tau_i}}(T^*_{z_{\tau_i}}M)$ if and only if the matrix $A_{\tau_i} Q_{\tau_i}^{-1} A_{\tau_i}^\top$ is degenerate. This is the matrix of the restriction of the nondegenerate quadratic form $v \mapsto v^\top Q_{\tau_i}^{-1} v$ to the image of the linear map $A_{\tau_i}^\top$. Hence $A_{\tau_i} Q_{\tau_i}^{-1} A_{\tau_i}^\top$

can be made nondegenerate by the arbitrary small perturbation of the map $A_{\tau_i} : T_{w_{\tau_i}}\mathcal{W} \to T_{z_{\tau_i}}M$. Such perturbations can be realized simultaneously for $i = 1, \dots, k$ [2] by passing to a continuous family $\tau \mapsto A'_\tau$, $t_0 \le \tau \le t_1$, arbitrary close and homotopic to the family $\tau \mapsto A_\tau$. In fact, A'_τ can be chosen equal to A_τ out of an arbitrarily small neighborhood of $\{\tau_1, \dots, \tau_k\}$. Putting now A'_τ instead of A_τ in the expression for $\Lambda(\tau)$ we obtain a family of Lagrangian subspaces $\Lambda'(\tau)$. This family is continuous (see the paragraph containing formula (20)) and homotopic to $\Lambda(\cdot)$. In particular, it has the same Maslov index as $\Lambda(\cdot)$. In other words, we can assume without lack of generality that $\Lambda(\tau_i) \cap T_{\ell_{\tau_i}}(T^*_{z_{\tau_i}}M) = 0$, $i = 0, 1, \dots, k+1$. Then $\mu(\Lambda(\cdot)) = \sum_{i=0}^{k} \mu\left(\Lambda(\cdot)\big|_{[\tau_i,\tau_{i+1}]}\right)$. Moreover, it follows from (21) and Lemma 1.2 that

$$\mu\left(\Lambda(\cdot)\big|_{[\tau_i,\tau_{i+1}]}\right) = \operatorname{ind}(A_{\tau_{i+1}} Q^{-1}_{\tau_{i+1}} A^\top_{\tau_{i+1}}) - \operatorname{ind}(A_{\tau_i} Q^{-1}_{\tau_i} A^\top_{\tau_i}).$$

Besides that, $\operatorname{ind} Q_{\tau_i} = \operatorname{ind} Q_{\tau_{i+1}}$ since Q_τ is nondegenerate for all $\tau \in [\tau_i, \tau_{i+1}]$ and continuously depends on τ.

Recall that $\operatorname{Hess}_{w_\tau}\left(J_\tau\big|_{\Phi^{-1}(z_\tau)}\right) = Q_\tau\big|_{\ker A_\tau}$. In order to complete proof of the theorem it remains to show that

$$\operatorname{ind} Q_\tau = \operatorname{ind}\left(Q_\tau\big|_{\ker A_\tau}\right) + \operatorname{ind}(A_\tau Q^{-1}_\tau A^\top_\tau) \tag{22}$$

for $\tau = \tau_i, \tau_{i+1}$.

Let us rearrange the second term in the right-hand side of (22). The change of variables $v = Q^{-1}_\tau A^\top_\tau z$, $z \in \mathbb{R}^n$, implies:

$$\operatorname{ind}(A_\tau Q^{-1}_\tau A^\top_\tau) = \operatorname{ind}\left(Q_\tau\big|_{\{Q^{-1}_\tau A^\top_\tau z \,:\, z \in \mathbb{R}^n\}}\right).$$

We have: $Q_\tau(v, \ker A_\tau) = 0$ if and only if $Q_\tau(v,\cdot) = z^\top A_\tau$ for some $z \in \mathbb{R}^n$, i.e., $v^\top Q_\tau = z^\top A_\tau$, $v = Q^{-1}_\tau A^\top_\tau z$. Hence the right-hand side of (22) takes the form

$$\operatorname{ind} Q_\tau = \operatorname{ind}\left(Q_\tau\big|_{\ker A_\tau}\right) + \operatorname{ind}\left(Q_\tau\big|_{\{v \,:\, Q_\tau(v, \ker A_\tau) = 0\}}\right)$$

and $Q_\tau\big|_{\{v : Q_\tau(v,\ker A_\tau)=0\}}$ is a nondegenerate form for $\tau = \tau_i, \tau_{i+1}$. Now equality (22) is reduced to the following elementary fact of linear algebra: If Q is a nondegenerate quadratic form on $\mathbb{R}^m$ and $E \subset \mathbb{R}^m$ is a linear subspace, then $\operatorname{ind} Q = \operatorname{ind}(Q|_E) + \operatorname{ind}\left(Q|_{E^\perp_Q}\right) + \dim(E \cap E^\perp_Q)$, where $E^\perp_Q = \{v \in \mathbb{R}^m : Q(v, E) = 0\}$ and $E \cap E^\perp_Q = \ker(Q|_E) = \ker\left(Q|_{E^\perp_Q}\right)$. □

Remark. Maslov index μ_Π is somehow more than just the intersection number with $\mathcal{M}_\Pi$. It can be extended, in a rather natural way, to all continuous

[2] We do not need to perturb A_{t_0} and $A_{t_{k+1}}$: assumption of the theorem and Lemma 1.1 guarantee the required nondegeneracy property.

curves in the Lagrange Grassmannian including those whose endpoint belong to $\mathcal{M}_\Pi$. This extension allows to get rid of the annoying nondegeneracy assumption for $\mathrm{Hess}_{w_{t_i}}(J_{t_i}|_{\Phi_{t_i}^{-1}(z_{t_i})})$ in the statement of Theorem 1.2. In general, Maslov index computes 1/2 of the difference of the signatures of the Hessians which is equal to the difference of the Morse indices in the degenerate case (see [3] for this approach).

1.7 Regular Extremals

A combination of the finite-dimensional Theorem 1.2 with the limiting procedure of Theorem 1.1 and with homotopy invariance of the Maslov index allows to efficiently compute Morse indices of the Hessians for numerous infinite-dimensional problems. Here we restrict ourselves to the simplest case of a regular extremal of the optimal control problem.

We use notations and definitions of Sects. 3 and 4. Let $h(\lambda, u)$ be the Hamiltonian of a smooth optimal control system and λ_t, $t_0 \le t \le t_1$, be an extremal contained in the regular domain $\mathcal{D}$ of h. Then λ_t is a solution of the Hamiltonian system $\dot\lambda = \mathbf{H}(\lambda)$, where $H(\lambda) = h(\lambda, \bar u(\lambda))$, $\frac{\partial h}{\partial u} h(\lambda, \bar u(\lambda)) = 0$.

Let $q(t) = \pi(\lambda_t)$, $t_0, \le t \le t_1$ be the extremal path. Recall that the pair $(\lambda_{t_0}, \lambda_t)$ is a Lagrange multiplier for the conditional minimum problem defined on an open subset of the space

$$M \times L_\infty([t_0, t_1], U) = \{(q_t, u(\cdot)) : q_t \in M, u(\cdot) \in L_\infty([t_0, t_1], U)\},$$

where $u(\cdot)$ is control and q_t is the value at t of the solution to the differential equation $\dot q = f(q, u(\tau))$, $\tau \in [t_0, t_1]$. In particular, $F_t(q_t, u(\cdot)) = q_t$. The cost is $J_{t_0}^t(q_t, u(\cdot))$ and constraints are $F_{t_0}(q_t, u(\cdot)) = q(t_0)$, $q_t = q(t)$.

Let us set $J_t(u) = J_{t_0}^t(q(t), u(\cdot))$, $\Phi_t(u) = F_{t_0}(q(t), u(\cdot))$. A covector $\lambda \in T^*M$ is a Lagrange multiplier for the problem (J_t, Φ_t) if and only if there exists an extremal $\hat\lambda_\tau$, $t_0 \le \tau \le t$, such that $\lambda_{t_0} = \lambda$, $\hat\lambda_t \in T^*_{q(t)}M$. In particular, λ_{t_0} is a Lagrange multiplier for the problem (J_t, Φ_t) associated to the control $u(\cdot) = \bar u(\lambda_\cdot)$. Moreover, all sufficiently close to λ_{t_0} Lagrange multipliers for this problem are values at t_0 of the solutions $\lambda(\tau)$, $t_0 \le \tau \le t$ to the Hamiltonian system $\dot\lambda = \mathbf{H}(\lambda)$ with the boundary condition $\lambda(t) \in T^*_{q(t)}M$.

We will use exponential notations for one-parametric groups of diffeomorphisms generated ordinary differential equations. In particular, $e^{\tau\mathbf{H}} : T^*M \to T^*M$, $\tau \in \mathbb{R}$, is a flow generated by the equation $\dot\lambda = \mathbf{H}(\lambda)$, so that $\lambda(\tau') = e^{(\tau'-\tau)\mathbf{H}}(\lambda(\tau), \tau, \tau' \in \mathbb{R}$, and Lagrange multipliers for the problem (J_t, Φ_t) fill the n-dimensional submanifold $e^{(t_0-t)\mathbf{H}}\left(T^*_{q(t)}M\right)$.

We set $\bar\Phi_t = (J_t, \Phi_t)$; it is easy to see that the $\mathcal{L}$-derivative $\mathcal{L}_{(\lambda_{t_0}, u)}(\bar\Phi_t)$ is the tangent space to $e^{(t_0-t)\mathbf{H}}\left(T^*_{q(t)}M\right)$, i.e., $\mathcal{L}_{(\lambda_{t_0}, u)}(\bar\Phi_t) = e_*^{(t_0-t)\mathbf{H}} T_{\lambda_t}\left(T^*_{q(t)}M\right)$. Indeed, let us recall the construction of the $\mathcal{L}$-derivative. First we linearize the equation for Lagrange multipliers at λ_{t_0}. Solutions of the linearized equation form an isotropic subspace $\mathcal{L}^0_{(\lambda_{t_0},u)}(\bar{\Phi}_t)$ of the symplectic space $T_{\lambda_{t_0}}(T^*M)$. If $\mathcal{L}^0_{(\lambda_{t_0},u)}(\bar{\Phi}_t)$ is a Lagrangian subspace (i.e., $\dim \mathcal{L}^0_{(\lambda_{t_0},u)}(\bar{\Phi}_t) = \dim M$), then $\mathcal{L}_{(\lambda^0_{t_0},u)}(\bar{\Phi}_t) = \mathcal{L}_{(\lambda_{t_0},u)}(\bar{\Phi}_t)$, otherwise we need a limiting procedure to complete the Lagrangian subspace. In the case under consideration, $\mathcal{L}^0_{(\lambda_{t_0},u)}(\bar{\Phi}_t) = e_*^{(t_0-t)\mathbf{H}} T_{\lambda_t}\left(T^*_{q(t)}M\right)$ has a proper dimension and thus coincides with $\mathcal{L}_{(\lambda_{t_0},u)}(\bar{\Phi}_t)$. We can check independently that $e_*^{(t_0-t)\mathbf{H}} T_{\lambda_t}\left(T^*_{q(t)}M\right)$ is Lagrangian: indeed, $T_{\lambda_t}\left(T^*_{q(t)}M\right)$ is Lagrangian and $e_*^{(t_0-t)\mathbf{H}} : T_{\lambda_t}(T^*M) \to T_{\lambda_{t_0}}(T^*M)$ is an isomorphism of symplectic spaces since Hamiltonian flows preserve the symplectic form.

So $t \mapsto \mathcal{L}_{(\lambda_{t_0},u)}(\bar{\Phi}_t)$ is a smooth curve in the Lagrange Grassmannian $L\left(T_{\lambda_{t_0}}(T^*M)\right)$ and we can try to compute Morse index of

$$\mathrm{Hess}_u \left(J_{t_1}\big|_{\Phi_{t_1}^{-1}(q(t_0))} \right) = \mathrm{Hess}_u \left(J_{t_0}^{t_1}\big|_{F_{t_0}^{-1}(q(t_0))\cap F_{t_1}^{-1}(q(t_1))} \right)$$

via the Maslov index of this curve. Of course, such a computation has no sense if the index is infinite.

Proposition 1.5. (Legendre condition) *If quadratic form $\frac{\partial^2 h}{\partial u^2}(\lambda_t, u(t))$ is negative definite for any $t \in [t_0, t_1]$, then* $\mathrm{ind}\,\mathrm{Hess}_u \left(J_{t_1}\big|_{\Phi_{t_1}^{-1}(q(t_0))} \right) < \infty$ *and* $\mathrm{Hess}_u \left(J_t\big|_{\Phi_t^{-1}(q(t_0))} \right)$ *is positive definite for any t sufficiently close to (and strictly greater than) t_0. If $\frac{\partial^2 h}{\partial u^2}(\lambda_t, u(t)) \not\leq 0$ for some $t \in [t_0, t_1]$, then* $\mathrm{ind}\,\mathrm{Hess}_u \left(J_{t_1}\big|_{\Phi_{t_1}^{-1}(q(t_0))} \right) = \infty$.

We do not give here the proof of this well-known result; you can find it in many sources (see, for instance, the textbook [7]). It is based on the fact that $\frac{\partial^2 h}{\partial u^2}(\lambda_t, u(t)) = \lambda(\frac{\partial^2 f}{\partial u^2}(q(t), u(t))) - \frac{\partial^2 \varphi}{\partial u^2}(q(t), u(t))$ is the infinitesimal (for the "infinitesimally small interval" at t) version of $\lambda_{t_0} D_u^2 \Phi_{t_1} - D_u^2 J_{t_1}$ while $\mathrm{Hess}_u \left(J_{t_1}\big|_{\Phi_{t_1}^{-1}(q(t_0))} \right) = \left(D_u^2 J_{t_1} - \lambda_{t_0} D_w^2 \Phi_{t_1}\right)\big|_{\ker D_u \Phi_{t_1}}$.

Next theorem shows that in the "regular" infinite dimensional situation of this section we may compute the Morse index similarly to the finite dimensional case. The proof of the theorem requires some information about second variation of optimal control problems which is out of the scope of these notes. The required information can be found in Chaps. 20 and 21 of [7]. Basically, it implies that finite dimensional arguments used in the proof of Theorem 1.2 are legal also in our infinite dimensional case.

We set: $\Lambda(t) = e_*^{(t_0-t)\mathbf{H}} T_{\lambda_t}\left(T^*_{q(t)}M\right)$.

Theorem 1.3. *Assume that $\frac{\partial^2 h}{\partial u^2}(\lambda_t, u(t))$ is a negative definite quadratic form and u is a regular point of Φ_t, $\forall t \in (t_0, t_1]$. Then:*

- *The form* $\mathrm{Hess}_u\left(J_{t_1}\big|_{\Phi_{t_1}^{-1}(q(t_0))}\right)$ *is degenerate if and only if* $\Lambda(t_1)\cap\Lambda(t_0)\neq 0$
- *If* $\Lambda(t_1)\cap\Lambda(t_0)=0$, *then there exists* $\bar t>t_0$ *such that*

$$\mathrm{ind}\,\mathrm{Hess}_u\left(J_{t_1}\big|_{\Phi_{t_1}^{-1}(q(t_0))}\right)=-\mu\left(\Lambda(\cdot)\big|_{[\tau,t_1]}\right),\quad\forall\tau\in(t_0,\bar t).\qquad\square$$

Note that Legendre condition implies monotonicity of the curve $\Lambda(\cdot)$; this property simplifies the evaluation of the Maslov index. Fix some local coordinates in M so that $T^*M\cong\{(p,q)\in\mathbb{R}^{n*}\times\mathbb{R}^n\}$.

Lemma 1.4. *Quadratic form* $\underline{\dot\Lambda}(t)$ *is equivalent (with respect to a linear change of variables) to the form* $-\frac{\partial^2 H}{\partial p^2}(\lambda_t)=\frac{\partial\bar u}{\partial p}^{\top}\frac{\partial^2 h}{\partial u^2}(\lambda_t,\bar u(\lambda_t))\frac{\partial\bar u}{\partial p}$.

Proof. Equality $\frac{\partial^2 H}{\partial p^2}=-\frac{\partial\bar u}{\partial p}^{*}\frac{\partial^2 h}{\partial u^2}\frac{\partial\bar u}{\partial p}$ is an easy corollary of the identities $H(p,q)=h(p,q,\bar u(p,q))$, $\frac{\partial h}{\partial u}\big|_{u=\bar u(p,q)}=0$. Indeed, $\frac{\partial^2 H}{\partial p^2}=2\frac{\partial^2 h}{\partial u\partial p}\frac{\partial\bar u}{\partial p}+\frac{\partial\bar u}{\partial p}^{\top}\frac{\partial^2 h}{\partial u^2}\frac{\partial\bar u}{\partial p}$ and $\frac{\partial}{\partial p}\left(\frac{\partial h}{\partial u}\right)=\frac{\partial^2 h}{\partial p\partial u}+\frac{\partial^2 h}{\partial u^2}\frac{\partial\bar u}{\partial p}=0$. Further, we have: $\frac{d}{dt}\Lambda(t)=\frac{d}{dt}e_*^{(t_0-t)\mathbf{H}}T_{\lambda_t}\left(T^*_{q(t)}M\right)=e_*^{(t_0-t)\mathbf{H}}\frac{d}{d\varepsilon}\Big|_{\varepsilon=0}e_*^{-\varepsilon\mathbf{H}}T_{\lambda_{t+\varepsilon}}\left(T^*_{q(t+\varepsilon)}M\right)$. Set $\Delta(\varepsilon)=e_*^{-\varepsilon\mathbf{H}}T_{\lambda_{t+\varepsilon}}\left(T^*_{q(t+\varepsilon)}M\right)\in L\left(T_{\lambda(t)}(T^*M)\right)$. It is enough to prove that $\underline{\dot\Delta}(0)$ is equivalent to $-\frac{\partial^2 H}{\partial p^2}(\lambda_t)$. Indeed, $\dot\Lambda(t)=e_*^{(t_0-t)\mathbf{H}}T_{\lambda_t}\dot\Delta(0)$, where

$$e_*^{(t_0-t)\mathbf{H}}:T_{\lambda_t}(T^*M)\to T_{\lambda_{t_0}}(T^*M)$$

is a symplectic isomorphism.
The association of the quadratic form $\underline{\dot\Lambda}(t)$ on the subspace $\Lambda(t)$ to the tangent vector $\dot\Lambda(t)\in L\left(T_{\lambda_{t_0}}(T^*M)\right)$ is intrinsic, i.e., depends only on the symplectic structure on $(T_{\lambda_{t_0}}(T^*M)$. Hence $\underline{\dot\Delta}(0)(\xi)=\underline{\dot\Lambda}(t)\left(e_*^{(t_0-t)\mathbf{H}}\xi\right)$, $\forall\xi\in\Delta(0)=T_{\lambda_t}\left(T^*_{q(t)}M\right)$.

What remains, is to compute $\underline{\dot\Delta}(0)$; we do it in coordinates. We have:

$$\Delta(\varepsilon)=\left\{(\xi(\varepsilon),\eta(\varepsilon)):\begin{array}{ll}\dot\xi(\tau)=\xi\frac{\partial^2 H}{\partial p\partial q}(\lambda_{t-\tau})+\eta^{\top}\frac{\partial^2 H}{\partial q^2}(\lambda_{t-\tau}), & \xi(0)\in\mathbb{R}^{n*}\\ \dot\eta(\tau)=-\frac{\partial^2 H}{\partial p^2}(\lambda_{t-\tau})\xi^{\top}-\frac{\partial^2 H}{\partial q\partial p}(\lambda_{t-\tau})\eta, & \eta(0)=0\end{array}\right\},$$

$$\underline{\dot\Delta}(0)(\xi(0))=\sigma\left((\xi(0),0),(\dot\xi(0),\dot\eta(0))\right)=\xi(0)\dot\eta(0)=-\xi(0)\frac{\partial^2 H}{\partial p^2}(\lambda_t)\xi(0)^{\top}.$$

$\square$

Now combining Lemma 1.4 with Theorem 1.3 and Corollary 1.1 we obtain the following version of the classical "Morse formula"

Corollary 1.2. *Under conditions of Theorem 1.3, if* $\{\tau\in(t_0,t_1]:\Lambda(\tau)\cap\Lambda(t_0)\neq 0\}$ *is a finite subset of* (t_0,t_1), *then*

$$\mathrm{ind}\,\mathrm{Hess}J_{t_1}\big|_{\Phi_{t_1}^{-1}(q(t_0))}=\sum_{\tau\in(t_0,t_1)}\dim(\Lambda(\tau)\cap\Lambda(t_0)).$$

2 Geometry of Jacobi Curves

2.1 Jacobi Curves

Computation of the $\mathcal{L}$-derivative for regular extremals in the last section has led us to the construction of curves in the Lagrange Grassmannians which works for all Hamiltonian systems on the cotangent bundles, independently on any optimal control problem. Set $\Delta_\lambda = T_\lambda(T_q^*M)$, where $\lambda \in T_q^*M$, $q \in M$. The curve $\tau \mapsto e_*^{-\tau\mathbf{H}}\Delta_{e^{\tau\mathbf{H}}(\lambda)}$ in the Lagrange Grassmannian $L\left(T_\lambda(T^*M)\right)$ is the result of the action of the flow $e^{t\mathbf{H}}$ on the vector distribution $\{\Delta_\lambda\}_{\lambda\in T^*M}$. Now we are going to study differential geometry of these curves; their geometry will provide us with a canonical connection on T^*M associated with the Hamiltonian system and with curvature-type invariants. All that gives a far going generalization (and a dynamical interpretation) of classical objects from Riemannian geometry.

In fact, construction of the basic invariants does not need symplectic structure and the Hamiltonian nature of the flow, we may deal with more or less arbitrary pairs (*vector field, rank n distribution*) on a $2n$-dimensional manifold N. The resulting curves belong to the usual Grassmannian of all n-dimensional subspaces in the $2n$-dimensional one. We plan to work for some time in this more general situation and then come back to the symplectic framework.

In these notes we mainly deal with the case of involutive distributions (i.e., with n-foliations) just because our main motivation and applications satisfy this condition. The reader can easily recover more general definitions and construction by himself.

So we consider a $2n$-dimensional smooth manifold N endowed with a smooth foliation of rank n. Let $z \in N$, by E_z we denote the passing through z leaf of the foliation; then E_z is an n-dimensional submanifold of N. Point z has a coordinate neighborhood O_z such that the restriction of the foliation to O_z is a (trivial) fiber bundle and the fibers $E_{z'}^{loc}$, $z' \in O_z$, of this fiber bundle are connected components of $E_{z'} \cap O_z$. Moreover, there exists a diffeomorphism $O_z \cong \mathbb{R}^n \times \mathbb{R}^n$, where $\mathbb{R}^n \times \{y\}$, $y \in \mathbb{R}^n$, are identified with the fibers so that both the typical fiber and the base are diffeomorphic to $\mathbb{R}^n$. We denote by O_z/E^{loc} the base of this fiber bundle and by $\pi : O_z \to O_z/E^{loc}$ the canonical projection.

Let ζ be a smooth vector field on N. Then $z' \mapsto \pi_*\zeta(z')$, $z' \in E_z^{loc}$ is a smooth mapping of E_z^{loc} to $T_{\pi(z)}(O_z/E^{loc})$. We denote the last mapping by $\Pi_z(\zeta) : E_z^{loc} \to T_{\pi(z)}(O_z/E^{loc})$.

Definition. We call ζ a *lifting* field if $\Pi_z(\zeta)$ is a constant mapping $\forall z \in N$; The field ζ is called *regular* if $\Pi_z(\zeta)$ is a submersion, $z \in N$.

The flow generated by the lifting field maps leaves of the foliation in the leaves, in other words it is leaves-wise. On the contrary, the flow generated by the regular field "smears" the fibers over O_z/E^{loc}; basic examples are second order differential equations on a manifold M treated as the vector fields on the tangent bundle $TM = N$.

Let us write things in coordinates: We fix local coordinates acting in the domain $O \subset N$, which turn the foliation into the Cartesian product of vector spaces: $O \cong \{(x,y) : x,y \in \mathbb{R}^n\}$, $\pi : (x,y) \mapsto y$. Then vector field ζ takes the form $\zeta = \sum_{i=1}^{n} \left(a^i \frac{\partial}{\partial x_i} + b^i \frac{\partial}{\partial y_i} \right)$, where a^i, b^i are smooth functions on $\mathbb{R}^n \times \mathbb{R}^n$. The coordinate representation of the map Π_z is: $\Pi_{(x,y)} : x \mapsto \left(b^1(x,y), \ldots, b^n(x,y)\right)^\top$. Field ζ is regular if and only if $\Pi_{(x,y)}$ are submersions; in other words, if and only if $\left(\frac{\partial b^i}{\partial x_j}\right)_{i,j=1}^{n}$ is a nondegenerate matrix. Field ζ is lifting if and only if $\frac{\partial b^i}{\partial x_j} \equiv 0$, $i,j = 1, \ldots, n$.

Now turn back to the coordinate free setting. The fibers E_z, $z \in N$ are integral manifolds of the involutive distribution $\mathcal{E} = \{T_z E_z : z \in N\}$. Given a vector field ζ on N, the (local) flow $e^{t\zeta}$ generated by ζ, and $z \in N$ we define the family of subspaces

$$J_z(t) = \left(e^{-t\zeta}\right)_* \mathcal{E}|_z \subset T_z N.$$

In other words, $J_z(t) = \left(e^{-t\zeta}\right)_* T_{e^{t\zeta}(z)} E_{e^{t\zeta}(z)}$, $J_z(0) = T_z E_z$.

$J_x(t)$ is an n-dimensional subspace of $T_z N$, i.e., an element of the Grassmannian $G_n(T_z N)$. We thus have (the germ of) a curve $t \mapsto J_z(t)$ in $G_n(T_z N)$ which is called a *Jacobi curve.*

Definition. We say that field ζ is *k-ample* for an integer k if $\forall z \in N$ and for any curve $t \mapsto \hat{J}_z(t)$ in $G_n(T_z N)$ with the same k-jet as $J_z(t)$ we have $\hat{J}_z(0) \cap \hat{J}_z(t) = 0$ for all t close enough but not equal to 0. The field is called *ample* if it is k-ample for some k.

It is easy to show that a field is 1-ample if and only if it is regular.

2.2 The Cross-Ratio

Let Σ be a $2n$-dimensional vector space, $v_0, v_1 \in G_n(\Sigma)$, $v_0 \cap v_1 = 0$. Then $\Sigma = v_0 + v_1$. We denote by $\pi_{v_0 v_1} : \Sigma \to v_1$ the projector of Σ onto v_1 parallel to v_0. In other words, $\pi_{v_0 v_1}$ is a linear operator on Σ such that $\pi_{v_0 v_1}\big|_{v_0} = 0$, $\pi_{v_0 v_1}\big|_{v_1} = \mathrm{id}$. Surely, there is a one-to-one correspondence between pairs of transversal n-dimensional subspaces of Σ and rank n projectors in $\mathrm{GL}(\Sigma)$.

Lemma 2.1. *Let $v_0 \in G_n(\Sigma)$; we set $v_0^\pitchfork = \{v \in G_n(\Sigma) : v \cap v_0 = 0\}$, an open dense subset of $G_n(\Sigma)$. Then $\{\pi_{v v_0} : v \in v_0^\pitchfork\}$ is an affine subspace of $GL(\Sigma)$.*

Indeed, any operator of the form $\alpha\pi_{vv_0} + (1-\alpha)\pi_{wv_0}$, where $\alpha \in \mathbb{R}$, takes values in v_0 and its restriction to v_0 is the identity operator. Hence $\alpha\pi_{vv_0} + (1-\alpha)\pi_{wv_0}$ is the projector of Σ onto v_0 along some subspace.

The mapping $v \mapsto \pi_{vv_0}$ thus serves as a local coordinate chart on $G_n(\Sigma)$. These charts indexed by v_0 form a natural atlas on $G_n(\Sigma)$.

Projectors π_{vw} satisfy the following basic relations:[3]

$$\pi_{v_0v_1} + \pi_{v_1v_0} = id, \quad \pi_{v_0v_2}\pi_{v_1v_2} = \pi_{v_1v_2}, \quad \pi_{v_0v_1}\pi_{v_0v_2} = \pi_{v_0v_1}, \tag{1}$$

where $v_i \in G_n(\Sigma)$, $v_i \cap v_j = 0$ for $i \neq j$. If $n = 1$, then $G_n(\Sigma)$ is just the projective line $\mathbb{RP}^1$; basic geometry of $G_n(\Sigma)$ is somehow similar to geometry of the projective line for arbitrary n as well. The group $\mathrm{GL}(\Sigma)$ acts transitively on $G_n(\Sigma)$. Let us consider its standard action on $(k+1)$-tuples of points in $G_n(\Sigma)$:

$$A(v_0, \ldots, v_k) \stackrel{def}{=} (Av_0, \ldots, Av_k), \quad A \in \mathrm{GL}(\Sigma),\ v_i \in G_n(\Sigma).$$

It is an easy exercise to check that the only invariants of a triple (v_0, v_1, v_2) of points of $G_n(\Sigma)$ for such an action are dimensions of the intersections: $\dim(v_i \cap v_j)$, $0 \leq i \leq 2$, and $\dim(v_0 \cap v_1 \cap v_2)$. Quadruples of points possess a more interesting invariant: a multidimensional version of the classical cross-ratio.

Definition. Let $v_i \in G_n(\Sigma)$, $i = 0, 1, 2, 3$, and $v_0 \cap v_1 = v_2 \cap v_3 = 0$. The cross-ratio of v_i is the operator $[v_0, v_1, v_2, v_3] \in \mathrm{gl}(v_1)$ defined by the formula:

$$[v_0, v_1, v_2, v_3] = \pi_{v_0v_1}\pi_{v_2v_3}\big|_{v_1}.$$

Remark. We do not lose information when restrict the product $\pi_{v_0v_1}\pi_{v_2v_3}$ to v_1; indeed, this product takes values in v_1 and its kernel contains v_0.

For $n = 1$, v_1 is a line and $[v_0, v_1, v_2, v_3]$ is a real number. For general n, the Jordan form of the operator provides numerical invariants of the quadruple v_i, $i = 0, 1, 2, 3$.

We will mainly use an infinitesimal version of the cross-ratio that is an invariant $[\xi_0, \xi_1] \in \mathrm{gl}(v_1)$ of a pair of tangent vectors $\xi_i \in T_{v_i}G_n(\Sigma)$, $i = 0, 1$, where $v_0 \cap v_1 = 0$. Let $\gamma_i(t)$ be curves in $G_n(\Sigma)$ such that $\gamma_i(0) = v_i$, $\frac{d}{dt}\gamma_i(t)\big|_{t=0} = \xi_i$, $i = 0, 1$. Then the cross-ratio: $[\gamma_0(t), \gamma_1(0), \gamma_0(\tau), \gamma_1(\theta)]$ is a well defined operator on $v_1 = \gamma_1(0)$ for all t, τ, θ close enough to 0. Moreover, it follows from (1) that $[\gamma_0(t), \gamma_1(0), \gamma_0(0), \gamma_1(0)] = [\gamma_0(0), \gamma_1(0), \gamma_0(t), \gamma_1(0)] = [\gamma_0(0), \gamma_1(0), \gamma_0(0), \gamma_1(t)] = id$. We set

$$[\xi_0, \xi_1] = \frac{\partial^2}{\partial t \partial \tau}[\gamma_0(t), \gamma_1(0), \gamma_0(0), \gamma_1(\tau)]\big|_{v_1}\Big|_{t=\tau=0}. \tag{2}$$

It is easy to check that the right-hand side of (2) depends only on ξ_0, ξ_1 and that $(\xi_0, \xi_1) \mapsto [\xi_0, \xi_1]$ is a bilinear mapping from $T_{v_0}G_n(\Sigma) \times T_{v_1}G_n(\Sigma)$ onto $gl(v_1)$.

Lemma 2.2. *Let $v_0, v_1 \in G_n(\Sigma)$, $v_0 \cap v_1 = 0$, $\xi_i \in T_{v_i}G_n(\Sigma)$, and $\xi_i = \frac{d}{dt}\gamma_i(t)\big|_{t=0}$, $i = 0, 1$. Then $[\xi_0, \xi_1] = \frac{\partial^2}{\partial t \partial \tau}\pi_{\gamma_1(t)\gamma_0(\tau)}\big|_{v_1}\Big|_{t=\tau=0}$ and v_1, v_0 are invariant subspaces of the operator $\frac{\partial^2}{\partial t \partial \tau}\pi_{\gamma_1(t)\gamma_0(\tau)}\big|_{v_1}\Big|_{t=\tau=0}$.*

[3] Numbering of formulas is separate in each of two parts of the paper.

Proof. According to the definition, $[\xi_0, \xi_1] = \frac{\partial^2}{\partial t \partial \tau}(\pi_{\gamma_0(t)\gamma_1(0)}\pi_{\gamma_0(0)\gamma_1(\tau)})\Big|_{v_1}\Big|_{t=\tau=0}$. The differentiation of the identities $\pi_{\gamma_0(t)\gamma_1(0)}\pi_{\gamma_0(t)\gamma_1(\tau)} = \pi_{\gamma_0(t)\gamma_1(0)}, \pi_{\gamma_0(t)\gamma_1(\tau)}\pi_{\gamma_0(0)\gamma_1(\tau)} = \pi_{\gamma_0(0)\gamma_1(\tau)}$ gives the equalities:

$$\frac{\partial^2}{\partial t \partial \tau}(\pi_{\gamma_0(t)\gamma_1(0)}\pi_{\gamma_0(0)\gamma_1(\tau)})\Big|_{t=\tau=0} = -\pi_{v_0 v_1}\frac{\partial^2}{\partial t \partial \tau}\pi_{\gamma_0(t)\gamma_1(\tau)}\Big|_{t=\tau=0}$$
$$= -\frac{\partial^2}{\partial t \partial \tau}\pi_{\gamma_0(t)\gamma_1(\tau)}\Big|_{t=\tau=0}\pi_{v_0 v_1}.$$

It remains to mention that $\frac{\partial^2}{\partial t \partial \tau}\pi_{\gamma_1(t)\gamma_0(\tau)} = -\frac{\partial^2}{\partial t \partial \tau}\pi_{\gamma_0(\tau)\gamma_1(t)}$. □

2.3 Coordinate Setting

Given $v_i \in G_n(\Sigma)$, $i = 0, 1, 2, 3$, we coordinatize $\Sigma = \mathbb{R}^n \times \mathbb{R}^n = \{(x, y) : x \in \mathbb{R}^n, y \in \mathbb{R}^n\}$ in such a way that $v_i \cap \{(0, y) : y \in \mathbb{R}^n\} = 0$. Then there exist $n \times n$-matrices S_i such that

$$v_i = \{(x, S_i x) : x \in \mathbb{R}^n\}, \quad i = 0, 1, 2, 3. \tag{3}$$

The relation $v_i \cap v_j = 0$ is equivalent to $\det(S_i - S_j) \neq 0$. If $S_0 = 0$, then the projector $\pi_{v_0 v_1}$ is represented by the $2n \times 2n$-matrix $\begin{pmatrix} 0 & S_1^{-1} \\ 0 & I \end{pmatrix}$. In general, we have

$$\pi_{v_0 v_1} = \begin{pmatrix} S_{01}^{-1} S_0 & -S_{01}^{-1} \\ S_1 S_{01}^{-1} S_0 & -S_1 S_{01}^{-1} \end{pmatrix},$$

where $S_{01} = S_0 - S_1$. Relation (3) provides coordinates $\{x\}$ on the spaces v_i. In these coordinates, the operator $[v_0, v_1, v_2, v_3]$ on v_1 is represented by the matrix:

$$[v_0, v_1, v_2, v_3] = S_{10}^{-1} S_{03} S_{32}^{-1} S_{21},$$

where $S_{ij} = S_i - S_j$.

We now compute the coordinate representation of the infinitesimal cross-ratio. Let $\gamma_0(t) = \{(x, S_t x) : x \in \mathbb{R}^n\}$, $\gamma_1(t) = \{(x, S_{1+t} x) : x \in \mathbb{R}^n\}$ so that $\xi_i = \frac{d}{dt}\gamma_i(t)\big|_{t=0}$ is represented by the matrix $\dot{S}_i = \frac{d}{dt}S_t\big|_{t=i}$, $i = 0, 1$. Then $[\xi_0, \xi_1]$ is represented by the matrix

$$\frac{\partial^2}{\partial t \partial \tau} S_{1t}^{-1} S_{t\tau} S_{\tau 0}^{-1} S_{01}\Big|_{\substack{t=0 \\ \tau=1}} = \frac{\partial}{\partial t} S_{1t}^{-1}\dot{S}_1\Big|_{t=0} = S_{01}^{-1}\dot{S}_0 S_{01}^{-1}\dot{S}_1.$$

So

$$[\xi_0, \xi_1] = S_{01}^{-1}\dot{S}_0 S_{01}^{-1}\dot{S}_1. \tag{4}$$

There is a canonical isomorphism $T_{v_0}G_n(\Sigma) \cong \mathrm{Hom}(v_0, \Sigma/v_0)$; it is defined as follows. Let $\xi \in T_{v_0}G_n(\Sigma)$, $\xi = \frac{d}{dt}\gamma(t)|_{t=0}$, and $z_0 \in v_0$. Take a smooth curve $z(t) \in \gamma(t)$ such that $z(0) = z_0$. Then the residue class $(\dot{z}(0) + v_0) \in \Sigma/v_0$ depends on ξ and z_0 rather than on a particular choice of $\gamma(t)$ and $z(t)$. Indeed, let $\gamma'(t)$ be another curve in $G_n(\Sigma)$ whose velocity at $t=0$ equals ξ. Take some smooth with respect to t bases of $\gamma(t)$ and $\gamma'(t)$: $\gamma(t) = span\{e_1(t),\dots,e_n(t)\}$, $\gamma'(t) = span\{e'_1(t),\dots,e'_n(t)\}$, where $e_i(0) = e'_i(0)$, $i=1,\dots,n$; then $(\dot e_i(0) - \dot e'_i(0)) \in v_0$, $i=1,\dots,n$. Let $z(t) = \sum\limits_{i=1}^{n} \alpha_i(t)e_i(t)$, $z'(t) = \sum\limits_{i=1}^{n} \alpha'_i(t)e'_i(t)$, where $\alpha_i(0) = \alpha'_i(0)$. We have:

$$\dot z(0) - \dot z'(0) = \sum_{i=1}^{n} \left((\dot\alpha_i(0) - \dot\alpha'_i(0))e_i(0) + \alpha'_i(0)(\dot e_i(0) - \dot e'_i(0))\right) \in v_0,$$

i.e., $\dot z(0) + v_0 = \dot z'(0) + v_0$.

We associate to ξ the mapping $\bar\xi : v_0 \to \Sigma/v_0$ defined by the formula $\bar\xi z_0 = \dot z(0) + v_0$. The fact that $\xi \to \bar\xi$ is an isomorphism of the linear spaces $T_{v_0}G_n(\Sigma)$ and $\mathrm{Hom}(v_0, \Sigma/v_0)$ can be easily checked in coordinates. The matrices $\dot S_i$ above are actually coordinate presentations of $\bar\xi_i$, $i = 0, 1$.

The standard action of the group $\mathrm{GL}(\Sigma)$ on $G_n(\Sigma)$ induces the action of $\mathrm{GL}(\Sigma)$ on the tangent bundle $TG_n(\Sigma)$. It is easy to see that the only invariant of a tangent vector ξ for this action is $\mathrm{rank}\,\bar\xi$ (tangent vectors are just "double points" or "pairs of infinitesimally close points" and number $(n - \mathrm{rank}\,\bar\xi)$ is the infinitesimal version of the dimension of the intersection for a pair of points in the Grassmannian). Formula (4) implies:

$$\mathrm{rank}[\xi_0, \xi_1] \le \min\{\mathrm{rank}\,\bar\xi_0, \mathrm{rank}\,\bar\xi_1\}.$$

2.4 Curves in the Grassmannian

Let $t \mapsto v(t)$ be a germ at $\bar t$ of a smooth curve in the Grassmannian $G_n(\Sigma)$.

Definition. We say that the germ $v(\cdot)$ is *ample* if $v(t) \cap v(\bar t) = 0$ $\forall t \ne \bar t$ and the operator-valued function $t \mapsto \pi_{v(t)v(\bar t)}$ has a pole at $\bar t$. We say that the germ $v(\cdot)$ is *regular* if the function $t \mapsto \pi_{v(t)v(\bar t)}$ has a simple pole at $\bar t$. A smooth curve in $G_n(\Sigma)$ is called ample (regular) if all its germs are ample (regular).

Assume that $\Sigma = \{(x,y) : x, y \in \mathbb{R}^n\}$ is coordinatized in such a way that $v(\bar t) = \{(x,0) : x \in \mathbb{R}^n\}$. Then $v(t) = \{(x, S_t x) : x \in \mathbb{R}^n\}$, where $S(\bar t) = 0$ and $\pi_{v(t)v(\bar t)} = \begin{pmatrix} I & -S_t^{-1} \\ 0 & 0 \end{pmatrix}$. The germ $v(\cdot)$ is ample if and only if the scalar function $t \mapsto \det S_t$ has a finite order root at $\bar t$. The germ $v(\cdot)$ is regular if and only if the matrix $\dot S_{\bar t}$ is not degenerate. More generally, the curve $\tau \mapsto \{(x, S_\tau x) : x \in \mathbb{R}^n\}$ is ample if and only if $\forall t$ the function $\tau \mapsto \det(S_\tau - S_t)$ has a finite order root at t. This curve is regular if and only if $\det \dot S_t \ne 0$, $\forall t$. The intrinsic version of this coordinate characterization of regularity reads: the curve $v(\cdot)$ is regular if and only if the map $\bar{\dot v}(t) \in \mathrm{Hom}(v(t), \Sigma/v(t))$ has rank n, $\forall t$.

Coming back to the vector fields and their Jacobi curves (see Sect. 2.1) one can easily check that a vector field is ample (regular) if and only if its Jacobi curves are ample (regular).

Let $v(\cdot)$ be an ample curve in $G_n(\Sigma)$. We consider the Laurent expansions at t of the operator-valued function $\tau \mapsto \pi_{v(\tau)v(t)}$,

$$\pi_{v(\tau)v(t)} = \sum_{i=-k_t}^{m} (\tau - t)^i \pi_t^i + O(\tau - t)^{m+1}.$$

Projectors of Σ on the subspace $v(t)$ form an affine subspace of $\mathrm{gl}(\Sigma)$ (cf. Lemma 2.1). This fact implies that π_t^0 is a projector of Σ on $v(t)$; in other words, $\pi_t^0 = \pi_{v^\circ(t)v(t)}$ for some $v^\circ(t) \in v(t)^\pitchfork$. We thus obtain another curve $t \mapsto v^\circ(t)$ in $G_n(\Sigma)$, where $\Sigma = v(t) \oplus v^\circ(t)$, $\forall t$. The curve $t \mapsto v^\circ(t)$ is called the *derivative curve* of the ample curve $v(\cdot)$.

The affine space $\{\pi_{wv(t)} : w \in v(t)^\pitchfork\}$ is a translation of the linear space $\mathfrak{N}(v(t)) = \{\mathfrak{n} : \Sigma \to v(t) \mid \mathfrak{n}|_{v(t)} = 0\} \subset \mathrm{gl}(\Sigma)\}$ containing only nilpotent operators. It is easy to see that $\pi_t^i \in \mathfrak{N}(v(t))$ for $i \neq 0$.

The derivative curve is not necessary ample. Moreover, it may be non-smooth and even discontinuous.

Lemma 2.3. *If $v(\cdot)$ is regular then $v^\circ(\cdot)$ is smooth.*

Proof. We will find the coordinate representation of $v^\circ(\cdot)$. Let $v(t) = \{(x, S_t x) : x \in \mathbb{R}^n\}$. Regularity of $v(\cdot)$ is equivalent to the nondegeneracy of $\dot{S}_t$. We have:

$$\pi_{v(\tau)v(t)} = \begin{pmatrix} S_{\tau t}^{-1} S_\tau & -S_{\tau t}^{-1} \\ S_t S_{\tau t}^{-1} S_\tau & -S_t S_{\tau t}^{-1} \end{pmatrix},$$

where $S_{\tau t} = S_\tau - S_t$. Then $S_{\tau t}^{-1} = (\tau - t)^{-1} \dot{S}_t^{-1} - \frac{1}{2} \dot{S}_t^{-1} \ddot{S}_t \dot{S}_t^{-1} + O(\tau - t)$ as $\tau \to t$ and

$$\pi_{v(\tau)v(t)} = (\tau - t)^{-1} \begin{pmatrix} \dot{S}_t^{-1} S_t & -\dot{S}_t^{-1} \\ S_t \dot{S}_t^{-1} S_t & -S_t \dot{S}_t^{-1} \end{pmatrix}$$

$$+ \begin{pmatrix} I - \frac{1}{2} \dot{S}_t^{-1} \ddot{S}_t \dot{S}_t^{-1} S_t & \frac{1}{2} \dot{S}_t^{-1} \ddot{S}_t \dot{S}_t^{-1} \\ S_t - \frac{1}{2} S_t \dot{S}_t^{-1} \ddot{S}_t \dot{S}_t^{-1} S_t & \frac{1}{2} S_t \dot{S}_t^{-1} \ddot{S}_t \dot{S}_t^{-1} \end{pmatrix} + O(\tau - t).$$

We set $A_t = -\frac{1}{2} \dot{S}_t^{-1} \ddot{S}_t \dot{S}_t^{-1}$; then $\pi_{v^\circ(t)v(t)} = \begin{pmatrix} I + A_t S_t & -A_t \\ S_t + S_t A_t S_t & -S_t A_t \end{pmatrix}$ is smooth with respect to t. Hence $t \mapsto v^\circ(t)$ is smooth. We obtain:

$$v^\circ(t) = \{(A_t y, y + S_t A_t y) : y \in \mathbb{R}^n\}. \tag{5}$$

2.5 The Curvature

Definition. Let v be an ample curve and v° be the derivative curve of v. Assume that v° is differentiable at t and set $R_v(t) = [\dot{v}^\circ(t), \dot{v}(t)]$. The operator $R_v(t) \in gl(v(t))$ is called the *curvature* of the curve v at t.

If v is a regular curve, then v° is smooth, the curvature is well-defined and has a simple coordinate presentation. To find this presentation, we will use formula (4) applied to $\xi_0 = \dot{v}^\circ(t)$, $\xi_1 = \dot{v}(t)$. As before, we assume that $v(t) = \{(x, S_t x) : x \in \mathbb{R}^n\}$; in particular, $v(t)$ is transversal to the subspace $\{(0,y) : y \in \mathbb{R}^n\}$. In order to apply (4) we need an extra assumption on the coordinatization of Σ: the subspace $v^\circ(t)$ has to be transversal to $\{(0,y) : y \in \mathbb{R}^n\}$ for given t. The last property is equivalent to the nondegeneracy of the matrix A_t (see (5)). It is important to note that the final expression for $R_v(t)$ as a differential operator of S must be valid without this extra assumption since the definition of $R_v(t)$ is intrinsic! Now we compute: $v^\circ(t) = \{(x, (A_t^{-1} + S_t)x) : x \in \mathbb{R}^n\}, R_v(t) = [\dot{v}^\circ(t), \dot{v}(t)] = A_t \frac{d}{dt}(A_t^{-1} + S_t)A_t\dot{S}_t = (A_t\dot{S}_t)^2 - \dot{A}_t\dot{S}_t = \frac{1}{4}(\dot{S}_t^{-1}\ddot{S}_t)^2 - \dot{A}_t\dot{S}_t$. We also have $\dot{A}\dot{S} = -\frac{1}{2}\frac{d}{dt}(\dot{S}^{-1}\ddot{S}\dot{S}^{-1})\dot{S} = (\dot{S}^{-1})^2 - \frac{1}{2}\dot{S}^{-1}\dddot{S}$. Finally,

$$R_v(t) = \frac{1}{2}\dot{S}_t^{-1}\dddot{S}_t - \frac{3}{4}(\dot{S}_t^{-1}\ddot{S}_t)^2 = \frac{d}{dt}\left((2\dot{S}_t)^{-1}\ddot{S}_t\right) - \left((2\dot{S}_t)^{-1}\ddot{S}_t\right)^2, \qquad (6)$$

the matrix version of the Schwartzian derivative.

Curvature operator is a fundamental invariant of the curve in the Grassmannian. One more intrinsic construction of this operator, without using the derivative curve, is provided by the following

Proposition 2.1. *Let v be a regular curve in $G_n(\Sigma)$. Then*

$$[\dot{v}(\tau), \dot{v}(t)] = (\tau - t)^{-2} id + \frac{1}{3}R_v(t) + O(\tau - t)$$

as $\tau \to t$.

Proof. It is enough to check the identity in some coordinates. Given t we may assume that

$$v(t) = \{(x, 0) : x \in \mathbb{R}^n\}, \quad v^\circ(t) = \{(0, y) : y \in \mathbb{R}^n\}.$$

Let $v(\tau) = \{(x, S_\tau x : x \in \mathbb{R}^n\}$, then $S_t = \ddot{S}_t = 0$ (see (5)). Moreover, we may assume that the bases of the subspaces $v(t)$ and $v^\circ(t)$ are coordinated in such a way that $\dot{S}_t = I$. Then $R_v(t) = \frac{1}{2}\dddot{S}_t$ (see (5)). On the other hand, formula (4) for the infinitesimal cross-ratio implies:

$$[\dot{v}(\tau), \dot{v}(t)] = S_\tau^{-1}\dot{S}_\tau S_\tau^{-1} = -\frac{d}{d\tau}(S_\tau^{-1})$$

$$= -\frac{d}{d\tau}\left((\tau - t)I + \frac{(\tau - t)^3}{6}\dddot{S}_t\right)^{-1} + O(\tau - t)$$

$$= -\frac{d}{d\tau}\left((\tau - t)^{-1}I - \frac{(\tau - t)}{6}\dddot{S}_t\right) + O(\tau - t) = (\tau - t)^{-2}I + \frac{1}{6}\dddot{S}_t + O(\tau - t).$$

□

Curvature operator is an invariant of the curves in $G_n(\Sigma)$ with fixed parametrizations. Asymptotic presentation obtained in Proposition 2.1 implies a nice chain rule for the curvature of the reparameterized curves.

Let $\varphi : \mathbb{R} \to \mathbb{R}$ be a regular change of variables, i.e., $\dot\varphi \neq 0$, $\forall t$. The standard imbedding $\mathbb{R} \subset \mathbb{RP}^1 = G_1(\mathbb{R}^2)$ makes φ a regular curve in $G_1(\mathbb{R}^2)$. As we know (see (6)), the curvature of this curve is the Schwartzian of φ:

$$R_\varphi(t) = \frac{\dddot\varphi(t)}{2\dot\phi(t)} - \frac{3}{4}\left(\frac{\ddot\varphi(t)}{\dot\varphi(t)}\right)^2.$$

We set $v_\varphi(t) = v(\varphi(t))$ for any curve v in $G_n(\Sigma)$.

Proposition 2.2. *Let v be a regular curve in $G_n(\Sigma)$ and $\varphi : \mathbb{R} \to \mathbb{R}$ be a regular change of variables. Then*

$$R_{v_\varphi}(t) = \dot\varphi^2(t) R_v(\varphi(t)) + R_\varphi(t). \tag{7}$$

Proof. We have

$$[\dot v_\varphi(\tau), \dot v_\varphi(t)] = (\tau - t)^{-2}\mathrm{id} + \frac{1}{3}R_{v_\varphi}(t) + O(\tau - t).$$

On the other hand,

$$[\dot v_\varphi(\tau), \dot v_\varphi(t)] = [\dot\varphi(\tau)\dot v(\varphi(\tau)), \dot\varphi(t)\dot v(\varphi(t))] = \dot\varphi(\tau)\dot\varphi(t)[\dot v(\varphi(\tau)), \dot v(\varphi(t))]$$

$$= \dot\varphi(\tau)\dot\varphi(t)\left((\varphi(\tau) - \varphi(t))^{-2}\mathrm{id} + \frac{1}{3}R_v(\varphi(t)) + O(\tau - t)\right)$$

$$= \frac{\dot\varphi(\tau)\dot\varphi(t)}{(\varphi(\tau) - \varphi(t))^2}\mathrm{id} + \frac{\dot\varphi^2(t)}{3}R_v(\varphi(t)) + O(\tau - t).$$

We treat φ as a curve in $\mathbb{RP}^1 = G_1(\mathbb{R}^2)$. Then $[\dot\varphi(\tau), \dot\varphi(t)] = \frac{\dot\varphi(\tau)\dot\varphi(t)}{(\varphi(\tau)-\varphi(t))^2}$, see (4). The one-dimensional version of Proposition 2.1 reads:

$$[\dot\varphi(\tau), \dot\varphi(t)] = (t - \tau)^{-2} + \frac{1}{3}R_\varphi(t) + O(\tau - t).$$

Finally,

$$[\dot v_\varphi(\tau), \dot v_\varphi(t)] = (t - \tau)^{-2} + \frac{1}{3}\left(R_\varphi(t) + \dot\varphi^2(t)R_v(\varphi(t))\right) + O(\tau - t). \quad \square$$

The following identity is an immediate corollary of Proposition 2.2:

$$\left(R_{v_\varphi} - \frac{1}{n}(\mathrm{tr}R_{v_\varphi})\mathrm{id}\right)(t) = \dot\varphi^2(t)\left(R_v - \frac{1}{n}(\mathrm{tr}R_v)\mathrm{id}\right)(\varphi(t)). \tag{8}$$

Definition. An ample curve v is called flat if $R_v(t) \equiv 0$.

It follows from Proposition 2.1 that any small enough piece of a regular curve can be made flat by a reparametrization if and only if the curvature of the curve is a scalar operator, i.e., $R_v(t) = \frac{1}{n}(\mathrm{tr} R_v(t))\mathrm{id}$. In the case of a nonscalar curvature, one can use equality (8) to define a distinguished parametrization of the curve and then derive invariants which do not depend on the parametrization.

Remark. In this paper we are mainly focused on the regular curves. See paper [6] for the version of the chain rule which is valid for any ample curve and for basic invariants of unparameterized ample curves.

2.6 Structural Equations

Assume that v and w are two smooth curves in $G_n(\Sigma)$ such that $v(t) \cap w(t) = 0$, $\forall t$.

Lemma 2.4. *For any t and any $e \in v(t)$ there exists a unique $f_e \in w(t)$ with the following property: $\exists$ a smooth curve $e_\tau \in v(\tau)$, $e_t = e$, such that $\frac{d}{d\tau} e_\tau \big|_{\tau=t} = f_e$. Moreover, the mapping $\Phi_t^{vw} : e \mapsto f_t$ is linear and for any $e_0 \in v(0)$ there exists a unique smooth curve $e(t) \in v(t)$ such that $e(0) = e_0$ and*

$$\dot{e}(t) = \Phi_t^{vw} e(t), \quad \forall t. \tag{9}$$

Proof. First we take any curve $\hat{e}_\tau \in v(\tau)$ such that $e_t = e$. Then $\hat{e}_\tau = a_\tau + b_\tau$ where $a_\tau \in v(t)$, $b_\tau \in w(t)$. We take $x_\tau \in v(\tau)$ such that $x_t = \dot{a}_t$ and set $e_\tau = \hat{e}_\tau + (t - \tau)x_\tau$. Then $\dot{e}_t = \dot{b}_t$ and we put $f_e = \dot{b}_t$.

Let us prove that $\dot{b}_t$ depends only on e and not on the choice of e_τ. Computing the difference of two admissible e_τ we reduce the lemma to the following statement: if $z(\tau) \in v(\tau)$, $\forall \tau$ and $z(t) = 0$, then $\dot{z}(t) \in v(t)$.

To prove the last statement we take smooth vector-functions $e_\tau^i \in v(\tau)$, $i = 1, \ldots, n$ such that $v(\tau) = span\{e_\tau^1, \ldots, e_\tau^n\}$. Then $z(\tau) = \sum_{i=1}^{n} \alpha_i(\tau) e_\tau^i$, $\alpha_i(t) = 0$. Hence $\dot{z}(t) = \sum_{i=1}^{n} \dot{\alpha}_i(t) e_t^i \in v_t$.

Linearity of the map Φ_t^{vw} follows from the uniqueness of f_e. Indeed, if $f_{e^i} = \frac{d}{d\tau} e_\tau^i \big|_{\tau=t}$, then $\frac{d}{d\tau}(\alpha_1 e_\tau^1 + \alpha_2 e_\tau^2)\big|_{\tau=t} = \alpha_1 f_{e^1} + \alpha_2 f_{e^2}$; hence $\alpha_1 f_{e^1} + \alpha_2 f_{e^2} = f_{\alpha_1 e^1 + \alpha_2 e^2}$, $\forall e^i \in v(t)$, $\alpha_i \in \mathbb{R}$, $i = 1, 2$.

Now consider the smooth submanifold $V = \{(t, e) : t \in \mathbb{R},\ e \in v(t)\}$ of $\mathbb{R} \times \Sigma$. We have $(1, \Phi_t^{vw} e) \in T_{(t,e)}V$ since $(1, \Phi_t^{vw} e)$ is the velocity of a curve $\tau \mapsto (\tau, e_\tau)$ in V. So $(t, e) \mapsto (1, \Phi_t^{vw} e)$, $(t, e) \in V$ is a smooth vector field on V. The curve $e(t) \in v(t)$ satisfies (9) if and only if $(t, e(t))$ is a trajectory of this vector field. Now the standard existence and uniqueness theorem for ordinary differential equations provides the existence of a unique solution to the Cauchy problem for small enough t while the linearity of the equation guarantees that the solution is defined for all t. □

It follows from the proof of the lemma that $\Phi_t^{vw} e = \pi_{v(t)w(t)} \dot{e}_\tau \big|_{\tau=t}$ for any $e_\tau \in v(\tau)$ such that $v_t = e$. Let $v(t) = \{(x, S_{vt}x) : x \in \mathbb{R}^n\}$, $w(t) = \{(x, S_{wt}x) : x \in \mathbb{R}^n\}$; the matrix presentation of Φ_t^{vw} in coordinates x is $(S_{wt} - S_{vt})^{-1}\dot{S}_{vt}$. Linear mappings Φ_t^{vw} and Φ_t^{wv} provide a factorization of the infinitesimal cross-ratio $[\dot{w}(t), \dot{v}(t)]$. Indeed, equality (4) implies:

$$[\dot{w}(t), \dot{v}(t)] = -\Phi_t^{wv}\Phi_t^{vw}. \tag{10}$$

Equality (9) implies one more useful presentation of the infinitesimal cross-ratio: if $e(t)$ satisfies (9), then

$$[\dot{w}(t), \dot{v}(t)]e(t) = -\Phi_t^{wv}\Phi_t^{vw}e(t) = -\Phi_t^{wv}\dot{e}(t) = -\pi_{w(t)v(t)}\ddot{e}(t). \tag{11}$$

Now let w be the derivative curve of v, $w(t) = v^\circ(t)$. It happens that $\ddot{e}(t) \in v(t)$ in this case and (11) is reduced to the *structural equation*:

$$\ddot{e}(t) = -[\dot{v}^\circ(t), \dot{v}(t)]e(t) = -R_v(t)e(t),$$

where $R_v(t)$ is the curvature operator. More precisely, we have the following

Proposition 2.3. *Assume that v is a regular curve in $G_n(\Sigma)$, v° is its derivative curve, and $e(\cdot)$ is a smooth curve in Σ such that $e(t) \in v(t)$, $\forall t$. Then $\dot{e}(t) \in v^\circ(t)$ if and only if $\ddot{e}(t) \in v(t)$.*

Proof. Given t, we take coordinates in such a way that $v(t) = \{(x, 0) : x \in \mathbb{R}^n\}$, $v^\circ(t) = \{(0, y) : y \in \mathbb{R}^n\}$. Then $v(\tau) = \{(x, S_\tau x) : x \in \mathbb{R}^n\}$ for τ close enough to t, where $S_t = \ddot{S}_t = 0$ (see (5)).

Let $e(\tau) = \{(x(\tau), S_\tau x(\tau))\}$. The inclusion $\dot{e}(t) \in v^\circ(t)$ is equivalent to the equality $\dot{x}(t) = 0$. Further,

$$\ddot{e}(t) = \{\ddot{x}(t), \ddot{S}_t x(t) + 2\dot{S}_t\dot{x}(t) + S_t\ddot{x}(t)\} = \{\ddot{x}(t), 2\dot{S}\dot{x}\} \in v(t).$$

Regularity of v implies the nondegeneracy of $\dot{S}(t)$. Hence $\ddot{e}(t) \in v(t)$ if and only if $\dot{x}(t) = 0$. □

Now equality (11) implies

Corollary 2.1. *If $\dot{e}(t) = \Phi_t^{vv^\circ}e(t)$, then $\ddot{e}(t) + R_v(t)e(t) = 0$.*

Let us consider invertible linear mappings $V_t : v(0) \to v(t)$ defined by the relations $V_t e(0) = e(t)$, $\dot{e}(\tau) = \Phi_\tau^{vv^\circ}e(\tau)$, $0 \le \tau \le t$. It follows from the structural equation that the curve v is uniquely reconstructed from $\dot{v}(0)$ and the curve $t \mapsto V_t^{-1}R_V(t)$ in $\mathrm{gl}(v(0))$. Moreover, let $v_0 \in G_n(\Sigma)$ and $\xi \in T_{v_0}G_n(\Sigma)$, where the map $\bar{\xi} \in \mathrm{Hom}(v_0, \Sigma/v_0)$ has rank n; then for any smooth curve $t \mapsto A(t)$ in $\mathrm{gl}(v_0)$ there exists a unique regular curve v such that $\dot{v}(0) = \xi$ and $V_t^{-1}R_v(t)V_t = A(t)$. Indeed, let $e_i(0)$, $i = 1, \ldots, n$, be a basis of v_0 and $A(t)e_i(0) = \sum_{j=1}^n a_{ij}(t)e_j(0)$. Then $v(t) = span\{e_1(t), \ldots, e_n(t)\}$, where

$$\ddot{e}_i(\tau) + \sum_{j=1}^n a_{ij}(\tau)e_j(\tau) = 0, \ 0 \le \tau \le t, \tag{12}$$

are uniquely defined by fixing the $\dot{v}(0)$.

The obtained classification of regular curves in terms of the curvature is particularly simple in the case of a scalar curvature operators $R_v(t) = \rho(t)\mathrm{id}$. Indeed, we have $A(t) = V_t^{-1}R_v(t)V_t = \rho(t)\mathrm{id}$ and system (12) is reduced to n copies of the Hill equation $\ddot{e}(\tau) + \rho(\tau)e(\tau) = 0$.

Recall that all $\xi \in TG_n(\Sigma)$ such that $\mathrm{rank}\,\xi = n$ are equivalent under the action of $\mathrm{GL}(\Sigma)$ on $TG_n(\Sigma)$ induced by the standard action on the Grassmannian $G_n(\Sigma)$. We thus obtain

Corollary 2.2. *For any smooth scalar function $\rho(t)$ there exists a unique, up to the action of $GL(\Sigma)$, regular curve v in $G_n(\Sigma)$ such that $R_v(t) = \rho(t)id$.*

Another important special class is that of symmetric curves.

Definition. A regular curve v is called *symmetric* if $V_tR_v(t) = R_v(t)V_t$, $\forall t$.

In other words, v is symmetric if and only the curve $A(t) = V_t^{-1}R_v(t)V_t$ in $\mathrm{gl}(v(0))$ is constant and coincides with $R_v(0)$. The structural equation implies

Corollary 2.3. *For any $n \times n$-matrix A_0, there exists a unique, up to the action of $GL(\Sigma)$, symmetric curve v such that $R_v(t)$ is similar to A_0.*

The derivative curve v° of a regular curve v is not necessary regular. The formula $R_v(t) = \Phi_t^{v^\circ v}\Phi_t^{vv^\circ}$ implies that v° is regular if and only if the curvature operator $R_v(t)$ is nondegenerate for any t. Then we may compute the second derivative curve $v^{\circ\circ} = (v^\circ)^\circ$.

Proposition 2.4. *A regular curve v with nondegenerate curvature operators is symmetric if and only if $v^{\circ\circ} = v$.*

Proof. Let us consider system (12). We are going to apply Proposition 2.3 to the curve v° (instead of v) and the vectors $\dot{e}_i(t) \in v^\circ(t)$. According to Proposition 2.3, $v^{\circ\circ} = v$ if and only if $\frac{d^2}{dt^2}\dot{e}_i(t) \in v^\circ(t)$. Differentiating (12) we obtain that $v^{\circ\circ} = v$ if and only if the functions $\alpha_{ij}(t)$ are constant. The last property is none other than a characterization of symmetric curves. □

2.7 Canonical Connection

Now we apply the developed theory of curves in the Grassmannian to the Jacobi curves $J_z(t)$ (see Sect. 2.1).

Proposition 2.5. *All Jacobi curves $J_z(\cdot)$, $z \in N$, associated to the given vector field ζ are regular (ample) if and only if the field ζ is regular (ample).*

Proof. The definition of the regular (ample) field is actually the specification of the definition of the regular (ample) germ of the curve in the Grassmannian: general definition is applied to the germs at $t = 0$ of the curves $t \mapsto J_z(t)$. What remains is to demonstrate that other germs of these curves are regular (ample) as soon as the germs at 0 are. The latter fact follows from the identity

$$J_z(t+\tau) = e_*^{-t\zeta} J_{e^{t\zeta}(z)}(\tau) \tag{13}$$

(which, in turn, is an immediate corollary of the identity $e_*^{-(t+\tau)\zeta} = e_*^{-t\zeta} \circ e_*^{-\tau\zeta}$). Indeed, (13) implies that the germ of $J_z(\cdot)$ at t is the image of the germ of $J_{e^{t\zeta}(\tau)}(\cdot)$ at 0 under the fixed linear transformation $e_*^{-t\zeta} : T_{e^{t\zeta}(z)}N \to T_z N$. The properties of the germs to be regular or ample survive linear transformations since they are intrinsic properties. □

Let ζ be an ample field. Then the derivative curves $J_z^{\circ}(t)$ are well-defined. Moreover, identity (13) and the fact that the construction of the derivative curve is intrinsic imply:

$$J_z^{\circ}(t) = e_*^{-t\zeta} J_{e^{t\zeta}(z)}^{\circ}(0). \tag{14}$$

The value at 0 of the derivative curve provides the splitting $T_z M = J_z(0) \oplus J_z^{\circ}(0)$, where the first summand is the tangent space to the fiber, $J_z(0) = T_z E_z$.

Now assume that $J_z^{\circ}(t)$ smoothly depends on z; this assumption is automatically fulfilled in the case of a regular ζ, where we have the explicit coordinate presentation for $J_z^{\circ}(t)$. Then the subspaces $J_z^{\circ}(0) \subset T_z N,\ z \in N$, form a smooth vector distribution, which is the direct complement to the vertical distribution $\mathcal{E} = \{T_z E_z : z \in N\}$. Direct complements to the vertical distribution are called Ehresmann connections (or just nonlinear connections, even if linear connections are their special cases). The Ehresmann connection $\mathcal{E}_\zeta = \{J_z^{\circ}(0) : z \in N\}$ is called the *canonical connection* associated with ζ and the correspondent splitting $TN = \mathcal{E} \oplus \mathcal{E}_\zeta$ is called the *canonical splitting.* Our nearest goal is to give a simple intrinsic characterization of $\mathcal{E}_\zeta$ which does not require the integration of the equation $\dot{z} = \zeta(z)$ and is suitable for calculations not only in local coordinates but also in moving frames.

Let $\mathcal{F} = \{F_z \subset T_z N : z \in N\}$ be an Ehresmann connection. Given a vector field ξ on E we denote $\xi_{ver}(z) = \pi_{F_z J_z(0)}\xi$, $\xi_{hor}(z) = \pi_{J_z(0) F_z}\xi$, the "vertical" and the "horizontal" parts of $\xi(z)$. Then $\xi = \xi_{ver} + \xi_{hor}$, where ξ_{ver} is a section of the distribution $\mathcal{E}$ and ξ_{hor} is a section of the distribution $\mathcal{F}$. In general, sections of $\mathcal{E}$ are called vertical fields and sections of $\mathcal{F}$ are called horizontal fields.

Proposition 2.6. *Assume that ζ is a regular field. Then $\mathcal{F} = \mathcal{E}_\zeta$ if and only if the equality*

$$[\zeta, [\zeta, \nu]]_{hor} = 2[\zeta, [\zeta, \nu]_{ver}]_{hor} \tag{15}$$

holds for any vertical vector field ν. Here $[\ ,\ ]$ is Lie bracket of vector fields.

Proof. The deduction of identity (15) is based on the following classical expression:

$$\frac{d}{dt} e_*^{-t\zeta} \xi = e_*^{-t\zeta} [\zeta, \xi], \tag{16}$$

for any vector field ξ.

Given $z \in N$, we take coordinates in T_zN in such a way that $T_zN = \{(x,y) : x, y \in \mathbb{R}^n\}$, where $J_z(0) = \{(x,0) : x \in \mathbb{R}^n\}$, $J_z^\circ(0) = \{(0,y) : y \in \mathbb{R}^n\}$. Let $J_z(t) = \{(x, S_t x) : x \in \mathbb{R}^n\}$, then $S_0 = \ddot{S}_0 = 0$ and $\det \dot{S}_0 \neq 0$ due to the regularity of the Jacobi curve J_z.

Let ν be a vertical vector field, $\nu(z) = (x_0, 0)$ and $\left(e_*^{-t\zeta}\nu\right)(z) = (x_t, y_t)$. Then $(x_t, 0) = \left(e_*^{-t\zeta}\nu\right)_{ver}(z)$, $(0, y_t) = \left(e_*^{-t\zeta}\nu\right)_{hor}(z)$. Moreover, $y_t = S_t x_t$ since $\left(e_*^{-t\zeta}\nu\right)(z) \in J_z(t)$. Differentiating the identity $y_t = S_t x_t$ we obtain: $\dot{y}_t = \dot{S}_t x_t + S_t \dot{x}_t$. In particular, $\dot{y}_0 = \dot{S}_0 x_0$. It follows from (16) that $(\dot{x}_0, 0) = [\zeta, \nu]_{ver}$, $(0, \dot{y}_0) = [\zeta, \nu]_{hor}$. Hence $(0, \dot{S}_0 x_0) = [\zeta, \nu]_{hor}(z)$, where, I recall, ν is any vertical field. Now we differentiate once more and evaluate the derivative at 0:

$$\ddot{y}_0 = \ddot{S}_0 x_0 + 2\dot{S}_0 \dot{x}_0 + S_0 \ddot{x}_0 = 2\dot{S}_0 \dot{x}_0. \tag{17}$$

The Lie bracket presentations of the left and right hand sides of (17) are: $(0, \ddot{y}_0) = [\zeta, [\zeta, \nu]]_{hor}$, $(0, \dot{S}_0 \dot{x}_0) = [\zeta, [\zeta, \nu]_{ver}]_{hor}$. Hence (17) implies identity (15).

Assume now that $\{(0,y) : y \in \mathbb{R}^n\} \neq J_z^\circ(0)$; then $\ddot{S}_0 x_0 \neq 0$ for some x_0. Hence $\ddot{y}_0 \neq 2\dot{S}_0 \dot{x}_0$ and equality (15) is violated. □

Inequality (15) can be equivalently written in the following form that is often more convenient for the computations:

$$\pi_*[\zeta, [\zeta, \nu]](z) = 2\pi_*[\zeta, [\zeta, \nu]_{ver}](z), \quad \forall z \in N. \tag{18}$$

Let $R_{J_z}(t) \in \mathrm{gl}(J_z(t))$ be the curvature of the Jacobi curve $J_z(t)$. Identity (13) and the fact that construction of the Jacobi curve is intrinsic imply that

$$R_{J_z}(t) = e_*^{-t\zeta} R_{J_{e^{t\zeta}(z)}}(0) e_*^{t\zeta}\big|_{J_z(t)}.$$

Recall that $J_z(0) = T_z E_z$; the operator $R_{J_z}(0) \in \mathrm{gl}(T_z E_z)$ is called *the curvature operator of the field* ζ at z. We introduce the notation: $R_\zeta(z) \stackrel{def}{=} R_{J_z}(0)$; then $R_\zeta = \{R_\zeta(z)\}_{z \in E}$ is an automorphism of the "vertical" vector bundle $\{T_z E_z\}_{z \in M}$.

Proposition 2.7. *Assume that ζ is an ample vector field and $J_z^\circ(0)$ is smooth with respect to z. Let $TN = \mathcal{E} \oplus \mathcal{E}_\zeta$ be the canonical splitting. Then*

$$R_\zeta \nu = -[\zeta, [\zeta, \nu]_{hor}]_{ver} \tag{19}$$

for any vertical field ν.

Proof. Recall that $R_{J_z}(0) = [\dot{J}_z^\circ(0), \dot{J}_z(0)]$, where $[\cdot,\cdot]$ is the infinitesimal cross-ratio (not the Lie bracket!). The presentation (10) of the infinitesimal cross-ratio implies:

$$R_\zeta(z) = R_{J_z}(0) = -\Phi_0^{J_z^\circ J_z} \Phi_0^{J_z J_z^\circ},$$

where $\Phi_0^{vw} e = \pi_{v(0)w(0)} \dot{e}_0$ for any smooth curve $e_\tau \in v(\tau)$ such that $e_0 = e$. Equalities (14) and (16) imply: $\Phi_0^{J_z J_z^\circ} \nu(z) = [\zeta, \nu]_{ver}(z), \ \forall z \in M$. Similarly, $\Phi_0^{J_z^\circ J_z} \mu(z) = [\zeta, \mu]_{hor}(z)$ for any horizontal field μ and any $z \in M$. Finally,

$$R_\zeta(z)\nu(z) = -\Phi_0^{J_z^\circ J_z} \Phi_0^{J_z J_z^\circ} = -[\zeta, [\zeta, \nu]_{hor}]_{ver}(z). \qquad \square$$

2.8 Coordinate Presentation

We fix local coordinates acting in the domain $\mathcal{O} \subset N$, which turn the foliation into the Cartesian product of vector spaces: $\mathcal{O} \cong \{(x, y) : x, y \in \mathbb{R}^n\}$, $\pi : (x, y) \mapsto y$. Then vector field ζ takes the form $\zeta = \sum_{i=1}^{n} \left(a^i \frac{\partial}{\partial x_i} + b^i \frac{\partial}{\partial y_i} \right)$, where a^i, b^i are smooth functions on $\mathbb{R}^n \times \mathbb{R}^n$. Below we use abridged notations: $\frac{\partial}{\partial x_i} = \partial_{x_i}$, $\frac{\partial \varphi}{\partial x_i} = \varphi_{x_i}$ etc. We also use the standard summation agreement for repeating indices.

Recall the coordinate characterization of the regularity property for the vector field ζ. Intrinsic definition of regular vector fields is done in Sect. 8; it is based on the mapping Π_z whose coordinate presentation is: $\Pi_{(x,y)} : x \mapsto \left(b^1(x, y), \ldots, b^n(x, y)\right)^\top$. Field ζ is regular if and only if Π_y are submersions; in other words, if and only if $\left(b^i_{x_j}\right)_{i,j=1}^{n}$ is a non degenerate matrix.

Vector fields ∂_{x_i}, $i = 1, \ldots, n$, provide a basis of the space of vertical fields. As soon as coordinates are fixed, any Ehresmann connection finds a unique basis of the form:

$$(\partial_{y_i})_{hor} = \partial_{y_i} + c_i^j \partial_{x_j},$$

where c_i^j, $i, j = 1, \ldots, n$, are smooth functions on $\mathbb{R}^n \times \mathbb{R}^n$. To characterize a connection in coordinates thus means to find functions c_i^j. In the case of the canonical connection of a regular vector field, the functions c_i^j can be easily recovered from identity (18) applied to $\nu = \partial_{x_i}$, $i = 1, \ldots, n$. We will do it explicitly for two important classes of vector fields: second order ordinary differential equations and Hamiltonian systems.

A second order ordinary differential equation

$$\dot{y} = x, \quad \dot{x} = f(x, y) \tag{20}$$

there corresponds to the vector field $\zeta = f^i \partial_{x_i} + x_i \partial_{y_i}$, where $f = (f_1, \ldots, f_n)^\top$. Let $\nu = \partial_{x_i}$; then

$$[\zeta, \nu] = -\partial_{y_i} - f^j_{x_i} \partial_{x_j}, \ [\zeta, \nu]_{ver} = (c_i^j - f^j_{x_i})\partial_{x_j},$$

$$\pi_*[\zeta, [\zeta, \nu]] = f^j_{x_i} \partial_{y_j}, \ \pi_*[\zeta, [\zeta, \nu]_{ver}] = (f^j_{x_i} - c_i^j)\partial_{y_j}.$$

Hence, in virtue of equality (18) we obtain that $c_i^j = \frac{1}{2} f^j_{x_i}$ for the canonical connection associated with the second order differential equation (20).

Now consider a Hamiltonian vector field $\zeta = -h_{y_i}\partial_{x_i} + h_{x_i}\partial_{y_i}$, where h is a smooth function on $\mathbb{R}^n \times \mathbb{R}^n$ (a Hamiltonian). The field ζ is regular if and only if the matrix $h_{xx} = \left(h_{x_i x_j}\right)_{i,j=1}^n$ is non degenerate. We are going to characterize the canonical connection associated with ζ. Let $C = \left(c_i^j\right)_{i,j=1}^n$; the straightforward computation similar to the computation made for the second order ordinary differential equation gives the following presentation for the matrix C:

$$2\left(h_{xx} C h_{xx}\right)_{ij} = h_{x_k} h_{x_i x_j y_k} - h_{y_k} h_{x_i x_j x_k} - h_{x_i y_k} h_{x_k x_j} - h_{x_i x_k} h_{y_k x_j}$$

or, in the matrix form:

$$2h_{xx} C h_{xx} = \{h, h_{xx}\} - h_{xy} h_{xx} - h_{xx} h_{yx},$$

where $\{h, h_{xx}\}$ is the Poisson bracket: $\{h, h_{xx}\}_{ij} = \{h, h_{x_i x_j}\} = h_{x_k} h_{x_i x_j y_k} - h_{y_k} h_{x_i x_j x_k}$.

Note that matrix C is symmetric in the Hamiltonian case (indeed, $h_{xx} h_{yx} = (h_{xy} h_{xx})^{\top}$). This is not occasional and is actually guaranteed by the fact that Hamiltonian flows preserve symplectic form $dx_i \wedge dy_i$. See Sect. 10 for the symplectic version of the developed theory.

As soon as we found the canonical connection, formula (19) gives us the presentation of the curvature operator although the explicit coordinate expression can be bulky. Let us specify the vector field more. In the case of the Hamiltonian of a natural mechanical system, $h(x,y) = \frac{1}{2}|x|^2 + U(y)$, the canonical connection is trivial: $c_i^j = 0$; the matrix of the curvature operator is just U_{yy}.

Hamiltonian vector field associated to the Hamiltonian $h(x,y) = g^{ij}(y) x_i x_j$ with a non degenerate symmetric matrix $\left(g^{ij}\right)_{i,j=1}^n$ generates a (pseudo-)Riemannian geodesic flow. Canonical connection in this case is classical Levi-Civita connection and the curvature operator is Ricci operator of (pseudo-)Riemannian geometry (see [4, Sect. 5] for details). Finally, Hamiltonian $h(x,y) = g^{ij}(y) x_i x_j + U(y)$ has the same connection as Hamiltonion $h(x,y) = g^{ij}(y) x_i x_j$ while its curvature operator is sum of Ricci operator and second covariant derivative of U.

2.9 Affine Foliations

Let $[\mathcal{E}]$ be the sheaf of germs of sections of the distribution $\mathcal{E} = \{T_z E_z : z \in N\}$ equipped with the Lie bracket operation. Then $[\mathcal{E}]_z$ is just the Lie algebra of germs at $z \in M$ of vertical vector fields. Affine structure on the foliation E is a sub-sheaf $[\mathcal{E}]^a \subset [\mathcal{E}]$ such that $[\mathcal{E}]^a_z$ is an Abelian sub-algebra of $[\mathcal{E}]_z$ and $\{\varsigma(z) : \varsigma \in [\mathcal{E}]^a_z\} = T_z E_z$, $\forall z \in N$. A foliation with a fixed affine structure is called the *affine foliation.*

The notion of the affine foliation generalizes one of the vector bundle. In the case of the vector bundle, the sheaf $[\mathcal{E}]^a$ is formed by the germs of vertical vector fields whose restrictions to the fibers are constant (i.e., translation invariant) vector fields on the fibers. In the next section we will describe an important class of affine foliations which is not reduced to the vector bundles.

Lemma 2.5. *Let $\mathcal{E}$ be an affine foliation, $\varsigma \in [\mathcal{E}]^a_z$ and $\varsigma(z) = 0$. Then $\varsigma|_{E_z} = 0$.*

Proof. Let $\varsigma_1, \dots, \varsigma_n \in [\mathcal{E}]^a_z$ be such that $\varsigma_1(z), \dots, \varsigma_n(z)$ form a basis of T_zE_z. Then $\varsigma = b_1\varsigma_1 + \cdots + b_n\varsigma_n$, where b_i are germs of smooth functions vanishing at z. Commutativity of $[\mathcal{E}]^a_z$ implies: $0 = [\varsigma_i, \varsigma] = (\varsigma_i b_1)\varsigma_1 + \cdots + (\varsigma_i b_n)\varsigma_n$. Hence functions $b_i|_{E_z}$ are constants, i.e., $b_i|_{E_z} = 0,\ i = 1, \dots, n$. $\square$

Lemma 2.5 implies that $\varsigma \in [\mathcal{E}]^a_z$ is uniquely reconstructed from $\varsigma(z)$. This property permits to define the *vertical derivative* of any vertical vector field ν on M. Namely, $\forall v \in T_zE_z$ we set

$$D_v\nu = [\varsigma, \nu](z), \text{ where } \varsigma \in [\mathcal{E}]^a_z,\ \varsigma(z) = v.$$

Suppose ζ is a regular vector field on the manifold N endowed with the affine n-foliation. The canonical Ehresmann connection $\mathcal{E}_\zeta$ together with the vertical derivative allow to define a canonical linear connection ∇ on the vector bundle $\mathcal{E}$. Sections of the vector bundle $\mathcal{E}$ are just vertical vector fields. We set

$$\nabla_\xi\nu = [\xi, \nu]_{ver} + D_\nu(\xi_{ver}),$$

where ξ is any vector field on N and ν is a vertical vector field. It is easy to see that ∇ satisfies axioms of a linear connection. The only non evident one is: $\nabla_{b\xi}\nu = b\nabla_\xi\nu$ for any smooth function b. Let $z \in N$,$\varsigma \in [\mathcal{E}]^a_z$, and $\varsigma(z) = \nu(z)$. We have

$$\nabla_{b\xi}\nu = [b\xi, \nu]_{ver} + [\varsigma, b\xi_{ver}] = b\left([\xi, \nu]_{ver} + [\varsigma, \xi_{ver}]\right) - (\nu b)\xi_{ver} + (\varsigma b)\xi_{ver}.$$

Hence

$$(\nabla_{b\xi}\nu)(z) = b(z)\left([\xi, \nu]_{ver}(z) + [\varsigma, \xi_{ver}](z)\right) = (b\nabla_\xi\nu)(z).$$

Linear connection ∇ gives us the way to express Pontryagin characteristic classes of the vector bundle $\mathcal{E}$ via the regular vector field ζ. Indeed, any linear connection provides an expression for Pontryagin classes. We are going to briefly recall the correspondent classical construction (see [13] for details). Let $R^\nabla(\xi, \eta) = [\nabla_\xi, \nabla_\eta] - \nabla_{[\xi,\eta]}$ be the curvature of linear connection ∇. Then $R^\nabla(\xi, \eta)\nu$ is $C^\infty(M)$-linear with respect to each of three arguments ξ, η, ν. In particular, $R^\nabla(\cdot,\cdot)\nu(z) \in \bigwedge^2(T_z^*N) \otimes T_zE_z,\ z \in N$. In other words, $R^\nabla(\cdot,\cdot) \in \mathrm{Hom}\left(\mathcal{E}, \bigwedge^2(T^*N) \otimes \mathcal{E}\right)$.

Consider the commutative exterior algebra

$$\bigwedge^{ev} N = C^\infty(N) \oplus \bigwedge^2 (T^*N) \oplus \cdots \oplus \bigwedge^{2n} (T^*N)$$

of the even order differential forms on N. Then R^∇ can be treated as an endomorphism of the module $\bigwedge^{ev} N \otimes \mathcal{E}$ over algebra $\bigwedge^{ev} N$, i.e., $R^\nabla \in \mathrm{End}_{\bigwedge^{ev} N}(\bigwedge^{ev} M \otimes \mathcal{E})$.

Now consider characteristic polynomial $\det(tI + \frac{1}{2\pi} R^\nabla) = t^n + \sum_{i=1}^{n} \phi_i t^{n-i}$, where the coefficient ϕ_i is an order $2i$ differential form on N. All forms ϕ_i are closed; the forms ϕ_{2k-1} are exact and the forms ϕ_{2k} represent the Pontryagin characteristic classes, $k = 1, \ldots, [\frac{n}{2}]$.

2.10 Symplectic Setting

Assume that N is a symplectic manifold endowed with a symplectic form σ. Recall that a symplectic form is just a closed non degenerate differential 2-form. Suppose E is a Lagrange foliation on the symplectic manifold (N, σ); this means that $\sigma|_{E_z} = 0$, $\forall z \in N$. Basic examples are cotangent bundles endowed with the standard symplectic structure: $N = T^*M$, $E_z = T^*_{\pi(z)} M$, where $\pi : T^*M \to M$ is the canonical projection. In this case $\sigma = d\tau$, where $\tau = \{\tau_z : z \in T^*M\}$ is the Liouville 1-form on T^*M defined by the formula: $\tau_z = z \circ \pi_*$. Completely integrable Hamiltonian systems provide another important class of Lagrange foliations. We will briefly recall the correspondent terminology. Details can be found in any introduction to symplectic geometry (for instance, in [10]).

Smooth functions on the symplectic manifold are called Hamiltonians. To any Hamiltonian there corresponds a Hamiltonian vector field $\mathbf{h}$ on M defined by the equation: $dh = \sigma(\cdot, \mathbf{h})$. The Poisson bracket $\{h_1, h_2\}$ of the Hamiltonians h_1 and h_2 is the Hamiltonian defined by the formula: $\{h_1, h_2\} = \sigma(\mathbf{h}_1, \mathbf{h}_2) = \mathbf{h}_1 h_2$. Poisson bracket is obviously anti-symmetric and satisfies the Jacobi identity: $\{h_1, \{h_2, h_3\}\} + \{h_3, \{h_1, h_2\}\} + \{h_2, \{h_3, h_1\}\} = 0$. This identity is another way to say that the form σ is closed. Jacobi identity implies one more useful formula: $\overrightarrow{\{h_1, h_2\}} = [\mathbf{h}_1, \mathbf{h}_2]$.

We say that Hamiltonians $h_1, \ldots, h_n$ are in involution if $\{h_i, h_j\} = 0$; then h_j is constant along trajectories of the Hamiltonian equation $\dot{z} = \mathbf{h}_i(z)$, $i, j = 1, \ldots, n$. We say that $h_1, \ldots, h_n$ are independent if $d_z h_1 \wedge \cdots \wedge d_z h_n \neq 0$, $z \in N$. n independent Hamiltonians in involution form a *completely integrable system*. More precisely, any of Hamiltonian equations $\dot{z} = \mathbf{h}_i(z)$ is completely integrable with first integrals $h_1, \ldots, h_n$.

Lemma 2.6. *Let Hamiltonians $h_1, \ldots, h_n$ form a completely integrable system. Then the n-foliation $E_z = \{z' \in M : h_i(z') = h_i(z),\ i = 1, \ldots, n\}$, $z \in N$, is Lagrangian.*

Proof. We have $\mathbf{h}_i h_j = 0$, $i,j = 1,\dots,n$, hence $\mathbf{h}_i(z)$ are tangent to E_z. Vectors $\mathbf{h}_1(z),\dots,\mathbf{h}_n(z)$ are linearly independent, hence

$$span\{\mathbf{h}_1(z),\dots,\mathbf{h}_n(z)\} = T_z E_z.$$

Moreover, $\sigma(\mathbf{h}_i,\mathbf{h}_j) = \{h_i,h_j\} = 0$, hence $\sigma|_{E_z} = 0$. □

Any Lagrange foliation possesses a canonical affine structure. Let $[\mathcal{E}]$ be the sheaf of germs of the distribution $\mathcal{E} = \{T_z E_z : z \in N\}$ as in Sect. 9; then $[\mathcal{E}]^a$ is the intersection of $[\mathcal{E}]$ with the sheaf of germs of Hamiltonian vector fields.

We have to check that Lie algebra $[\mathcal{E}]^a_z$ is Abelian and generates $T_z E_z$, $\forall z \in N$. First check the Abelian property. Let $\mathbf{h}_1,\mathbf{h}_2 \in [\mathcal{E}]^a_z$; we have $[\mathbf{h}_1,\mathbf{h}_2] = \overrightarrow{\{h_1,h_2\}}$, $\{h_1,h_2\} = \sigma(\mathbf{h}_1,\mathbf{h}_2) = 0$, since $\mathbf{h}_i$ are tangent to E_z and $\sigma|_{E_z} = 0$. The second property follows from the Darboux–Weinstein theorem (see [10]) which states that all Lagrange foliations are locally equivalent. More precisely, this theorem states that any $z \in M$ possesses a neighborhood O_z and local coordinates which turn the restriction of the Lagrange foliation E to O_z into the trivial bundle $\mathbb{R}^n \times \mathbb{R}^n = \{(x,y) : x,y \in \mathbb{R}^n\}$ and, simultaneously, turn $\sigma|_{O_z}$ into the form $\sum\limits_{i=1}^{n} dx_i \wedge dy_i$. In this special coordinates, the fibers become coordinate subspaces $\mathbb{R}^n \times \{y\}$, $y \in \mathbb{R}^n$, and the required property is obvious: vector fields $\frac{\partial}{\partial x_i}$ are Hamiltonian fields associated to the Hamiltonians $-y_i$, $i = 1,\dots,n$.

Suppose ζ is a Hamiltonian field on the symplectic manifold endowed with the Lagrange foliation, $\zeta = \mathbf{h}$. Let $\varsigma \in [\mathcal{E}]^a_z$, $\varsigma = \mathbf{s}$; then $\varsigma h = \{s,h\}$. The field $\mathbf{h}$ is regular if and only if the quadratic form $s \mapsto \{s,\{s,h\}\}(z)$ has rank n. Indeed, in the "Darboux–Weinstein coordinates" this quadratic form has the matrix $\{\frac{\partial^2 h}{\partial x_i \partial x_j}\}_{i,j=1}^n$.

Recall that the tangent space $T_z N$ to the symplectic manifold N is a symplectic space endowed with the symplectic structure σ_z. An n-dimensional subspace $\upsilon \subset T_z N$ is a Lagrangian subspace if $\sigma_z|_\upsilon = 0$. The set

$$L(T_z N) = \{\upsilon \in G_n(T_z M) : \sigma_z|_\upsilon = 0\}$$

of all Lagrange subspaces of $T_z M$ is a Lagrange Grassmannian.

Hamiltonian flow $e^{t\mathbf{h}}$ preserves the symplectic form, $\left(e^{t\mathbf{h}}\right)^* \sigma = \sigma$. Hence $\left(e^{t\mathbf{h}}\right)_* : T_z N \to T_{e^{t\mathbf{h}}(z)} N$ transforms Lagrangian subspaces in the Lagrangian ones. It follows that the Jacobi curve $J_z(t) = \left(e^{-t\mathbf{h}}\right)_* T_{e^{t\mathbf{h}}(z)} E_{e^{t\mathbf{h}}(z)}$ consists of Lagrangian subspaces, $J_z(t) \in L(T_z N)$.

We need few simple facts on Lagrangian Grassmannians (see Sect. 1.6 for the basic information and [3, Sect. 4] for a consistent description of their geometry). Let $(\Sigma,\bar\sigma)$ be a $2n$-dimensional symplectic space and $\upsilon_0,\upsilon_1 \in L(\Sigma)$ be a pair of transversal Lagrangian subspaces, $\upsilon_0 \cap \upsilon_1 = 0$. Bilinear form $\bar\sigma$ induces a non degenerate pairing of the spaces υ_0 and υ_1 by the rule

$(e,f) \mapsto \bar\sigma(e,f)$, $e \in v_0, f \in v_1$. To any basis $e_1,\dots,e_n$ of v_0 we may associate a unique dual basis $f_1,\dots,f_n$ of v_1 such that $\bar\sigma(e_i,f_j) = \delta_{ij}$. The form $\bar\sigma$ is totally normalized in the basis $e_1,\dots,e_n,f_1,\dots,f_n$ of Σ, since $\sigma(e_i,e_j) = \sigma(f_i,f_j) = 0$. It follows that symplectic group

$$\mathrm{Sp}(\Sigma) = \{A \in \mathrm{GL}(\Sigma) : \bar\sigma(Ae,Af) = \bar\sigma(e,f),\ e,f \in \Sigma\}$$

acts transitively on the pairs of transversal Lagrangian subspaces.

Next result is a "symplectic specification" of Lemma 2.1 from Sect. 9.

Lemma 2.7. *Let $v_0 \in L(\Sigma)$; then $\{\pi_{vv_0} : v \in v_0^{\pitchfork} \cap L(\Sigma)\}$ is an affine subspace of the affine space $\{\pi_{vv_0} : v \in v_0^{\pitchfork}\}$ characterized by the relation:*

$$v \in v_0^{\pitchfork} \cap L(\Sigma) \ \Leftrightarrow\ \bar\sigma(\pi_{vv_0}\cdot,\cdot) + \bar\sigma(\cdot,\pi_{vv_0}\cdot) = \bar\sigma(\cdot,\cdot).$$

Proof. Assume that $v_1 \in v_0^{\pitchfork} \cap L(\Sigma)$. Let $e,f \in \Sigma$, $e = e_0 + e_1$, $f = f_0 + f_1$ where $e_i, f_i \in v_i$, $i = 0,1$; then

$$\bar\sigma(e,f) = \bar\sigma(e_0+e_1, f_0+f_1) = \bar\sigma(e_0,f_1) + \bar\sigma(e_1,f_0)$$
$$= \bar\sigma(e_0,f) + \bar\sigma(e,f_0) = \bar\sigma(\pi_{v_1v_0}e,f) + \bar\sigma(e,\pi_{v_1v_0}f).$$

Conversely, let $v \in v_0^{\pitchfork}$ is not a Lagrangian subspace. Then there exist $e,f \in v$ such that $\bar\sigma(e,f) \neq 0$, while $\bar\sigma(\pi_{vv_0}e,f) = \bar\sigma(e,\pi_{vv_0}f) = 0$. □

Corollary 2.4. *Let $v(\cdot)$ be an ample curve in $G_n(\Sigma)$ and $v^\circ(\cdot)$ be the derivative curve of $v(\cdot)$. If $v(t) \in L(\Sigma)$, $\forall t$, then $v^\circ(t) \in L(\Sigma)$.*

Proof. The derivative curve v° was defined in Sect. 11. Recall that $\pi_{v^\circ(t)v(t)} = \pi_t^0$, where π_t^0 is the free term of the Laurent expansion

$$\pi_{v(\tau)v(t)} \approx \sum_{i=-k_t}^{\infty} (\tau - t)^i \pi_t^i.$$

The free term π_t^0 belongs to the affine hull of $\pi_{v(\tau)v(t)}$, when τ runs a neighborhood of t. Since $\pi_{v(\tau)v(t)}$ belongs to the affine space $\{\pi_{vv_0} : v \in v_0^{\pitchfork} \cap L(\Sigma)\}$, then π_t^0 belongs to this affine space as well. □

We call a *Lagrange distribution* any rank n vector distribution $\{\Lambda_z \subset T_zN : z \in N\}$ on the symplectic manifold N such that $\Lambda_z \in L(T_zN)$, $z \in N$.

Corollary 2.5. *Canonical Ehresmann connection $\mathcal{E}_\zeta = \{J_z^\circ(0) : z \in N\}$ associated to an ample Hamiltonian field $\zeta = \vec{h}$ is a Lagrange distribution.* □

It is clearly seeing in coordinates how Lagrange Grassmanian is sitting in the usual one. Let $\Sigma = \mathbb{R}^{n*} \times \mathbb{R}^n = \{(\eta,y) : \eta \in \mathbb{R}^{n*}, y \in \mathbb{R}^n\}$. Then any $v \in (\{0\} \times \mathbb{R}^n)^{\pitchfork}$ has a form $v = \{(y^\top, Sy) : y \in \mathbb{R}^n\}$, where S is an $n \times n$-matrix. It is easy to see that v is a Lagrangian subspace if and only if S is a symmetric matrix, $S = S^\top$.

2.11 Monotonicity

We continue to study curves in the Lagrange Grassmannian $L(T_zN)$, in particular, the Jacobi curves $t \mapsto \left(e^{-t\mathbf{H}}\right)_* T_{e^{t\mathbf{H}}(z)} E_{e^{t\mathbf{H}}(z)}$. In Sect. 6 we identified the velocity $\dot{\Lambda}(t)$ of any smooth curve $\Lambda(\cdot)$ in $L(T_zN)$ with a quadratic form $\underline{\dot{\Lambda}}(t)$ on the subspace $\Lambda(t) \subset T_zN$. Recall that the curve $\Lambda(\cdot)$ was called monotone increasing if $\underline{\dot{\Lambda}}(t) \geq 0,\ \forall t$; it is called monotone decreasing if $\underline{\dot{\Lambda}}(t) \leq 0$. It is called monotone in both cases.

Proposition 2.8. *Set* $\Lambda(t) = \left(e^{-t\mathbf{H}}\right)_* T_{e^{t\mathbf{H}}(z)} E_{e^{t\mathbf{H}}(z)}$; *then quadratic form* $\underline{\dot{\Lambda}}(t)$ *is equivalent (up to a linear change of variables) to the form*

$$\varsigma \mapsto -(\varsigma \circ \varsigma H)(e^{t\mathbf{H}}(z)), \quad \varsigma \in [\mathcal{E}]^a_{e^{t\mathbf{H}}(z)}, \tag{21}$$

on $E_{e^{t\mathbf{H}}(z)}$.

Proof. Let $z_t = e^{t\mathbf{H}}(z)$, then

$$\frac{d}{dt}\Lambda(t) = \frac{d}{dt} e_*^{(t_0-t)\mathbf{H}} T_{z_t} E_{z_t} = e_*^{(t_0-t)\mathbf{H}} \frac{d}{d\varepsilon}\Big|_{\varepsilon=0} e_*^{-\varepsilon\mathbf{H}} T_{z_{t+\varepsilon}} E_{z_{t+\varepsilon}}.$$

Set $\Delta(\varepsilon) = e_*^{-\varepsilon\mathbf{H}} T_{z_{t+\varepsilon}} E_{z_{t+\varepsilon}} \in L\left(T_{z_t}N\right)$. It is enough to prove that $\underline{\dot{\Delta}}(0)$ is equivalent to form (21). Indeed, $\dot{\Lambda}(t) = e_*^{(t_0-t)\mathbf{H}} T_{z_t}\dot{\Delta}(0)$, where

$$e_*^{(t_0-t)\mathbf{H}} : T_{z_t}N \to T_{z_{t_0}}N$$

is a symplectic isomorphism. The association of the quadratic form $\underline{\dot{\Lambda}}(t)$ on the subspace $\Lambda(t)$ to the tangent vector $\dot{\Lambda}(t) \in L\left(T_{z_{t_0}}N\right)$ is intrinsic, i.e., depends only on the symplectic structure on $T_{z_{t_0}}N$. Hence $\underline{\dot{\Delta}}(0)(\xi) = \underline{\dot{\Lambda}}(t)\left(e_*^{(t_0-t)\mathbf{H}}\xi\right)$, $\forall \xi \in \Delta(0) = T_{z_t}E_{z_t}$.

What remains, is to compute $\underline{\dot{\Delta}}(0)$; we do it in the Darboux–Weinstein coordinates $z = (x, y)$. We have: $\Delta(\varepsilon) =$

$$\left\{ (\xi(\varepsilon), \eta(\varepsilon)) : \begin{array}{l} \dot{\xi}(\tau) = \xi(\tau)\frac{\partial^2 H}{\partial x \partial y}(z_{t-\tau}) + \eta(\tau)^\top \frac{\partial^2 H}{\partial y^2}(z_{t-\tau}), \quad \xi(0) = \xi \in \mathbb{R}^{n*} \\ \dot{\eta}(\tau) = -\frac{\partial^2 H}{\partial x^2}(z_{t-\tau})\xi(\tau)^\top - \frac{\partial^2 H}{\partial y \partial x}(z_{t-\tau})\eta(\tau),\ \eta(0) = 0 \in \mathbb{R}^n \end{array} \right\},$$

$$\underline{\dot{\Delta}}(0)(\xi) = \sigma\left((\xi, 0), (\dot{\xi}(0), \dot{\eta}(0))\right) = \xi\dot{\eta}(0) = -\xi\frac{\partial^2 H}{\partial x^2}(z_t)\xi^\top.$$

Recall now that form (21) has matrix $\frac{\partial^2 H}{\partial x^2}(z_t)$ in the Darboux–Weinstein coordinates. □

This proposition clearly demonstrates the importance of monotone curves. Indeed, monotonicity of Jacobi curves is equivalent to the convexity (or concavity) of the Hamiltonian on each leaf of the Lagrange foliation. In the case of a cotangent bundle this means the convexity or concavity of the Hamiltonian with respect to the impulses. All Hamiltonians (energy functions) of mechanical systems are like that! This is not an occasional fact but a corollary of the list action principle. Indeed, trajectories of the mechanical Hamiltonian system are extremals of the least action principle and the energy function itself is the Hamiltonian of the correspondent regular optimal control problem as it was considered in Sect. 7. Moreover, it was stated in Sect. 7 that convexity of the Hamiltonian with respect to the impulses is necessary for the extremals to have finite Morse index. It turns out that the relation between finiteness of the Morse index and monotonicity of the Jacobi curve has a fundamental nature. A similar property is valid for any, not necessary regular, extremal of a finite Morse index. Of course, to formulate this property we have first to explain what are Jacobi curve for non regular extremals. To do that, we come back to the very beginning; indeed, Jacobi curves appeared first as the result of calculation of the $\mathcal{L}$-derivative at the regular extremal (see Sects. 7 and 8). On the other hand, $\mathcal{L}$-derivative is well-defined for any extremal of the finite Morse index as it follows from Theorem 1.1. We thus come to the following construction in which we use notations and definitions of Sects. 3 and 4.

Let $h(\lambda, u)$ be the Hamiltonian of a smooth optimal control system, λ_t, $t_0 \le t \le t_1$, an extremal, and $q(t) = \pi(\lambda_t)$, $t_0, \le t \le t_1$ the extremal path. Recall that the pair $(\lambda_{t_0}, \lambda_t)$ is a Lagrangian multiplier for the conditional minimum problem defined on an open subset of the space

$$M \times L_\infty([t_0, t_1], U) = \{(q_t, u(\cdot)) : q \in M, u(\cdot) \in L_\infty([t_0, t_1], U)\},$$

where $u(\cdot)$ is control and q_t is the value at t of the solution to the differential equation $\dot{q} = f(q, u(\tau))$, $\tau \in [t_0, t_1]$. In particular, $F_t(q_t, u(\cdot)) = q_t$. The cost is $J_{t_0}^{t_1}(q_t, u(\cdot))$ and constraints are $F_{t_0}(q_t, u(\cdot)) = q(t_0)$, $q_t = q(t)$.

Let us set $J_t(u) = J_{t_0}^t(q(t), u(\cdot))$, $\Phi_t(u) = F_{t_0}(q(t), u(\cdot))$. A covector $\lambda \in T^*M$ is a Lagrange multiplier for the problem (J_t, Φ_t) if and only if there exists an extremal $\hat{\lambda}_\tau$, $t_0 \le \tau \le t$, such that $\lambda_{t_0} = \lambda$, $\hat{\lambda}_t \in T^*_{q(t)}M$. In particular, λ_{t_0} is a Lagrange multiplier for the problem (J_t, Φ_t) associated to the control $u(\cdot) = \bar{u}(\lambda_.)$.

Assume that $\operatorname{ind}\operatorname{Hess}_u \left(J_{t_1}\big|_{\Phi_{t_1}^{-1}(q(t_0))}\right) < \infty$, $t_0 \le t \le t_1$ and set $\bar{\Phi}_t = (J_t, \Phi_t)$. The curve

$$t \mapsto \mathcal{L}_{(\lambda_{t_0}, u)}(\bar{\Phi}_t), \quad t_0 \le t \le t_1$$

in the Lagrange Grassmannian $L\left(T_{\lambda_{t_0}}(T^*M)\right)$ is called the Jacobi curve associated to the extremal λ_t, $t_0 \le t \le t_1$.

In general, the Jacobi curve $t \mapsto \mathcal{L}_{(\lambda_{t_0}, u)}(\bar{\Phi}_t)$ is not smooth, it may even be discontinues, but it is monotone decreasing in a sense we are going to briefly describe now. You can find more details in [2] (just keep in mind that similar quantities may have opposite signs in different papers; sign agreements vary from paper to paper that is usual for symplectic geometry). Monotone curves in the Lagrange Grassmannian have analytic properties similar to scalar monotone functions: no more than a countable set of discontinuity points, right and left limits at every point, and differentiability almost everywhere with semi-definite derivatives (nonnegative for monotone increasing curves and nonpositive for decreasing ones). True reason for such a monotonicity is a natural monotonicity of the family $\bar{\Phi}_t$. Indeed, let $\tau < t$, then $\bar{\Phi}_\tau$ is, in fact, the restriction of $\bar{\Phi}_t$ to certain subspace: $\bar{\Phi}_\tau = \bar{\Phi}_t \circ \mathfrak{p}_\tau$, where $\mathfrak{p}_\tau(u)(s) = \begin{cases} u(s)\ , s < \tau \\ \tilde{u}(s)\ , s > \tau \end{cases}$. One can define the Maslov index of a (maybe discontinues) monotone curve in the Lagrange Grassmannian and the relation between the Morse and Maslov index indices from Theorem 1.3 remains true.

In fact, Maslov index is a key tool in the whole construction. The starting point is the notion of a *simple curve.* A smooth curve $\Lambda(\tau)$, $\tau_0 \le \tau \le \tau_1$, in the Lagrange Grassmannian $L(\Sigma)$ is called simple if there exists $\Delta \in L(\Sigma)$ such that $\Delta \cap \Lambda(\tau) = 0$, $\forall \tau \in [\tau_0, \tau_1]$; in other words, the entire curve is contained in one coordinate chart. It is not hard to show that any two points of $L(\Sigma)$ can be connected by a simple monotone increasing (as well as monotone decreasing) curve. An important fact is that the Maslov index $\mu(\Lambda_\Pi(\cdot))$ of a simple monotone increasing curve $\Lambda(\tau)$, $\tau_0 \le \tau \le \tau_1$ is uniquely determined by the triple $(\Pi, \Lambda(\tau_0), \Lambda(\tau_1))$; i.e., it has the same value for all simple monotone increasing curves connecting $\Lambda(\tau_0)$ with $\Lambda(\tau_1)$. A simple way to see this is to find an intrinsic algebraic expression for the Maslov index preliminary computed for some simple monotone curve in some coordinates. We can use Lemma 1.2 for this computation since the curve is simple. The monotonic increase of the curve implies that $S_{\Lambda(t_1)} > S_{\Lambda(t_0)}$.

Exercise. Let S_0, S_1 be nondegenerate symmetric matrices and $S_1 \ge S_0$. Then $\mathrm{ind} S_0 - \mathrm{ind} S_1 = \mathrm{ind}\left(S_0^{-1} - S_1^{-1}\right)$.

Let $x \in (\Lambda(\tau_0) + \Lambda(\tau_1) \cap \Pi$ so that $x = x_0 + x_1$, where $x_i \in \Lambda(\tau_i)$, $i = 0, 1$. We set $\mathfrak{q}(x) = \sigma(x_1, x_0)$. If $\Lambda(\tau_0) \cap \Lambda(\tau_1) = 0$, then $\Lambda(\tau_0) + \Lambda(\tau_1) = \Sigma$, x is any element of Π and x_0, x_1 are uniquely determined by x. This is not true if $\Lambda(\tau_0) \cap \Lambda(\tau_1) \ne 0$ but $\mathfrak{q}(x)$ is well-defined anyway: $\sigma(x_1, x_2)$ depends only on $x_0 + x_1$ since σ vanishes on $\Lambda(\tau_i)$, $i = 0, 1$.

Now we compute $\mathfrak{q}$ in coordinates. Recall that

$$\Lambda(\tau_i) = \{(y^\top, S_{\Lambda(\tau_i)} y) : y\mathbb{R}^n\},\ i = 0, 1,\ \Pi = \{y^\top, 0) : y \in \mathbb{R}^n\}.$$

We have

$$\mathfrak{q}(x) = y_1^\top S_{\Lambda(\tau_0)} y_0 - y_0^\top S_{\Lambda(\tau_1)} y_1,$$

where $x = (y_0^\top + y_1^\top, 0)$, $S_{\Lambda(\tau_0)} y_0 + S_{\Lambda(\tau_1)} y_1 = 0$. Hence $y_1 = -S^{-1}_{\Lambda(tau_1)} S_{\Lambda(\tau_0)} y_0$ and

$$\mathfrak{q}(x) = -y_0^\top S_{\Lambda(\tau_0)} y_0 - \left(S_{\Lambda(\tau_0)} y_0\right)^\top S^{-1}_{\Lambda(\tau_1)} S_{\Lambda(\tau_0)} y_0 = y^\top \left(S^{-1}_{\Lambda(\tau_0)} - S^{-1}_{\Lambda(\tau_1)}\right) y,$$

where $y = S_{\Lambda(\tau_0)} y_0$. We see that the form $\mathfrak{q}$ is equivalent, up to a linear change of coordinates, to the quadratic form defined by the matrix $S^{-1}_{\Lambda(\tau_0)} - S^{-1}_{\Lambda(\tau_1)}$. Now we set

$$\mathrm{ind}_\Pi(\Lambda(\tau_0), \Lambda(\tau_1)) \stackrel{def}{=} \mathrm{ind}\, \mathfrak{q}.$$

The above exercise and Lemma 1.2 imply the following:

Lemma 2.8. *If $\Lambda(\tau)$, $\tau_0 \leq \tau \leq \tau_1$, is a simple monotone increasing curve, then*

$$\mu(\Lambda(\cdot)) = \operatorname{ind}_\Pi(\Lambda(\tau_0), \Lambda(\tau_1)).$$

Note that definition of the form $\mathfrak{q}$ does not require transversality of $\Lambda(\tau_i)$ to Π. It is convenient to extend definition of $\operatorname{ind}_\Pi(\Lambda(\tau_0), \Lambda(\tau_1))$ to this case. General definition is as follows:

$$\operatorname{ind}_\Pi(\Lambda_0, \Lambda_1) = \operatorname{ind} \mathfrak{q} + \frac{1}{2}(\dim(\Pi \cap \Lambda_0) + \dim(\Pi \cap \Lambda_1)) - \dim(\Pi \cap \Lambda_0 \cap \Lambda_1).$$

The Maslov index also has appropriate extension (see [3, Sect. 4]) and Lemma 2.8 remains true.

Index $\operatorname{ind}_\Pi(\Lambda_0, \Lambda_1)$ satisfies the triangle inequality:

$$\operatorname{ind}_\Pi(\Lambda_0, \Lambda_2) \leq \operatorname{ind}_\Pi(\Lambda_0, \Lambda_1) + \operatorname{ind}_\Pi(\Lambda_1, \Lambda_2).$$

Indeed, the right-hand side of the inequality is equal to the Maslov index of a monotone increasing curve connecting Λ_0 with Λ_2, i.e., of the concatenation of two simple monotone increasing curves. Obviously, the Maslov index of a simple monotone increasing curve is not greater than the Maslov index of any other monotone increasing curve connecting the same endpoints.

The constructed index gives a nice presentation of the Maslov index of any (not necessary simple) monotone increasing curve $\Lambda(t)$, $t_0 \leq t \leq t_1$:

$$\mu_\Pi(\Lambda(\cdot)) = \sum_{i=0}^{l} \operatorname{ind}_\Pi(\Lambda(\tau_i), \Lambda(\tau_{i+1})), \tag{22}$$

where $t_0 = \tau_0 < \tau_1 < \cdots < \tau_l < \tau_{l+1} = t_1$ and $\Lambda\big|_{[\tau_i, \tau_{i+1}]}$ are simple pieces of the curve $\Lambda(\cdot)$. If the pieces are not simple, then the right-hand side of (22) gives a low bound for the Maslov index (due to the triangle inequality).

Let now $\Lambda(t)$, $t_0 \leq t \leq t_1$, be a smooth curve which is *not* monotone increasing. Take any subdivision $t_0 = \tau_0 < \tau_1 < \cdots < \tau_l < \tau_{l+1} = t_1$ and compute the sum $\sum\limits_{i=0}^{l} \operatorname{ind}_\Pi(\Lambda(\tau_i), \Lambda(\tau_{i+1}))$. This sum inevitably goes to infinity when the subdivision becomes finer and finer. The reason is as follows: $\operatorname{ind}_\Pi(\Lambda(\tau_i), \Lambda(\tau_{i+1})) > 0$ for any simple piece $\Lambda\big|_{[\tau_i, \tau_{i+1}]}$ such that $\underline{\dot{\Lambda}}(\tau) \ngeq 0$, $\forall \tau \in [\tau_i, \tau_{i+1}]$ and $\mu_\Pi(\Lambda\big|_{[\tau_i, \tau_{i+1}]}) = 0$. I advise reader to play with the one-dimensional case of the curve in $L(\mathbb{R}^2) = S^1$ to see better what's going on.

This should now be clear how to manage in the general nonsmooth case. Take a curve $\Lambda(\cdot)$ (an arbitrary mapping from $[t_0, t_1]$ into $L(\Sigma)$). For any finite subset $\mathcal{T} = \{\tau_1 \ldots, \tau_l\} \subset (t_0, t_1)$, where $t_0 = \tau_0 < \tau_1 < \cdots < \tau_l < \tau_{l+1} = t_1$, we compute the sum $I_\Pi^{\mathcal{T}} = \sum\limits_{i=0}^{l} \operatorname{ind}_\Pi(\Lambda(\tau_i), \Lambda(\tau_{i+1}))$ and then find supremum of these sums for all finite subsets: $I_\Pi(\Lambda(\cdot)) = \sup\limits_{\mathcal{T}} I_\Pi^{\mathcal{T}}$. The curve $\Lambda(\cdot)$ is called monotone increasing if $I_\Pi < \infty$; it is not hard to show that the last property does not depend on Π and that monotone increased curves enjoy listed above analytic properties. A curve $\Lambda(\cdot)$ is called monotone decreasing if inversion of the parameter $t \mapsto t_0 + t_1 - t$ makes it monotone increasing.

We set $\mu(\Lambda(\cdot)) = I_\Pi(\Lambda(\cdot))$ for any monotone increasing curve and $\mu(\Lambda(\cdot)) = -I_\Pi(\hat{\Lambda}(\cdot))$ for a monotone decreasing one, where $\hat{\Lambda}(t) = \Lambda(t_0 + t_1 - t)$. The defined in this way Maslov index of a discontinues monotone curve equals the Maslov index of the continues curve obtained by gluing all discontinuities with simple monotone curves of the same direction of monotonicity.

If $\Lambda(t) = \mathcal{L}_{(\lambda_{t_0},u)}(\bar{\Phi}_t)$ is the Jacobi curve associated to the extremal with a finite Morse index, then $\Lambda(\cdot)$ is monotone decreasing and its Maslov index computes $\operatorname{ind}\operatorname{Hess}_u\left(J_{t_1}\big|_{\Phi_{t_1}^{-1}(q(t_0))}\right)$ in the way similar to Theorem 1.3. Of course, these nice things have some value only if we can effectively find Jacobi curves for singular extremals: their definition was too abstract. Fortunately, this is not so hard; see [5] for the explicit expression of Jacobi curves for a wide class of singular extremals and, in particular, for singular curves of rank 2 vector distributions (these last Jacobi curves have found important applications in the geometry of distributions, see [11,14]).

One more important property of monotonic curves is as follows.

Lemma 2.9. *Assume that $\Lambda(\cdot)$ is monotone and right-continues at t_0, i.e., $\Lambda(t_0) = \lim_{t \searrow t_0} \Lambda(t)$. Then $\Lambda(t_0) \cap \Lambda(t) = \bigcap_{t_0 \le \tau \le t} \Lambda(t)$ for any t sufficiently close to (and greater than) t_0.*

Proof. We may assume that $\Lambda(\cdot)$ is monotone increasing. Take centered at $\Lambda(t_0)$ local coordinates in the Lagrange Grassmannian; the coordinate presentation of $\Lambda(t)$ is a symmetric matrix $S_{\Lambda(t)}$, where $S_{\Lambda(t_0)} = 0$ and $t \mapsto y^\top S_{\Lambda(t)} y$ is a monotone increasing scalar function $\forall y \in \mathbb{R}^n$. In particular, $\ker S_{\Lambda(t)} = \Lambda(t) \cap \Lambda(t_0)$ is a monotone decreasing family of subspaces. □

We set $\Gamma_t = \bigcap_{t_0 \le \tau \le t} \Lambda(\tau)$, a monotone decreasing family of isotropic subspaces. Let $\Gamma = \max_{t > t_0} \Gamma_t$, then $\Gamma_t = \Gamma$ for all $t > t_0$ sufficiently close to t_0. We have: $\Lambda(t) = \Lambda(t)^\angle$ and $\Lambda(t) \supset \Gamma$ for all $t > t_0$ close enough to t_0; hence $\Gamma_t^\angle \supset \Lambda(t)$. In particular, $\Lambda(t)$ can be treated as a Lagragian subspace of the symplectic space $\Gamma^\angle/\Gamma$. Moreover, Lemma 2.9 implies that $\Lambda(t) \cap \Lambda(t_0) = \Gamma$. In other words, $\Lambda(t)$ is transversal to $\Lambda(t_0)$ in $\Gamma^\angle/\Gamma$. In the case of a real-analytic monotone curve $\Lambda(\cdot)$ this automatically implies that $\Lambda(\cdot)$ is an ample curve in $\Gamma^\angle/\Gamma$. Hence any nonconstant monotone analytic curve is reduced to an ample monotone curve. It becomes ample after the factorization by a fixed (motionless) subspace.

2.12 Comparison Theorem

We come back to smooth regular curves after the deviation devoted to a more general perspective.

Lemma 2.10. *Let $\Lambda(t)$, $t \in [t_0, t_1]$ be a regular monotone increasing curve in the Lagrange Grassmannian $L(\Sigma)$. Then $\{t \in [t_0, t_1] : \Lambda(t) \cap \Pi \neq 0\}$ is a finite subset of $[t_0, t_1]$ $\forall \Pi \in L(\Sigma)$. If t_0 and t_1 are out of this subset, then*

$$\mu_\Pi(\Lambda(\cdot)) = \sum_{t \in (t_0, t_1)} \dim(\Lambda(t) \cap \Pi).$$

Proof. We have to proof that $\Lambda(t)$ may have a nontrivial intersection with Π only for isolated values of t; the rest is Lemma 1.1. Assume that $\Lambda(t) \cap \Pi \neq 0$ and take a centered at Π coordinate neighborhood in $L(\Sigma)$ which contains $\Lambda(t)$. In these coordinates, $\Lambda(\tau)$ is presented by a symmetric matrix $S_\Lambda(\tau)$ for any τ sufficiently close to t and $\Lambda(\tau) \cap \Pi = \ker S_{\Lambda(\tau)}$. Monotonicity and regularity properties are equivalent to the inequality $\dot{S}_{\Lambda(\tau)} > 0$. In particular, $y^\top \dot{S}_{\Lambda(t)} y > 0$ $\forall y \in \ker S_{\Lambda(t)} \setminus \{0\}$. The last inequality implies that $S_{\Lambda(\tau)}$ is a nondegenerate for all τ sufficiently close and not equal to t.

Definition. Parameter values τ_0, τ_1 are called conjugate for the continues curve $\Lambda(\cdot)$ in the Lagrange Grassmannian if $\Lambda(\tau_0) \cap \Lambda(\tau_1) \neq 0$; the dimension of $\Lambda(\tau_0) \cap \Lambda(\tau_1)$ is the *multiplicity* of the conjugate parameters.

If $\Lambda(\cdot)$ is a regular monotone increasing curve, then, according to Lemma 2.9, conjugate points are isolated and the Maslov index $\mu_{\Lambda(t_0)}\left(\Lambda\big|_{[t,t_1]}\right)$ equals the sum of multiplicities of the conjugate to t_0 parameter values located in (t, t_1). If $\Lambda(\cdot)$ is the Jacobi curve of an extremal of an optimal control problem, then this Maslov index equals the Morse index of the extremal; this is why conjugate points are so important.

Given a regular monotone curve $\Lambda(\cdot)$, the quadratic form $\underline{\dot{\Lambda}}(t)$ defines an Euclidean structure $\langle \cdot, \cdot \rangle_{\dot{\Lambda}(t)}$ on $\Lambda(t)$ so that $\underline{\dot{\Lambda}}(t)(x) = \langle x, x \rangle_{\dot{\Lambda}(t)}$. Let $R_\Lambda(t) \in$ gl$(\Lambda(t))$ be the curvature operator of the curve $\Lambda(\cdot)$; we define the *curvature quadratic form* $r_\Lambda(t)$ on $\Lambda(t)$ by the formula:

$$r_\Lambda(t)(x) = \langle R_\Lambda(t)x, x \rangle_{\dot{\Lambda}(t)}, \quad x \in \Lambda(t).$$

Proposition 2.9. *The curvature operator $R_\Lambda(t)$ is a self-adjoint operator for the Euclidean structure $\langle \cdot, \cdot \rangle_{\dot{\Lambda}(t)}$. The form $r_\Lambda(t)$ is equivalent (up to linear changes of variables) to the form $\underline{\dot{\Lambda}}^\circ(t)$, where $\Lambda^\circ(\cdot)$ is the derivative curve.*

Proof. The statement is intrinsic and we may check it in any coordinates. Fix t and take Darboux coordinates $\{(\eta, y) : \eta \in \mathbb{R}^{n*}, y \in \mathbb{R}^n\}$ in Σ in such a way that $\Lambda(t) = \{(y^\top, 0) : y \in \mathbb{R}^n\}$, $\Lambda^\circ(t) = \{(0, y) : y \in \mathbb{R}^n\}$, $\underline{\dot{\Lambda}}(t)(y) = y^\top y$. Let $\Lambda(\tau) = \{(y^\top, S_\tau y) : y \in \mathbb{R}^n\}$, then $S_t = 0$. Moreover, $\dot{S}(t)$

is the matrix of the form $\underline{\dot{\Lambda}}(t)$ in given coordinates, hence $\dot{S}_t = I$. Recall that $\Lambda^\circ(\tau) = \{(y^\top A_\tau y, y + S_\tau A_\tau y) : y \in \mathbb{R}^n\}$, where $A_\tau = -\frac{1}{2}\dot{S}_\tau^{-1}\ddot{S}_\tau\dot{S}_\tau^{-1}$ (see (5)). Hence $\ddot{S}_t = 0$. We have: $R_\Lambda(t) = \frac{1}{2}\dddot{S}_t$, $r_\Lambda(t)(y) = \frac{1}{2}y^\top \dddot{S}_t\, y$,

$$\underline{\dot{\Lambda}}^\circ(t)(y) = \sigma\Big((0, y), (y^\top \dot{A}_t, 0)\Big) = -y^\top \dot{A}_t y = \frac{1}{2} y^\top \dddot{S}_t\, y.$$

So $r_\Lambda(t)$ and $\underline{\dot{\Lambda}}^\circ(t)$ have equal matrices for our choice of coordinates in $\Lambda(t)$ and $\Lambda^\circ(t)$. The curvature operator is self-adjoint since it is presented by a symmetric matrix in coordinates where form $\underline{\dot{\Lambda}}(t)$ is the standard inner product.

□

Proposition 2.9 implies that the curvature operators of regular monotone curves in the Lagrange Grassmannian are diagonalizable and have only real eigenvalues.

Theorem 2.1. *Let $\Lambda(\cdot)$ be a regular monotone curve in the Lagrange Grassmannian $L(\Sigma)$, where $\dim \Sigma = 2n$:*

- *If all eigenvalues of $R_\Lambda(t)$ do not exceed a constant $c \geq 0$ for any t from the domain of $\Lambda(\cdot)$, then $|\tau_1 - \tau_0| \geq \frac{\pi}{\sqrt{c}}$ for any pair of conjugate parameter values τ_0, τ_1. In particular, If all eigenvalues of $R_\Lambda(t)$ are nonpositive $\forall t$, then $\Lambda(\cdot)$ does not possess conjugate parameter values*
- *If $\mathrm{tr} R_\Lambda(t) \geq nc$ for some constant $c > 0$ and $\forall t$, then, for arbitrary $\tau_0 \leq t$, the segment $[t, t + \frac{\pi}{\sqrt{c}}]$ contains a conjugate to τ_0 parameter value as soon as this segment is contained in the domain of $\Lambda(\cdot)$.*

Both estimates are sharp.

Proof. We may assume without lack of generality that $\Lambda(\cdot)$ is ample monotone increasing. We start with the case of nonpositive eigenvalues of $R_\Lambda(t)$. The absence of conjugate points follows from Proposition 2.9 and the following

Lemma 2.11. *Assume that $\Lambda(\cdot)$ is an ample monotone increasing (decreasing) curve and $\Lambda^\circ(\cdot)$ is a continues monotone decreasing (increasing) curve. Then $\Lambda(\cdot)$ does not possess conjugate parameter values and there exists a $\lim\limits_{t\to+\infty} \Lambda(t) = \Lambda_\infty$.*

Proof. Take some value of the parameter τ_0; then $\Lambda(\tau_0)$ and $\Lambda^\circ(\tau_0)$ is a pair of transversal Lagrangian subspaces. We may choose coordinates in the Lagrange Grassmannian in such a way that $S_{\Lambda(\tau))} = 0$ and $S_{\Lambda^\circ(\tau_0)} = I$, i.e., $\Lambda(\tau_0)$ is represented by zero $n \times n$-matrix and $\Lambda^\circ(\tau_0)$ by the unit matrix. Monotonicity assumption implies that $t \mapsto S_{\Lambda(t)}$ is a monotone increasing curve in the space of symmetric matrices and $t \mapsto S_{\Lambda^\circ(t)}$ is a monotone decreasing curve. Moreover, transversality of $\Lambda(t)$ and $\Lambda^\circ(t)$ implies that $S_{\Lambda^\circ(t)} - S_{\Lambda(t)}$ is a nondegenerate matrix. Hence $0 < S_{\Lambda(t)} < S_{\Lambda^\circ(t)} \leq I$ for any $t > \tau_0$. In particular, $\Lambda(t)$ never leaves the coordinate neighborhood under consideration for $T > \tau_0$, the subspace $\Lambda(t)$ is always transversal to $\Lambda(\tau_0)$ and has a limit Λ_∞, where $S_{\Lambda_\infty} = \sup\limits_{t\geq\tau_0} S_{\Lambda(t)}$. □

Now assume that the eigenvalues of $R_\Lambda(t)$ do not exceed a constant $c > 0$. We are going to reparametrize the curve $\Lambda(\cdot)$ and to use the chain rule (7). Take some $\bar{t}$ in the domain of $\Lambda(\cdot)$ and set

$$\varphi(t) = \frac{1}{\sqrt{c}}\left(\arctan(\sqrt{c}t) + \frac{\pi}{2}\right) + \bar{t}, \quad \Lambda_\varphi(t) = \Lambda(\varphi(t)).$$

We have: $\varphi(\mathbb{R}) = \left(\bar{t}, \bar{t} + \frac{\pi}{\sqrt{c}}\right)$, $\dot{\varphi}(t) = \frac{1}{ct^2+1}$, $R_\varphi(t) = -\frac{c}{(ct^2+1)^2}$. Hence, according to the chain rule (7), the operator

$$R_{\Lambda_\varphi}(t) = \frac{1}{(ct^2+1)^2}\left(R_\Lambda(\varphi(t)) - cI\right)$$

has only nonpositive eigenvalues. Already proved part of the theorem implies that Λ_φ does not possess conjugate values of the parameter. In other words, any length $\frac{\pi}{\sqrt{c}}$ interval in the domain of $\Lambda(\cdot)$ is free of conjugate pairs of the parameter values.

Assume now that $\mathrm{tr}R_\Lambda(t) \geq nc$. We will prove that the existence of $\Delta \in L(\Sigma)$ such that $\Delta \cap \Lambda(t) = 0$ for all $t \in [\bar{t}, \tau]$ implies that $\tau - \bar{t} < \frac{\pi}{\sqrt{c}}$. We will prove it by contradiction. If there exists such a Δ, then $\Lambda\big|_{[\bar{t},\tau]}$ is completely contained in a fixed coordinate neighborhood of $L(\Sigma)$, therefore the curvature operator $R_\Lambda(t)$ is defined by the formula (6). Put $B(t) = (2\dot{S}_t)^{-1}\ddot{S}_t$, $b(t) = \mathrm{tr}B(t)$, $t \in [\bar{t}, \tau]$. Then

$$\dot{B}(t) = B^2(t) + R_\Lambda(t), \quad \dot{b}(t) = \mathrm{tr}B^2(t) + \mathrm{tr}R_\Lambda(t).$$

Since for an arbitrary symmetric $n \times n$-matrix A we have $\mathrm{tr}A^2 \geq \frac{1}{n}(\mathrm{tr}A)^2$, the inequality $\dot{b} \geq \frac{b^2}{n} + nc$ holds. Hence $b(t) \geq \beta(t)$, $\bar{t} \leq t \leq \tau$, where $\beta(\cdot)$ is a solution of the equation $\dot{\beta} = \frac{\beta^2}{n} + nc$, i.e., $\beta(t) = n\sqrt{c}\tan(\sqrt{c}(t - t_0))$. The function $b(\cdot)$ together with $\beta(\cdot)$ are bounded on the segment $[\bar{t}, \tau]$. Hence $\tau - t \leq \frac{\pi}{\sqrt{c}}$.

To verify that the estimates are sharp, it is enough to consider regular monotone curves of constant curvature. □

2.13 Reduction

We consider a Hamiltonian system on a symplectic manifold N endowed with a fixed Lagrange foliation E. Assume that $g : N \to \mathbb{R}$ is a first integral of our Hamiltonian system, i.e., $\{h, g\} = 0$.

Lemma 2.12. *Let $z \in N$, $g(z) = c$. The leaf E_z is transversal to $g^{-1}(c)$ at z if and only if $\mathbf{g}(z) \notin T_zE_z$.*

Proof. Hypersurface $g^{-1}(c)$ is not transversal to $g^{-1}(c)$ at z if and only if

$$d_z g(T_z E_z) = 0 \Leftrightarrow \sigma(\mathbf{g}(z), T_z E_z) = 0 \Leftrightarrow \mathbf{g}(z) \in (T_z E_z)^{\angle} = T_z E_z. \quad \square$$

If all points of some level $g^{-1}(c)$ satisfy conditions of Lemma 2.12, then $g^{-1}(c)$ is a (2n-1)-dimensional manifold foliated by $(n-1)$-dimensional submanifolds $E_z \cap g^{-1}(c)$. Note that $\mathbb{R}\mathbf{g}(z) = \ker \sigma\big|_{T_z g^{-1}(c)}$, hence $\Sigma_z^g \stackrel{def}{=} T_z g^{-1}(c)/\mathbb{R}\mathbf{g}(z)$ is a $2(n-1)$-dimensional symplectic space and $\Delta_z^g \stackrel{def}{=} T_z\left(E_z \cap g^{-1}(c)\right)$ is a Lagrangian subspace in L_z^g, i.e., $\Delta_z^g \in L(\Sigma_z^g)$.

The submanifold $g^{-1}(c)$ is invariant for the flow $e^{t\mathbf{h}}$. Moreover, $e_*^{t\mathbf{h}}\mathbf{g} = \mathbf{g}$. Hence $e_*^{t\mathbf{h}}$ induces a symplectic transformation $e_*^{t\mathbf{h}} : \Sigma_z^g \to \Sigma_{e^{t\mathbf{h}}(z)}^g$. Set $J_z^g(t) = e_*^{-t\mathbf{h}} \Delta_{e^{t\mathbf{h}}(z)}^g$. The curve $t \mapsto J_z^g(t)$ in the Lagrange Grassmannian $L(\Sigma_z^g)$ is called a *reduced Jacobi curve* for the Hamiltonian field $\mathbf{h}$ at $z \in N$.

The reduced Jacobi curve can be easily reconstructed from the Jacobi curve $J_z(t) = e_*^{-t\mathbf{h}}\left(T_{e^{t\mathbf{h}}(z)} E_{e^{t\mathbf{h}}(z)}\right) \in L(T_z N)$ and vector $\mathbf{g}(z)$. An elementary calculation shows that

$$J_z^g(t) = J_z(t) \cap \mathbf{g}(z)^{\angle} + \mathbb{R}\mathbf{g}(z).$$

Now we can temporary forget the symplectic manifold and Hamiltonians and formulate everything in terms of the curves in the Lagrange Grassmannian. So let $\Lambda(\cdot)$ be a smooth curve in the Lagrange Grassmannian $L(\Sigma)$ and γ a one-dimensional subspace in Σ. We set $\Lambda^\gamma(t) = \Lambda(t) \cap \gamma^{\angle} + \gamma$, a Lagrange subspace in the symplectic space $\gamma^{\angle}/\gamma$. If $\gamma \not\subset \Lambda(t)$, then $\Lambda^\gamma(\cdot)$ is smooth and $\underline{\dot{\Lambda}}^\gamma(t) = \underline{\dot{\Lambda}}(t)\big|_{\Lambda(t)\cap\gamma^{\angle}}$ as it easily follows from the definitions. In particular, monotonicity of $\Lambda(\cdot)$ implies monotonicity of $\Lambda^\gamma(\cdot)$; if $\Lambda(\cdot)$ is regular and monotone, then $\Lambda^\gamma(\cdot)$ is also regular and monotone. The curvatures and the Maslov indices of $\Lambda(\cdot)$ and $\Lambda^\gamma(\cdot)$ are related in a more complicated way. The following result is proved in [9].

Theorem 2.2. *Let $\Lambda(t)$, $t \in [t_0, t_1]$ be a smooth monotone increasing curve in $L(\Sigma)$ and γ a one-dimensional subspace of Σ such that $\gamma \not\subset \Lambda(t)$, $\forall t \in [t_0, t_1]$. Let $\Pi \in L(\Sigma)$, $\gamma \not\subset \Pi$, $\Lambda(t_0) \cap \Pi = \Lambda(t_1) \cap \Pi = 0$. Then:*

- $\mu_\Pi(\Lambda(\cdot)) \le \mu_{\Pi^\gamma}(\Lambda^\gamma(\cdot)) \le \mu_\Pi(\Lambda(\cdot)) + 1$
- *If $\Lambda(\cdot)$ is regular, then $r_{\Lambda^\gamma}(t) \ge r_\Lambda(t)\big|_{\Lambda(t)\cap\gamma^{\angle}}$ and*
  $\operatorname{rank}\left(r_{\Lambda^\gamma}(t) - r_\Lambda(t)\big|_{\Lambda(t)\cap\gamma^{\angle}}\right) \le 1$

The inequality $r_{\Lambda^\gamma}(t) \ge r_\Lambda(t)\big|_{\Lambda(t)\cap\gamma^{\angle}}$ turns into the equality if $\gamma \subset \Lambda^\circ(t)$, $\forall t$. Then $\gamma \subset \ker \underline{\dot{\Lambda}}^\circ(t)$. According to Proposition 2.9, to γ there corresponds a one-dimensional subspace in the kernel of $r_\Lambda(t)$; in particular, $r_\Lambda(t)$ is degenerate.

Return to the Jacobi curves $J_z(t)$ of a Hamiltonian field $\mathbf{h}$. There always exists at least one first integral: the Hamiltonian h itself. In general,

$\mathbf{h}(z) \notin J_z^\circ(0)$ and the reduction procedure has a nontrivial influence on the curvature (see [8,9] for explicit expressions). Still, there is an important class of Hamiltonians and Lagrange foliations for which the relation $\mathbf{h}(z) \in J_z^\circ(0)$ holds $\forall z$. These are homogeneous on fibers Hamiltonians on cotangent bundles. In this case the generating homotheties of the fibers Euler vector field belongs to the kernel of the curvature form.

2.14 Hyperbolicity

Definition. We say that a Hamiltonian function h on the symplectic manifold N is regular with respect to the Lagrange foliation E if the functions $h\big|_{E_z}$ have nondegenerate second derivatives at z, $\forall z \in N$ (second derivative is well-defined due to the canonical affine structure on E_z). We say that h is monotone with respect to E if $h\big|_{E_z}$ is a convex or concave function $\forall z \in N$.

Typical examples of regular monotone Hamiltonians on the cotangent bundles are energy functions of natural mechanical systems. Such a function is the sum of the kinetic energy whose Hamiltonian system generates the Riemannian geodesic flow and a "potential" that is a constant on the fibers function. Proposition 2.8 implies that Jacobi curves associated to the regular monotone Hamiltonians are also regular and monotone. We will show that negativity of the curvature operators of such a Hamiltonian implies the hyperbolic behavior of the Hamiltonian flow. This is a natural extension of the classical result about Riemannian geodesic flows.

Main tool is the structural equation derived in Sect. 13. First we will show that this equation is well coordinated with the symplectic structure. Let $\Lambda(t)$, $t \in \mathbb{R}$, be a regular curve in $L(\Sigma)$ and $\Sigma = \Lambda(t) \oplus \Lambda^\circ(t)$ the correspondent canonical splitting. Consider the structural equation

$$\ddot{e}(t) + R_\Lambda(t)e(t) = 0, \quad \text{where } e(t) \in \Lambda(t),\ \dot{e}(t) \in \Lambda^\circ(t), \tag{23}$$

(see Corollary 2.1).

Lemma 2.13. *The mapping $e(0) \oplus \dot{e}(0) \mapsto e(t) \oplus \dot{e}(t)$, where $e(\cdot)$ and $\dot{e}(\cdot)$ satisfies (23), is a symplectic transformation of Σ.*

Proof. We have to check that $\sigma(e_1(t), e_2(t))$, $\sigma(\dot{e}_1(t), \dot{e}_2(t))$, $\sigma(e_1(t), \dot{e}_2(t))$ do not depend on t as soon as $e_i(t), \dot{e}_i(t)$, $i = 1, 2$, satisfy (23). First two quantities vanish since $\Lambda(t)$ and $\Lambda^\circ(t)$ are Lagrangian subspaces. The derivative of the third quantity vanishes as well since $\ddot{e}_i(t) \in \Lambda(t)$. □

Let h be a regular monotone Hamiltonian on the symplectic manifold N equipped with a Lagrange foliation E. As before, we denote by $J_z(t)$ the Jacobi curves of $\mathbf{h}$ and by $J_z^h(t)$ the reduced to the level of h Jacobi curves (see previous section). Let $R(z) = R_{J_z}(0)$ and $R^h(z) = R_{J_z^h}(0)$ be the curvature operators of $J_z(\cdot)$ and $J_z^h(\cdot)$ correspondently. We say that the Hamiltonian field $\mathbf{h}$ has a negative curvature at z with respect to E if all eigenvalues of $R(z)$ are negative. We say that $\mathbf{h}$ has a negative reduced curvature at z if all eigenvalues of R_z^h are negative.

Proposition 2.10. *Let $z_0 \in N$, $z_t = e^{t\mathbf{h}}(z)$. Assume that that $\overline{\{z_t : t \in \mathbb{R}\}}$ is a compact subset of N and that N is endowed with a Riemannian structure. If* $\mathbf{h}$ *has a negative curvature at any $z \in \overline{\{z_t : t \in \mathbb{R}\}}$, then there exists a constant $\alpha > 0$ and a splitting $T_{z_t}N = \Delta^+_{z_t} \oplus \Delta^-_{z_t}$, where $\Delta^\pm_{z_t}$ are Lagrangian subspaces of $T_{z_t}N$ such that $e^{\tau\mathbf{h}}_*(\Delta^\pm_{z_t}) = \Delta^\pm_{z_{t+\tau}}$ $\forall t, \tau \in \mathbb{R}$ and*

$$\|e^{\pm\tau\mathbf{h}}_* \zeta_\pm\| \geq e^{\alpha\tau}\|\zeta_\pm\| \quad \forall \zeta \in \Delta^\pm_{z_t},\ \tau \geq 0,\ t \in \mathbb{R}. \tag{24}$$

Similarly, if $\mathbf{h}$ *has a negative reduced curvature at any $z \in \overline{\{z_t : t \in \mathbb{R}\}}$, then there exists a splitting $T_{z_t}(h^{-1}(c)/\mathbb{R}h(z_t)) = \hat{\Delta}^+_{z_t} \oplus \hat{\Delta}^-_{z_t}$, where $c = h(z_0)$ and $\hat{\Delta}^\pm_{z_t}$ are Lagrangian subspaces of $T_{z_t}(h^{-1}(c)/\mathbb{R}h(z_t))$ such that $e^{\tau\mathbf{h}}_*(\hat{\Delta}^\pm_{z_t}) = \hat{\Delta}^\pm_{z_{t+\tau}}$ $\forall t, \tau \in \mathbb{R}$ and $\|e^{\pm\tau\mathbf{h}}_* \zeta_\pm\| \geq e^{\alpha\tau}\|\zeta_\pm\|$ $\forall \zeta \in \hat{\Delta}^\pm_{z_t}$, $\tau \geq 0$, $t \in \mathbb{R}$.*

Proof. Obviously, the desired properties of $\Delta^\pm_{z_t}$ and $\hat{\Delta}^\pm_{z_t}$ do not depend on the choice of the Riemannian structure on N. We will introduce a special Riemannian structure determined by h. The Riemannian structure is a smooth family of inner products $\langle\cdot,\cdot\rangle_z$ on T_zN, $z \in N$. We have $T_zN = J_z(0) \oplus J^\circ_z(0)$, where $J_z(0) = T_zE_z$. Replacing h with $-h$ if necessary we may assume that $h\big|_{E_z}$ is a strongly convex function. First we define $\langle\cdot,\cdot\rangle_z\big|_{J_z(0)}$ to be equal to the second derivative of $h\big|_{E_z}$. Symplectic form σ induces a nondegenerate pairing of $J_z(0)$ and $J^\circ_z(0)$. In particular, for any $\zeta \in J_z(0)$ there exists a unique $\zeta^\circ \in J^\circ_z(0)$ such that $\sigma(\zeta^\circ,\cdot)\big|_{J_z(0)} = \langle\zeta,\cdot\rangle_z\big|_{J_z(0)}$. There exists a unique extension of the inner product $\langle\cdot,\cdot\rangle_z$ from $J_z(0)$ to the whole T_zN with the following properties:

- $J^\circ_z(0)$ is orthogonal to $J_z(0)$ with respect to $\langle\cdot,\cdot\rangle_z$
- $\langle\zeta_1,\zeta_2\rangle_z = \langle\zeta^\circ_1,\zeta^\circ_2\rangle_z$ $\forall \zeta_1,\zeta_2 \in J_z(0)$

We will need the following classical fact from Hyperbolic Dynamics (see, for instance, [12, Sect. 17.6]).

Lemma 2.14. *Let $A(t)$, $t \in \mathbb{R}$, be a bounded family of symmetric $n \times n$-matrices whose eigenvalues are all negative and uniformly separated from 0. Let $\Gamma(t,\tau)$ be the fundamental matrix of the $2n$-dimensional linear system $\dot{x} = -y$, $\dot{y} = A(t)x$, where $x, y \in \mathbb{R}^n$, i.e.,*

$$\frac{\partial}{\partial t}\Gamma(t,\tau) = \left(\begin{smallmatrix} 0 & -I \\ A & 0 \end{smallmatrix}\right)\Gamma(t,\tau), \quad \Gamma(\tau,\tau) = \left(\begin{smallmatrix} I & 0 \\ 0 & I \end{smallmatrix}\right). \tag{25}$$

Then there exist closed conic neighborhoods C^+_Γ, C^-_Γ, where $C^+_\Gamma \cap C^-_\Gamma = 0$, of some n-dimensional subspaces of $\mathbb{R}^{2n}$ and a constant $\alpha > 0$ such that

$$\Gamma(t,\tau)C^+_\Gamma \subset C^+_\Gamma, \quad |\Gamma(t,\tau)\xi_+| \geq e^{\alpha(\tau-t)}|\xi_+|,\ \forall \xi_+ \in C^+_\Gamma,\ t \leq \tau,$$

and

$$\Gamma(t,\tau)C^-_\Gamma \subset C^-_\Gamma, \quad |\Gamma(t,\tau)\xi_-| \geq e^{\alpha(t-\tau)}|\xi_-|,\ \forall \xi_- \in C^-_\Gamma,\ t \geq \tau.$$

The constant α depends only on upper and lower bounds of the eigenvalues of $A(t)$. $\square$

Corollary 2.6. *Let $C^{\pm}_{\Gamma}$ be the cones described in Lemma 2.14; then $\Gamma(0,\pm t)C^{\pm}_{\Gamma} \subset \Gamma(0;\pm\tau)C^{\pm}_{\Gamma}$ for any $t \geq \tau \geq 0$ and the subsets $K^{\pm}_{\Gamma} = \bigcap\limits_{t\geq 0} \Gamma(0,t)C^{\pm}_{\Gamma}$ are Lagrangian subspaces of $\mathbb{R}^n \times \mathbb{R}^n$ equipped with the standard symplectic structure.*

Proof. The relations $\Gamma(\tau,t)C^{+}_{\Gamma} \subset C^{+}_{\Gamma}$ and $\Gamma(\tau,t)C^{-}_{\Gamma} \subset C^{-}_{\Gamma}$ imply:

$$\Gamma(0,\pm t)C^{\pm}_{\Gamma} = \Gamma(0,\pm\tau)\Gamma(\pm\tau,\pm t)C^{\pm}_{\Gamma} \subset \Gamma(0,\pm\tau)C^{\pm}_{\Gamma}.$$

In what follows we will study K^{+}_{Γ}; the same arguments work for K^{-}_{Γ}. Take vectors $\zeta, \zeta' \in K^{+}_{\Gamma}$; then $\zeta = \Gamma(0,t)\zeta_t$ and $\zeta' = \Gamma(0,t)\zeta'_t$ for any $t \geq 0$ and some $\zeta_t, \zeta'_t \in C^{+}_{\Gamma}$. Then, according to Lemma 2.14, $|\zeta_t| \leq e^{-\alpha t}|\zeta|$, $|\zeta'_t| \leq e^{-\alpha t}|\zeta'|$, i.e., ζ_t and ζ'_t tend to 0 as $t \to +\infty$. On the other hand,

$$\sigma(\zeta,\zeta') = \sigma(\Gamma(0,t)\zeta_t, \Gamma(0,t)\zeta'_t) = \sigma(\zeta_t,\zeta'_t) \quad \forall t \geq 0$$

since $\Gamma(0,t)$ is a symplectic matrix. Hence $\sigma(\zeta,\zeta') = 0$.

We have shown that K^{+}_{Γ} is an isotropic subset of $\mathbb{R}^n \times \mathbb{R}^n$. On the other hand, K^{+}_{Γ} contains an n-dimensional subspace since C^{+}_{Γ} contains one and $\Gamma(0,t)$ are invertible linear transformations. Isotropic n-dimensional subspace is equal to its skew-orthogonal complement, therefore K^{+}_{Γ} is a Lagrangian subspace. □

Take now a regular monotone curve $\Lambda(t)$, $t \in \mathbb{R}$ in the Lagrange Grassmannian $L(\Sigma)$. We may assume that $\Lambda(\cdot)$ is monotone increasing, i.e., $\dot{\Lambda}(t) > 0$. Recall that $\underline{\dot{\Lambda}}(t)(e(t)) = \sigma(e(t),\dot{e}(t))$, where $e(\cdot)$ is an arbitrary smooth curve in Σ such that $e(\tau) \in \Lambda(\tau)$, $\forall\tau$. Differentiation of the identity $\sigma(e_1(\tau),e_2(\tau)) = 0$ implies: $\sigma(e_1(t),\dot{e}_2(t)) = -\sigma(\dot{e}_1(t),e_2(t)) = \sigma(e_2(t),\dot{e}_1(t))$ if $e_i(\tau) \in \Lambda(\tau)$, $\forall\tau$, $i = 1,2$. Hence the Euclidean structure $\langle\cdot,\cdot\rangle_{\dot{\Lambda}(t)}$ defined by the quadratic form $\underline{\dot{\Lambda}}(t)$ reads: $\langle e_1(t),e_2(t)\rangle_{\dot{\Lambda}(t)} = \sigma(e_1(t),\dot{e}_2(t))$.

Take a basis $e_1(0),\ldots,e_n(0)$ of $\Lambda(0)$ such that the form $\underline{\dot{\Lambda}}(t)$ has the unit matrix in this basis, i.e., $\sigma(e_i(0),\dot{e}_j(0)) = \delta_{ij}$. In fact, vectors $\dot{e}_j(0)$ are defined modulo $\Lambda(0)$; we can normalize them assuming that $\dot{e}_i(0) \in \Lambda^\circ(0)$, $i = 1,\ldots,n$. Then $e_1(0),\ldots,e_n(0),\dot{e}_1(0),\ldots,\dot{e}_n(0)$ is a Darboux basis of Σ. Fix coordinates in Σ using this basis: $\Sigma = \mathbb{R}^n \times \mathbb{R}^n$, where $\left(\begin{smallmatrix} x \\ y \end{smallmatrix}\right) \in \mathbb{R}^n \times \mathbb{R}^n$ is identified with $\sum\limits_{j=1}^{n} \left(x^j e_j(0) + y^j \dot{e}_j(0)\right) \in \Sigma$, $x = (x^1,\ldots,x^n)^\top$, $y = (y^1,\ldots,y^n)^\top$.

We claim that there exists a smooth family $A(t)$, $t \in \mathbb{R}$, of symmetric $n \times n$ matrices such that $A(t)$ has the same eigenvalues as $R_\Lambda(t)$ and

$$\Lambda(t) = \Gamma(0,t)\left(\begin{smallmatrix} \mathbb{R}^n \\ 0 \end{smallmatrix}\right), \quad \Lambda^\circ(t) = \Gamma(0,t)\left(\begin{smallmatrix} 0 \\ \mathbb{R}^n \end{smallmatrix}\right), \quad \forall t \in \mathbb{R}$$

in the fixed coordinates, where $\Gamma(t,\tau)$ satisfies (25). Indeed, let $e_i(t)$, $i = 1,\ldots,n$, be solutions to the structural equations (23). Then

$$\Lambda(t) = span\{e_1(t),\ldots,e_n(t)\}, \quad \Lambda^\circ(t) = span\{\dot{e}_1(t),\ldots,\dot{e}_n(t)\}.$$

Moreover, $\ddot{e}_i(t) = -\sum_{i=1}^{n} a_{ij}(t)e_j(t)$, where $A(t) = \{a_{ij}(t)\}_{i,j=1}^{n}$ is the matrix of the operator $R_\Lambda(t)$ in the "moving" basis $e_1(t), \dots, e_n(t)$. Lemma 1.13 implies that $\langle e_i(t), e_j(t)\rangle_{\dot{\Lambda}(t)} = \sigma(e_i(t), \dot{e}_j(t)) = \delta_{ij}$. In other words, the Euclidean structure $\langle\cdot,\cdot\rangle_{\dot{\Lambda}(t)}$ has unit matrix in the basis $e_1(t), \dots, e_n(t)$. Operator $R_\Lambda(t)$ is self-adjoint for the Euclidean structure $\langle\cdot,\cdot\rangle_{\dot{\Lambda}(t)}$ (see Proposition 2.9). Hence matrix $A(t)$ is symmetric.

Let $e_i(t) = \begin{pmatrix} x_i(t) \\ y_i(t) \end{pmatrix} \in \mathbb{R}^n \times \mathbb{R}^n$ in the fixed coordinates. Make up $n \times n$-matrices $X(t) = (x_1(t), \dots, x_n(t))$, $Y(t) = (y_1(t), \dots, y_n(t))$, and a $2n \times 2n$-matrix $\begin{pmatrix} X(t) & \dot{X}(t) \\ Y(t) & \dot{Y}(t) \end{pmatrix}$. We have

$$\frac{d}{dt}\begin{pmatrix} X & \dot{X} \\ Y & \dot{Y} \end{pmatrix}(t) = \begin{pmatrix} X & \dot{X} \\ Y & \dot{Y} \end{pmatrix}(t)\begin{pmatrix} 0 & -A(t) \\ I & 0 \end{pmatrix}, \quad \begin{pmatrix} X & \dot{X} \\ Y & \dot{Y} \end{pmatrix}(0) = \begin{pmatrix} I & 0 \\ 0 & I \end{pmatrix}.$$

Hence $\begin{pmatrix} X & \dot{X} \\ Y & \dot{Y} \end{pmatrix}(t) = \Gamma(t,0)^{-1} = \Gamma(0,t)$.

Let now $\Lambda(\cdot)$ be the Jacobi curve, $\Lambda(t) = J_{z_0}(t)$. Set $\xi_i(z_t) = e_*^{t\mathbf{h}} e_i(t)$, $\eta_i(z_t) = e_*^{t\mathbf{h}}\dot{e}_i(t)$; then

$$\xi_1(z_t), \dots, \xi_n(z_t), \eta_1(z_t), \dots, \eta_n(z_t) \tag{26}$$

is a Darboux basis of $T_{z_t}N$, where $J_{z_t}(0) = span\{\xi_1(z_t), \dots, \xi_n(z_t)\}$, $J^\circ_{z_t}(0) = span\{\eta_1(z_t), \dots, \eta_n(z_t)\}$. Moreover, the basis (26) is orthonormal for the inner product $\langle\cdot,\cdot\rangle_{z_t}$ on $T_{z_t}N$.

The intrinsic nature of the structural equation implies the translation invariance of the construction of the frame (26): if we would start from z_s instead of z_0 and put $\Lambda(t) = J_{z_s}(t)$, $e_i(0) = \xi_i(z_s)$, $\dot{e}_i(0) = \eta_i(z_s)$ for some $s \in \mathbb{R}$, then we would obtain $e_*^{t\mathbf{h}} e_i(t) = \xi_i(z_{s+t})$, $e_*^{t\mathbf{h}}\dot{e}_i(t) = \eta_i(z_{s+t})$.

The frame (26) gives us fixed orthonormal Darboux coordinates in $T_{z_s}N$ for $\forall\, s \in \mathbb{R}$ and the correspondent symplectic $2n \times 2n$-matrices $\Gamma_{z_s}(\tau, t)$. We have: $\Gamma_{z_s}(\tau,t) == \Gamma_{z_0}(s+\tau, s+t)$; indeed, $\Gamma_{z_s}(\tau,t)\begin{pmatrix} x \\ y \end{pmatrix}$ is the coordinate presentation of the vector

$$e_*^{(\tau-t)\mathbf{h}}\left(\sum_i x^i \xi^i(z_{s+t}) + y^i \eta_i(z_{s+t})\right)$$

in the basis $\xi_i(z_{s+\tau})$, $\eta_i(z_{s+\tau})$. In particular,

$$|\Gamma_{z_s}(0,t)\begin{pmatrix} x \\ y \end{pmatrix}| = \left\| e_*^{-t\mathbf{h}}\left(\sum_i x^i \xi^i(z_{s+t}) + y^i \eta_i(z_{s+t})\right)\right\|_{z_s}. \tag{27}$$

Recall that $\xi_1(z_\tau), \dots, \xi_n(z_\tau), \eta_1(z_\tau), \dots, \eta_n(z_\tau)$ is an orthonormal frame for the scalar product $\langle\cdot,\cdot\rangle_{z_\tau}$ and $\|\zeta\|_{z_\tau} = \sqrt{\langle\zeta,\zeta\rangle_{z_\tau}}$.

We introduce the notation:

$$\lfloor W \rfloor_{z_s} = \left\{ \sum_i x^i \xi^i(z_s) + y^i \eta_i(z_s) : \binom{x}{y} \in W \right\},$$

for any $W \subset \mathbb{R}^n \times \mathbb{R}^n$. Let $C^{\pm}_{\Gamma_{z_0}}$ be the cones from Lemma 2.14. Then

$$e_*^{-\tau \mathbf{h}} \lfloor \Gamma_{z_s}(0,t) C^{\pm}_{\Gamma_{z_0}} \rfloor_{z_{s-\tau}} = \lfloor \Gamma_{z_{s-\tau}}(0,t+\tau) C^{\pm}_{\Gamma_{z_0}} \rfloor_{z_{s-\tau}}, \quad \forall t, \tau, s. \tag{28}$$

Now set $K^+_{\Gamma_{z_s}} = \bigcap_{t \geq 0} C^+_{\Gamma_{z_0}}$, $K^-_{\Gamma_{z_s}} = \bigcap_{t \leq 0} C^-_{\Gamma_{z_0}}$ and $\Delta^{\pm}_{z_s} = \lfloor K^{\mp}_{\Gamma_{z_s}} \rfloor_{z_s}$. Corollary 2.6 implies that $\Delta^{\pm}_{z_s}$ are Lagrangian subspaces of $T_{z_s}N$. Moreover, it follows from (28) that $e_*^{t\mathbf{h}} \Delta^{\pm}_{z_s} = \Delta^{\pm}_{z_{s+t}}$, while (28) and (27) imply inequalities (24).

This finishes the proof of the part of Proposition 2.10 which concerns Jacobi curves $J_z(t)$. We leave to the reader a simple adaptation of this proof to the case of reduced Jacobi curves $J^h_z(t)$. □

Remark. Constant α depends, of course, on the Riemannian structure on N. In the case of the special Riemannian structure defined at the beginning of the proof of Proposition 2.10 this constant depends only on the upper and lower bounds for the eigenvalues of the curvature operators and reduced curvature operators correspondently (see Lemma 2.14 and further arguments).

Let e^{tX}, $t \in \mathbb{R}$ be the flow generated by the the vector field X on a manifold M. Recall that a compact invariant subset $W \subset M$ of the flow e^{tX} is called a hyperbolic set if there exists a Riemannian structure in a neighborhood of W, a positive constant α, and a splitting $T_zM = E^+_z \oplus E^-_z \oplus \mathbb{R}X(z)$, $z \in W$, such that $X(z) \neq 0$, $e_*^{tX} E^{\pm}_z = E^{\pm}_{e^{tX}(z)}$, and $\|e_*^{\pm tX}\zeta^{\pm}\| \geq e^{\alpha t}\|\zeta^{\pm}\|$, $\forall t \geq 0$, $\zeta^{\pm} \in E^{\pm}_z$. Just the fact some invariant set is hyperbolic implies a rather detailed information about asymptotic behavior of the flow in a neighborhood of this set (see [12] for the introduction to Hyperbolic Dynamics). The flow e^{tX} is called an Anosov flow if the entire manifold M is a hyperbolic set.

The following result is an immediate corollary of Proposition 2.10 and the above remark.

Theorem 2.3. *Let h be a regular monotone Hamiltonian on N, $c \in \mathbb{R}$, $W \subset h^{-1}(c)$ a compact invariant set of the flow $e^{t\mathbf{h}}$, $t \in \mathbb{R}$, and $d_z h \neq 0$, $\forall z \in W$. If* $\mathbf{h}$ *has a negative reduced curvature at every point of W, then W is a hyperbolic set of the flow $e^{t\mathbf{h}}\big|_{h^{-1}(c)}$.* □

This theorem generalizes a classical result about geodesic flows on compact Riemannian manifolds with negative sectional curvatures. Indeed, if N is the cotangent bundle of a Riemannian manifold and $e^{t\mathbf{h}}$ is the geodesic flow, then negativity of the reduced curvature of $\mathbf{h}$ means simply negativity of the sectional Riemannian curvature. In this case, the Hamiltonian h is homogeneous on the fibers of the cotangent bundle and the restrictions $e^{t\mathbf{h}}\big|_{h^{-1}(c)}$ are equivalent for all $c > 0$.

The situation changes if h is the energy function of a general natural mechanical system on the Riemannian manifold. In this case, the flow and the reduced curvature depend on the energy level. Still, negativity of the sectional curvature implies negativity of the reduced curvature at $h^{-1}(c)$ for all sufficiently big c. In particular, $e^{t\mathbf{h}}\big|_{h^{-1}(c)}$ is an Anosov flow for any sufficiently big c; see [8,9] for the explicit expression of the reduced curvature in this case.

Theorem 2.3 concerns only the reduced curvature while the next result deals with the (not reduced) curvature of $\mathbf{h}$.

Theorem 2.4. *Let h be a regular monotone Hamiltonian and W a compact invariant set of the flow $e^{t\mathbf{h}}$. If $\mathbf{h}$ has a negative curvature at any point of W, then W is a finite set and each point of W is a hyperbolic equilibrium of the field $\mathbf{h}$.*

Proof. Let $z \in W$; the trajectory $z_t = e^{t\mathbf{h}}(z)$, $t \in \mathbb{R}$, satisfies conditions of Proposition 2.10. Take the correspondent splitting $T_{z_t}N = \Delta^+_{z_t} \oplus \Delta^-_{z_t}$. In particular, $\mathbf{h}(z_t) = \mathbf{h}^+(z_t) + \mathbf{h}^-(z_t)$, where $\mathbf{h}^\pm(z_t) \in \Delta^\pm_{z_t}$.

We have $e^{\tau\mathbf{h}}_*\mathbf{h}(z_t) = \mathbf{h}(z_{t+\tau})$. Hence

$$\begin{aligned}\|\mathbf{h}(z_{t+\tau})\| = \|e^{\tau\mathbf{h}}_*\mathbf{h}(z_t)\| &\geq \|e^{\tau\mathbf{h}}_*\mathbf{h}^+(z_t)\| - \|e^{\tau\mathbf{h}}_*\mathbf{h}^-(z_t)\| \\ &\geq e^{\alpha\tau}\|\mathbf{h}^+(z_t)\| - e^{-\alpha\tau}\|\mathbf{h}^-(z_t)\|, \quad \forall \tau \geq 0.\end{aligned}$$

Compactness of $\overline{\{z_t : t \in \mathbb{R}\}}$ implies that $\mathbf{h}^+(z_t)$ is uniformly bounded; hence $\mathbf{h}^+(z_t) = 0$. Similarly, $\|\mathbf{h}(z_{t-\tau}\| \geq e^{\alpha\tau}\|\mathbf{h}^-(z_t)\| - e^{-\alpha\tau}\|\mathbf{h}^+(z_t)\|$ that implies the equality $\mathbf{h}^-(z_t) = 0$. Finally, $\mathbf{h}(z_t) = 0$. In other words, $z_t \equiv z$ is an equilibrium of $\mathbf{h}$ and $T_zN = \Delta^+_z \oplus \Delta^-_z$ is the splitting of T_zN into the repelling and attracting invariant subspaces for the linearization of the flow $e^{t\mathbf{h}}$ at z. Hence z is a hyperbolic equilibrium; in particular, z is an isolated equilibrium of $\mathbf{h}$. □

We say that a subset of a finite dimensional manifold is bounded if it has a compact closure.

Corollary 2.7. *Assume that h is a regular monotone Hamiltonian and $\mathbf{h}$ has everywhere negative curvature. Then any bounded semi-trajectory of the system $\dot{z} = \mathbf{h}(z)$ converges to an equilibrium with the exponential rate while another semi-trajectory of the same trajectory must be unbounded.* □

Typical Hamiltonians which satisfy conditions of Corollary 2.7 are energy functions of natural mechanical systems in $\mathbb{R}^n$ with a strongly concave potential energy. Indeed, in this case, the second derivative of the potential energy is equal to the matrix of the curvature operator in the standard Cartesian coordinates (see Sect. 2.8).

References

1. A. A. Agrachev, *Topology of quadratic maps and Hessians of smooth maps.* Itogi Nauki; Algebra, Topologiya, Geometriya, 1988, v.26, 85–124 (in Russian). English. transl.: J. Soviet Math., Plenum Publ. Corp., 1990, 990–1013

2. A. A. Agrachev, R. V. Gamkrelidze, *Symplectic geometry and necessary conditions for optimality.* Matem. Sbornik, 1991, v.182 (in Russian). English transl.: Math. USSR Sbornik, 1992, v.72, 29–45
3. A. A. Agrachev, R. V. Gamkrelidze, *Symplectic methods in optimization and control.* In the book: Geometry of Feedback and Optimal Control. B. Jakubczyk, W. Respondek, Eds. Marcel Dekker, 1998, 19–77
4. A. A. Agrachev, R. V. Gamkrelidze, *Feedback–invariant optimal control theory and differential geometry, I. Regular extremals.* J. Dynamical and Control Systems, 1997, v.3, 343–389
5. A. A. Agrachev, *Feedback–invariant optimal control theory and differential geometry, II. Jacobi curves for singular extremals.* J. Dynamical and Control Systems, 1998, v.4, 583–604
6. A. A. Agrachev, I. Zelenko, *Geometry of Jacobi curves, I, II.* J. Dynamical and Control Systems, 2002, v.8, 93–140; 167–215
7. A. A. Agrachev, Yu. L. Sachkov, *Control theory from the geometric viewpoint.* Springer Verlag, 2004, xiv+412pp.
8. A. A. Agrachev, N. Chtcherbakova, *Hamiltonian systems of negative curvature are hyperbolic.* Russian Math. Dokl., 2005, v.400, 295–298
9. A. A. Agrachev, N. Chtcherbakova, I. Zelenko, *On curvatures and focal points of dynamical Lagrangian distributions and their reductions by first integrals.* J. Dynamical and Control Systems, 2005, v.11, 297–327
10. V. I. Arnold, A. B. Givental, *Symplectic geometry.* Springer Verlag, Encyclopedia of Mathematical Sciences, v.4, 1988, 1–136
11. B. Dubrov, I. Zelenko, *Canonical frame for rank 2 distributions of maximal class.* C. R. Acad. Sci. Paris, v. 342, 2006, 589-594
12. A. Katok, B. Hasselblatt, *Introduction to the modern theory of dynamical systems.* Cambridge Univ. Press, 1998
13. J. W. Milnor, J. D. Stasheff, *Characteristic classes.* Princeton Univ. Press, 1974
14. I. Zelenko, *On variational approach to differential invariants of rank two distributions*, Differential Geom. Appl., 2006, v. 24, 235–259

Lecture Notes on Logically Switched Dynamical Systems

A.S. Morse*

Department of Electrical Engineering, Yale University, P.O. Box 208267,
New Haven, CT 06520-8284, USA
morse@sysc.eng.yale.edu

Introduction

The subject of logically switched dynamical systems is a large one which overlaps with many areas including hybrid system theory, adaptive control, optimal control, cooperative control, etc. Ten years ago we presented a lecture, documented in [1], which addressed several of the areas of logically switched dynamical systems which were being studied at the time. Since then there have been many advances in many directions, far to many too adequately address in these notes. One of the most up to date and best written books on the subject is the monograph by Liberzon [2] to which we refer the reader for a broad but incisive perspective as well an extensive list of references.

In these notes we will deal with two largely disconnected topics, namely switched adaptive control (sometimes called supervisory control) and "flocking" which is about the dynamics of reaching a consensus in a rapidly changing environment. In the area of adaptive control we focus mainly on one problem which we study in depth. Our aim is to give a thorough analysis under realistic assumptions of the adaptive version of what is perhaps the most important design objective in all of feedback control, namely set-point control of a single-input, single output process admitting a linear model. While the non-adaptive version the set-point control problem is very well understood and has been so for more than a half century, the adaptive version still is not because there is no credible counterpart in an adaptive context of the performance theories which address the non-adaptive version of the problem. In fact, even just the stabilization question for the adaptive version of the problem did not really get ironed out until ten years ago, except under unrealistic assumptions which ignored the effects of noise and/or un-modeled dynamics.

As a first step we briefly discuss the problem of adaptive disturbance rejection. Although the switching logic we consider contains no logic or discrete

*This research was supported by the US Army Research Office, the US National Science Foundation and by a gift from the Xerox Corporation.

event sub-system, the problem nonetheless sets the stage for what follows. One of the things which turns out (in retrospect) to have impeded progress with adaptive control has been the seemingly benign assumption that the parameters of the (nominal) model of the process to be controlled are from a *continuum* of possible values. In Chap. 4 we briefly discuss the unpleasant consequences of this assumption and outline some preliminary ideas which might be used to deal with them.

In Chap. 5 we turn to detailed discussion of a switched adaptive controller capable of causing the output of an imprecisely modeled process to approach and track a constant reference signal. The material in this chapter provides a clear example of what is meant by a logically switched dynamical system. Finally in Chap. 6 we consider several switched dynamical systems which model the behavior of a group of mobile autonomous agents moving in a rapidly changing environment using distributed controls. We begin with what most would agree is the quintessential problem in the area of switched dynamical systems.

1 The Quintessential Switched Dynamical System Problem

The quintessential problem in the area of switched dynamical systems is this: Given a compact subset $\mathcal{P}$ of a finite dimensional space, a parameterized family of $n \times n$ matrices $\mathcal{A} = \{A_p : p \in \mathcal{P}\}$, and a family $\mathcal{S}$ of piecewise-constant switching signals $\sigma : [0, \infty) \to \mathcal{P}$, determine necessary and sufficient conditions for A_σ to be exponentially stable for every $\sigma \in \mathcal{S}$. There is a large literature on this subject. Probably its most comprehensive treatment to date is in the monograph [2] by Liberzon mentioned before. The most general version of the problem is known to be undecidable [3]. These notes deal with two special versions of this problem. Each arises in a specific context and thus is much more structured than the general problem just formulated.

1.1 Dwell-Time Switching

In the first version of the problem, $\mathcal{S}$ consists of all switching signals whose switching times are separated by τ_D times unit where τ_D is a pre-specified positive number called a *dwell time*. More precisely, $\sigma \in \mathcal{S}$ is said to have dwell time τ_D if and only if σ switches values at most once, or if it switches more that once, the set of time differences between any two successive switches is bounded below τ_D. In Sect. 5.1 we will encounter a switching logic which generates such signals.

Note that the class $\mathcal{S}$ just defined contains constant switching signal $\sigma(t) = p,\ t \geq 0$ for any value of $p \in \mathcal{P}$. A necessary condition for A_σ to be exponentially stable for every $\sigma \in \mathcal{S}$, is therefore that each $A_p \in \mathcal{A}$ is exponentially stable. In other words, if A_σ to be exponentially stable for every

$\sigma \in \mathcal{S}$, then for each $p \in \mathcal{P}$ there must exist non-negative numbers t_p and λ_p, with λ_p positive such that $|e^{A_p t}| \leq e^{\lambda_p(t_p - t)}, \quad t \geq 0$. Here and elsewhere through the end of Chap. 5, the symbol $|\cdot|$ denotes any norm on a finite dimensional linear space. It is quite easy to show by example that this condition is not sufficient unless τ_D is large. An estimate of how large τ_D has to be in order to guarantee exponential stability, is provided by the following lemma.

Lemma 1.1. *Let $\{A_p : p \in \mathcal{P}\}$ be a set of real, $n \times n$ matrices for which there are non-negative numbers t_p and λ_p with λ_p positive such that*

$$|e^{A_p t}| \leq e^{\lambda_p(t_p - t)}, \quad t \geq 0 \tag{1.1}$$

Suppose that τ_D is a finite number satisfying

$$\tau_D > t_p, \quad p \in \mathcal{P} \tag{1.2}$$

For any switching signal $\sigma : [0, \infty) \rightarrow \mathcal{P}$ with dwell time τ_D, the state transition matrix of A_σ satisfies

$$|\Phi(t, \mu)| \leq e^{\lambda(T - (t - \mu))}, \quad \forall\, t \geq \mu \geq 0 \tag{1.3}$$

where λ is a positive number defined by

$$\lambda = \inf_{p \in \mathcal{P}} \left\{ \lambda_p \left(1 - \frac{t_p}{\tau_D} \right) \right\} \tag{1.4}$$

and

$$T = \frac{2}{\lambda} \sup_{p \in \mathcal{P}} \{\lambda_p t_p\} \tag{1.5}$$

Moreover,

$$\lambda \in (0, \lambda_p], \quad p \in \mathcal{P}. \tag{1.6}$$

The estimate given by this lemma can be used to make sure that the "slow switching assumption" discussed in Sect. 5.2 is satisfied.

Proof of Lemma 1.1: Since $\mathcal{P}$ is a closed, bounded set, $\sup_{p \in \mathcal{P}} t_p < \infty$. Thus a finite τ_D satisfying (1.2) exists. Clearly $\lambda_p(1 - \frac{t_p}{\tau_D}) > 0$, $p \in \mathcal{P}$. From this and the definition of λ it follows that (1.6) holds and that

$$e^{\lambda_p(t_p - \tau_D)} \leq e^{-\lambda \tau_D}, \quad p \in \mathcal{P}$$

This and (1.1) imply that for $t \geq \tau_D$

$$\begin{aligned} |e^{A_p t}| &\leq e^{\lambda_p(t_p - t)} = e^{\lambda_p(t_p - \tau_D)} e^{-\lambda_p(t - \tau_D)} \leq e^{-\lambda \tau_D} e^{-\lambda_p(t - \tau_D)} \\ &\leq e^{-\lambda \tau_D} e^{-\lambda(t - \tau_D)} \leq e^{-\lambda t}, \quad t \geq \tau_D,\, p \in \mathcal{P} \end{aligned} \tag{1.7}$$

It also follows from (1.1) and the definition of T that

$$|e^{A_p t}| \leq e^{\lambda(\frac{T}{2}-t)}, \quad t \in [0, \tau_D), \quad p \in \mathcal{P} \tag{1.8}$$

Set $t_0 = 0$ and let $t_1, t_2 \ldots$ denote the times at which σ switches. Write p_i for the value of σ on $[t_{i-1}, t_i)$. Note that for $t_{j-1} \leq \mu \leq t_j \leq t_i \leq t \leq t_{i+1}$,

$$\Phi(t,\mu) = e^{A_{p_{i+1}}(t-t_i)} \left(\prod_{q=j+1}^{i} e^{A_{p_q}(t_q - t_{q-1})} \right) e^{A_{p_j}(t_j - \mu)}$$

In view of (1.7) and (1.8)

$$\begin{aligned}
|\Phi(t,\mu)| &\leq |e^{A_{p_{i+1}}(t-t_i)}| \left(\prod_{q=j+1}^{i} |e^{A_{p_q}(t_q - t_{q-1})}| \right) |e^{A_{p_j}(t_j - \mu)}| \\
&\leq e^{\lambda(\frac{T}{2}-(t-t_i))} \left(\prod_{q=j+1}^{i} e^{-\lambda(t_q - t_{q-1})} \right) e^{\lambda(\frac{T}{2}-(t_j-\mu))} \\
&= e^{\lambda(T-(t-\mu))}
\end{aligned}$$

On the other hand, for $i > 0$, $t_{i-1} \leq \mu \leq t \leq t_i$, (1.8) implies that

$$|\Phi(t,\mu)| \leq e^{\lambda(\frac{T}{2}-(t-\mu))} \leq e^{\lambda(T-(t-\mu))}$$

and so (1.3) is true. □

Input–Output Gains of Switched Linear Systems

In deriving stability margins and systems gains for the supervisory control systems discussed in Sect. 5 we will make use of induced "gains" (i.e. norms) of certain types of switched linear systems. Quantification of stability margins and the devising of a much needed performance theory of adaptive control, thus relies heavily on our ability to characterize these induced gains. In this section we make precise what types of induced gains we are referring to and we direct the reader to some recent work aimed at their characterization.

To begin, suppose that $\{(A_p, B_p, C_p, D_p) : p \in \mathcal{P}\}$ is a family of coefficient matrices of m-input, r-output, n-dimensional, exponentially stable linear systems. Then any $\sigma \in \mathcal{S}$ determines a switched linear system of the form

$$\Sigma_\sigma \stackrel{\Delta}{=} \left\{ \begin{array}{l} \dot{x} = A_\sigma x + B_\sigma u \\ y = C_\sigma x + D_\sigma u \end{array} \right\} \tag{1.9}$$

Thus if $x(0) \stackrel{\Delta}{=} 0$, then $y = Y_\sigma \circ u$, where Y_σ is the input–output mapping

$$u \longmapsto \int_0^t C_{\sigma(t)} \Phi(t,\tau) D_{\sigma(\tau)} B_\sigma(\tau) u(\tau) d\tau + D_\sigma u,$$

and Φ is the state transition matrix of A_σ. Let prime denotes transpose and, for any integrable, vector-valued signal v on $[0, \infty)$, let $||\cdot||$ denotes the two-norm

$$||v|| \stackrel{\Delta}{=} \sqrt{\int_0^\infty v'(t)v(t)dt}$$

The *input–output gain* of Σ_σ is then the induced two-norm

$$\gamma(\sigma) \stackrel{\Delta}{=} \inf\{g : ||Y \circ u|| \leq g||u||, \ \forall u \in \mathcal{L}_2\}$$

where $\mathcal{L}_2$ is the space of all signals with finite two-norms. Define the gain $\mathfrak{g}$ of the *multi-system* $\{(A_p, B_p, C_p, D_p), p \in \mathcal{P}\}$ to be

$$\mathfrak{g} \stackrel{\Delta}{=} \sup_{\sigma \in \mathcal{S}} \gamma(\sigma)$$

Thus $\mathfrak{g}$ is the worst case input–output gain of (1.9) as σ ranges over all switching signals in $\mathcal{S}$. Two problems arise:

1. Derive conditions in terms of τ_D and the multi-system $\{(A_p, B_p, C_p, D_p), p \in \mathcal{P}\}$ under which $\mathfrak{g}$ is a finite number.
2. Assuming these conditions hold, characterize $\mathfrak{g}$ in terms of $\{(A_p, B_p, C_p, D_p), p \in \mathcal{P}\}$ and τ_D.

The first of the two problems just posed implicitly contains as a subproblem the quintessential switched dynamical system problem posed at the beginning of this section. A sufficient condition for $\mathfrak{g}$ to be finite is that τ_D satisfies condition (1.2) of Lemma 1.1. For the second problem, what would be especially useful would be a characterization of $\mathfrak{g}$ which is coordinate-independent; that is a characterization which depends only on the transfer matrices $C_p(sI - A_p)^{-1}B_p + D_p$, $p \in \mathcal{P}$ and not on the specific realizations of these transfer matrices which define $\{(A_p, B_p, C_p, D_p), p \in \mathcal{P}\}$. For example, it is reasonable to expect that there might be a characterization of $\mathfrak{g}$ in terms of the $\mathcal{H}^\infty$ norms of the $C_p(sI - A_p)^{-1}B_p + D_p$, $p \in \mathcal{P}$, at least for τ_D sufficiently large.

The problem of characterizing $\mathfrak{g}$ turns out to be a good deal more difficult that one might at first suspect, even if all one wants is a characterization of the limiting value of $\mathfrak{g}$ as $\tau_D \to \infty$ [4]. In fact, contrary to intuition, one can show by example that this limiting value may, in some cases, be strictly greater than the supremum over $\mathcal{P}$ of the $\mathcal{H}_\infty$ norms of the $C_p(sI - A_p)^{-1}B_p + D_p$. We refer the interested reader to [4] for a more detailed discussion of this subject. It is results along these lines which will eventually lead to a bona fide performance theory for adaptive control.

1.2 Switching Between Stabilizing Controllers

In many applications, including those discussed in Chap. 5, the matrix A_σ arises within a linear system which models the closed loop connection consisting of fixed linear system in feedback with a switched linear system. For

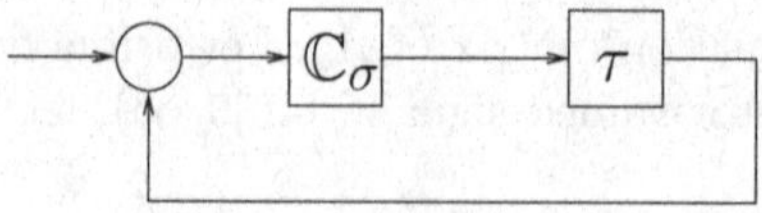

Fig. 1. A switched control system

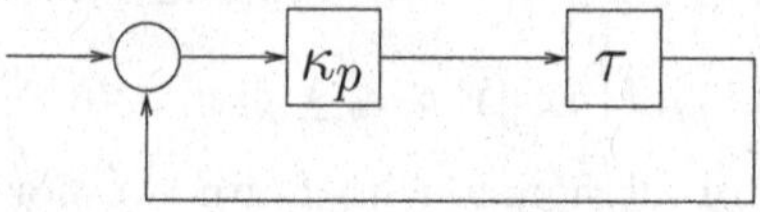

Fig. 2. A Linear Control System

example, it is possible to associate with a given family of controller transfer functions $\{\kappa_p : p \in \mathcal{P}\}$ together with a given process model transfer function τ, a switched control system of the form shown in Fig. 1 where $\mathbb{C}_\sigma$ is a switched controller with instantaneous transfer function κ_σ.

Not always appreciated, but nonetheless true is the fact that the input–output properties of such a system depends on the specific realizations of the κ_p. Said differently, it is really not possible to talk about switched linear systems from a strictly input–output point of view. A good example, which makes this point, occurs when for each $p \in \mathcal{P}$, the closed loop systems shown in Fig. 2 is stable. Under these conditions, the system shown in Fig. 1 turns out to be exponentially stable for every possible piecewise continuous switching signal, no matter how fast the switching, but only for certain realizations of the κ_p [5]. The key idea is to first use a Youla parameterization to represent the entire class of controller transfer functions and second to realize the family in such a way so that what is actually being switched within $\mathbb{C}_\sigma$ are suitably defined realizations of the Youla parameters, one for each κ_p. We refer the reader to [5] for a detailed discussion of this idea.

1.3 Switching Between Graphs

The system just discussed is a switched dynamical system because the controller within the system is a switched controller. Switched dynamical systems can arise for other reasons. An interesting example of this when the overall system under consideration models the motions of a group of mobile autonomous agents whose specific movements are governed by strategies which depend on the movements of their nearby agents. A switched dynamical model can arise in this context because each agent's neighbors may change over time. What is especially interesting about this type of system is the interplay between the underlying graphs which characterize neighbor relationships, and the evolution of the system over time. Chapter 6 discusses in depth several example of this type of system.

2 Switching Controls with Memoryless Logics

2.1 Introduction

Classical relay control can be thought of as a form of switching control in which the logic generating the switching signal is a memoryless system. A good example of this is the adaptive disturbance rejector devised by I. M. Lie Ying at the High Altitude Control Laboratory at the Tibet Institute of Technology in Lhasa. Ying's work provides a clear illustration of what the use of switching can accomplish in a control setting, even if there is no logic or memory involved.

2.2 The Problem

In [6], Ying gives a definitive solution to the long-standing problem of constructing an adaptive feedback control for a one-dimensional siso linear system with an unmeasurable, bounded, exogenous disturbance input so as to cause the system's output to go to zero asymptotically. More specifically he considered the problem of constructing an adaptive feedback control $u = f(y)$ for the one-dimensional linear system

$$\dot{y} = -y + u + d$$

so as to cause the system's output y to go to zero no matter what exogenous disturbance $d : [0, \infty) \to \mathbb{R}$ might be, so long as d is bounded and piecewise-continuous. Up until the time of Ying's work, solutions to this long standing problem had been shown to exist only under the unrealistic assumption that d could be measured [7]. Ying made no such assumption.

2.3 The Solution

The adaptive control he devised is described by the equations

$$u = -k\sigma(y)$$
$$\dot{k} = |y|$$

where

$$\sigma(y) = \begin{cases} 1 & y \geq 0 \\ -1 & y < 0 \end{cases} \tag{2.1}$$

The closed-loop system is thus

$$\dot{y} = -y - k\sigma(y) + d, \tag{2.2}$$
$$\dot{k} = |y| \tag{2.3}$$

2.4 Analysis

Concerned that skeptical readers might doubt the practicality of his idea, Ying carried out a full analysis of the system. Here is his reasoning.

To study this system's behavior, let b be any positive number for which

$$|d(t)| \leq b, \quad t \geq 0 \tag{2.4}$$

and let V denote the Lyapunov function

$$V = \frac{y^2}{2} + \frac{(k-b)^2}{2} \tag{2.5}$$

In view of the definition of σ in (2.1), the rate of change of V along a solution to (2.2) and (2.3) can be written as

$$\dot{V} = \dot{y}y + \dot{k}(k-b) = -y^2 - k|y| + dy + |y|(k-b)$$

Since (2.4) implies that $dy \leq b|y|$, $\dot{V}$ must satisfy

$$\dot{V} \leq -y^2 - k|y| + b|y| + |y|(k-b)$$

Therefore

$$\dot{V} \leq -y^2.$$

From this and (2.5) it follows that y and k are bounded and that y has a bounded $\mathcal{L}^2$ norm. In addition, since (2.2) implies that $\dot{y}$ is bounded, it must be true that $y \rightarrow 0$ as $t \rightarrow \infty$ (cf. [8]) which is what is desired. The practical significance of this result, has been firmly established by computer simulation performed be numerous graduate students all over the world.

3 Collaborations

Much of the material covered in these notes has appeared in one form or another in published literature. There are however several notable exceptions in Chap. 6. Among these are the idea of composing directed graphs and the interrelationships between rooted graphs, Sarymsakov graphs, and neighbor shared graphs discussed in Sect. 6.1. The entire section on measurement delays (Sect. 6.3) is also new. All of the new topics addressed in Chap. 6 were developed in collaboration with Ming Cao and Brian Anderson. Daniel Spielman also collaborated with us on most of the convergence results and Jia Fang helped with the development as well. Most of this material will be published elsewhere as one or more original research papers.

4 The Curse of the Continuum

Due to its roots in nonlinear estimation theory, parameter identification theory generally focuses on problems in which unknown model parameters are assumed to lie within given continuums. Parameter-adaptive control, which is largely an outgrowth of identification theory, also usually addresses problems in which unknown process model parameters are assumed to lie within continuums. The continuum assumption comes with a large price tag because a typical parameter estimation problem over a continuum is generally not tractable unless the continuum is convex and the dependence on parameters is linear. These practical limitations have deep implications. For example, a linearly parameterized transfer matrix on a convex set can easily contain points in its parameter spaces at which the transfer matrix has an unstable pole and zero in common. For nonlinear systems, the problem can be even worse – for those process model parameterizations in which parameters cannot be separated from signals, it is usually impossible to construct a finite-dimensional multi-estimator needed to carry out the parameter estimation process. We refer to these and other unfortunate consequences of the continuum assumption as the *Curse of the Continuum*. An obvious way to avoid the curse, is to formulate problems in such a way so that the parameter space of interest is finite or perhaps countable. But many problems begin with parameter spaces which are continuums. How is one to reformulate such a problem using a finite parameter space, without serious degradation in expected performance? How should a parameter search be carried out in a finite parameter space so that one ends up with a provably correct overall adaptive algorithm? It is these questions to which this brief chapter and Chap. 5 are addressed.

4.1 Process Model Class

Let $\mathbb{P}$ be a process to be controlled and suppose for simplicity that $\mathbb{P}$ is a siso system admitting a linear model. Conventional linear feedback theory typically assumes that $\mathbb{P}$'s transfer function lies in a known open ball

$$\mathbb{B}(\nu, r)$$

of radius r centered at nominal transfer function ν in a metric space $\mathcal{T}$. In contrast, main-stream adaptive control typically assumes $\mathbb{P}$'s transfer function lies in a known set of the form

$$\mathcal{M} = \bigcup_{p \in \mathcal{P}} \mathbb{B}(\nu_p, r_p)$$

where $\mathcal{P}$ is a compact continuum within a finite dimensional space and $p \longmapsto r_p$ is at least bounded.

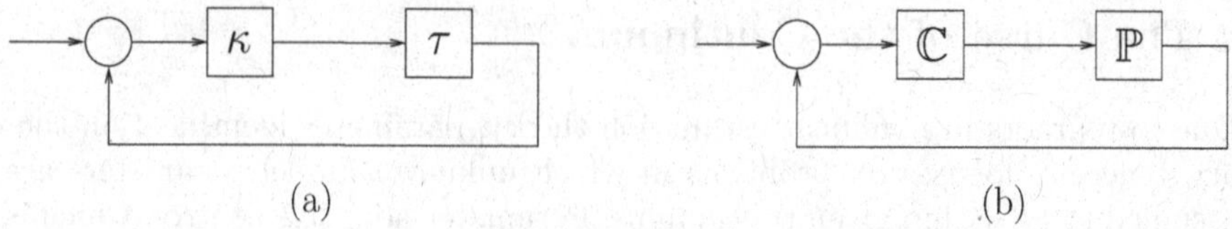

Fig. 3. Non-adaptive control

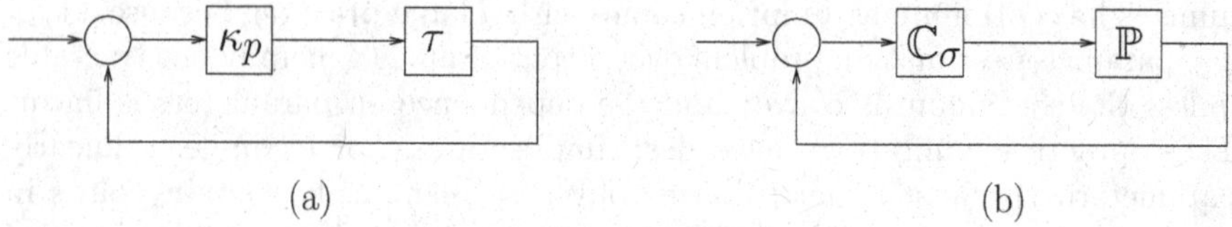

Fig. 4. Adaptive control

In a conventional non-adaptive control problem, a controller transfer function κ is chosen to endow the feedback system shown in Fig. 3a with stability and other prescribed properties for each candidate transfer function $\tau \in \mathbb{B}(\nu, r)$. Control of $\mathbb{P}$ is then carried out by applying to $\mathbb{P}$ a controller $\mathbb{C}$ with transfer function κ as shown in Fig. 3b.

In the adaptive control case, for each $p \in \mathcal{P}$, controller transfer function κ_p is chosen to endow the feedback system shown in Fig. 4a with stability and other prescribed properties for each $\tau \in \mathbb{B}(\nu_p, r_p)$. Adaptive control of $\mathbb{P}$ is then carried, in accordance with the idea of "certainty equivalence by using a parameter-varying controller or *multi-controller* $\mathbb{C}_\sigma$ with instantaneous transfer function κ_σ where σ is the index in $\mathcal{P}$ of the "best" current estimate of the ball within which $\mathbb{P}$'s transfer function resides.[1]

Since $\mathcal{P}$ is a continuum and κ_p is to be defined for each $p \in \mathcal{P}$, the actual construction of κ_p is at best a challenging problem, especially if the construction is based on LQG or $\mathcal{H}^\infty$ techniques. Moreover, because of the continuum, the associated estimation of the index of the ball within which $\mathbb{P}$'s transfer function resides will be intractable unless demanding conditions are satisfied. Roughly speaking, $\mathcal{P}$ must be convex and the dependence of candidate

[1] *Certainty equivalence* is a heuristic idea which advocates that the feedback controller applied to an imprecisely modeled process should, at each instant of time, be designed of the basis of a current estimate of what the process is, with the understanding that each such estimate is to be viewed as correct even though it may not be. The term is apparently due to Herbert Simon [9] who used in a 1956 paper [10] to mean something somewhat different then what's meant here and throughout the field of parameter adaptive control. The consequence of using this idea in an adaptive context is to cause the interconnection of the controlled process, the multi-controller and the parameter estimator to be detectable through the error between the output of the process and its estimate for every frozen parameter estimate – and this is true whether the three subsystems involved are linear or not [11].

process models on p must be linear. And in the more general case of nonlinear process models, a certain separability condition [11, 12] would have to hold which models as simple as

$$\dot{y} = \sin(py) + u$$

fail to satisfy. The totality of these implications is rightly called *the curse of the continuum.*

Example

The following example illustrates one of the difficulties which arises as a consequence of the continuum assumption. Ignoring un-modeled dynamics, suppose that $\mathcal{M}$ is simply the parameterized family of candidate process model transfer functions

$$\mathcal{M} = \left\{ \frac{s - \frac{1}{6}(p+2)}{s^2 + ps - \frac{2}{9}p(p+2)} : \quad p \in \mathcal{P} \right\}$$

where $\mathcal{P} = \{p : -1 \le p \le 1\}$. Note that there is no transfer function in $\mathcal{M}$ with a common pole and zero because the polynomial function

$$s^2 + ps - \frac{2}{9}p(p+2)|_{s=\frac{1}{6}(p+2)}$$

is nonzero for all $p \in \mathcal{P}$. The parameterized transfer function under consideration can be written as

$$\frac{s - \frac{1}{6}(p+2)}{s^2 + ps + q}$$

where

$$q = -\frac{2}{9}p(p+2). \tag{4.1}$$

Thus $\mathcal{M}$ is also the set of transfer functions

$$\mathcal{M} = \left\{ \frac{s - \frac{1}{6}(p+2)}{s^2 + ps + q} : \quad (p, q) \in \mathcal{Q} \right\}, \tag{4.2}$$

where $\mathcal{Q}$ is the two-parameter space

$$\mathcal{Q} = \{(p, q) : q + \frac{2}{9}p(p+2) = 0, \ -1 \le p \le 1\}.$$

The set of points in $\mathcal{Q}$ form a parabolic curve segment as shown in Fig. 5. Although the parameterized transfer function defining $\mathcal{M}$ in (4.2) depends linearly on p and q, the parameter space $\mathcal{Q}$ is not convex. Thus devising a provably parameter estimation algorithm for this parameterization would be difficult. In a more elaborate example of this type, where more parameters would be involved, the parameter estimation problem would typically be intractable.

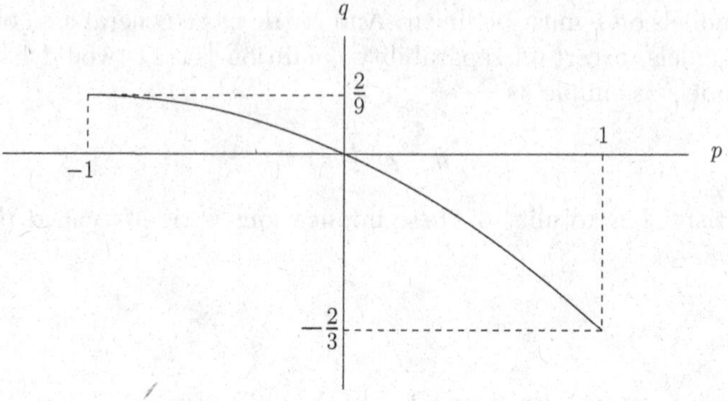

Fig. 5. Parameter space $\mathcal{Q}$

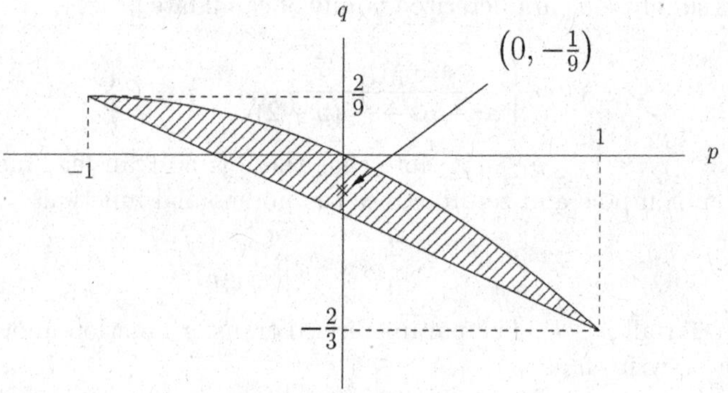

Fig. 6. Parameter space $\bar{\mathcal{Q}}$

There is a natural way to get around this problem, if the goal is identification as opposed to adaptive control. The idea is to embed $\mathcal{Q}$ in a larger parameter space which is convex. The smallest convex space $\bar{\mathcal{Q}}$ containing $\mathcal{Q}$ is the convex hull of $\mathcal{Q}$ as shown in Fig. 6.

The corresponding set of transfer functions is

$$\bar{\mathcal{M}} = \left\{ \frac{s - \frac{1}{6}(p+2)}{s^2 + ps + q}, \quad (p,q) \in \bar{\mathcal{Q}} \right\}$$

It is easy to see that *any* linearly parameterized family of transfer functions containing $\mathcal{M}$ which is defined on a convex parameter space, must also contain $\bar{\mathcal{M}}$. While this procedure certainly makes tractably the problem of estimating parameters, the procedure introduces a new problem. Note that if the newly parameterized transfer function is evaluated at the point $(0, -\frac{1}{9}) \in \bar{\mathcal{Q}}$, what results is the transfer function

$$\left. \frac{s - \frac{1}{6}(p+2)}{s^2 + ps + q} \right|_{(p,\,q)=(0,-\frac{1}{9})} = \frac{s - \frac{1}{3}}{s^2 - \frac{1}{9}} .$$

This transfer function has a common pole and zero in the right half plane at $s = \frac{1}{3}$. In summary, the only way to embed $\mathcal{M}$ in a larger set which is linearly parameterized on a convex parameter space, is to introduce candidate process model transfer functions which have right half plane pole-zero cancellations. For any such candidate process model transfer function τ, it is impossible to construct a controller which stabilizes the feedback loop shown in Fig. 4a. Thus the certainty equivalence based approach we have outlined for defining $\mathbb{C}_\sigma$ cannot be followed here.

There is a way to deal with this problem if one is willing to use a different paradigm to construct $\mathbb{C}_\sigma$ [13]; the method relies on an alternative to certainty equivalence to define $\mathbb{C}_\sigma$ for values of σ which are "close" to points on parameter space at which such pole zero cancellations occur. Although the method is systematic and provably correct, the multi-controller which results is more complicated than the simple certainty-equivalence based multi-controller described above. There is another way to deal with this problem [14] which we discuss next.

4.2 Controller Covering Problem

As before let $\mathbb{P}$ be a siso process to be controlled and suppose that $\mathbb{P}$ has a model in

$$\mathcal{M} = \bigcup_{p\in\mathcal{P}} \mathbb{B}(\nu_p, r_p)$$

where $\mathbb{B}(\nu_p, r_p)$ is a ball of radius r_p centered at nominal transfer function ν_p in a metric space $\mathcal{T}$ with metric μ. Suppose in addition that $\mathcal{P}$ is a compact continuum within a finite dimensional space and $p \longmapsto r_p$ is at least bounded. Instead of trying re-parameterize as we did in the above example, suppose instead we try to embed $\mathcal{M}$ is a larger class of transfer functions which is the union of a *finite set* of balls in $\mathcal{T}$, each ball being small enough so that it can be adequately controlled with a single conventional linear controller. We pursue this idea as follows.

Let us agree to say that a finite set of controller transfer functions $\mathcal{K}$ is a *control cover* of $\mathcal{M}$ if for each transfer function $\tau \in \mathcal{M}$ there is at least one transfer function $\kappa \in \mathcal{K}$ which endows the closed-loop system shown in Fig. 7 with at least stability and possible other prescribed properties.

The *controller covering problem* for a given $\mathcal{M}$ is to find a control cover, if one exists.

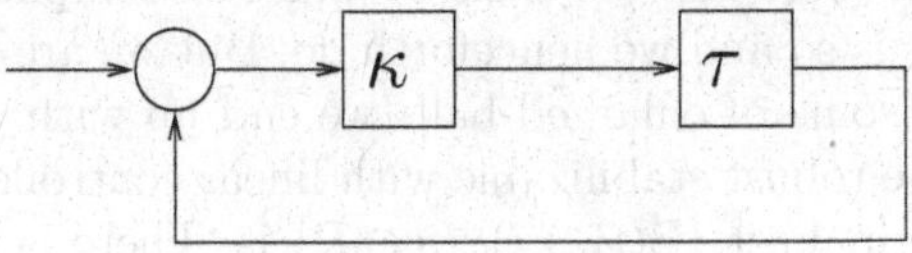

Fig. 7. Feedback loop

4.3 A Natural Approach

There is a natural way to try to solve the controller covering problem if each of the balls $\mathbb{B}(\nu_i, r_i)$ is small enough so that each transfer function in any given ball can be adequately controlled by the same fixed linear controller. In particular, if we could cover $\mathcal{M}$ with a finite set of balls – say $\{\mathbb{B}(\nu_p, r_p) : p \in \mathcal{Q}\}$ where $\mathcal{Q}$ is a finite subset of $\mathcal{P}$ – then we could find controller transfer functions κ_p, $p \in \mathcal{Q}$, such that for each $p \in \mathcal{Q}$ and each $\tau \in \mathbb{B}(\nu_p, r_p)$, κ_p provides the system shown in Fig. 4a with at least stability and possibly other prescribed properties. The problem with this approach is that $\mathcal{M}$ is typically not compact in which case no such finite cover will exist. It is nevertheless possible to construct a finite cover of $\mathcal{M}$ using enlarged versions of some of the balls in the set

$$\{\mathbb{B}(\nu_p, r_p) : p \in \mathcal{P}\}$$

In fact, $\mathcal{M}$ can be covered by any one such ball. For example, for any fixed $q \in \mathcal{P}$, $\mathcal{M} \subset \mathbb{B}(\nu_p, s_q)$ where

$$s_q = \sup_{p \in \mathcal{P}} r_p + \sup_{p \in \mathcal{P}} \mu(\nu_p, \nu_q)$$

This can be checked by simply applying the triangle inequality. The real problem then is to construct a cover using balls which are small enough so that all transfer functions in any one ball can be adequately controlled by the same linear controller. The following lemma [15] takes a step in this direction.

Lemma 4.1. *If $\mathcal{Q}$ is a finite subset of $\mathcal{P}$ such that*

$$\{\nu_p : p \in \mathcal{P}\} \subset \bigcup_{q \in \mathcal{Q}} \mathbb{B}(\nu_q, r_q)$$

then

$$\mathcal{M} \subset \bigcup_{q \in \mathcal{Q}} \mathbb{B}(\nu_q, r_q + s_q)$$

where for $q \in \mathcal{Q}$,

$$s_q = \sup_{p \in \mathcal{P}_q} r_p$$

and $\mathcal{P}_q$ is the set of all $p \in \mathcal{P}$ such that $\nu_p \in \mathbb{B}(\nu_q, r_q)$.

In other words, if we can cover the set of nominal process model transfer functions $\mathcal{N} = \{\nu_p : p \in \mathcal{P}\}$ with a finite set of balls from the set $\{\mathbb{B}(\nu_p, r_p) : p \in \mathcal{P}\}$, then by enlarging these balls as the lemma suggests, we can over $\mathcal{M}$. Of course for such a cover of $\mathcal{N}$ to exist, $\mathcal{N}$ must be compact. It is reasonable to assume that this is so and we henceforth do. But we are not yet out of the woods because the some of enlarged balls we end up with may still turn out to be too large to be robust stabilizable with linear controllers, even if each of the balls in the original set $\{\mathbb{B}(\nu_p, r_p) : p \in \mathcal{P}\}$ is. There is a different way to proceed which avoids this problem if certain continuity conditions apply. We discuss this next.

4.4 A Different Approach

Let us assume that

1. $p \longmapsto r_p$ is continuous on $\mathcal{P}$.
2. For each $p \in \mathcal{P}$ and each positive number ϵ_p there is a number δ_p for which
$$\mu(\nu_q, \nu_p) < \epsilon_p$$
whenever $|q-p| < \delta_p$ where $|\cdot|$ is the norm in the finite-dimensional space within which $\mathcal{P}$ resides.

The following lemma is from [14].

Lemma 4.2. *Let the preceding assumptions hold. For each function $p \longmapsto s_p$ which is positive on $\mathcal{P}$, there is a finite subset $\mathcal{Q} \subset \mathcal{P}$ such that*

$$\mathcal{M} \subset \bigcap_{q \in \mathcal{Q}} \mathbb{B}(\nu_q, r_q + s_q)$$

The key point here is that if the continuity assumptions hold, then it is possible to cover $\mathcal{M}$ with a finite set of balls which are arbitrarily close in size to the originals in the set $\{\mathbb{B}(\nu_p, r_p) : p \in \mathcal{Q}\}$. Thus if each of the original balls is robustly stabilizable with a linear controller, then so should be the expanded balls if they are chosen close enough in size to the originals. Of course small enlargements may require the use of lots of balls to cover $\mathcal{M}$.

4.5 Which Metric?

In order to make use of the preceding to construct a control cover of $\mathcal{M}$, we need a metric which at least guarantees that if κ stabilizes ν, {i.e., if $1 + \kappa\nu$ has all its zeros in the open left-half plane}, then κ also stabilizes any transfer function in the ball $\mathbb{B}(\nu, r)$ for r sufficiently small. Picking a metric with this *robust stabilization property* is not altogether trivial. For example, although for sufficiently small g the transfer function

$$\tau_g = \frac{s - 1 + g}{(s+1)(s-1)}$$

can be made arbitrarily close to the transfer function

$$\nu = \tau_g\big|_{g=0} = \frac{1}{(s+1)}$$

in the metric space of normed differences between transfer function coefficients, for any controller transfer function κ which stabilizes ν one can always find a non-zero value of g sufficiently small such that κ does not stabilize τ_g at this value. This metric clearly does not have the property we seek. Two metrics which do are the gap metric [16] and the v-metric [17]. Moreover the v-metric is known to satisfy the continuity assumption stated just above Lemma 4.2 [14]. Thus we are able to construct a controller cover as follows.

4.6 Construction of a Control Cover

Suppose that the admissible process model transfer function class

$$\mathcal{M} = \bigcup_{p \in \mathcal{P}} \mathbb{B}(\nu_p, r_p)$$

is composed of balls $\mathbb{B}(\nu_p, r_p)$ which are open neighborhoods in the metric space of transfer functions with the v-metric μ. Suppose that these balls are each small enough so that for some sufficiently small continuous, positive function $p \longmapsto \rho_p$ we can construct for each $p \in \mathcal{P}$, a controller transfer function κ_p which stabilizes each transfer function in $\mathbb{B}(\nu_p, r_p + \rho_p)$. By Lemma 4.2, we can then construct a finite subset $\mathcal{Q} \subset \mathcal{P}$ such that

$$\mathcal{M} \subset \bigcup_{p \in \mathcal{Q}} \mathbb{B}(\nu_p, r_p + \rho_p)$$

By construction, κ_p stabilizes each transfer function in $\mathbb{B}(\nu_p, r_p + \rho_p)$. Thus for each $\tau \in \mathcal{M}$ there must be at least one value of $q \in \mathcal{Q}$ such that κ_q stabilizes τ. Since $\mathcal{Q}$ is a finite set, $\{\kappa_q : q \in \mathcal{Q}\}$ is a controller cover of $\mathcal{M}$.

5 Supervisory Control

Much has happened in adaptive control in the last 40 years. The solution to the classical model reference problem is by now very well understood. Provably correct algorithms exist which, at least in theory, are capable of dealing with un-modeled dynamics, noise, right-half-plane zeros, and even certain types of nonlinearities. However despite these impressive gains, there remain many important, unanswered questions: Why, for example, is it still so difficult to explain to a novice why a particular algorithm is able to functions correctly in the face of un-modeled process dynamics and $\mathcal{L}^\infty$ bounded noise? How much un-modeled dynamics can a given algorithm tolerate before loop-stability is lost? How do we choose an adaptive control algorithm's many design parameters to achieve good disturbance rejection, transient response, etc.?

There is no doubt that there will eventually be satisfactory answers to all of these questions, that adaptive control will become much more accessible to non-specialists, that we will be able to much more clearly and concisely quantify un-modeled dynamics norm bounds, disturbance-to-controlled output gains, and so on and that because of this we will see the emergence of a bona fide computer-aided adaptive control design methodology which relies much more on design principals then on trial and error techniques. The aim of this chapter is to take a step towards these ends.

The intent of the chapter is to provide a relatively uncluttered analysis of the behavior of a set-point control system consisting of a poorly modeled process, an integrator and a multi-controller supervised by an estimator-based algorithm employing dwell-time switching. The system has been considered previously in [18, 19] where many of the ideas which follow were first presented. Similar systems have been analyzed in one form or another in [20–22] and elsewhere under various assumptions. It has been shown in [19] that the system's supervisor can successfully orchestrate the switching of a sequence of candidate set-point controllers into feedback with the system's imprecisely modeled siso process so as (i) to cause the output of the process to approach and track a constant reference input despite norm-bounded un-modeled dynamics, and constant disturbances and (ii) to insure that none of the signals within the overall system can grow without bound in response to bounded disturbances, be they constant or not. The objective of this chapter is to derive the same results in a much more straight forward manner. In fact this has already been done in [23] and [24] for a supervisory control system in which the switching between candidate controllers is constrained to be "slow." This restriction not only greatly simplified the analysis in comparison with that given in [19], but also made it possible to derive reasonably explicit upper bounds for the process's allowable un-modeled dynamics as well as for the system's disturbance-to-tracking error gain. In these notes we also constrain switching to be slow.

Adaptive set-point control systems typically consist of at least a process to be controlled, an integrator, a parameter tunable controller or "multi-controller," a parameter estimator or "multi-estimator," and a tuner or "switching logic." In sharp contrast with non-adaptive linear control systems where subsystems are typically analyzed together using one overall linear model, in the adaptive case the sub-systems are not all linear and cannot be easily analyzed as one big inter-connected non-linear system. As a result, one needs to keep track of lots of equations, which can be quite daunting. One way to make things easier is to use carefully defined block diagrams which summarize equations and relations between signals in a way no set of equations can match. We make extensive use of such diagrams in this chapter.

5.1 The System

This section describes the overall structure of the supervisory control system to be considered. We begin with a description of the process.

Process = $\mathbb{P}$

The problem of interest is to construct a control system capable of driving to and holding at a prescribed set-point r, the output of a process modeled by a dynamical system with "large" uncertainty. The process $\mathbb{P}$ is presumed to admit the model of a siso linear system whose transfer function from control input u to measured output y is a member of a continuously parameterized class of admissible transfer functions of the form

$$\mathcal{M} = \bigcup_{p \in \mathcal{P}} \{\nu_p + \delta : \|\delta\| \leq \epsilon_p\}$$

where $\mathcal{P}$ is a compact subset of a finite dimensional space,

$$\nu_p \stackrel{\Delta}{=} \frac{\alpha_p}{\beta_p}$$

is a pre-specified, strictly proper, *nominal transfer function*, ϵ_p is a real non-negative number, δ is a proper stable transfer function whose poles all have real parts less than the negative of a pre-specified *stability margin* $\lambda > 0$, and $\|\cdot\|$ is the shifted infinity norm

$$\|\delta\| = \sup_{\omega \in \mathbb{R}} |\delta(j\omega - \lambda)|$$

It is assumed that the coefficients of α_p and β_p depend continuously on p and for each $p \in \mathcal{P}$, that β_p is monic and that α_p and β_p are co-prime. All transfer functions in $\mathcal{M}$ are thus proper, but not necessarily stable rational functions. Prompted by the requirements of set-point control, it is further assumed that the numerator of each transfer function in $\mathcal{M}$ is non-zero at $s = 0$. The specific model of the process to be controlled is shown in Fig. 8.

Here y is the process's measured output, $\mathbf{d}$ is a disturbance, and p^* is the index of the nominal process model transfer function which models $\mathbb{P}$.

We will consider the case when $\mathcal{P}$ is a finite set and also the case when $\mathcal{P}$ contains a continuum. Although both cases will be treated more or less simultaneously, there are two places where the continuum demands special consideration. The first is when one tries to construct stabilizable and detectable realizations of continuously parameterized transfer functions, which are continuous functions of p. In particular, unless the transfer functions in question all have the same McMillan degree, constructing realizations with all of these properties can be quite a challenge. The second place where the continuum requires special treatment, is when one seeks to characterize the key property implied by dwell time switching. Both of these matters will be addressed later in this chapter.

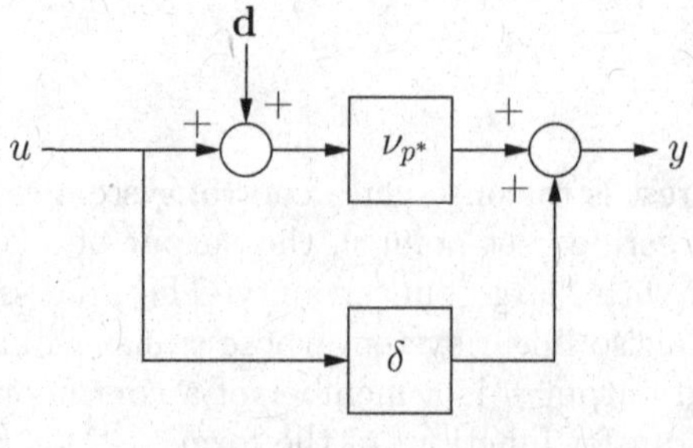

Fig. 8. Process model

In the sequel, we define one by one, the component subsystems of the overall supervisory control system under consideration. We begin with the "multi-controller."

Multi-Controller $= \mathbb{C}_\sigma$

We take as given a continuously parameterized family of "off-the-shelf" loop controller transfer functions $\mathcal{K} \stackrel{\Delta}{=} \{\kappa_p : p \in \mathcal{P}\}$ with at least the following property:

Stability margin property: *For each $p \in \mathcal{P}$, $-\lambda$ is greater than the real parts of all of the closed-loop poles*[2] *of the feedback interconnection* (Fig. 9).

We emphasize that stability margin property is only a minimal requirement on the κ_p. Actual design of the κ_p could be carried out using any one of a number of techniques, including linear quadratic or $\mathcal{H}^\infty$ methods, parameter-varying techniques, pole placement, etc.

We will also take as given an integer $n_C \geq 0$ and a continuously parameterized family of n_C-dimensional realizations $\{A_C(p), b_C(p), f_C(p), g_C(p)\}$, one for each $\kappa_p \in \mathcal{K}$. These realizations are required to be chosen so that for each $p \in \mathcal{P}$, $(f_C(p), \lambda I + A_C(p))$ is detectable and $(\lambda I + A_C(p), b_C(p))$ is stabilizable. There are a great many different ways to construct such realizations, once one has in hand an upper bound n_κ on the McMillan Degrees of the κ_p. One is the $2n_\kappa$-dimensional *identifier-based* realization

$$\left\{ \begin{pmatrix} A_I & 0 \\ 0 & A_I \end{pmatrix} + \begin{pmatrix} b_I \\ 0 \end{pmatrix} f_p, \begin{pmatrix} g_p b_I \\ b_I \end{pmatrix}, f_p, g_p \right\} \tag{5.1}$$

where (A_I, b_I) is any given parameter-independent, n_κ-dimensional siso, controllable pair with A_I stable and f_p are g_p are respectively a parameter-dependent $1 \times 2n_\kappa$ matrix and a parameter dependent scalar. Another is the n_κ-dimensional *observer-based* realization

$$\{A_O + k_p f_O, b_p, f_O, g_p\}$$

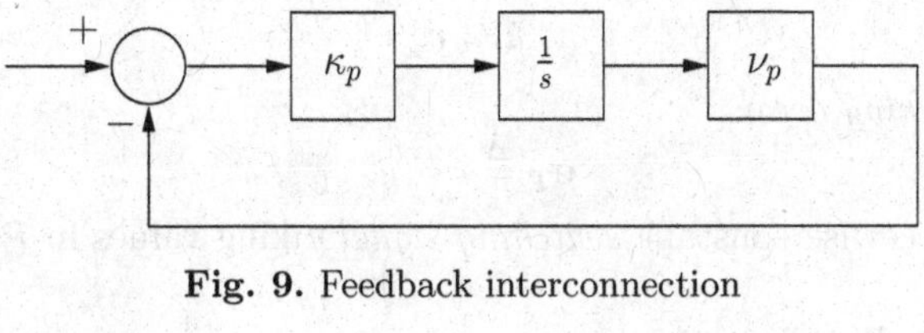

Fig. 9. Feedback interconnection

[2] By the closed-loop poles are meant the zeros of the polynomial $s\rho_p\beta_p + \gamma_p\alpha_p$, where $\frac{\alpha_p}{\beta_p}$ and $\frac{\gamma_p}{\rho_p}$ are the reduced transfer functions ν_p and κ_p respectively.

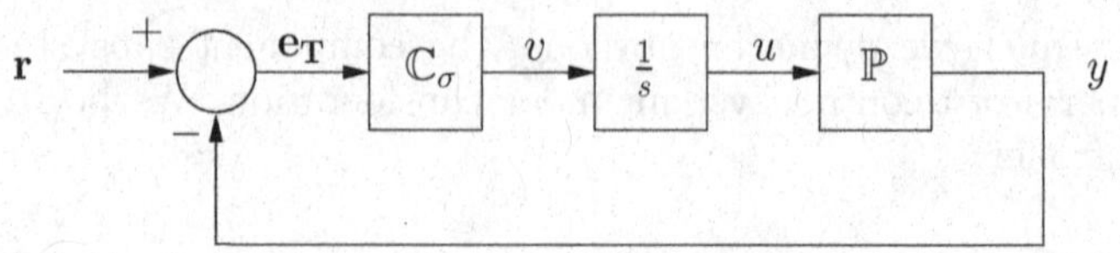

Fig. 10. Supervised sub-system

where (f_O, A_O) is any given n_κ-dimensional, parameter-independent, observable pair with A_O stable and k_p and g_p are respectively a parameter-dependent $n_\kappa \times 1$ matrix and a parameter dependent scalar. Thus for the identifier-based realization $n_C = 2n_\kappa$ whereas for the observer-based realization $n_C = n_\kappa$. In either case, linear systems theory dictates that for a these realizations to exist, it is necessary to be able to represent each transfer function in $\mathcal{K}$ as a rational function $\rho(s)$ with a monic denominator of degree n_κ. Moreover, $\rho(s)$ must be defined so that the greatest common divisor of its numerator and denominator is a factor of the characteristic polynomial of A_I or A_O depending on which type of realization is being used. For the case when $\mathcal{K}$ is a finite set, this is easily accomplished by simply multiplying both the numerator and denominator of each reduced transfer function κ_p in $\mathcal{K}$ by an appropriate monic polynomial μ_p of sufficiently high degree so that the rational function $\rho(s)$ which results has a denominator of degree n_κ. Carry out this step for the case when $\mathcal{P}$ contains a continuum is more challenging because to obtain a realization which depends continuously on p, one must choose the coefficients of μ_p to depend continuously on p as well. One obvious way to side-step this problem is to deal only with the case when all of the transfer functions in $\mathcal{K}$ have the same McMillan degree, because in this case μ_p can always be chosen to be a fixed polynomial not depending on p. We will not pursue this issue in any greater depth in these notes. Instead, we will simply assume that such a continuously parameterized family of realizations of the κ_p has been constructed.

Given such a family of realizations, the sub-system to be supervised is of the form shown in Fig. 10 where $\mathbb{C}_\sigma$ is the n_C-dimensional switchable dynamical system

$$\dot{x}_C = A_C(\sigma)x_C + b_C(\sigma)\mathbf{e_T} \qquad v = f_C(\sigma)x_C + g_C(\sigma)\mathbf{e_T}, \tag{5.2}$$

called a *multi-controller*, v is the input to the *integrator*

$$\dot{u} = v, \tag{5.3}$$

$\mathbf{e_T}$ is the *tracking error*

$$\mathbf{e_T} \stackrel{\Delta}{=} r - y, \tag{5.4}$$

and σ is a piecewise constant *switching signal* taking values in $\mathcal{P}$.

Supervisor = $\mathbb{E} + \mathbb{W} + \mathbb{D}$

Our aim here is to define a "supervisor" which is capable of generating σ in real time so as to ensure both:

1. *Global boundedness* of all system signals in the face of an arbitrary but bounded disturbance inputs
2. *Set-point regulation* {i.e., $\mathbf{e_T} \to 0$} in the event that the disturbance signal is constant

As we shall see, the kind of supervisor we will define will deliver even more – a form of exponential stability which will ensure that neither bounded measurement noise nor bounded system noise entering the system at any point can cause any signal in the system to grow without bound.

Parallel Realization of an Estimator-Based Supervisor

To understand the basic idea behind the type of supervisor we ultimately intend to discuss, it is helpful to first consider what we shall call a *parallel realized estimator-based supervisor.* This type of supervisor is applicable only when $\mathcal{P}$ is a finite set. So for the moment assume that $\mathcal{P}$ contains m elements $p_1, p_2, \ldots, p_m$ and consider the system shown in Fig. 11.

Here each y_p is a suitably defined estimate of y which would be asymptotically correct if ν_p were the process model's transfer function and there were no noise or disturbances. The system which generates y_p would typically be an observer for estimating the output of a linear realization of nominal process transfer function ν_p. For each $p \in \mathcal{P}$, $e_p = y_p - y$ denotes the pth output estimation error and μ_p is a suitably defined norm-squared value of e_p called a *monitoring signal* which is used by the supervisor to assess the potential performance of controller p. $\mathbb{S}$ is a switching logic whose function is to determine

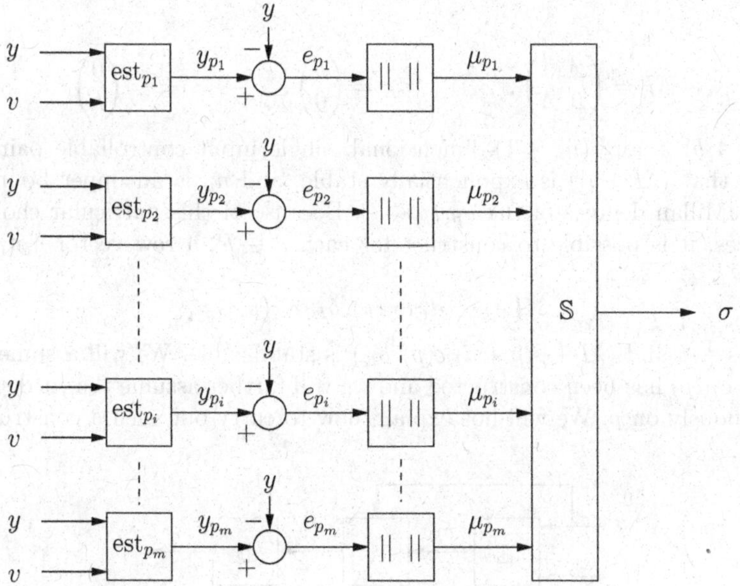

Fig. 11. Parallel realization of an estimator-based supervisor

σ on the basis of the current values of the μ_p. The underlying decision making strategy used by such a supervisor is basically this: From time to time select for σ, that candidate control index q whose corresponding monitoring signal μ_q is the smallest among the μ_p, $p \in \mathcal{P}$. The justification for this heuristic idea is that the nominal process model whose associated monitoring signal is the smallest, "best" approximates what the process is and thus the candidate controller designed on the basis of this model ought to be able to do the best job of controlling the process.

Estimator-Based Supervisor

The supervisor shown in Fig. 11 is a hybrid dynamical system whose inputs are v and y and whose output is σ. One shortcoming of this particular architecture is that it is only applicable when $\mathcal{P}$ is a finite set. The supervisor we will now define is functionally the same as the supervisor shown in Fig. 11 but has a different realization, one which can be applied even when $\mathcal{P}$ contains a continuum of points. The supervisor consists of three subsystems: a *multi-estimator* $\mathbb{E}$, a *weight generator* $\mathbb{W}$, and a specific switching logic $\mathbb{S} \stackrel{\Delta}{=} \mathbb{D}$ called *dwell-time switching* (Fig. 12).

We now describe each of these subsystems in greater detail.

Multi-Estimator $= \mathbb{E}$

By a *multi-estimator* $\mathbb{E}$ for $\frac{1}{s}\mathcal{N}$, is meant an n_E-dimensional linear system

$$\dot{x}_E = A_E x_E + d_E y + b_E v \tag{5.5}$$

where

$$A_E = \begin{pmatrix} A & 0 \\ 0 & A \end{pmatrix}, \qquad d_E = \begin{pmatrix} b \\ 0 \end{pmatrix}, \qquad b_E = \begin{pmatrix} 0 \\ b \end{pmatrix}$$

Here (A, b) is any $(n_\nu + 1)$-dimensional, single input controllable pair chosen so that $(\lambda I + A)$ is exponentially stable, and n_ν is an upper bound on the McMillan degrees of the ν_p, $p \in \mathcal{P}$. Because of this particular choice of matrices, it is possible to construct for each $p \in \mathcal{P}$, a row vector $c_E(p)$ for which

$$\{A_E + d_E c_E(p), b_E, c_E(p)\}$$

realizes $\frac{1}{s}\nu_p$ and $(\lambda I + A_E + d_E c(p), b_E)$ is stabilizable. We will assume that such a $c_E(p)$ has been constructed and we will further assume that it depends continuously on p. We will not explain how to carry out such a construction

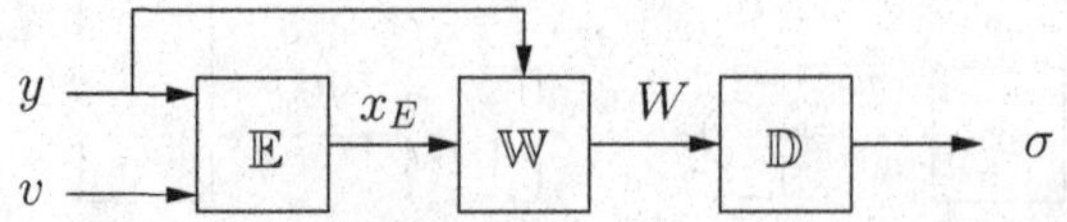

Fig. 12. Estimator-based supervisor

here, even though the procedure is straight forward if $\mathcal{P}$ is a finite set or if all the transfer functions in $\mathcal{N}$ have the same McMillan degree.

The parameter-dependent row vector $c_E(p)$ is used in the definition of $\mathbb{W}$ which will be given in Sect. 5.1. With $c_E(p)$ in hand it is also possible to define *output estimation errors*

$$e_p \stackrel{\Delta}{=} c_E(p)x_E - y, \quad p \in \mathcal{P} \tag{5.6}$$

While these error signals are not actually generated by the supervisor, they play an important role in explaining how the supervisor functions. It should be mentioned that for the case when $\mathcal{P}$ contains a continuum, the same issues arise in defining the quadruple $\{A_E + d_E c_E(p), b_E, c_E(p)\}$ to realize the $\frac{1}{s}\nu_p$, as were raised earlier when we discussed the problem of realizing the κ_p using the identifier-based quadruple in (5.1). Like before, we will sidestep these issues by assuming for the case when $\mathcal{P}$ is not finite, that all nominal process model transfer functions have the same McMillan degree.

Weight Generator $= \mathbb{W}$

The supervisor's second subsystem, $\mathbb{W}$, is a causal dynamical system whose inputs are x_E and y and whose state and output W is a symmetric "weighting matrix" which takes values in a linear space $\mathcal{W}$ of symmetric matrices. W together with a suitably defined *monitoring function* $M : \mathcal{W} \times \mathcal{P} \rightarrow \mathbb{R}$ determine a scalar-valued *monitoring signal* of the form

$$\mu_p \stackrel{\Delta}{=} M(W, p) \tag{5.7}$$

which is viewed by the supervisor as a measure of the expected performance of controller p. $\mathbb{W}$ and M are defined by

$$\dot{W} = -2\lambda W + \begin{pmatrix} x_E \\ y \end{pmatrix}\begin{pmatrix} x_E \\ y \end{pmatrix}', \tag{5.8}$$

and

$$M(W, p) = \left(c_E(p)\ -1\right) W \left(c_E(p)\ -1\right)' \tag{5.9}$$

respectively, where $W(0)$ may be chosen to be any matrix in $\mathcal{W}$. The definitions of $\mathbb{W}$ and M are prompted by the observation that if μ_p are given by (5.7), then

$$\dot{\mu}_p = -2\lambda\mu_p + e_p^2, \quad p \in \mathcal{P}$$

because of (5.6), (5.8) and (5.9). Note that this implies that

$$\mu_p(T) = e^{-2\lambda T}||e_p||_T^2 + e^{-2\lambda T}M(W(0), p), \qquad T \geq 0,\ p \in \mathcal{P}$$

where, for any piecewise-continuous signal $z : [0, \infty) \rightarrow \mathbb{R}^n$, and any time $T > 0$, $||z||_T$ is the *exponentially weighted 2-norm*

$$||z||_T \triangleq \sqrt{\int_0^T e^{2\lambda t}|z(t)|^2 dt}$$

Thus if $W(0) = 0$, $\mu_p(t)$ is simply a scaled version of the square of the exponentially weighted 2-norm of e_p.

Dwell-time Switching Logic = $\mathbb{D}$

The supervisor's third subsystem, called a *dwell-time switching logic* $\mathbb{D}$, is a hybrid dynamical system whose input and output are W and σ respectively, and whose state is the ordered triple $\{X, \tau, \sigma\}$. Here X is a discrete-time matrix which takes on sampled values of W, and τ is a continuous-time variable called a *timing signal*. τ takes values in the closed interval $[0, \tau_D]$, where τ_D is a pre-specified positive number called a *dwell time*. Also assumed pre-specified is a *computation time* $\tau_C \leq \tau_D$ which bounds from above for any $X \in \mathcal{W}$, the time it would take a supervisor to compute a value $p \in \mathcal{P}$ which minimizes $M(X, p)$. Between "event times," τ is generated by a reset integrator according to the rule $\dot{\tau} = 1$. Event times occur when the value of τ reaches either $\tau_D - \tau_C$ or τ_D; at such times τ is reset to either 0 or $\tau_D - \tau_C$ depending on the value of $\mathbb{D}$'s state. $\mathbb{D}$'s internal logic is defined by the flow diagram shown in Fig. 13 where p_x denotes a value of $p \in \mathcal{P}$ which minimizes $M(X, p)$.

Note that implementation of the supervisor just described can be accomplished when $\mathcal{P}$ contains either finitely or infinitely many points. However when $\mathcal{P}$ is a continuum, for the required minimization of $M(X, p)$ to be tractable, it will typically be necessary to make assumptions about both $c_E(p)$ and $\mathcal{P}$. For example, if $c_E(p)$ is an affine linear function and $\mathcal{P}$ is a finite union of convex sets, the minimization of $M(X, p)$ will be a finite family of finite-dimensional convex programming problems.

In the sequel we call a piecewise-constant signal $\bar{\sigma} : [0, \infty) \to \mathcal{P}$ *admissible* if it either switches values at most once, or if it switches more than once and the set of time differences between each two successive switching times is bounded below by τ_D. We write $\mathcal{S}$ for the set of all admissible switching signals. Because of the definition of $\mathbb{D}$, it is clear its output σ will be admissible. This means that switching cannot occur infinitely fast and thus that existence and uniqueness of solutions to the differential equations involved is not an issue.

Closed-Loop Supervisory Control System

The overall system just described, admits a block diagram description of the form shown in Fig. 14. The basic properties of this system are summarized by the following theorem.

Theorem 5.1. *Let $\tau_C \geq 0$ be fixed. Let τ_D be any positive number no smaller than τ_C. There are positive numbers ϵ_p, $p \in \mathcal{P}$, for which the following statements are true provided the process $\mathbb{P}$ has a transfer function in $\mathcal{M}$.*

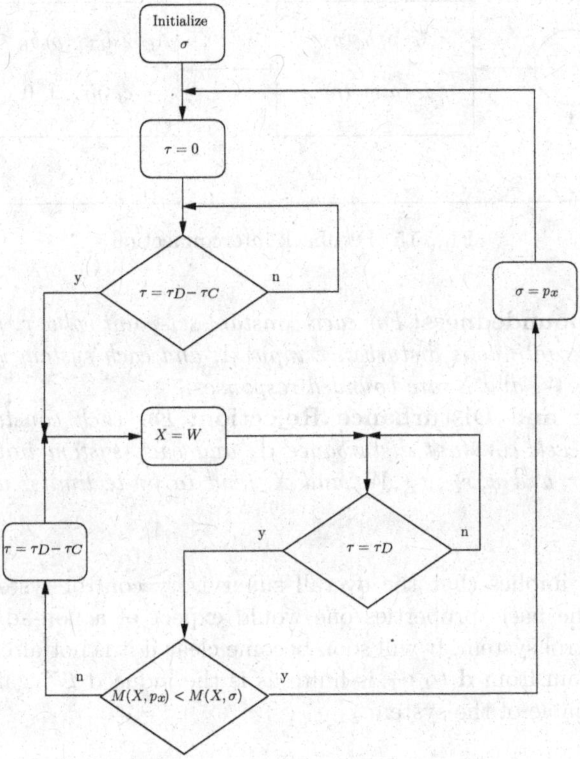

Fig. 13. Dwell-time switching logic $\mathbb{D}$

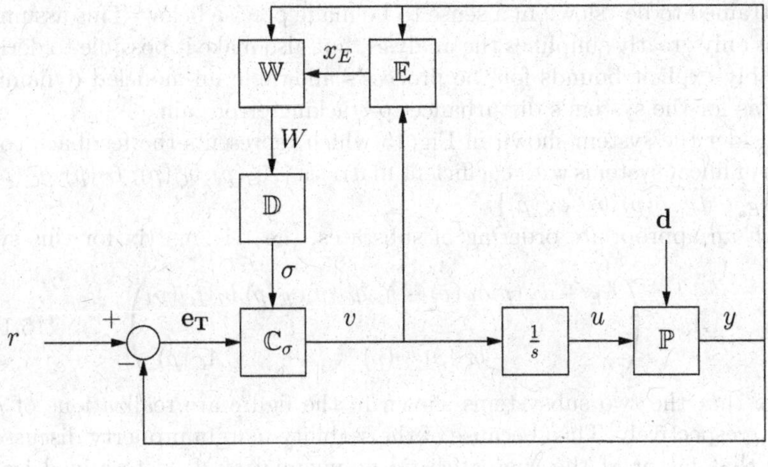

Fig. 14. Supervisory control system

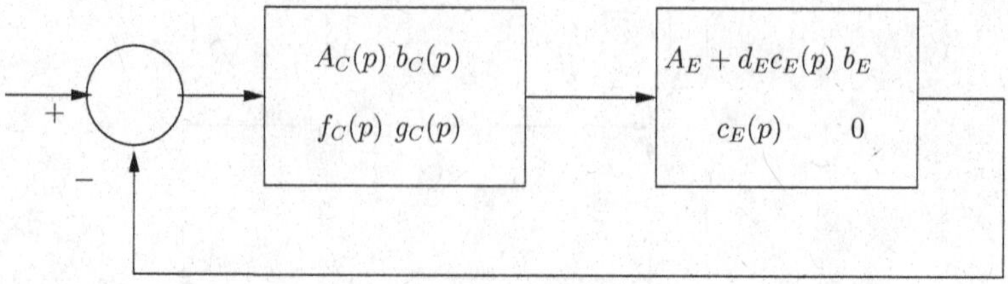

Fig. 15. Feedback interconnection

1. **Global Boundedness:** *For each constant set-point value* r, *each bounded piecewise-continuous disturbance input* $\mathbf{d}$, *and each system initialization,* u, x_C, x_E, W, *and* X *are bounded responses.*
2. **Tracking and Disturbance Rejection:** *For each constant set-point value* r, *each constant disturbance* $\mathbf{d}$, *and each system initialization,* y *tends to* r *and* u, x_C, x_E, W, *and* X *tend to finite limits, all as fast as* $e^{-\lambda t}$.

The theorem implies that the overall supervisory control system shown in Fig. 14 has the basic properties one would expect of a non-adaptive linear set-point control system. It will soon become clear if it is not already that the induced $\mathcal{L}^2$ gain from $\mathbf{d}$ to $\mathbf{e_T}$ is finite as is the induced $\mathcal{L}^\infty$ gain from $\mathbf{d}$ to any state variable of the system.

5.2 Slow Switching

Although it is possible to establish correctness of the supervisory control system just described without any further qualification [19], in these notes we will only consider the case when the switching between candidate controllers is constrained to be "slow" in a sense to be made precise below. This assumption not only greatly simplifies the analysis, but also make it possible to derive reasonably explicit bounds for the process's allowable un-modeled dynamics as well as for the system's disturbance-to-tracking-error gain.

Consider the system shown in Fig. 15 which represents the feedback connection of linear systems with coefficient matrices $\{A_C(p), b_C(p), f_C(p), g_C(p)\}$ and $\{A_E + d_E c_E(p), b_E, c_E(p)\}$.

With an appropriate ordering of substates, the "A" matrix for this system is

$$A_p = \begin{pmatrix} A_E + d_E(p)c_E(p) - b_E g_C(p)c_E(p) & b_E f_C(p) \\ -b_C(p)c_E(p) & A_C(p) \end{pmatrix}. \qquad (5.10)$$

Observe that the two subsystems shown in the figure are realizations of κ_p and $\frac{1}{s}\nu_p$ respectively. Thus because of the stability margin property discussed earlier, that factor of the characteristic polynomial of A_p determined by κ and $\frac{1}{s}\nu$ must have all its roots to the left of the vertical line $s = -\lambda$ in the complex plane. Any remaining eigenvalues of A_p must also line to the left of the line $s = -\lambda$ because $\{\lambda I + A_C(p), b_C(p), f_C(p), g_C(p)\}$ and $\{\lambda I + A_E + d_E c_E(p), b_E, c_E(p)\}$ are stabilizable and detectable systems. In other words, $\lambda I + A_p$ is an exponentially stable matrix and this is true for all $p \in \mathcal{P}$. In the sequel we assume the following.

Slow switching assumption: *Dwell time τ_D is large enough so that for each admissible switching signal $\sigma : [0, \infty) \rightarrow \mathcal{P}$, $\lambda I + A_\sigma$ is an exponentially stable matrix.*

Using Lemma 1.1 from Sect. 1.1, it is possible to compute an explicit lower bound for τ_D for which this assumption holds. Here's how. Since each $\lambda I + A_p$ is exponentially stable and $p \longmapsto A_p$ is continuous, it is possible to compute continuous, non-negative and positive functions $p \longmapsto t_p$ and $p \longmapsto \lambda_p$ respectively, such that

$$|e^{(\lambda I + A_p t)}| \leq e^{\lambda_p(t_p - t)}, \quad t \geq 0$$

It follows from Lemma 1.1 that if τ_D is chosen to satisfy

$$\tau_D > \sup_{p \in \mathcal{P}}\{t_p\}$$

then for each admissible switching signal σ, $\lambda I + A_\sigma$ will be exponentially stable.

5.3 Analysis

Our aim here is to establish a number of basic properties of the supervisory control system under consideration. We assume that $\mathbf{r}$ is an arbitrary but constant set-point value. In addition we invariably ignore initial condition dependent terms which decay to zero as fast as $e^{-\lambda t}$, as this will make things much easier to follow. A more thorough analysis which would take these terms into account can carried out in essentially the same manner.

Output Estimation Error e_{p^*}

Assume that the diagram in Fig. 8 correctly models the process and consequently that p^* is the index of the correct nominal model transfer function ν_{p^*}. In this section we develop a useful formula for e_{p^*} where for $p \in \mathcal{P}$, e_p is the output estimation error

$$e_p = C_E x_E - y \tag{5.11}$$

defined previously by (5.6). In the sequel, for any signal w and polynomial $\alpha(s)$, we use the notation $\alpha(s)w$ to denote the action of the differential operator polynomial $\alpha(s)|_{s=\frac{d}{dt}}$ on w. For the sake of conciseness, we proceed formally, ignoring questions of differentiability of w. We will need the following easily verifiable fact.

Lemma 5.1. *For any triple of real matrices* $\{A_{n\times n},\ b_{n\times 1},\ c_{i\times n}\}$

$$c(sI - A)^{-1}b = \frac{\pi(s) - \bar{\pi}(s)}{\pi(s)}$$

where $\pi(s)$ *and* $\bar{\pi}(s)$ *are the characteristic polynomials of* A *and* $A + bc$ *respectively.*

A proof will not be given.

The process model depicted in Fig. 8 implies that

$$\beta_{p^*}y = (\alpha_{p^*} + \beta_{p^*}\delta)u + \alpha_{p^*}\mathbf{d}$$

This and the fact that $\dot{u} = v$ enable us to write

$$s\beta_{p^*}y = (\alpha_{p^*} + \beta_{p^*}\delta)v + s\alpha_{p^*}\mathbf{d} \tag{5.12}$$

In view of (5.11)

$$e_{p^*} = c_E(p^*)x_E - y \tag{5.13}$$

Using this, is possible to re-write estimator equation $\dot{x}_E = A_E x_E + d_E y + b_E v$ defined by (5.5), as

$$\dot{x}_E = (A_E + d_E c_E(p^*))x_E - d_E e_{p^*} + b_E v \tag{5.14}$$

Since $\{A_E + d_E c_E(p^*), b_E, c_E(p^*)\}$ realizes $\frac{1}{s}\nu_{p^*}$ and $\nu_{p^*} = \frac{\alpha_{p^*}(s)}{\beta_{p^*}(s)}$ it must be true that

$$c_E(p^*)(sI - A_E - d_E c_E(p^*))^{-1}b_E = \frac{\alpha_{p^*}\theta(s)}{s\beta_{p^*}(s)\theta(s)}$$

where $s\beta_{p^*}(s)\theta(s)$ is the characteristic polynomial of $A_E + d_E c_E(p^*)$ and θ is a polynomial of unobservable-uncontrollable eigenvalues of $\{A_E + d_E c_E(p^*), b_E, c_E(p^*)\}$. By assumption, $(s+\lambda)\theta$ is thus a stable polynomial. By Lemma 5.1,

$$c_E(p^*)(sI - A_E - d_E c_E(p^*))^{-1}d_E = \frac{\omega_E(s)\theta(s) - s\beta_{p^*}(s)\theta(s)}{s\beta_{p^*}(s)\theta(s)}$$

where $\omega_E(s)\theta(s)$ is the characteristic polynomial of A_E. These formulas and (5.14) imply that

$$s\beta_{p^*}\theta c_E(p^*)x_E = -(\omega_E\theta - s\beta_{p^*}\theta)e_{p^*} + \alpha_{p^*}\theta v$$

Therefore

$$s\beta_{p^*}c_E(p^*)x_E = -(\omega_E - s\beta_{p^*})e_{p^*} + \alpha_{p^*}v$$

This, (5.12) and (5.13) thus imply that

$$s\beta_{p^*}e_{p^*} = -(\omega_E - s\beta_{p^*})e_{p^*} - \beta_{p^*}\delta v - s\alpha_{p^*}\mathbf{d}$$

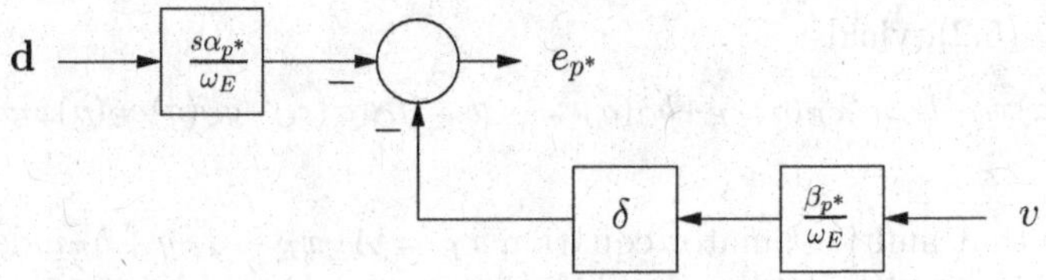

Fig. 16. Output estimation error e_{p^*}

and consequently that

$$\omega_E e_{p^*} = -\beta_{p^*}\delta v - s\alpha_{p^*}\mathbf{d}$$

In summary, the assumption that $\mathbb{P}$ is modeled by the system shown in Fig. 8, implies that the relationship between v, and e_{p^*} is as shown in Fig. 16.

Note that because of what has been assumed about δ and about the spectrum of A_E, the poles of all three transfer functions shown in this diagram lie to the left of the line $s = -\lambda$ in the complex plane.

Multi-Estimator/Multi-Controller Equations

Our next objective is to combine into a single model, the equations which describe $\mathbb{C}_\sigma$ and $\mathbb{E}$. As a first step, write $\bar{x}_E$ for the shifted state

$$\bar{x}_E = x_E + A_E^{-1}b_E r \tag{5.15}$$

Note that because of Lemma 5.1

$$c_E(p)(sI - A_E)^{-1}d_E = \frac{\omega(s) - s\beta_p(s)\theta_p(s)}{\omega(s)}, \quad p \in \mathcal{P}$$

where ω is the characteristic polynomial of A_E and for $p \in \mathcal{P}$, $s\beta_p\theta_p$ is the characteristic polynomial of $A_E + d_E c_E(p)$. Evaluation of this expression at $s = 0$ shows that

$$c_E(p)A_E^{-1}d_E = -1, \quad p \in \mathcal{P}$$

Therefore the pth output estimation error $e_p = c_E(p)x_E - y$ can be written as $e_p = c_E(p)\bar{x}_E + r - y$. But by definition, the tracking error is $e_{\mathbf{T}} = r - y$, so

$$e_p = c_E(p)\bar{x}_E + e_{\mathbf{T}}, \quad p \in \mathcal{P} \tag{5.16}$$

By evaluating this expression at $p = \sigma$, then solving for $e_{\mathbf{T}}$, one obtains

$$\mathbf{e_T} = e_\sigma - c_E(\sigma)\bar{x}_E \tag{5.17}$$

Substituting this expression for $\mathbf{e_T}$ into the multi-controller equations

$$\dot{x}_C = A_C(\sigma)x_C + b_C(\sigma)\mathbf{e_T} \qquad v = f_C(\sigma)x_C + g_C(\sigma)\mathbf{e_T}$$

defined by (5.2), yields

$$\dot{x}_C = A_C(\sigma)x_C - b_C(\sigma)c_E(\sigma)\bar{x}_E + b_C(\sigma)e_\sigma \qquad v = f_C(\sigma)x_C - g_C(\sigma)c_E(\sigma)\bar{x}_E + g_C(\sigma)e_\sigma \tag{5.18}$$

Note next that multi-estimator equation $\dot{x}_E = A_E x_E + d_E y + b_E v$ defined by (5.5) can be re-written using the shifted state $\bar{x}_E$ defined by (5.15) as

$$\dot{\bar{x}}_E = A_E \bar{x}_E - d_E(r - y) + b_E v$$

Therefore

$$\dot{\bar{x}}_E = A_E \bar{x}_E - d_E \mathbf{e_T} + b_E v$$

Substituting in the expression for $\mathbf{e_T}$ in (5.17) and the formula for v in (5.18) one gets

$$\dot{\bar{x}}_E = (A_E + d_E c_E(\sigma) - b_E g_C(\sigma) c_E(\sigma))\bar{x}_E + b_E f_C(\sigma) x_C + (b_E g_C(\sigma) - d_E) e_\sigma \tag{5.19}$$

Finally if we define the composite state

$$x = \begin{pmatrix} \bar{x}_E \\ x_C \end{pmatrix} \tag{5.20}$$

then it is possible to combine (5.18) and (5.19) into a single model

$$\dot{x} = A_\sigma x + b_\sigma e_\sigma \tag{5.21}$$

where for $p \in \mathcal{P}$, A_p is the matrix defined previously by (5.10) and

$$b_p = \begin{pmatrix} b_E g_C(p) - d_E \\ \\ b_C(p) \end{pmatrix}$$

The expressions for $\mathbf{e_T}$ and v in (5.17) and (5.18) can also be written in terms of x as

$$\mathbf{e_T} = e_\sigma + c_\sigma x \tag{5.22}$$

and

$$v = f_\sigma x + g_\sigma e_\sigma \tag{5.23}$$

respectively, where for $p \in \mathcal{P}$

$$c_p = -\begin{pmatrix} c_E(p) & 0 \end{pmatrix} \qquad f_p = \begin{pmatrix} -g_C(p)c_E(p) & f_C(p) \end{pmatrix} \qquad g_p = g_C(p)$$

Moreover, in view of (5.16),

$$e_p = e_q + c_{pq} x, \quad p, q \in \mathcal{P} \tag{5.24}$$

where

$$c_{pq} = \begin{pmatrix} c_E(p) - c_E(q) & 0 \end{pmatrix}, \quad p, q \in \mathcal{P} \tag{5.25}$$

Equations (5.20)–(5.24) can be thought of as an alternative description of $\mathbb{C}_\sigma$ and $\mathbb{E}$. We will make use of these equations a little later.

Exponentially Weighted 2-Norm

In Sect. 5.1 we noted that each monitoring signal $\mu_p(t)$ could be written as

$$\mu_p(t) = e^{-2\lambda t}||e_p||_t^2 + e^{-2\lambda t}M(W(0), p), \qquad t \geq 0, \; p \in \mathcal{P}$$

where, for any piecewise-continuous signal $z : [0, \infty) \to \mathbb{R}^n$, and any time $t > 0$, $||z||_t$ is the *exponentially weighted 2-norm*

$$||z||_t \triangleq \sqrt{\int_0^t e^{2\lambda\tau}|z(\tau)|^2 d\tau}$$

Since we are considering the case when $W(0) = 0$, the expression for μ simplifies to

$$\mu_p(t) = e^{-2\lambda t}||e_p||_t^2$$

Thus

$$||e_p||_t^2 = e^{2\lambda t}\mu_p(t), \quad p \in \mathcal{P} \tag{5.26}$$

The analysis which follows will be carried out using this exponentially weighted norm. In addition, for any time-varying siso linear system Σ of the form $y = c(t)x + d(t)u$, $\dot{x} = A(t)x + b(t)u$ we write

$$\left\|\begin{matrix} A & b \\ c & d \end{matrix}\right\|$$

for the induced norm

$$\sup\{||y_u||_\infty : u \in \mathcal{U}\}$$

where y_u is Σ's zero initial state, output response to u and $\mathcal{U}$ is the space of all piecewise continuous signals u such that $||u||_\infty = 1$. The induced norm of Σ is finite whenever $\lambda I + A(t)$ is {uniformly} exponentially stable.

We note the following easily verifiable facts about the norm we are using. If $e^{-\lambda t}||u||_t$ is bounded on $[0, \infty)$ {in the $\mathcal{L}^\infty$ sense}, then so is y_u provided $d = 0$ and $\lambda I - A(t)$ is exponentially stable. If u is bounded on $[0, \infty)$ in the $\mathcal{L}^\infty$ sense, then so is $e^{-\lambda t}||u||_{\{0,t\}}$. If $u \to 0$ as $t \to \infty$, then so does $e^{-\lambda t}||u||_t$.

$\mathcal{P}$ Is a Finite Set

It turns out that at this point the analysis for the case when $\mathcal{P}$ is a finite set, proceeds along a different path that the path to be followed in the case when $\mathcal{P}$ contains infinitely many points. In this section we focus exclusively on the case when $\mathcal{P}$ is finite.

As a first step, let us note that the relationships between $\mathbf{e_T}$, v and e_σ given by (5.22)–(5.24) can be conveniently represented by block diagrams which, in turn, can be added to the block diagram shown in Fig. 16. What results in the block diagram shown in Fig. 17.

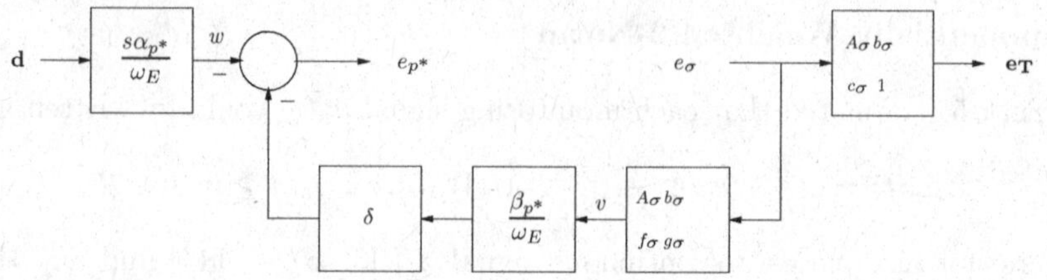

Fig. 17. A representation of $\mathbb{P}$, $\mathbb{C}_\sigma$, $\mathbb{E}$ and $\frac{1}{s}$

In drawing this diagram we have set $p = \sigma$ and $q = p^*$ in (5.24) and we have represented the system defined by (5.21), (5.22) and (5.23) as two separate exponentially stable subsystems, namely

$$\begin{aligned} \dot{x}_1 &= A_\sigma x_1 + b_\sigma e_\sigma \\ v &= f_\sigma x_1 + g_\sigma e_\sigma \end{aligned} \qquad\qquad \begin{aligned} \dot{x}_2 &= A_\sigma x_2 + b_\sigma e_\sigma \\ \mathbf{e_T} &= c_\sigma x_2 + e_\sigma \end{aligned}$$

where $x_1 = x_2 = x$. Note that the signal in the block diagram labeled w, will tend to zero if $\mathbf{d}$ is constant because of the zero at $s = 0$ in the numerator of the transfer function in the block driven by $\mathbf{d}$.

To proceed beyond this point, we will need to "close the loop" in the sense that we will need to relate e_σ to e_{p^*}. To accomplish this we need to address the consequences of dwell time switching, which up until now we have not considered.

Dwell-Time Switching

Note that each of the five blocks in Fig. 17 represents an exponentially stable linear system with stability margin λ. Thus if it happened to be true that $e_\sigma = g e_{p^*}$ for some sufficiently small constant gain g then we would be able to deduce stability in the sense that the induced norm from d to $\mathbf{e_T}$ would be finite. Although no such gain exists, it nonetheless turns out to be true that there is a constant gain for which $||e_\sigma||_t \leq g||e_{p^*}||_t$ for all $t \geq 0$. To explain why this is so we will need the following proposition.

Proposition 5.1. *Suppose that $\mathcal{P}$ contains $m > 0$ elements that W is generated by (5.8), that the μ_p, $p \in \mathcal{P}$, are defined by (5.7) and (5.9), that $W(0) = 0$, and that σ is the response of $\mathbb{D}$ to W. Then for each time $T > 0$, there exists a piecewise constant function $\psi : [0, \infty) \to \{0, 1\}$ such that for all $q \in \mathcal{P}$,*

$$\int_0^\infty \psi(t)dt \leq m(\tau_D + \tau_C) \tag{5.27}$$

and

$$||(1 - \psi)e_\sigma + \psi e_q||_T \leq \sqrt{m}||e_q||_T \tag{5.28}$$

Proposition 5.1 highlights the essential consequences of dwell time switching needed to analyze the system under consideration for the case when $\mathcal{P}$ is finite. The proposition is proved in Sect. 5.4.

A Snapshot at Time T

Fix $T > 0$ and let ψ be as in Proposition 5.1. In order to make use of (5.28), it is convenient to introduce the signal

$$z = (1 - \psi)e_\sigma + \psi e_{p^*} \tag{5.29}$$

since, with $q \stackrel{\Delta}{=} p^*$, (5.28) then becomes

$$||z||_T \leq \sqrt{m}||e_{p^*}||_T \tag{5.30}$$

Note that (5.24) implies that $e_\sigma = e_{p^*} + c_{\sigma p^*}x$. Because of this, the expression for z in (5.29) can be written as

$$z = (1 - \psi)e_\sigma + \psi(e_\sigma - c_{\sigma p^*}x)$$

Therefore after cancellation

$$z = e_\sigma - \psi c_{\sigma p^*}x$$

or

$$e_\sigma = \psi c_{\sigma p^*}x + z \tag{5.31}$$

Recall that

$$\dot{x} = A_\sigma x + b_\sigma e_\sigma \tag{5.32}$$

The point here is that (5.31) and (5.32) define a linear system with input z and output e_σ. We refer to this system as the *injected sub-system* of the overall supervisory control system under consideration. Adding a block diagram representation of this sub-system to the block diagram in Fig. 17 results is the block diagram shown in Fig. 18 which can be though of as a snapshot of the entire supervisory control system at time T. Of course the dashed block shown in the diagram is not really a block in the usual sense of signal flow. Nonetheless its inclusion in the diagram is handy for deriving norm bound inequalities, since in the sense of norms, the dashed block does provide the correct inequality, namely (5.30).

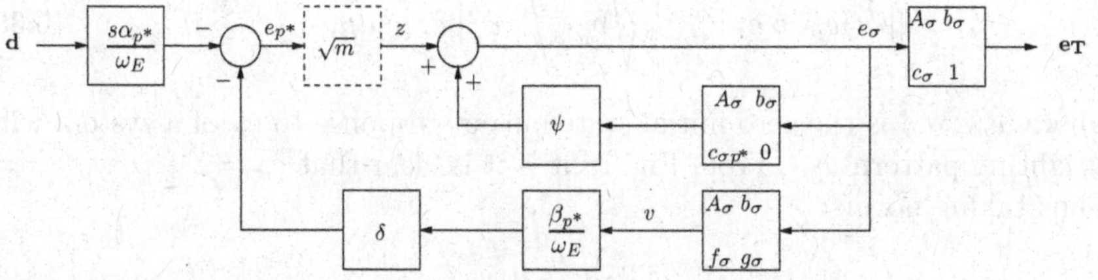

Fig. 18. A snapshot of the complete system at time T

System Gains

Let us note that for any given admissible switching signal σ, each of the six blocks in Fig. 18, excluding the dashed block and the block for ψ, represents an exponentially stable linear system with stability margin λ. It is convenient at this point to introduce certain worst case "system gains" associated with these blocks. In particular, let us define for $p \in \mathcal{P}$

$$\mathfrak{a}_p \triangleq \sqrt{2}\left|\!\left|\!\left| \frac{s\alpha_p}{\omega_E} \right|\!\right|\!\right|, \qquad \mathfrak{b}_p \triangleq \sqrt{2}\left|\!\left|\!\left| \frac{\beta_p}{\omega_E} \right|\!\right|\!\right| \sup_{\sigma\in\mathcal{S}} \left\|\begin{matrix} A_\sigma & b_\sigma \\ f_\sigma & g_\sigma \end{matrix}\right\|, \qquad \mathfrak{c} \triangleq \sup_{\sigma\in\mathcal{S}} \left\|\begin{matrix} A_\sigma & b_\sigma \\ c_\sigma & 1 \end{matrix}\right\|$$

where, as defined earlier, $|\!|\!| \cdot |\!|\!|$ is the shifted infinity norm and $\mathcal{S}$ is the set of all admissible switching signals. In the light of Fig. 18, it is easy to see that

$$||\mathbf{e_T}||_t \leq \mathfrak{c}||e_\sigma||_t, \quad t \geq 0 \tag{5.33}$$

and that

$$||e_{p^*}||_t \leq \epsilon_{p^*}\frac{\mathfrak{b}_{p^*}}{\sqrt{2}}|||e_\sigma||_t + \frac{\mathfrak{a}_{p^*}}{\sqrt{2}}||d||_t, \qquad t \geq 0, \tag{5.34}$$

where ϵ_{p^*} is the norm bound on δ.

To proceed we need an inequality which relates the norm of e_σ to the norm of z. For this purpose we introduce one more system gain, namely

$$\mathfrak{v}_p \triangleq \sup_{\sigma\in\mathcal{S}} \sup_{t\geq 0} \int_0^t |c_{\sigma(t)p}\Phi(t,\tau)b_{\sigma(\tau)}e^{\lambda(t-\tau)}|^2 d\tau, \quad p \in \mathcal{P}$$

where $\Phi(t,\tau)$ is the state transition matrix of A_σ. Note that each $\mathfrak{v}_p$ is finite because of the Slow Switching Assumption.

Analysis of the injected sub-system in Fig. 18 can now be carried out as follows. Set

$$w_p(t,\tau) = c_{\sigma(t)p}\Phi(t,\tau)b_{\sigma(\tau)}$$

Using Cauchy-Schwartz

$$||\psi(w_{p^*} \circ e_\sigma)||_t \leq \sqrt{\mathfrak{v}_{p^*}\int_0^t \psi^2||e_\sigma||_\mu^2 d\mu}, \qquad t \geq 0 \tag{5.35}$$

where $w_{p^*} \circ e_\sigma$ is the zero initial state output response to e_σ of a system with weighting pattern w_{p^*}. From Fig. 18 it is it is clear that $e_\sigma = z + \psi(w_{p^*} \circ e_\sigma)$. Thus taking norms

$$||e_\sigma||_t \leq ||z||_t + ||\psi(w_{p^*} \circ e_\sigma||_t$$

Therefore

$$||e_\sigma||_t^2 \leq 2||z||_t^2 + 2||\psi(w_{p^*} \circ e_\sigma||_t^2$$

Thus using (5.35)

$$||e_\sigma||_t^2 \leq 2||z||_t^2 + 2\mathfrak{v}_{p^*}\int_0^t \psi^2||e_\sigma||_\mu^2 d\mu, \quad 0 \leq t \leq T$$

Hence by the Bellman–Gronwall Lemma

$$||e_\sigma||_T^2 \leq \left(2e^{2\mathfrak{v}_{p^*}\int_0^T \psi^2 dt}\right)||z||_T^2$$

so

$$||e_\sigma||_T \leq \left(\sqrt{2}e^{\mathfrak{v}_{p^*}\int_0^T \psi^2 dt}\right)||z||_T$$

From this, (5.27), and the fact that $\psi^2 = \psi$, we arrive at

$$||e_\sigma||_T \leq \left(\sqrt{2}e^{\mathfrak{v}_{p^*}m(\tau_D+\tau_C)}\right)||z||_T \tag{5.36}$$

Thus the induced gain from z to e_σ of the injected sub-system shown in Fig. 18 is bounded above by $\sqrt{2}e^{\mathfrak{v}_{p^*}m(\tau_D+\tau_C)}$. We emphasize that this is a finite number, not depending on T.

Stability Margin

We have developed four key inequalities, namely (5.30), (5.33), (5.34) and (5.36) which we repeat below for ease of reference.

$$||z||_T \leq \sqrt{m}||e_{p^*}||_T \tag{5.37}$$

$$||\mathbf{e_T}||_T \leq \mathfrak{c}||e_\sigma||_T \tag{5.38}$$

$$||e_{p^*}||_T \leq \epsilon_{p^*}\frac{\mathfrak{b}_{p^*}}{\sqrt{2}}||e_\sigma||_T + \frac{\mathfrak{a}_{p^*}}{\sqrt{2}}||d||_T \tag{5.39}$$

$$||e_\sigma||_T \leq \left(\sqrt{2}e^{\mathfrak{v}_{p^*}m(\tau_D+\tau_C)}\right)||z||_T \tag{5.40}$$

Inequalities (5.37), (5.39) and (5.40) imply that

$$||e_\sigma||_T \leq \sqrt{m}e^{\mathfrak{v}_{p^*}m(\tau_D+\tau_C)})(\epsilon_{p^*}\mathfrak{b}_{p^*}||e_\sigma||_T + \mathfrak{a}_{p^*}||d||_T).$$

Thus if ϵ_{p^*} satisfies the *small gain condition*

$$\boxed{\epsilon_{p^*} < \frac{e^{-\mathfrak{v}_{p^*}m(\tau_D+\tau_C)}}{\mathfrak{b}_{p^*}\sqrt{m}}} \tag{5.41}$$

then

$$||e_\sigma||_T \leq \frac{\mathfrak{a}_{p^*}}{\frac{e^{-\mathfrak{v}_{p^*}m(\tau_D+\tau_C)}}{\sqrt{m}} - \epsilon_{p^*}\mathfrak{b}_{p^*}}||\mathbf{d}||_T. \tag{5.42}$$

The inequality in (5.41) provides an explicit upper bound for the norm of allowable un-modeled process dynamics, namely $|\delta|$.

A Bound on the Disturbance-to-Tracking-Error Gain

Note that (5.38) and (5.42) can be combined to provide an inequality of the form

$$||\mathbf{e_T}||_T \leq \mathfrak{g}_{p^*}||\mathbf{d}||_T, \quad T \geq 0 \tag{5.43}$$

where

$$\boxed{\mathfrak{g}_{p^*} = \frac{\mathfrak{c}\mathfrak{a}_{p^*}}{\frac{e^{-\mathfrak{v}_{p^*}m(\tau_D+\tau_C)}}{\sqrt{m}} - \epsilon_{p^*}\mathfrak{b}_{p^*}}} \tag{5.44}$$

The key point here is that $\mathfrak{g}_{p^*}$ does *not* depend on T even though the block diagram in Fig. 18 does. Because of this, (5.43) must hold for all T. In other words, even though we have carried out an analysis at a fixed time T and in the process have had to define a several signals {e.g., ψ} which depended on T, in the end we have obtained an inequality namely (5.43), which is valid for all T. Because of this we can conclude that

$$||\mathbf{e_T}||_\infty \leq \mathfrak{g}_{p^*}||\mathbf{d}||_\infty \tag{5.45}$$

Thus $\mathfrak{g}_{p^*}$ bounds from above the overall disturbance-to-tracking-error gain of the system we have been studying.

Global Boundedness

The global boundedness condition of Theorem 5.1 can now easily be justified as follows. Suppose $\mathbf{d}$ is bounded on $[0, \infty)$ in the $\mathcal{L}^\infty$ sense. Then so must be $e^{-\lambda t}||\mathbf{d}||_t$. Hence by (5.42), $e^{-\lambda t}||e_\sigma||_t$ must be bounded on $[0, \infty)$ as well. This, the differential equation for x in (5.32), and the exponential stability of $\lambda I + A_\sigma$ then imply that x is also bounded on $[0, \infty)$. In view of (5.20) and (5.15), x_E and x_C must also be bounded. Next recall that the zeros of ω_E {i.e., the eigenvalues of A_E} have negative real parts less than $-\lambda$, and that the transfer function $\frac{\beta_{p^*}}{\omega_E}\delta$ in Fig. 17 is strictly proper. From these observations, the fact that $e^{-\lambda t}||e_\sigma||_t$ is bounded on $[0, \infty)$, and the block diagram in Fig. 17 one readily concludes that e_{p^*} is bounded on $[0, \infty)$. From (5.24), $e_\sigma = e_{p^*} + c_{\sigma p^*}x$. Therefore e_σ is bounded on $[0, \infty)$. Boundedness of $\mathbf{e_T}$ and v follow at once from (5.22) and (5.23) respectively. In view of (5.4), y must be bounded. Thus W must be bounded because of (5.8). Finally note that u must be bounded because of the boundedness of y and v and because of our standing assumption that the transfer function of $\mathbb{P}$ is non-zero at $s = 0$. This, in essence, proves Claim 1 of Theorem 5.1.

Convergence

Now suppose that $\mathbf{d}$ is a constant. Examination of Fig. 17 reveals that w must tend to zero as fast as $e^{-\lambda t}$ because of the zero at $s = 0$ in the numerator of the transfer function from $\mathbf{d}$ to w. Thus $||w||_\infty < \infty$. Figure 17 also implies that

$$||e_{p^*}||_T \leq ||w||_T + \epsilon_{p^*}\frac{b_{p^*}}{\sqrt{2}}||e_\sigma||_T$$

This (5.37), (5.40) and (5.41) implies that

$$||e_\sigma||_T \leq \frac{\sqrt{2}}{\frac{e^{-\mathfrak{v}_{p^*}m(\tau_D+\tau_C)}}{\sqrt{m}} - \epsilon_{p^*}\mathfrak{b}_{p^*}}||w||_T$$

Since this inequality holds for all $T \geq 0$, it must be true that $||e_\sigma||_\infty < \infty$. Hence e_σ must tend to zero as fast as $e^{-\lambda t}$. So therefore must x because of the differential equation for x in (5.21). In view of (5.20) $\bar{x}_E$ and x_C must tend to zero. Thus x_E must tend to $A_E^{-1}b_E r$ because of (5.15). Moreover e_{p^*} must tend to zero as can be plainly seen from Fig. 17. Hence from the formulas (5.22) and(5.23) for $\mathbf{e_T}$ and v respectively one concludes that these signals must tend to zero as well. In view of (5.4), y must tent to r. Thus W must approach a finite limit because of (5.8). Finally note that u tend to a finite limit because y and v do and because of our standing assumption that the transfer function of $\mathbb{P}$ is non-zero at $s = 0$. This, in essence, proves Claim 2 of Theorem 5.1.

$\mathcal{P}$ Is Not a Finite Set

We now consider the more general case when $\mathcal{P}$ is a compact but not necessarily finite subset of a finite dimensional linear space. The following proposition replaces Proposition 5.1 which is clearly not applicable to this case. The proposition relies on the fact that every nominal transfer function in $\mathcal{N}$ can be modeled by a linear system of dimension at most n_E.

Dwell-Time Switching

Proposition 5.2. *Suppose that $\mathcal{P}$ is a compact subset of a finite dimensional space, that $p \longmapsto c_E(p)$ is a continuous function taking values in $\mathbb{R}^{1\times n_E}$, that $c_{pq} = \big(c_E(p) - c_E(q)\ 0\big)$ as in (5.25), that W is generated by (5.8), that the μ_p, $p \in \mathcal{P}$, are defined by (5.7) and (5.9), that $W(0) = 0$, and that σ is the response of $\mathbb{D}$ to W. For each $q \in \mathcal{P}$, each real number $\rho > 0$ and each fixed time $T > 0$, there exists piecewise-constant signals $h : [0, \infty) \to \mathbb{R}^{1\times(n_E+n_C)}$ and $\psi : [0, \infty) \to \{0, 1\}$ such that*

$$|h(t)| \leq \rho, \quad t \geq 0, \tag{5.46}$$

$$\int_0^\infty \psi(t)dt \leq n_E(\tau_D + \tau_C), \tag{5.47}$$

and

$$||(1-\psi)(e_\sigma - hx) + \psi e_q||_T \leq \left\{1 + 2n_E\left(\frac{1 + \sup_{p\in\mathcal{P}}|c_{pq}|}{\rho}\right)^{n_E}\right\}||e_q||_T \tag{5.48}$$

This Proposition is proved in Sect. 5.4. Proposition 5.2 summarizes the key consequences of dwell time switching which are needed to analyze the system under consideration. The term involving h in (5.48) present some minor difficulties which we will deal with next.

Let us note that for any piece-wise continuous matrix-valued signal $h : [0, \infty) \rightarrow \mathbb{R}^{1\times(n_E+n_C)}$, it is possible to re-write (5.21)–(5.22) as

$$\dot{x} = (A_\sigma + b_\sigma h)x + b_\sigma \bar{e} \tag{5.49}$$

$$\mathbf{e_T} = \bar{e} + (c_\sigma + h)x \tag{5.50}$$

and

$$v = (f_\sigma + g_\sigma h)x + g_\sigma \bar{e} \tag{5.51}$$

respectively where

$$\bar{e} = e_\sigma - hx. \tag{5.52}$$

Note also that the matrix $\lambda I + A_\sigma + b_\sigma h$ will be exponentially stable for any $\sigma \in \mathcal{S}$ if $|h| \leq \rho, t \geq 0$, where ρ is a sufficiently small positive number. Such a value of ρ exists because $p \longmapsto A_p$ and $p \longmapsto b_p$ are continuous and bounded functions on $\mathcal{P}$ and because $\lambda I + A_\sigma$ is exponentially stable for every admissible switching signal. In the sequel we will assume that ρ is such a number and that $\mathcal{H}$ is the set of all piece-wise continuous signals h satisfying $|h| \leq \rho, t \geq 0$.

A Snapshot at Time T

Now fix $T > 0$ and let ψ and h be signals for which (5.46)–(5.48) hold with $q = p^*$. To account for any given h in (5.49)–(5.51), we will use in place of the diagram in Fig. 17, the diagram shown in Fig. 19. As with the representation in Fig. 17, we are representing the system defined by (5.49)–(5.51) as two separate subsystems, namely

$$\begin{aligned} \dot{x}_1 &= (A_\sigma + b_\sigma h)x_1 + b_\sigma \bar{e} \\ v &= (f_\sigma + g_\sigma h)x_1 + g_\sigma \bar{e} \end{aligned} \qquad \begin{aligned} \dot{x}_2 &= (A_\sigma + b_\sigma h)x_2 + b_\sigma \bar{e} \\ \mathbf{e_T} &= (c_\sigma + h)x_2 + \bar{e} \end{aligned}$$

where $x_1 = x_2 = x$.

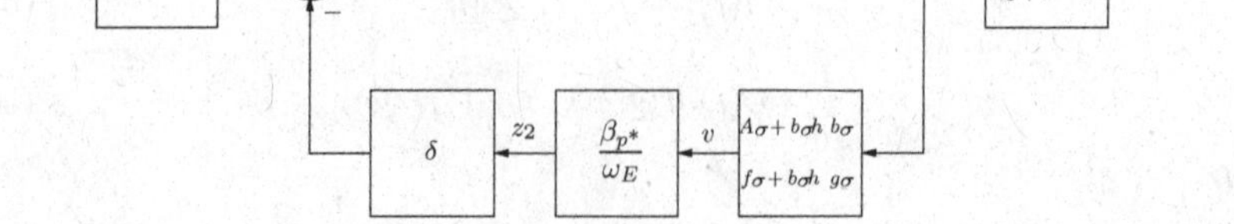

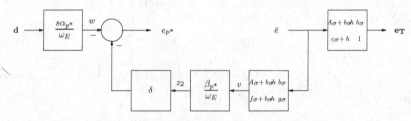

Fig. 19. A Representation of $\mathbb{P}$, $\mathbb{C}_\sigma$, $\mathbb{E}$ and $\frac{1}{s}$

Note that each of the five blocks in Fig. 19 represents an exponentially stable linear system with stability margin λ.

In order to make use of (5.48), it is helpful to introduce the signal

$$\bar{z} = (1 - \psi)(e_\sigma - hx) + \psi e_{p^*} \tag{5.53}$$

since (5.48) then becomes

$$||\bar{z}||_T \leq \gamma_{p^*} ||e_{p^*}||_T \tag{5.54}$$

where

$$\gamma_{p^*} \triangleq \left\{ 1 + 2n_E \left(\frac{1 + \sup_{p \in \mathcal{P}} |c_{pp^*}|}{\rho} \right)^{n_E} \right\}$$

Note that

$$e_\sigma = e_{p^*} + c_{\sigma p^*} x$$

because of (5.24). Solving for e_{p^*} in substituting the result into (5.53) gives

$$\bar{z} = (1 - \psi)(e_\sigma - hx) + \psi(e_\sigma - c_{\sigma p^*} x)$$

Thus

$$\bar{z} = e_\sigma - hx - \psi(c_{\sigma p^*} - h)x$$

In view of (5.52) we can therefore write

$$\bar{z} = \bar{e} - \psi(c_{\sigma p^*} - h)x$$

or

$$\bar{e} = \bar{z} + \psi(c_{\sigma p^*} - h)x \tag{5.55}$$

Recall that (5.49) states that

$$\dot{x} = (A_\sigma + b_\sigma h)x + b_\sigma \bar{e}. \tag{5.56}$$

Observe that (5.55) and (5.56) define a linear system with input $\bar{z}$ and output $\bar{e}$ which we refer to as the *injected system* for the problem under consideration. Adding a block diagram representation of this sub-system to the block diagram in Fig. 19 results in the block diagram shown in Fig. 20. Just as in the case when $\mathcal{P}$ is finite, the dashed block is not really a block in the sense of signal flow.

System Gains

Let us note that for any given admissible switching signal σ, each of the six blocks in Fig. 20, excluding the dashed block and the block for ψ, represents an exponentially stable linear system with stability margin λ. It is convenient at this point to introduce certain worst case "system gains" associated with these blocks. In particular, let us define for $p \in \mathcal{P}$

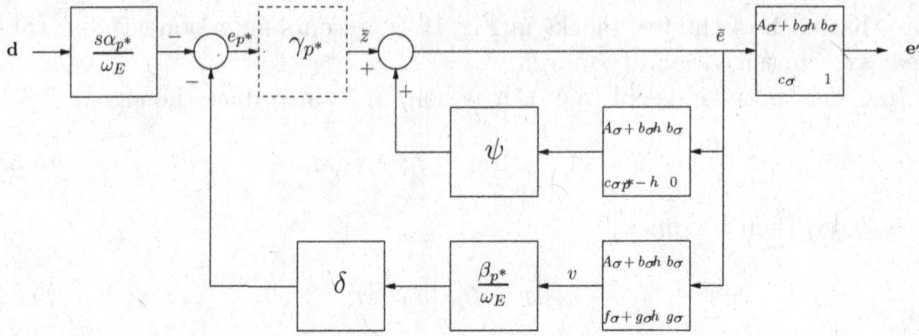

Fig. 20. A Snapshot of the complete system at time T

$$\bar{\mathfrak{b}}_p \triangleq \sqrt{2}\left|\frac{\beta_p}{\omega_E}\right|\left\{\sup_{h\in\mathcal{H}}\sup_{\sigma\in\mathcal{S}}\left\|\begin{matrix} A_\sigma + b_\sigma h & b_\sigma \\ f_\sigma + g_\sigma h & g_\sigma \end{matrix}\right\|\right\} \qquad \bar{\mathfrak{c}} \triangleq \sup_{h\in\mathcal{H}}\sup_{\sigma\in\mathcal{S}}\left\|\begin{matrix} A_\sigma + b_\sigma h & b_\sigma \\ c_\sigma + h & 1 \end{matrix}\right\|$$

In the light of Fig. 20, it is easy to see that

$$||\mathbf{e_T}||_t \leq \bar{\mathfrak{c}}||\bar{e}||_t \quad t \geq 0, \tag{5.57}$$

and that

$$||e_{p^*}||_t \leq \epsilon_{p^*}\frac{\bar{\mathfrak{b}}_{p^*}}{\sqrt{2}}||\bar{e}||_t + \frac{\mathfrak{a}_{p^*}}{\sqrt{2}}||d||_t, \qquad t \geq 0 \tag{5.58}$$

where ϵ_{p^*} is the norm bound on δ and $\mathfrak{a}_p$ is as defined in Sect. 5.3.

To proceed we need an inequality which related the norm of $\bar{e}$ to the norm of $\bar{z}$. For this purpose we introduce the additional system gain

$$\bar{\mathfrak{v}}_q \triangleq \sup_{h\in\mathcal{H}}\sup_{\sigma\in\mathcal{S}}\sup_{t\geq 0}\int_0^t |(c_{\sigma(t)q} - h(t))\Phi(t,\tau)b_{\sigma(\tau)}e^{\lambda(t-\tau)}|^2 d\tau$$

where $\Phi(t,\tau)$ is the state transition matrix of $A_\sigma + b_\sigma h$. Note that each $\bar{\mathfrak{v}}_q$ is finite because of the Slow Switching Assumption.

Analysis of the injected sub-system shown in Fig. 20 is the same as in the case when $\mathcal{P}$ is finite. Instead of (5.36), what one obtains in this case is the inequality

$$||\bar{e}||_T \leq \left(\sqrt{2}e^{\bar{\mathfrak{v}}_{p^*}n_E(\tau_D+\tau_C)}\right)||\bar{z}||_T \tag{5.59}$$

Stability Margin

We have developed four key inequalities for the problem at hand, namely (5.54), (5.57), (5.58) and (5.59) which we repeat for ease of reference.

$$||\bar{z}||_T \leq \gamma_{p^*}||e_{p^*}||_T \tag{5.60}$$
$$||\mathbf{e_T}||_T \leq \bar{\mathfrak{c}}||\bar{e}||_T \tag{5.61}$$
$$||e_{p^*}||_T \leq \epsilon_{p^*}\frac{\bar{\mathfrak{b}}_{p^*}}{\sqrt{2}}|||\bar{e}||_T + \frac{\mathfrak{a}_{p^*}}{\sqrt{2}}||d||_T \tag{5.62}$$
$$||\bar{e}||_T \leq \left(\sqrt{2}e^{\bar{\mathfrak{v}}_{p^*}m(\tau_D+\tau_C)}\right)||\bar{z}||_T \tag{5.63}$$

Observe that except for different symbols, these inequalities are exactly the same as those in (5.37)–(5.40) respectively. Because of this, we can state at once that if ϵ_{p^*} satisfies the small gain condition

$$\boxed{\epsilon_{p^*} < \frac{e^{-\bar{\mathfrak{v}}_{p^*}n_E(\tau_D+\tau_C)}}{\bar{\mathfrak{b}}_{p^*}\gamma_{p^*}}} \tag{5.64}$$

then

$$||\bar{e}||_T \leq \frac{\mathfrak{a}_{p^*}}{\frac{e^{-\bar{\mathfrak{v}}_{p^*}n_E(\tau_D+\tau_C)}}{\gamma_{P^*}} - \epsilon_{p^*}\bar{\mathfrak{b}}_{p^*}}||\mathbf{d}||_T, \tag{5.65}$$

As in the case when $\mathcal{P}$ is finite, inequality in (5.64) provides an explicit bound for the norm of allowable process dynamics.

A Bound on the Disturbance-to-Tracking-Error Gain

Note that (5.61) and (5.65) can be combined to provide an inequality of the form

$$||\mathbf{e_T}||_T \leq \bar{\mathfrak{g}}_{p^*}||\mathbf{d}||_T, \quad T \geq 0$$

where

$$\boxed{\bar{\mathfrak{g}}_{p^*} = \frac{\bar{\mathfrak{c}}\mathfrak{a}_{p^*}}{\frac{e^{-\bar{\mathfrak{v}}_{p^*}n_E(\tau_D+\tau_C)}}{\gamma_{P^*}} - \epsilon_{p^*}\bar{\mathfrak{b}}_{p^*}}} \tag{5.66}$$

Moreover, because the preceding inequality holds for all $T > 0$ and $\mathfrak{g}_{p^*}$ is independent of T, it must be true that

$$||\mathbf{e_T}||_\infty \leq \bar{\mathfrak{g}}_{p^*}||\mathbf{d}||_\infty$$

Thus for the case when $\mathcal{P}$ contains infinitely many points, $\bar{\mathfrak{g}}_{p^*}$ bounds from above the overall system's disturbance-to-tracking-error gain.

Global Boundedness and Convergence

It is clear from the preceding that the reasoning for the case when $\mathcal{P}$ contains infinitely many points parallels more or less exactly the reasoning used for the case when $\mathcal{P}$ contains only finitely many points. Thus for example, the claims of Theorem 5.1 regarding global boundedness and exponential convergence for the case when $\mathcal{P}$ contains infinitely many points, can be established in essentially the same way as which they were established earlier in these notes for the case when $\mathcal{P}$ is a finite set. For this reason, global boundedness and convergence arguments will not be given here.

5.4 Analysis of the Dwell Time Switching Logic

We now turn to the analysis of dwell time switching. In the sequel, $T > 0$ is fixed, σ is a given switching signal, $t_0 \stackrel{\Delta}{=} 0$, t_i denotes the ith time at which σ switches and p_i is the value of σ on $[t_{i-1}, t_i)$; if σ switches at most $n < \infty$ times then $t_{n+1} \stackrel{\Delta}{=} \infty$ and p_{n+1} denotes σ's value on $[t_n, \infty)$. Any time X takes on the current value of W is called a *sample time*. We use the notation $\lfloor t \rfloor$ to denote the sample time just preceding time t, if $t > \tau_D - \tau_C$, and the number zero otherwise. Thus, for example, $\lfloor t_0 \rfloor = 0$ and $\lfloor t_i \rfloor = t_i - \tau_C$, $i > 0$. We write k for that integer for which $T \in [t_{k-1}, t_k)$. For each $j \in \{1, 2, \ldots, k\}$ define

$$\bar{t}_j = \begin{cases} t_j & \text{if } j < k \\ T & \text{if } j = k \end{cases},$$

and let $\phi_j : [0, \infty) \rightarrow \{0, 1\}$ be that piecewise-constant signal which is zero everywhere except on the interval

$$[\lfloor t_j \rfloor,\ \bar{t}_j), \qquad \text{if } \bar{t}_j - t_{j-1} \leq \tau_D$$

or

$$[\lfloor \bar{t}_j \rfloor - \tau_C,\ \bar{t}_j), \ \text{if } \bar{t}_j - t_{j-1} > \tau_D$$

In either case ϕ_j has support no greater than $\tau_D + \tau_C$ and is idempotent {i.e., $\phi_j^2 = \phi_j$}. The following lemma describes the crucial consequence of dwell time switching upon which the proofs of Proposition 5.1 and 5.2 depend.

Lemma 5.2. *For each* $j \in \{1, 2, \ldots, k\}$

$$||(1 - \phi_j)e_{p_j}||_{\bar{t}_j} \leq ||(1 - \phi_j)e_q||_T, \qquad \forall q \in \mathcal{P}$$

Proof of Lemma 5.2: The definition of dwell time switching implies that

$$\mu_{p_j}(\lfloor t_{j-1} \rfloor) \leq \mu_q(\lfloor t_{j-1} \rfloor), \quad \forall q \in \mathcal{P},$$

$$\mu_{p_j}(\lfloor \bar{t}_j \rfloor - \tau_C) \leq \mu_q(\lfloor \bar{t}_j \rfloor - \tau_C), \forall q \in \mathcal{P} \text{ if } \bar{t}_j - t_{j-1} > \tau_D$$

As noted earlier, for all $t \geq 0$

$$\mu_p(t) = e^{-2\lambda t}||e_p||_t^2 \ , \ p \in \mathcal{P}$$

Therefore

$$\left.\begin{aligned} ||e_{p_j}||^2_{\lfloor t_{j-1} \rfloor} &\leq ||e_q||^2_{\lfloor t_{j-1} \rfloor}, \qquad \forall q \in \mathcal{P} \\ ||e_{p_j}||^2_{(\lfloor \bar{t}_j \rfloor - \tau_C)} &\leq ||e_q||^2_{(\lfloor \bar{t}_j \rfloor - \tau_C)}, \ \forall q \in \mathcal{P}, \text{ if } \bar{t}_j - t_j > \tau_D \end{aligned}\right\} \tag{5.67}$$

The definitions of ϕ_j implies that for $l \in \mathcal{P}$

$$||(1-\phi_j)e_l||^2_{\bar{t}_j} = \begin{cases} ||(1-\phi_j)e_l||^2_{\lfloor t_{j-1}\rfloor} & \text{if } \bar{t}_j - t_{j-1} \leq \tau_D, \\ ||(1-\phi_j)e_l||^2_{(\lfloor \bar{t}_j \rfloor - \tau_C)} & \text{if } \bar{t}_j - t_{j-1} > \tau_D \end{cases}$$

From this and (5.67) we obtain for all $q \in \mathcal{P}$

$$||(1-\phi_j)e_{p_j}||^2_{\bar{t}_j} = ||e_{p_j}||^2_{\lfloor t_{j-1}\rfloor} \leq ||e_q||^2_{\lfloor t_{j-1}\rfloor} \leq ||e_q||^2_{\bar{t}_j} = ||(1-\phi_j)e_q||^2_{\bar{t}_j}$$

if $\bar{t}_j - t_{j-1} \leq \tau_D$ and

$$||(1-\phi_j)e_{p_j}||^2_{\bar{t}_j} = ||e_{p_j}||^2_{\lfloor \bar{t}_j-\tau_C\rfloor} \leq ||e_q||^2_{\lfloor \bar{t}_j-\tau_C\rfloor} \leq ||e_q||^2_{\bar{t}_j} = ||(1-\phi_j)e_q||^2_{\bar{t}_j}$$

if $\bar{t}_j - t_{j-1} > \tau_D$. From this and the fact that

$$||(1-\phi_j)e_q||^2_{\bar{t}_j} \leq ||(1-\phi_j)e_q||^2_T, \; q \in \mathcal{P},$$

there follows

$$||(1-\phi_j)e_{p_j}||_{\bar{t}_j} \leq ||(1-\phi_j)e_q||_T, \;\; \forall \; q \in \mathcal{P}$$

□

Implication of Dwell-Time Switching When $\mathcal{P}$ Is a Finite Set

The proof of Proposition 5.1 makes use of the following lemma.

Lemma 5.3. *For all* $\mu_i \in [0,1]$, $i \in \{1,2,\ldots,m\}$

$$\sum_{i=1}^{m}(1-\mu_i) \leq (m-1) + \prod_{i=1}^{m}(1-\mu_i) \tag{5.68}$$

Proof of Lemma 5.3: Set $x_i = 1-\mu_i$, $i \in \{1,2,\ldots,\}$. It is enough to show that for $x_i \in [0,1]$, $i \in \{1,2,\ldots,\}$

$$\sum_{i=1}^{j} x_i \leq (j-1) + \prod_{i=1}^{j} x_j \tag{5.69}$$

for $j \in \{1,2,\ldots,m\}$. Clearly (5.69) is true if $j=1$. Suppose therefore that for some $k>0$, (5.69) holds for $j \in \{1,2,\ldots,k\}$. Then

$$\begin{aligned} \sum_{i=1}^{k+1} x_i &= x_{k+1} + \sum_{i=1}^{k} x_i \leq x_{k+1} + (k-1) + \prod_{i=1}^{k} x_i \\ &\leq (1-x_{k+1})\left(1-\prod_{i=1}^{k} x_i\right) + x_{k+1} + (k-1) + \prod_{i=1}^{k} x_i \\ &= k + \prod_{i=1}^{k+1} x_i \end{aligned}$$

By induction, (5.69) thus holds for $j \in \{1, 2, \dots, m\}$. □

Proof of Proposition 5.1: For each distinct $p \in \{p_1, p_2, \dots, p_k\}$, let $\mathcal{I}_p$ denote the set of nonnegative integers i such that $p_i = p$ and write j_p for the largest integer in $\mathcal{I}_p$. Note that $j_{p_k} = k$. Let $\mathcal{J}$ denote the set of all such j. since $\mathcal{P}$ contains m elements, m bounds from above the number of elements in $\mathcal{J}$. For each $j \in \mathcal{J}$, define

$$\psi \triangleq 1 - \prod_{j \in \mathcal{J}} (1 - \phi_j) \tag{5.70}$$

Since each ϕ_p has co-domain $\{0, 1\}$, support no greater than $\tau_D + \tau_C$ and is idempotent, it must be true that ψ has co-domain $\{0, 1\}$, support no greater than $m(\tau_D + \tau_C)$ and is idempotent as well. Therefore, (5.27) holds.

Now

$$||(1-\psi)e_\sigma||_T^2 = \sum_{j \in \mathcal{J}} \sum_{i \in \mathcal{I}_{p_j}} (||(1-\psi)e_{p_j}||_{\bar{t}_i}^2 - ||(1-\psi)e_{p_j}||_{\bar{t}_{i-1}}^2) \leq \sum_{j \in \mathcal{J}} ||(1-\psi)e_{p_j}||_{\bar{t}_j}^2 \tag{5.71}$$

In view of (5.70) we can write

$$\sum_{j \in \mathcal{J}} ||(1-\psi)e_{p_j}||_{\bar{t}_j}^2 = \sum_{j \in \mathcal{J}} \left\| \left\{ \prod_{l \in \mathcal{J}} (1 - \phi_l) \right\} e_{p_j} \right\|_{\bar{t}_j}^2 \tag{5.72}$$

But

$$\sum_{j \in \mathcal{J}} \left\| \left\{ \prod_{l \in \mathcal{J}} (1 - \phi_l) \right\} e_{p_j} \right\|_{\bar{t}_j}^2 \leq \sum_{j \in \mathcal{J}} ||(1 - \phi_j)e_{p_j}||_{\bar{t}_j}^2$$

From this, Lemma 5.2, (5.71), and (5.72) it follows that

$$||(1 - \psi)e_\sigma||_T^2 \leq \sum_{j \in \mathcal{J}} ||(1 - \phi_j)e_q||_T^2, \quad \forall\ q \in \mathcal{P}$$

Thus for $q \in \mathcal{P}$

$$\begin{aligned} ||(1 - \psi)e_\sigma||_T^2 &\leq \sum_{j \in \mathcal{J}} \int_0^T \{e_q e^{\lambda t}\}^2 (1 - \phi_j)^2 dt \\ &= \int_0^T \{e_q e^{\lambda t}\}^2 \left\{ \sum_{j \in \mathcal{J}} (1 - \phi_j)^2 \right\} dt \\ &= \int_0^T \{e_q e^{\lambda t}\}^2 \left\{ \sum_{j \in \mathcal{J}} (1 - \phi_j) \right\} dt \end{aligned}$$

This, Lemma 5.3 and (5.70) imply that

$$\begin{aligned}||(1-\psi)e_\sigma||_T^2 &\leq \int_0^T \{e_q e^{\lambda t}\}^2 \left\{ m - 1 + \prod_{p\in\mathcal{P}_T} (1-\phi_p) \right\} dt \\ &= \int_0^T \{e_q e^{\lambda t}\}^2 \{m-\psi\} dt \\ &= \int_0^T \{e_q e^{\lambda t}\}^2 \{m-\psi^2\} dt\end{aligned}$$

Hence

$$||(1-\psi)e_\sigma||_T^2 \leq m||e_q||_T^2 - ||\psi e_q||_T^2 \tag{5.73}$$

Now

$$||(1-\psi)e_\sigma||_T^2 + ||\psi e_q||_T^2 = ||(1-\psi)e_\sigma + \psi e_q||_T^2$$

because $\psi(1-\psi)=0$. From this and (5.73) it follows that

$$||(1-\psi)e_\sigma + \psi e_q||_T^2 \leq m||e_q||_T^2$$

and thus that (5.28) is true □

Implication of Dwell-Time Switching When $\mathcal{P}$ Is Not a Finite Set

To prove Proposition 5.2 we will need the following result which can be easily deduced from the discussion about strong bases in Sect. 5.4.

Lemma 5.4. *Let ϵ be a positive number and suppose that $\mathcal{X} = \{x_1, x_2, \ldots x_m\}$ is any finite set of vectors in a real n-dimensional space such that $|x_m| > \epsilon$. There exists a subset of $\bar{m} \leq n$ positive integers $\mathcal{N} = \{i_1, i_2, \ldots, i_{\bar{m}}\}$, each no larger than m, and a set of real numbers a_{ij}, $i \in \mathcal{M} = \{1, 2, \ldots m\}$, $j \in \mathcal{N}$ such that*

$$\left| x_i - \sum_{j\in\mathcal{N}} a_{ij}x_j \right| \leq \epsilon, \quad i \in \mathcal{M}$$

where

$$\begin{aligned} a_{ij} &= 0, \quad i\in\mathcal{M}, \quad j\in\mathcal{N}, \quad i>j, \\ |a_{ij}| &\leq \frac{(1+\sup\mathcal{X})^n}{\epsilon}, \quad i\in\mathcal{M}, \quad j\in\mathcal{N}\end{aligned}$$

Proof of Proposition 5.2: There are two cases to consider:
Case I: Suppose that $|c_{p_i q}| \leq \rho$ for $i \in \{1, 2, \ldots, k\}$. In this case set $\psi(t) = 0$, $t \geq 0$, $h(t) = c_{\sigma(t)q}$ for $t \in [0, t_k)$, and $h(t) = 0$ for $t > t_k$. Then (5.47) and (5.46) hold and $e_\sigma = hx + e_q$ for $t \in [0, T)$. Therefore $||e_\sigma - hx||_T = ||e_q||_T$ and (5.48) follows.

Case II: Suppose the assumption of Case I is not true in which case there is a largest integer $m \in \{1, 2, \dots, k\}$ such that $|c_{p_m q}| > \rho$. We claim that there is a non-negative integer $\bar{m} \leq n_E$, a set of $\bar{m}$ positive integers $\mathcal{J} = \{i_1, i_2, \dots, i_{\bar{m}}\}$, each no greater than k, and a set of piecewise constant signals $\gamma_j : [0, \infty) \to$ IR, $j \in \mathcal{J}$, such that

$$\left| c_{\sigma(t)q} - \sum_{j \in \mathcal{J}} \gamma_j(t) c_{p_j q} \right| \leq \rho, \quad 0 \leq t \leq T \tag{5.74}$$

where for all $j \in \mathcal{J}$

$$\gamma_j(t) = 0, \quad t \in (t_j, \infty), \tag{5.75}$$

$$|\gamma_j(t)| \leq \left(\frac{1 + \sup_{p \in \mathcal{P}} |c_{pq}|}{\rho} \right)^{n_E}, \quad t \in [0, t_j) \tag{5.76}$$

To establish this claim, we first note that $\{c_{p_1 q}, c_{p_2 q}, \dots, c_{p_m q}\} \subset \{c_{pq} : p \in \mathcal{P}\}$ and that $\{c_{pq} : p \in \mathcal{P}\}$ is a bounded subset of an n_E dimensional space. By Lemma 5.4 we thus know that there must exist a subset of $\bar{m} \leq n_E$ integers $\mathcal{J} = \{i_1, i_2, \dots, i_{\bar{m}}\}$, each no greater than m, and a set of real numbers g_{ij}, $i \in \mathcal{M} = \{1, 2, \dots, m\}$, $j \in \mathcal{J}$ such that

$$\left| c_{p_i q} - \sum_{j \in \mathcal{J}} g_{ij} c_{p_j q} \right| \leq \rho, \quad i \in \mathcal{M} \tag{5.77}$$

where

$$g_{ij} = 0, \quad i \in \mathcal{M},\ j \in \mathcal{J},\ i > j, \tag{5.78}$$

$$|g_{ij}| \leq \frac{(1 + \sup_{p \in \mathcal{P}} |c_{pq}|)^{n_E}}{\rho}, \quad i \in \mathcal{M},\ j \in \mathcal{N} \tag{5.79}$$

Thus if for each $j \in \mathcal{J}$, we define $\gamma_j(t) = g_{ij}$, $t \in [t_{i-1}, t_i)$, $i \in \mathcal{M}$, and $\gamma_j(t) = 0$, $t > t_m$ then (5.74)–(5.76) will all hold.

To proceed, define $h(t)$ for $t \in [0, t_m)$ so that

$$h(t) = c_{p_i q} - \sum_{j \in \mathcal{J}} g_{ij} c_{p_j q}, \ t \in [t_{i-1}, t_i),\ i \in \mathcal{M}$$

and for $t > t_m$ so that

$$h(t) = \begin{cases} c_{p_i q} \ t \in [t_{i-1}, t_i) \ i \in \{m+1, \dots, k\} \\ 0 \quad t > t_k \end{cases}$$

Then (5.46) holds because of (5.74) and the assumption that $|c_{p_i q}| \leq \rho$ for $i \in \{m+1, \dots, k\}$. The definition of h implies that

$$c_{\sigma(t)q} - h(t) = \sum_{j\in\mathcal{J}} \gamma_j(t)c_{p_jq}, \quad t \in [0, T)$$

and thus that

$$e_{\sigma(t)}(t) - e_q(t) - h(t)x(t) = \sum_{j\in\mathcal{J}} \gamma_j(t)(e_{p_j}(t) - e_q(t)), \quad t \in [0, T) \tag{5.80}$$

For each $j \in \mathcal{J}$, let

$$\psi \stackrel{\Delta}{=} 1 - \prod_{j\in\mathcal{J}} (1 - \phi_j) \tag{5.81}$$

Since each ϕ_j has co-domain $\{0, 1\}$, support no greater than $\tau_D + \tau_C$ and is idempotent, it must be true that ψ also has co-domain $\{0, 1\}$, is idempotent, and has support no greater than $\bar{m}(\tau_D + \tau_C)$. In view of the latter property and the fact that $\bar{m} \leq n_E$, (5.47) must be true.

By Lemma 5.2

$$||(1 - \phi_j)e_{p_j}||_{\bar{t}_j} \leq ||e_q||_T, \quad \forall j \in \mathcal{J}, \ q \in \mathcal{P}$$

From this and the triangle inequality

$$||(1 - \phi_j)(e_{p_j} - e_q)||_{\bar{t}_j} \leq 2||e_q||_T, \quad \forall j \in \mathcal{J}, \ q \in \mathcal{P}. \tag{5.82}$$

From (5.80)

$$\begin{aligned} ||(1 - \psi)(e_\sigma - e_q - \bar{f}x)||_T &= \left\| \textstyle\sum_{j\in\mathcal{N}} (1 - \psi)\gamma_j(e_{p_j} - e_q) \right\|_T \\ &\leq \textstyle\sum_{j\in\mathcal{N}} ||(1 - \psi)\gamma_j(e_{p_j} - e_q)||_T \end{aligned} \tag{5.83}$$

But

$$||(1 - \psi)\gamma_j(e_{p_j} - e_q)||_T = ||(1 - \psi)\gamma_j(e_{p_j} - e_q)||_{t_j}$$

because of (5.75). In view of (5.76)

$$||(1 - \psi)\gamma_j(e_{p_j} - e_q)||_{\{0,T\}} \leq \bar{\gamma}||(1 - \psi)(e_{p_j} - e_q)||_{t_j} \tag{5.84}$$

where

$$\bar{\gamma} \stackrel{\Delta}{=} \frac{(1 + \sup_{p\in\mathcal{P}} |c_{pq}|)^{n_E}}{\rho}$$

Now $||(1 - \psi)(e_{p_j} - e_q)||_{t_j} \leq ||(1 - \phi_j)(e_{p_j} - e_q)||_{t_j}$ because of (5.70). From this, (5.82) and (5.84) it follows that

$$||(1 - \psi)\gamma_j(e_{p_j} - e_q)||_T \leq 2\bar{\gamma}||e_q||_T$$

In view of (5.83) and the fact that $\bar{m} \leq n_E$, it follows that

$$||(1 - \psi)(e_\sigma - e_q - hx)||_T \leq 2n_E\bar{\gamma}||e_q||_T$$

But

$$(1-\psi)(e_\sigma - hx)) + \psi e_q = (1-\psi)(e_\sigma - e_q - hx) - e_q$$

so by the triangle inequality

$$||(1-\psi)(e_\sigma - hx)) + \psi e_q||_T \leq ||(1-\psi)(e_\sigma - e_q - hx)||_T + ||e_q||_T$$

Therefore

$$||(1-\psi)(e_\sigma - hx) + \psi e_q||_T \leq (1 + 2n_E\bar{\gamma})||e_q||_T$$

and (5.48) is true. □

Strong Bases

Let $\mathcal{X}$ be a subset of a real, finite dimensional linear space with norm $|\cdot|$ and let ϵ be a positive number. A nonempty list of vectors $\{x_1,\ x_2,\ \ldots,\ x_{\bar{n}}\}$ in $\mathcal{X}$ is *ϵ-independent* if

$$|x_{\bar{n}}| \geq \epsilon, \tag{5.85}$$

and, for $k \in \{1, 2, \ldots, \bar{n}-1\}$,

$$\left| x_k + \sum_{j=k+1}^{\bar{n}} \mu_j x_j \right| \geq \epsilon, \quad \forall \mu_j \in \mathbb{R} \tag{5.86}$$

$\{x_1,\ x_2, ldots, x_{\bar{n}}\}$ *ϵ-spans* $\mathcal{X}$ if for each $x \in \mathcal{X}$ there is a set of real numbers $\{b_1, b_2, \ldots, b_{\bar{n}}\}$, called *$\epsilon$-coordinates*, such that

$$\left| x - \sum_{i=1}^{\bar{n}} b_i x_i \right| \leq \epsilon \tag{5.87}$$

The following lemma gives an estimate on how large these ϵ-coordinates can be assuming $\mathcal{X}$ is abounded subset.

Lemma 5.5. *Let $\mathcal{X}$ be a bounded subset which is ϵ-spanned by an ϵ-independent list $\{x_1, x_2, \ldots, x_{\bar{n}}\}$. Suppose that x is a vector in $\mathcal{X}$ and that $b_1, b_2, \ldots b_{\bar{n}}$ is a set of ϵ-coordinates of x with respect to $\{x_1,\ x_2,\ \ldots,\ x_{\bar{n}}\}$. Then*

$$|b_i| \leq \left(1 + \frac{\sup \mathcal{X}}{\epsilon}\right)^{\bar{n}}, \quad i \in \{1, 2, \ldots, \bar{n}\} \tag{5.88}$$

This lemma will be proved in a moment.

Now suppose that $\mathcal{X}$ is an finite list of vectors $x_1,\ x_2,\ \ldots,\ x_m$ in a real n-dimensional vector space. Suppose, in addition, that $|x_m| \geq \epsilon$. It is possible to extract from $\mathcal{X}$ an ordered subset $\{x_{i_1}, x_{i_2}, \ldots, x_{i_{\bar{n}}}\}$, with $\bar{n} \leq n$, which is ϵ-independent and which ϵ-spans $\mathcal{X}$. Moreover the i_j can always be chosen so that

$$i_1 < i_2 < i_3 < \cdots < i_{\bar{n}} = m \tag{5.89}$$

and also so that for suitably defined $b_{ij} \in \mathbb{R}$

$$\left| x_i - \sum_{j=k+1}^{\bar{n}} b_{ij} x_{i_j} \right| \leq \epsilon, i \in \{i_k + 1, i_k + 2, \ldots, i_{k+1}\}, k \in \{1, 2, \ldots, \bar{n} - 1\}, \tag{5.90}$$

$$\left| x_i - \sum_{j=1}^{\bar{n}} b_{ij} x_{i_j} \right| \leq \epsilon, i \in \{1, 2, \ldots, k_1\} \tag{5.91}$$

In fact, the procedure for doing this is almost identical to the familiar procedure for extracting from $\{x_1,\ x_2, \ldots,\ x_m\}$, an ordered subset which is linearly independent {in the usual sense} and which spans the span of $\{x_1,\ x_2,\ \ldots,\ x_m\}$. The construction of interest here begins by defining an integer $j_1 \stackrel{\Delta}{=} m$. j_2 is then defined to be the greatest integer $j < j_1$ such that

$$|x_j - \mu x_{j_1}| \geq \epsilon \quad \forall \mu \in \mathbb{R},$$

if such an integer exists. If not, one defines $\bar{n} \stackrel{\Delta}{=} 1$ and $i_1 \stackrel{\Delta}{=} j_1$ and the construction is complete. If j_2 exists, j_3 is then defined to be the greatest integer $j < j_2$ such that

$$|x_j - \mu_1 x_{j_1} - \mu_2 x_{j_2}| \geq \epsilon \quad \forall \mu_i \in \mathbb{R},$$

if such an integer exists. If not, one defines $\bar{n} \stackrel{\Delta}{=} 2$ and $i_k \stackrel{\Delta}{=} j_{\bar{m}+1-k},\ k \in \{1, 2\}$.... and so on. By this process one thus obtains an ϵ-independent, ϵ-spanning subset of $\mathcal{X}$ for which there exist numbers a_{ij} such that (5.89)–(5.91) hold. Since such $b_{ij},\ j \in \{1, 2, \ldots, \bar{n}\}$, are ϵ-coordinates of $x_i,\ i \in \{1, 2, \ldots, \bar{n}\}$, each coordinate must satisfy the same bound inequality as the b_i in (5.88). Moreover, because $\bar{n}$ cannot be larger than the dimension of the smallest linear space containing $\mathcal{X}$, $\bar{n} \leq n$.

Proof of Lemma 5.5: For $k \in \{1, 2, \ldots, \bar{n}\}$ let

$$y_k \stackrel{\Delta}{=} \sum_{i=k}^{\bar{n}} b_i x_i \tag{5.92}$$

We claim that

$$|b_k| \leq \frac{|y_k|}{\epsilon}, \quad k \in \{1, 2, \ldots, \bar{n}\} \tag{5.93}$$

Now (5.93) surely holds for $k = \bar{n}$, because of (5.85) and the formula $|y_{\bar{n}}| = |b_{\bar{n}}||x_{\bar{n}}|$ which, in turn, is a consequence of (5.92). Next fix $k \in \{1, 2, , \ldots, \bar{n} - 1\}$. Now (5.93) is clearly true if $b_k = 0$. Suppose $b_k \neq 0$ in which case

$$y_k = b_k \left(x_k + \sum_{j=k+1}^{\bar{n}} \mu_j x_j \right)$$

where $\mu_j \triangleq \frac{b_j}{b_k}$. From this and (5.86) it follows that $|y_k| \geq |b_k|\epsilon$, $k \in \{1, 2, \ldots, \bar{n}\}$, so (5.93) is true.

Next write $y_1 = (y_1 - x) + x$. Then $|y_1| \leq |y_1 - x| + |x|$. But $|x| \leq \sup \mathcal{X}$ because $x \in \mathcal{X}$ and $|y_1 - x| \leq \epsilon$ because of (5.87) and the definition of y_1 in (5.92). Therefore

$$\frac{|y_1|}{\epsilon} \leq \left(1 + \frac{\sup \mathcal{X}}{\epsilon}\right) \tag{5.94}$$

From (5.92) we have that $y_{k+1} = y_k - b_k x_k$, $k \in \{1, 2, \ldots, \bar{n} - 1\}$. Thus $|y_{k+1}| \leq |y_k| + |b_k||x_k|$, $k \in \{1, 2, \ldots, \bar{n} - 1\}$. Dividing both sides of this inequality by ϵ and then using (5.93) and $|x_k| \leq \sup \mathcal{X}$, we obtain the inequality

$$\frac{|y_{k+1}|}{\epsilon} \leq \left(1 + \frac{\sup \mathcal{X}}{\epsilon}\right) \frac{|y_k|}{\epsilon}, \quad k \in \{1, 2, \ldots, \bar{n} - 1\}$$

This and (5.94) imply that

$$\frac{|y_k|}{\epsilon} \leq \left(1 + \frac{\sup \mathcal{X}}{\epsilon}\right)^k, \quad k \in \{1, 2, \ldots, \bar{n}\}$$

In view of (5.93), it follows that (5.88) is true. □

6 Flocking

Current interest in cooperative control of groups of mobile autonomous agents has led to the rapid increase in the application of graph theoretic ideas together with more familiar dynamical systems concepts to problems of analyzing and synthesizing a variety of desired group behaviors such as maintaining a formation, swarming, rendezvousing, or reaching a consensus. While this in-depth assault on group coordination using a combination of graph theory and system theory is in its early stages, it is likely to significantly expand in the years to come. One line of research which "graphically" illustrates the combined use of these concepts, is the recent theoretical work by a number of individuals which successfully explains the heading synchronization phenomenon observed in simulation by Vicsek [25], Reynolds [26] and others more than a decade ago. Vicsek and co-authors consider a simple discrete-time model consisting of n autonomous agents or particles all moving in the plane with the same speed but with different headings. Each agent's heading is updated using a local rule based on the average of its own heading plus the current headings of its "neighbors." Agent i's *neighbors* at time t, are those agents which are either in or on a circle of pre-specified radius r_i centered at agent i's current position. In their paper, Vicsek et al. provide a variety of interesting simulation results which demonstrate that the nearest neighbor rule they are studying can cause all agents to eventually move in the same direction despite the absence of centralized coordination and despite the fact that each agent's set of nearest neighbors can change with time. A theoretical explanation for this observed behavior has recently been given in [27]. The explanation exploits ideas from graph theory [28] and from the theory of non-homogeneous Markov chains [29–31]. With the benefit of hind-sight it is now reasonably clear that it is more the graph theory than the Markov chains which will prove key as this line of research advances. An illustration of this is the recent extension of the findings of [27] which explain the behavior of Reynolds' full nonlinear "boid" system [32]. By appealing to the concept of *graph composition*, we side-step most issues involving products of stochastic matrices and present in this chapter a variety of graph theoretic results which explain how convergence to a common heading is achieved.

Since the writing of [27] many important papers have appeared which extend the Vicsek problem in many directions and expand the results obtained [33–37]. Especially noteworthy among these are recent papers by Moreau [33] and Beard [34] which address the modified versions of the Vicsek problem in which different agents use different sensing radii r_i. The asymmetric neighbor relationships which result necessitate the use of directed graphs rather than undirected graphs to represent neighbor relation. We will use directed graphs in this chapter, not only because we want to deal with different sensing radii, but also because working with directed graphs enables us to give convergence conditions for the symmetric version of Vicsek's problem which are less restrictive than those originally presented in [27].

Vicsek's problem is what in computer science is called a "consensus problem" or an "agreement problem." Roughly speaking, one has a group of agents which are all trying to agree on a specific value of some quantity. Each agent initially has only limited information available. The agents then try to reach a consensus by passing what they know between them either just once or repeatedly, depending on the specific problem of interest. For the Vicsek problem, each agent always knows only its own heading and the headings of its neighbors. One feature of the Vicsek problem which sharply distinguishes it from other consensus problems, is that each agent's neighbors change with time, because all agents are in motion for the problems considered in these notes. The theoretical consequence of this is profound: it renders essentially useless a large body of literature appropriate to the convergence analysis of "nearest neighbor" algorithms with fixed neighbor relationships. Said differently, for the linear heading update rules considered in this chapter, understanding the difference between fixed neighbor relationships and changing neighbor relationships is much the same as understanding the difference between the stability of time – invariant linear systems and time – varying linear systems.

6.1 Leaderless Coordination

The system to be studied consists of n autonomous agents, labeled 1 through n, all moving in the plane with the same speed but with different headings. Each agent's heading is updated using a simple local rule based on the average of its own heading plus the headings of its "neighbors." Agent i's *neighbors* at time t, are those agents, including itself, which are either in or on a circle of pre-specified radius r_i centered at agent i's current position. In the sequel $\mathcal{N}_i(t)$ denotes the set of labels of those agents which are neighbors of agent i at time t. Agent i's heading, written θ_i, evolves in discrete-time in accordance with a model of the form

$$\theta_i(t+1) = \frac{1}{n_i(t)}\left(\sum_{j\in\mathcal{N}_i(t)} \theta_j(t)\right) \tag{6.1}$$

where t is a discrete-time index taking values in the non-negative integers $\{0, 1, 2, \ldots\}$, and $n_i(t)$ is the number of neighbors of agent i at time t.

The explicit form of the update equations determined by (6.1) depends on the relationships between neighbors which exist at time t. These relationships can be conveniently described by a directed graph $\mathbb{G}$ with vertex set $\mathcal{V} = \{1, 2, \ldots n\}$ and "arc set" $\mathcal{A}(\mathbb{G}) \subset \mathcal{V} \times \mathcal{V}$ which is defined in such a way so that (i, j) is an *arc* or directed edge from i to j just in case agent i is a neighbor of agent j. Thus $\mathbb{G}$ is a directed graph on n vertices with at most one arc from any vertex to another and with exactly one self-arc at each vertex. We write $\mathcal{G}$ for the set of all such graphs and $\bar{\mathcal{G}}$ for the set of all directed graphs with vertex set $\mathcal{V}$. We use the symbol $\bar{\mathcal{P}}$ to denote a suitably defined set indexing $\bar{\mathcal{G}}$ and we write $\mathcal{P}$ for the subset of $\bar{\mathcal{P}}$ which indexes $\mathcal{G}$. Thus $\mathcal{G} = \{\mathbb{G}_p : p \in \mathcal{P}\}$ where for $p \in \bar{\mathcal{P}}$, $\mathbb{G}_p$ denotes the pth graph in $\bar{\mathcal{G}}$. It is natural to call to a vertex i a *neighbor* of vertex j in $\mathbb{G}$ if (i, j) is and arc in $\mathbb{G}$. In addition we sometimes refer to a vertex k as a *observer* of vertex j in $\mathbb{G}$ if (j, k) is and arc in $\mathbb{G}$. Thus every vertex of $\mathbb{G}$ can *observe* its neighbors, which with the interpretation of vertices as agents, is precisely the kind of relationship $\mathbb{G}$ is suppose to represent.

The set of agent heading update rules defined by (6.1) can be written in state form. Toward this end, for each $p \in \mathcal{P}$, define *flocking matrix*

$$F_p = D_p^{-1} A_p' \tag{6.2}$$

where A_p' is the transpose of the "adjacency matrix" of the graph $\mathbb{G}_p$ and D_p the diagonal matrix whose jth diagonal element is the "in-degree" of vertex j within the graph.[3] Then

$$\theta(t+1) = F_{\sigma(t)}\theta(t), \quad t \in \{0, 1, 2, \ldots\} \tag{6.3}$$

where θ is the heading vector $\theta = \begin{pmatrix}\theta_1 \ \theta_2 \ \ldots \ \theta_n\end{pmatrix}'$ and $\sigma : \{0, 1, \ldots\} \to \mathcal{P}$ is a switching signal whose value at time t, is the index of the graph representing

[3] By the *adjacency matrix* of a directed graph $\mathbb{G} \in \bar{\mathcal{G}}$ is meant an $n \times n$ matrix whose ijth entry is a 1 if (i, j) is an arc in $\mathcal{A}(\mathbb{G})$ and 0 if it is not. The *in-degree* of vertex j in $\mathbb{G}$ is the number of arcs in $\mathcal{A}(\mathbb{G})$ of the form (i, j); thus j's in-degree is the number of *incoming* arcs to vertex j.

the agents' neighbor relationships at time t. A complete description of this system would have to include a model which explains how σ changes over time as a function of the positions of the n agents in the plane. While such a model is easy to derive and is essential for simulation purposes, it would be difficult to take into account in a convergence analysis. To avoid this difficulty, we shall adopt a more conservative approach which ignores how σ depends on the agent positions in the plane and assumes instead that σ might be any switching signal in some suitably defined set of interest.

Our ultimate goal is to show for a large class of switching signals and for any initial set of agent headings that the headings of all n agents will converge to the same steady state value θ_{ss}. Convergence of the θ_i to θ_{ss} is equivalent to the state vector θ converging to a vector of the form $\theta_{ss}\mathbf{1}$ where $\mathbf{1} \triangleq \begin{pmatrix}1 \; 1 \; \dots \; 1\end{pmatrix}'_{n\times 1}$. Naturally there are situations where convergence to a common heading cannot occur. The most obvious of these is when one agent – say the ith – starts so far away from the rest that it never acquires any neighbors. Mathematically this would mean not only that $\mathbb{G}_{\sigma(t)}$ is never strongly connected[4] at any time t, but also that vertex i remains an isolated vertex of $\mathbb{G}_{\sigma(t)}$ for all t in the sense that within each $\mathbb{G}_{\sigma(t)}$, vertex i has no incoming arcs other than its own self-arc. This situation is likely to be encountered if the r_i are very small. At the other extreme, which is likely if the r_i are very large, all agents might remain neighbors of all others for all time. In this case, σ would remain fixed along such a trajectory at that value in $p \in \mathcal{P}$ for which $\mathbb{G}_p$ is a complete graph. Convergence of θ to $\theta_{ss}\mathbf{1}$ can easily be established in this special case because with σ so fixed, (6.3) is a linear, time-invariant, discrete-time system. The situation of perhaps the greatest interest is between these two extremes when $\mathbb{G}_{\sigma(t)}$ is not necessarily complete or even strongly connected for any $t \geq 0$, but when no strictly proper subset of $\mathbb{G}_{\sigma(t)}$'s vertices is isolated from the rest for all time. Establishing convergence in this case is challenging because σ changes with time and (6.3) is not time-invariant. It is this case which we intend to study.

Strongly Rooted Graphs

In the sequel we will call a vertex i of a directed graph $\mathbb{G} \in \bar{\mathcal{G}}$, a *root* of $\mathbb{G}$ if for each other vertex j of $\mathbb{G}$, there is a path from i to j. Thus i is a root of $\mathbb{G}$, if it is the root of a directed spanning tree of $\mathbb{G}$. We will say that $\mathbb{G}$ is *rooted at* i if i is in fact a root. Thus $\mathbb{G}$ is rooted at i just in case each other vertex of $\mathbb{G}$ is *reachable* from vertex i along a path within the graph. $\mathbb{G}$ is *strongly rooted at* i if each other vertex of $\mathbb{G}$ is reachable from vertex i along a path of

[4] A directed graph $\mathbb{G} \in \bar{\mathcal{G}}$ with arc set $\mathcal{A}$ is *strongly connected* if has a "path" between each distinct pair of its vertices i and j; by a *path* {of *length* m} between vertices i and j is meant a sequence of arcs in $\mathcal{A}$ of the form $(i, k_1), (k_1, k_2), \dots (k_m, j)$ where $i, k_1, \dots, k_m$, and j are distinct vertices. $\mathbb{G}$ is *complete* if has a path of length one {i.e., an arc} between each distinct pair of its vertices.

length 1. Thus $\mathbb{G}$ is strongly rooted at i if i is a neighbor of every other vertex in the graph. By a *rooted graph* $\mathbb{G} \in \bar{\mathcal{G}}$ is meant a graph which possesses at least one root. Finally, a *strongly rooted graph* is a graph which at least one vertex at which it is strongly rooted. It is now possible to state the following elementary convergence result.

Theorem 6.1. *Let $\mathcal{Q}$ denote the subset of $\mathcal{P}$ consisting of those indices q for which $\mathbb{G}_q \in \mathcal{G}$ is strongly rooted. Let $\theta(0)$ be fixed and let $\sigma : \{0,1,2,\ldots\} \rightarrow \mathcal{P}$ be a switching signal satisfying $\sigma(t) \in \mathcal{Q}$, $t \in \{0,1,\ldots\}$. Then there is a constant steady state heading θ_{ss} depending only on $\theta(0)$ and σ for which*

$$\lim_{t \to \infty} \theta(t) = \theta_{ss}\mathbf{1} \tag{6.4}$$

where the limit is approached exponentially fast.

In order to explain why this theorem is true, we will make use of certain structural properties of the F_p. As defined, each F_p is square and non-negative, where by a *non-negative* matrix is meant a matrix whose entries are all non-negative. Each F_p also has the property that its row sums all equal 1 {i.e., $F_p\mathbf{1} = \mathbf{1}$}. Matrices with these two properties are called {row} *stochastic* [38]. Because each vertex of each graph in $\mathcal{G}$ has a self-arc, the F_p have the additional property that their diagonal elements are all non-zero. Let $\mathcal{S}$ denote the set of all $n \times n$ row stochastic matrices whose diagonal elements are all positive. $\mathcal{S}$ is closed under multiplication because the class of all $n \times n$ stochastic matrices is closed under multiplication and because the class of $n \times n$ non-negative matrices with positive diagonals is also.

In the sequel we write $M \geq N$ whenever $M - N$ is a non-negative matrix. We also write $M > N$ whenever $M - N$ is a positive matrix where by a *positive matrix* is meant a matrix with all positive entries.

Products of Stochastic Matrices

Stochastic matrices have been extensively studied in the literature for a long time largely because of their connection with Markov chains [29–31]. One problem studied which is of particular relevance here, is to describe the asymptotic behavior of products of $n \times n$ stochastic matrices of the form

$$S_j S_{j-1} \cdots S_1$$

as j tends to infinity. This is equivalent to looking at the asymptotic behavior of all solutions to the recursion equation

$$x(j+1) = S_j x(j) \tag{6.5}$$

since any solution $x(j)$ can be written as

$$x(j) = (S_j S_{j-1} \cdots S_1)\, x(1), \quad j \geq 1$$

One especially useful idea, which goes back at least to [39] and more recently to [40], is to consider the behavior of the scalar-valued non-negative function $V(x) = \lceil x \rceil - \lfloor x \rfloor$ along solutions to (6.5) where $x = (x_1 \, x_2 \cdots x_n)'$ is a non-negative n vector and $\lceil x \rceil$ and $\lfloor x \rfloor$ are its largest and smallest elements respectively. The key observation is that for any $n \times n$ stochastic matrix S, the ith entry of Sx satisfies

$$\sum_{j=1}^{n} s_{ij}x_j \geq \sum_{j=1}^{n} s_{ij}\lfloor x \rfloor = \lfloor x \rfloor$$

and

$$\sum_{j=1}^{n} s_{ij}x_j \leq \sum_{j=1}^{n} s_{ij}\lceil x \rceil = \lceil x \rceil$$

Since these inequalities hold for all rows of Sx, it must be true that $\lfloor Sx \rfloor \geq \lfloor x \rfloor$, $\lceil Sx \rceil \leq \lceil x \rceil$ and, as a consequence, that $V(Sx) \leq V(x)$. These inequalities and (6.5) imply that the sequences

$$\lfloor x(1) \rfloor, \lfloor x(2) \rfloor, \ldots \qquad \lceil x(1) \rceil, \lceil x(2) \rceil, \ldots \qquad V(x(1)), V(x(2)), \ldots$$

are each monotone. Thus because each of these sequences is also bounded, the limits

$$\lim_{j \to \infty} \lfloor x(j) \rfloor, \qquad \lim_{j \to \infty} \lceil x(j) \rceil, \qquad \lim_{j \to \infty} V(x(j))$$

each exist. Note that whenever the limit of $V(x(j))$ is zero, all components of $x(j)$ must tend to the same value and moreover this value must be a constant equal to the limiting value of $\lfloor x(j) \rfloor$.

There are various different ways one might approach the problem of developing conditions under which $S_j S_{j-1} \cdots S_1$ converges to a constant matrix of the form $\mathbf{1}c$ or equivalently $x(j)$ converges to some scalar multiple of $\mathbf{1}$. For example, since for any $n \times n$ stochastic matrix S, $S\mathbf{1} = \mathbf{1}$, it must be true that span $\{\mathbf{1}\}$ is an S-invariant subspace for any such S. From this and standard existence conditions for solutions to linear algebraic equations, it follows that for any $(n-1) \times n$ matrix P with kernel spanned by $\mathbf{1}$, the equations $PS = \tilde{S}P$ has unique solutions $\tilde{S}$, and moreover that

$$\text{spectrum } S = \{1\} \cup \text{spectrum } \tilde{S} \tag{6.6}$$

As a consequence of the equations $PS_j = \tilde{S}_j P, \;\; j \geq 1$, it can easily be seen that

$$\tilde{S}_j \tilde{S}_{j-1} \cdots \tilde{S}_1 P = P S_j S_{j-1} \cdots S_1$$

Since P has full row rank and $P\mathbf{1} = 0$, the convergence of a product of the form $S_j S_{j-1} \cdots S_1$ to $\mathbf{1}c$ for some constant row vector c, is equivalent to convergence of the corresponding product $\tilde{S}_j \tilde{S}_{j-1} \cdots \tilde{S}_1$ to the zero matrix. There are two problems with this approach. First, since P is not unique, neither are the

$\tilde{S}_i$. Second it is not so clear how to going about picking P to make tractable the problem of proving that the resulting product $\tilde{S}_j\tilde{S}_{j-1}\cdots\tilde{S}_1$ tends to zero. Tractability of the latter problem generally boils down to choosing a norm for which the $\tilde{S}_i$ are all contractive. For example, one might seek to choose a suitably weighted 2-norm. This is in essence the same thing choosing a common quadratic Lyapunov function. Although each $\tilde{S}_i$ can easily be shown to be discrete-time stable, it is known that there are classes of S_i which give rise to $\tilde{S}_i$ for which no such common Lyapunov matrix exists [41] regardless of the choice of P. Of course there are many other possible norms to choose from other than two norms. In the end, success with this approach requires one to simultaneously choose *both* a suitable P and an appropriate norm with respect to which the $\tilde{S}_i$ are all contractive. In the sequel we adopt a slightly different, but closely related approach which ensures that we can work with what is perhaps the most natural norm for this type of convergence problem, the infinity norm.

To proceed, we need a few more ideas concerned with non-negative matrices. For any non-negative matrix R of any size, we write $||R||$ for the largest of the row sums of R. Note that $||R||$ is the induced infinity norm of R and consequently is sub-multiplicative. Note in addition that $||x|| = \lceil x \rceil$ for any non-negative n vector x. Moreover, $||M_1|| \leq ||M_2||$ if $M_1 \leq M_2$. Observe that for any $n \times n$ stochastic matrix S, $||S|| = 1$ because the row sums of a stochastic matrix all equal 1. We extend the domain of definitions of $\lfloor \cdot \rfloor$ and $\lceil \cdot \rceil$ to the class of all non-negative $n \times m$ matrix M, by letting $\lfloor M \rfloor$ and $\lceil M \rceil$ now denote the $1 \times m$ row vectors whose jth entries are the smallest and largest elements respectively, of the jth column of M. Note that $\lfloor M \rfloor$ is the largest $1 \times m$ non-negative row vector c for which $M - \mathbf{1}c$ is non-negative and that $\lceil M \rceil$ is the smallest non-negative row vector c for which $\mathbf{1}c - M$ is non-negative. Note in addition that for any $n \times n$ stochastic matrix S,

$$S = \mathbf{1}\lfloor S \rfloor + \lfloor\!\lfloor S \rfloor\!\rfloor \quad \text{and} \quad S = \mathbf{1}\lceil S \rceil - \lceil\!\lceil S \rceil\!\rceil \tag{6.7}$$

where $\lfloor\!\lfloor S \rfloor\!\rfloor$ and $\lceil\!\lceil S \rceil\!\rceil$ are the non-negative matrices

$$\lfloor\!\lfloor S \rfloor\!\rfloor = S - \mathbf{1}\lfloor S \rfloor \quad \text{and} \quad \lceil\!\lceil S \rceil\!\rceil = \mathbf{1}\lceil S \rceil - S \tag{6.8}$$

respectively. Moreover the row sums of $\lfloor\!\lfloor S \rfloor\!\rfloor$ are all equal to $1 - \lfloor S \rfloor \mathbf{1}$ and the row sums of $\lceil\!\lceil S \rceil\!\rceil$ are all equal to $\lceil S \rceil \mathbf{1} - 1$ so

$$||\lfloor\!\lfloor S \rfloor\!\rfloor|| = 1 - \lfloor S \rfloor \mathbf{1} \quad \text{and} \quad ||\lceil\!\lceil S \rceil\!\rceil|| = \lceil S \rceil \mathbf{1} - 1 \tag{6.9}$$

In the sequel we will also be interested in the matrix

$$[\![S]\!] = \lfloor\!\lfloor S \rfloor\!\rfloor + \lceil\!\lceil S \rceil\!\rceil \tag{6.10}$$

This matrix satisfies

$$[\![S]\!] = \mathbf{1}(\lceil S \rceil - \lfloor S \rfloor) \tag{6.11}$$

because of (6.7).

For any infinite sequence of $n \times n$ stochastic matrices $S_1, S_2, \ldots$, we henceforth use the symbol $\lfloor \cdots S_j \cdots S_1 \rfloor$ to denote the limit

$$\lfloor \cdots S_j \cdots S_2 S_1 \rfloor = \lim_{j \to \infty} \lfloor S_j \cdots S_2 S_1 \rfloor \tag{6.12}$$

From the preceding discussion it is clear that this limit exists whether or not the product $S_j \cdots S_2 S_1$ itself has a limit. Two situations can occur. Either the product $S_j \cdots S_2 S_1$ converges to a rank one matrix or it does not. It is quite possible for such a product to converge to a matrix which is not rank one. An example of this would be a sequence in which S_1 is any stochastic matric of rank greater than 1 and for all $i > 1$, $S_i = I_{n \times n}$. In the sequel we will develop necessary conditions for $S_j \cdots S_2 S_1$ to converge to a rank one matrix as $j \to \infty$. Note that if this occurs, then the limit must be of the form $\mathbf{1}c$ where $c\mathbf{1} = 1$ because stochastic matrices are closed under multiplication.

In the sequel we will say that a matrix product $S_j S_{j-1} \cdots S_1$ *converges to* $\mathbf{1}\lfloor \cdots S_j \cdots S_1 \rfloor$ *exponentially fast at a rate no slower than* λ if there are non-negative constants b and λ with $\lambda < 1$, such that

$$||(S_j \cdots S_1) - \mathbf{1}\lfloor \cdots S_j \cdots S_2 S_1 \rfloor|| \leq b\lambda^j, \qquad j \geq 1 \tag{6.13}$$

The following proposition implies that such a stochastic matrix product will so converge if the matrix product $\lfloor S_j \cdots S_1 \rceil$ converges to 0.

Proposition 6.1. *Let $\bar{b}$ and λ be non-negative numbers with $\lambda < 1$. Suppose that $S_1, S_2, \ldots,$ is an infinite sequence of $n \times n$ stochastic matrices for which*

$$||\lfloor S_j \cdots S_1 \rceil|| \leq \bar{b}\lambda^j, \quad j \geq 0 \tag{6.14}$$

Then the matrix product $S_j \cdots S_2 S_1$ converges to $\mathbf{1}\lfloor \cdots S_j \cdots S_1 \rfloor$ exponentially fast at a rate no slower than λ.

The proof of Proposition 6.1 makes use of the first of the two inequalities which follow.

Lemma 6.1. *For any two $n \times n$ stochastic matrices S_1 and S_2,*

$$\lfloor S_2 S_1 \rfloor - \lfloor S_1 \rfloor \leq \lceil S_2 \rceil \lfloor S_1 \rceil \tag{6.15}$$

$$\lfloor S_2 S_1 \rceil \leq \lfloor S_2 \rceil \lfloor S_1 \rceil \tag{6.16}$$

Proof of Lemma 6.1: Since $S_2 S_1 = S_2(\mathbf{1}\lfloor S_1 \rfloor + \lfloor S_1 \rceil) = \mathbf{1}\lfloor S_1 \rfloor + S_2\lfloor S_1 \rceil$ and $S_2 = \mathbf{1}\lceil S_2 \rceil - \lceil S_2 \rfloor$, it must be true that $S_2 S_1 = \mathbf{1}(\lfloor S_1 \rfloor + \lceil S_2 \rceil \lfloor S_1 \rceil) - \lceil S_2 \rfloor \lfloor S_1 \rceil$. But $\lceil S_2 S_1 \rceil$ is the smallest non-negative row vector c for which $\mathbf{1}c - S_2 S_1$ is non-negative. Therefore

$$\lceil S_2 S_1 \rceil \leq \lfloor S_1 \rfloor + \lceil S_2 \rceil \lfloor S_1 \rceil \tag{6.17}$$

Moreover $\lfloor S_2S_1 \rfloor \leq \lceil S_2S_1 \rceil$ because of (6.11). This and (6.17) imply $\lfloor S_2S_1 \rfloor \leq \lfloor S_1 \rfloor + \lceil S_2 \rceil \lfloor S_1 \rfloor$ and thus that (6.15) is true.

Since $S_2S_1 = S_2(\mathbf{1}\lfloor S_1 \rfloor + \lfloor S_1 \rfloor) = \mathbf{1}\lfloor S_1 \rfloor + S_2\lfloor S_1 \rfloor$ and $S_2 = \lfloor S_2 \rfloor + \lfloor S_2 \rfloor$, it must be true that $S_2S_1 = \mathbf{1}(\lfloor S_1 \rfloor + \lfloor S_2 \rfloor \lfloor S_1 \rfloor) + \lfloor S_2 \rfloor \lfloor S_1 \rfloor$. But $\lfloor S_2S_1 \rfloor$ is the largest non-negative row vector c for which $S_2S_1 - \mathbf{1}c$ is non-negative so

$$S_2S_1 \leq \mathbf{1}\lfloor S_2S_1 \rfloor + \lfloor S_2 \rfloor \lfloor S_1 \rfloor \tag{6.18}$$

Now it is also true that $S_2S_1 = \mathbf{1}\lfloor S_2S_1 \rfloor + \lfloor S_2S_1 \rfloor$. From this and (6.18) it follows that (6.16) is true. □

Proof of Proposition 6.1: Set $X_j = S_j \cdots S_1, \quad j \geq 1$ and note that each X_j is a stochastic matrix. In view of (6.15),

$$\lfloor X_{j+1} \rfloor - \lfloor X_j \rfloor \leq \lceil S_{j+1} \rceil \lfloor X_j \rfloor, \quad j \geq 1$$

By hypothesis, $|||\lfloor X_j \rfloor||| \leq \bar{b}\lambda^j, \quad j \geq 1$. Moreover $|||\lceil S_{j+1} \rceil||| \leq n$ because all entries in S_{j+1} are bounded above by 1. Therefore

$$|||\lfloor X_{j+1} \rfloor - \lfloor X_j \rfloor||| \leq n\bar{b}\lambda^j, \quad j \geq 1 \tag{6.19}$$

Clearly

$$\lfloor X_{j+i} \rfloor - \lfloor X_j \rfloor = \sum_{k=1}^{i} (\lfloor X_{i+j+1-k} \rfloor - \lfloor X_{i+j-k} \rfloor), \quad i, j \geq 1$$

Thus, by the triangle inequality

$$|||\lfloor X_{j+i} \rfloor - \lfloor X_j \rfloor||| \leq \sum_{k=1}^{i} |||\lfloor X_{i+j+1-k} \rfloor - \lfloor X_{i+j-k} \rfloor|||, \quad i, j \geq 1$$

This and (6.19) imply that

$$|||\lfloor X_{j+i} \rfloor - \lfloor X_j \rfloor||| \leq n\bar{b} \sum_{k=1}^{i} \lambda^{(i+j-k)}, \quad i, j \geq 1$$

Now

$$\sum_{k=1}^{i} \lambda^{(i+j-k)} = \lambda^j \sum_{k=1}^{i} \lambda^{(i-k)} = \lambda^j \sum_{q=1}^{i} \lambda^{q-1} \leq \lambda^j \sum_{q=1}^{\infty} \lambda^{q-1}$$

But $\lambda < 1$ so

$$\sum_{q=1}^{\infty} \lambda^{q-1} = \frac{1}{(1 - \lambda)}$$

Therefore

$$|||\lfloor X_{i+j}\rfloor - \lfloor X_j\rfloor||| \leq n\bar{b}\frac{\lambda^j}{(1-\lambda)}, \quad i,j \geq 1. \tag{6.20}$$

Set $c = \lfloor \cdots S_j \cdots S_1 \rfloor$ and note that

$$\begin{aligned} |||\lfloor X_j\rfloor - c|| &= |||\lfloor X_j\rfloor - \lfloor X_{i+j}\rfloor + \lfloor X_{i+j}\rfloor - c|| \\ &\leq |||\lfloor X_j\rfloor - \lfloor X_{i+j}\rfloor||| + |||\lfloor X_{i+j}\rfloor - c||, \quad i,j \geq 1 \end{aligned}$$

In view of (6.20)

$$|||\lfloor X_j\rfloor - c|| \leq n\bar{b}\frac{\lambda^j}{(1-\lambda)} + |||\lfloor X_{i+j}\rfloor - c||, \quad i,j \geq 1$$

Since

$$\lim_{i\to\infty} |||\lfloor X_{i+j}\rfloor - c|| = 0$$

it must be true that

$$|||\lfloor X_j\rfloor - c|| \leq n\bar{b}\frac{\lambda^j}{(1-\lambda)}, \quad j \geq 1$$

But $||\mathbf{1}(\lfloor X_j\rfloor - c)|| = |||\lfloor X_j\rfloor - c||$ and $X_j = S_j \cdots S_1$. Therefore

$$||\mathbf{1}(\lfloor S_j \cdots S_1\rfloor - c)|| \leq n\bar{b}\frac{\lambda^j}{(1-\lambda)}, \quad j \geq 1 \tag{6.21}$$

In view of (6.7)

$$S_j \cdots S_1 = \mathbf{1}\lfloor S_j \cdots S_1\rfloor + \lfloor S_j \cdots S_1\rceil, \ j \geq 1$$

Therefore

$$\begin{aligned} ||(S_j \cdots S_1) - \mathbf{1}c|| &= ||\mathbf{1}\lfloor S_j \cdots S_1\rfloor + \lfloor S_j \cdots S_1\rceil - \mathbf{1}c|| \\ &\leq ||\mathbf{1}\lfloor S_j \cdots S_1\rfloor - \mathbf{1}c|| + |||\lfloor S_j \cdots S_1\rceil|||, \quad j \geq 1 \end{aligned}$$

From this, (6.14) and (6.21) it follows that

$$||S_j \cdots S_1 - \mathbf{1}c|| \leq \bar{b}\left(1 + \frac{n}{(1-\lambda)}\right)\lambda^j, \quad j \geq 1$$

and thus that (6.13) holds with $b = \bar{b}\left(1 + \frac{n}{(1-\lambda)}\right)$. □

Convergence

We are now in a position to make some statements about the asymptotic behavior of a product of $n \times n$ stochastic matrices of the form $S_j S_{j-1} \cdots S_1$ as

j tends to infinity. Note first that (6.16) generalizes to sequences of stochastic matrices of any length. Thus

$$\lfloor S_j S_{j-1} \cdots S_2 S_1 \rfloor \leq \lfloor S_j \rfloor \lfloor S_{j-1} \rfloor \cdots \lfloor S_1 \rfloor \tag{6.22}$$

It is therefore clear that condition (6.14) of Proposition 6.1 will hold with $\bar{b} = 1$ if

$$|| \lfloor S_j \rfloor \cdots \lfloor S_1 \rfloor || \leq \lambda^j \tag{6.23}$$

for some nonnegative number $\lambda < 1$. Because $|| \cdot ||$ is sub-multiplicative, this means that a product of stochastic matrices $S_j \cdots S_1$ will converge to a limit of the form $\mathbf{1}c$ for some constant row-vector c if for each of the matrices S_i in the sequence $S_1, S_2, \ldots$ satisfies the norm bound $|| \lfloor S_i \rfloor || < \lambda$. We now develop a condition, tailored to our application, for this to be so. For any $n \times n$ stochastic matrix, let $\gamma(S)$ denote that graph $\mathbb{G} \in \bar{\mathcal{G}}$ whose adjacency matrix is the transpose of the matrix obtained by replacing all of S's non-zero entries with 1s.

Lemma 6.2. *For each $n \times n$ stochastic matrix S whose graph $\gamma(S)$ is strongly rooted*

$$|| \lfloor S \rfloor || < 1 \tag{6.24}$$

Proof: Let A be the adjacency matrix of $\gamma(S)$. Since $\gamma(S)$ is strongly rooted, its adjacency matrix A must have a positive row i for every i which is the label of a root of $\gamma(S)$. Since the positions of the non-zero entries of S and A' are the same, this means that S's ith column s_i will be positive if i is a root. Clearly $\lfloor S \rfloor$ will have its ith entry non-zero if vertex i is a root of $\gamma(S)$. Since $\gamma(S)$ is strongly rooted, at least one such root exists which implies that $\lfloor S \rfloor$ is non-zero, and thus that $1 - \lfloor S \rfloor \mathbf{1} < 1$. From this and (6.9) it follows that (6.24) is true. □

It can be shown very easily that if (6.24) holds, then $\gamma(S)$ must be strongly rooted. We will not need this fact so the proof is omitted.

Proposition 6.2. *Let $\mathcal{S}_{sr}$ be any closed set of $n \times n$ stochastic matrices whose graphs $\gamma(S)$, $S \in \mathcal{S}_{sr}$ are all strongly rooted. Then any product $S_j \cdots S_1$ of matrices from $\mathcal{S}_{sr}$ converges to $\mathbf{1}\lfloor \cdots S_j \cdots S_1 \rfloor$ exponentially fast as $j \to \infty$ at a rate no slower than λ, where λ is a non-negative constant depending on $\mathcal{S}_{sr}$ and satisfying $\lambda < 1$.*

Proof of Proposition 6.2: In view of Lemma 6.2, $|| \lfloor S \rfloor || < 1$, $S \in \mathcal{S}_{sr}$. Let

$$\lambda = \max_{S \in \mathcal{S}_{sr}} || \lfloor S \rfloor ||$$

Because $\mathcal{S}_{sr}$ is closed and bounded and $|| \lfloor \cdot \rfloor ||$ is continuous, $\lambda < 1$. Clearly $|| \lfloor S \rfloor || \leq \lambda, \ i \geq 1$ so (6.23) must hold for any sequence of matrices $S_1, S_2, \ldots$ from $\mathcal{S}_{sr}$. Therefore for any such sequence $|| \lfloor S_j \cdots S_1 \rfloor || \leq \lambda^j, \ j \geq 0$. Thus by

Proposition 6.1, the product $\Pi(j) = S_j S_{j-1} \cdots S_1$ converges to $\mathbf{1}\lfloor \cdots S_j \cdot S_1 \rfloor$ exponentially fast at a rate no slower than λ. $\square$

Proof of Theorem 6.1: By definition, the graph $\mathbb{G}_p$ of each matrix F_p in the finite set $\{F_p : p \in \mathcal{Q}\}$ is strongly rooted. By assumption, $F_{\sigma(t)} \in \{F_p : p \in \mathcal{Q}\}$, $t \geq 0$. In view of Proposition 6.2, the product $F_{\sigma(t)} \cdots F_{\sigma(0)}$ converges to $\mathbf{1}\lfloor \cdots F_{\sigma(t)} \cdots F_{\sigma(0)} \rfloor$ exponentially fast at a rate no slower than

$$\lambda = \max_{p \in \mathcal{Q}} || \lfloor F_p \rfloor ||$$

But it is clear from (6.3) that

$$\theta(t) = F_{\sigma(t-1)} \cdots F_{\sigma(1)} F_{\sigma(0)} \theta(0), \quad t \geq 1$$

Therefore (6.4) holds with $\theta_{ss} = \lfloor \cdots F_{\sigma(t)} \cdots F_{\sigma(0)} \rfloor \theta(0)$ and the convergence is exponential. $\square$

Convergence Rate

Using (6.9) it is possible to calculate a worst case value for the convergence rate λ used in the proof of Theorem 6.1. Fix $p \in \mathcal{Q}$ and consider the flocking matrix F_p and its associated graph $\mathbb{G}_p$. Because $\mathbb{G}_p$ is strongly rooted, at least one vertex – say the kth – must be a root with arcs to each other vertex. In the context of (6.1), this means that agent k must be a neighbor of every agent. Thus θ_k must be in each sum in (6.1). Since each n_i in (6.1) is bounded above by n, this means that the smallest element in column k of F_p, is bounded below by $\frac{1}{n}$. Since (6.9) asserts that $|| \lfloor F_p \rfloor || = 1 - \lfloor F_p \rfloor \mathbf{1}$, it must be true that $|| \lfloor F_p \rfloor || \leq 1 - \frac{1}{n}$. This holds for all $p \in \mathcal{Q}$. Moreover in the worst case when $\mathbb{G}_p$ is strongly rooted at just one vertex and all vertices are neighbors of at least one common vertex, $|| \lfloor F_p \rfloor || = 1 - \frac{1}{n}$. It follows that the worst case convergence rate is

$$\max_{p \in \mathcal{Q}} || \lfloor F_p \rfloor || = 1 - \frac{1}{n} \tag{6.25}$$

Rooted Graphs

The proof of Theorem 6.1 depends crucially on the fact that the graphs encountered along a trajectory of (6.3) are all strongly rooted. It is natural to ask if this requirement can be relaxed and still have all agents' headings converge to a common value. The aim of this section is to show that this can indeed be accomplished. To do this we need to have a meaningful way of "combining" sequences of graphs so that only the combined graph need be strongly rooted, but not necessarily the individual graphs making up the combination. One possible notion of combination of a sequence $\mathbb{G}_{p_1}, \mathbb{G}_{p_2}, \ldots, \mathbb{G}_{p_k}$ would be that graph in $\mathcal{G}$ whose arc set is the union of the arc sets of the graphs in the sequence. It turns out that because we are interested in *sequences* of graphs rather than mere *sets* of graphs, a simple union is not quite the appropriate notion for our purposes because a union does not take into account the order in which the graphs are encountered along a trajectory. What is appropriate is a slightly more general notion which we now define.

Composition of Graphs

Let us agree to say that the *composition* of a directed graph $\mathbb{G}_{p_1} \in \bar{\mathcal{G}}$ with a directed graph $\mathbb{G}_{p_2} \in \bar{\mathcal{G}}$, written $\mathbb{G}_{p_2} \circ \mathbb{G}_{p_1}$, is the directed graph with vertex set $\{1, \ldots, n\}$ and arc set defined in such a way so that (i, j) is an arc of the composition just in case there is a vertex q such that (i, q) is an arc of $\mathbb{G}_{p_1}$ and (q, j) is an arc of $\mathbb{G}_{p_2}$. Thus (i, j) is an arc of $\mathbb{G}_{p_2} \circ \mathbb{G}_{p_1}$ if and only if i has an observer in $\mathbb{G}_{p_1}$ which is also a neighbor of j in $\mathbb{G}_{p_2}$. Note that $\bar{\mathcal{G}}$ is closed under composition and that composition is an associative binary operation; because of this, the definition extend unambiguously to any finite sequence of directed graphs $\mathbb{G}_{p_1}, \mathbb{G}_{p_2}, \ldots, \mathbb{G}_{p_k}$.

If we focus exclusively on graphs in $\mathcal{G}$, more can be said. In this case the definition of composition implies that the arcs of $\mathbb{G}_{p_1}$ and $\mathbb{G}_{p_2}$ are arcs of $\mathbb{G}_{p_2} \circ \mathbb{G}_{p_1}$. The definition also implies in this case that if $\mathbb{G}_{p_1}$ has a directed path from i to k and $\mathbb{G}_{p_2}$ has a directed path from k to j, then $\mathbb{G}_{p_2} \circ \mathbb{G}_{p_1}$ has a directed path from i to j. Both of these implications are consequences of the requirement that the vertices of the graphs in $\mathcal{G}$ all have self arcs. Note in addition that $\mathcal{G}$ is closed under composition. It is worth emphasizing that the union of the arc sets of a sequence of graphs $\mathbb{G}_{p_1}, \mathbb{G}_{p_2}, \ldots, \mathbb{G}_{p_k}$ in $\mathcal{G}$ must be contained in the arc set of their composition. However the converse is not true in general and it is for this reason that composition rather than union proves to be the more useful concept for our purposes.

Suppose that $A_p = \big(a_{ij}(p)\big)$ and $A_q = \big(a_{ij}(q)\big)$ are the adjacency matrices of $\mathbb{G}_p \in \bar{\mathcal{G}}$ and $\mathbb{G}_q \in \bar{\mathcal{G}}$ respectively. Then the adjacency matrix of the composition $\mathbb{G}_q \circ \mathbb{G}_p$ must be the matrix obtained by replacing all non-zero elements in $A_p A_q$ with ones. This is because the ijth entry of $A_p A_q$, namely

$$\sum_{k=1}^{n} a_{ik}(p) a_{kj}(q),$$

will be non-zero just in case there is at least one value of k for which both $a_{ik}(p)$ and $a_{kj}(q)$ are non-zero. This of course is exactly the condition for the ijth element of the adjacency matrix of the composition $\mathbb{G}_q \circ \mathbb{G}_p$ to be non-zero. Note that if S_1 and S_2 are $n \times n$ stochastic matrices for which $\gamma(S_1) = \mathbb{G}_p$ and $\gamma(S_2) = \mathbb{G}_q$, then the matrix which results by replacing by ones, all non-zero entries in the stochastic matrix $S_2 S_1$, must be the transpose of the adjacency matrix of $\mathbb{G}_q \circ \mathbb{G}_p$. In view of the definition of $\gamma(\cdot)$, it therefore must be true that $\gamma(S_2 S_1) = \gamma(S_2) \circ \gamma(S_1)$. This obviously generalizes to finite products of stochastic matrices.

Lemma 6.3. *For any sequence of $n \times n$ stochastic matrices $S_1, S_2, \dots, S_j$,*

$$\gamma(S_j \cdots S_1) = \gamma(S_j) \circ \cdots \circ \gamma(S_1)$$

Compositions of Rooted Graphs

We now give several different conditions under which the composition of a sequence of graphs is strongly rooted.

Proposition 6.3. *Suppose $n > 1$ and let $\mathbb{G}_{p_1}, \mathbb{G}_{p_2}, \dots, \mathbb{G}_{p_m}$ be a finite sequence of rooted graphs in $\mathcal{G}$.*

1. *If $m \geq n^2$, then $\mathbb{G}_{p_m} \circ \mathbb{G}_{p_{m-1}} \circ \cdots \circ \mathbb{G}_{p_1}$ is strongly rooted.*
2. *If $\mathbb{G}_{p_1}, \mathbb{G}_{p_2}, \dots, \mathbb{G}_{p_m}$ are all rooted at v and $m \geq n-1$, then $\mathbb{G}_{p_m} \circ \mathbb{G}_{p_{m-1}} \circ \cdots \circ \mathbb{G}_{p_1}$ is strongly rooted at v.*

The requirement that all the graphs in the sequence be rooted at a single vertex v is obviously more restrictive than the requirement that all the graphs be rooted, but not necessarily at the same vertex. The price for the less restrictive assumption, is that the bound on the number of graphs needed in the more general case is much higher than the bound given in the case which all the graphs are rooted at v. It is undoubtedly true that the bound n^2 for the more general case is too conservative. The more special case when all graphs share a common root is relevant to the leader follower version of the problem which will be discussed later in these notes. Proposition 6.3 will be proved in a moment.

Note that a strongly connected graph is the same as a graph which is rooted at every vertex and that a complete graph is the same as a graph which is strongly rooted at every vertex. In view of these observations and Proposition 6.3 we can state the following

Proposition 6.4. *Suppose $n > 1$ and let $\mathbb{G}_{p_1}, \mathbb{G}_{p_2}, \dots, \mathbb{G}_{p_m}$ be a finite sequence of strongly connected graphs in $\mathcal{G}$. If $m \geq n-1$, then $\mathbb{G}_{p_m} \circ \mathbb{G}_{p_{m-1}} \circ \cdots \circ \mathbb{G}_{p_1}$ is complete.*

To prove Proposition 6.3 we will need some more ideas. We say that a vertex $v \in \mathcal{V}$ is a *observer* of a subset $\mathcal{S} \subset \mathcal{V}$ in a graph $\mathbb{G} \in \bar{\mathcal{G}}$, if v is an observer of at least one vertex in $\mathcal{S}$. By the *observer function* of a graph $\mathbb{G} \in \bar{\mathcal{G}}$, written $\alpha(\mathbb{G}, \cdot\,)$ we mean the function $\alpha(\mathbb{G}, \cdot\,) : 2^{\mathcal{V}} \to 2^{\mathcal{V}}$ which assigns to each subset $\mathcal{S} \subset \mathcal{V}$, the subset of vertices in $\mathcal{V}$ which are observers of $\mathcal{S}$ in $\mathbb{G}$. Thus $j \in \alpha(\mathbb{G}, i)$ just in case $(i, j) \in \mathcal{A}(\mathbb{G})$. Note that if $\mathbb{G}_p \in \bar{\mathcal{G}}$ and $\mathbb{G}_q$ in $\mathcal{G}$, then

$$\alpha(\mathbb{G}_p, \mathcal{S}) \subset \alpha(\mathbb{G}_q \circ \mathbb{G}_p, \mathcal{S}), \quad \mathcal{S} \in 2^{\mathcal{V}} \tag{6.26}$$

because $\mathbb{G}_q \in \mathcal{G}$ implies that the arcs in $\mathbb{G}_p$ are all arcs in $\mathbb{G}_q \circ \mathbb{G}_p$. Observer functions have the following important property.

Lemma 6.4. *For all $\mathbb{G}_p, \mathbb{G}_q \in \bar{\mathcal{G}}$ and any non-empty subset $\mathcal{S} \subset \mathcal{V}$,*

$$\alpha(\mathbb{G}_q, \alpha(\mathbb{G}_p, \mathcal{S})) = \alpha(\mathbb{G}_q \circ \mathbb{G}_p, \mathcal{S}) \tag{6.27}$$

Proof: Suppose first that $i \in \alpha(\mathbb{G}_q, \alpha(\mathbb{G}_p, \mathcal{S}))$. Then (j,i) is an arc in $\mathbb{G}_q$ for some $j \in \alpha(\mathbb{G}_p, \mathcal{S})$. Hence (k,j) is an arc in $\mathbb{G}_p$ for some $k \in \mathcal{S}$. In view of the definition of composition, (k,i) is an arc in $\mathbb{G}_q \circ \mathbb{G}_p$ so $i \in \alpha(\mathbb{G}_q \circ \mathbb{G}_p, \mathcal{S})$. Since this holds for all $i \in \mathcal{V}$, $\alpha(\mathbb{G}_q, \alpha(\mathbb{G}_p, \mathcal{S})) \subset \alpha(\mathbb{G}_q \circ \mathbb{G}_p, \mathcal{S})$.

For the reverse inclusion, fix $i \in \alpha(\mathbb{G}_q \circ \mathbb{G}_p, \mathcal{S})$ in which case (k,i) is an arc in $\mathbb{G}_q \circ \mathbb{G}_p$ for some $k \in \mathcal{S}$. By definition of composition, there exists an $j \in \mathcal{V}$ such that (k,j) is an arc in $\mathbb{G}_p$ and (j,i) is an arc in $\mathbb{G}_q$. Thus $j \in \alpha(\mathbb{G}_p, \mathcal{S})$. Therefore $i \in \alpha(\mathbb{G}_q, \alpha(\mathbb{G}_p, \mathcal{S}))$. Since this holds for all $i \in \mathcal{V}$, $\alpha(\mathbb{G}_q, \alpha(\mathbb{G}_p, \mathcal{S})) \supset \alpha(\mathbb{G}_q \circ \mathbb{G}_p, \mathcal{S})$. Therefore (6.27) is true. □

To proceed, let us note that each subset $\mathcal{S} \subset \mathcal{V}$ induces a unique subgraph of $\mathbb{G}$ with vertex set $\mathcal{S}$ and arc set $\mathcal{A}$ consisting of those arcs (i,j) of $\mathbb{G}$ for which both i and j are vertices of $\mathcal{S}$. This together with the natural partial ordering of $\mathcal{V}$ by inclusion provides a corresponding partial ordering of $\bar{\mathcal{G}}$. Thus if $\mathcal{S}_1$ and $\mathcal{S}_2$ are subsets of $\mathcal{V}$ and $\mathcal{S}_1 \subset \mathcal{S}_2$, then $\mathbb{G}_1 \subset \mathbb{G}_2$ where for $i \in \{1,2\}$, $\mathbb{G}_i$ is the subgraph of $\mathbb{G}$ induced by $\mathcal{S}_i$. For any $v \in \mathcal{V}$, there is a unique largest subgraph rooted at v, namely the graph induced by the vertex set $\mathcal{V}(v) = \{v\} \cup \alpha(\mathbb{G}, v) \cup \cdots \cup \alpha^{n-1}(\mathbb{G}, v)$ where $\alpha^i(\mathbb{G}, \cdot)$ denotes the composition of $\alpha(\mathbb{G}, \cdot)$ with itself i times. We call this graph, the *rooted graph generated by* v. It is clear that $\mathcal{V}(v)$ is the smallest $\alpha(\mathbb{G}, \cdot)$-invariant subset of $\mathcal{V}$ which contains v.

The proof of Propositions 6.3 depends on the following lemma.

Lemma 6.5. *Let $\mathbb{G}_p$ and $\mathbb{G}_q$ be graphs in $\mathcal{G}$. If $\mathbb{G}_q$ is rooted at v and $\alpha(\mathbb{G}_p, v)$ is a strictly proper subset of $\mathcal{V}$, then $\alpha(\mathbb{G}_p, v)$ is also a strictly proper subset of $\alpha(\mathbb{G}_q \circ \mathbb{G}_p, v)$*

Proof of Lemma 6.5: In general $\alpha(\mathbb{G}_p, v) \subset \alpha(\mathbb{G}_q \circ \mathbb{G}_p, v)$ because of (6.26). Thus if $\alpha(\mathbb{G}_p, v)$ is not a strictly proper subset of $\alpha(\mathbb{G}_q \circ \mathbb{G}_p, v)$, then $\alpha(\mathbb{G}_p, v) = \alpha(\mathbb{G}_q \circ \mathbb{G}_p, v)$ so $\alpha(\mathbb{G}_q \circ \mathbb{G}_p, v) \subset \alpha(\mathbb{G}_p, v)$. In view of (6.27), $\alpha(\mathbb{G}_q \circ \mathbb{G}_p, v) = \alpha(\mathbb{G}_q, \alpha(\mathbb{G}_p, v))$. Therefore $\alpha(\mathbb{G}_q, \alpha(\mathbb{G}_p, v)) \subset \alpha(\mathbb{G}_p, v)$. Moreover, $v \in \alpha(\mathbb{G}_p, v)$ because v has a self-arc in $\mathbb{G}_p$. Thus $\alpha(\mathbb{G}_p, v)$ is a strictly proper subset of $\mathcal{V}$ which contains v and is $\alpha(\mathbb{G}_q, \cdot)$-invariant. But this is impossible because $\mathbb{G}_q$ is rooted at v. □

Proof of Proposition 6.3: Suppose $m \geq n^2$. In view of (6.26), $\mathcal{A}(\mathbb{G}_{p_k} \circ \mathbb{G}_{p_{k-1}} \circ \cdots \circ \mathbb{G}_{p_1}) \subset \mathcal{A}(\mathbb{G}_{p_m} \circ \mathbb{G}_{p_{m-1}} \circ \cdots \circ \mathbb{G}_{p_1})$ for any positive integer $k \leq m$. Thus $\mathbb{G}_{p_m} \circ \mathbb{G}_{p_{m-1}} \circ \cdots \circ \mathbb{G}_{p_1}$ will be strongly rooted if there exists an integer $k \leq n^2$ such that $\mathbb{G}_{p_k} \circ \mathbb{G}_{p_{k-1}} \circ \cdots \circ \mathbb{G}_{p_1}$ is strongly rooted. It will now be shown that such an integer exists.

If $\mathbb{G}_{p_1}$ is strongly rooted, set $k = 1$. If $\mathbb{G}_{p_1}$ is not strongly rooted, then let $i > 1$ be the least positive integer not exceeding n^2 for which $\mathbb{G}_{p_{i-1}} \circ \cdots \circ \mathbb{G}_{p_1}$ is not strongly rooted. If $i < n^2$, set $k = i$ in which case $\mathbb{G}_{p_k} \circ \cdots \circ \mathbb{G}_{p_1}$ is strongly rooted. Therefore suppose $i = n^2$ in which case $\mathbb{G}_{p_{j-1}} \circ \cdots \circ \mathbb{G}_{p_1}$ is not strongly rooted for $j \in \{2, 3, \ldots, n^2\}$. Fix $j \in \{2, 3, \ldots, n^2\}$ and let v_j be any root of $\mathbb{G}_{p_j}$. Since $\mathbb{G}_{p_{j-1}} \circ \cdots \circ \mathbb{G}_{p_1}$ is not strongly rooted, $\alpha(\mathbb{G}_{p_{j-1}} \circ \cdots \circ \mathbb{G}_{p_1}, v_j)$ is a strictly proper subset of $\mathcal{V}$. Hence by Lemma 6.5, $\alpha(\mathbb{G}_{p_{j-1}} \circ \cdots \circ \mathbb{G}_{p_1}, v_j)$ is also a strictly proper subset of $\alpha(\mathbb{G}_{p_j} \circ \cdots \circ \mathbb{G}_{p_1}, v_j)$. Thus $\mathcal{A}(\mathbb{G}_{p_{j-1}} \circ \cdots \circ \mathbb{G}_{p_1})$ is a strictly proper subset of $\mathcal{A}(\mathbb{G}_{p_j} \circ \cdots \circ \mathbb{G}_{p_1})$. Since this holds for all $j \in \{2, 3, \ldots, n^2\}$ each containment in the ascending chain

$$\mathcal{A}(\mathbb{G}_{p_1}) \subset \mathcal{A}(\mathbb{G}_{p_2} \circ \mathbb{G}_{p_1}) \subset \cdots \subset \mathcal{A}(\mathbb{G}_{p_{n^2}} \circ \cdots \circ \mathbb{G}_{p_1})$$

is strict. Since $\mathcal{A}(\mathbb{G}_{p_1})$ must contain at least one arc, and there are at most n^2 arcs in any graph in $\mathcal{G}$, $\mathbb{G}_{p_k} \circ \mathbb{G}_{p_{k-1}} \circ \cdots \circ \mathbb{G}_{p_1}$ must be strongly rooted if $k = n^2$.

Now suppose that $m \geq n-1$ and $\mathbb{G}_{p_1}, \mathbb{G}_{p_2}, \ldots, \mathbb{G}_{p_m}$ are all rooted at v. In view of (6.26), $\mathcal{A}(\mathbb{G}_{p_k} \circ \mathbb{G}_{p_{k-1}} \circ \cdots \circ \mathbb{G}_{p_1}) \subset \mathcal{A}(\mathbb{G}_{p_m} \circ \mathbb{G}_{p_{m-1}} \circ \cdots \circ \mathbb{G}_{p_1})$ for any positive integer $k \leq m$. Thus $\mathbb{G}_{p_m} \circ \mathbb{G}_{p_{m-1}} \circ \cdots \circ \mathbb{G}_{p_1}$ will be strongly rooted at v if there exists an integer $k \leq n-1$ such that

$$\alpha(\mathbb{G}_{p_k} \circ \mathbb{G}_{p_{k-1}} \circ \cdots \circ \mathbb{G}_{p_1}, v) = \mathcal{V} \tag{6.28}$$

It will now be shown that such an integer exists.

If $\alpha(\mathbb{G}_{p_1}, v) = \mathcal{V}$, set $k = 1$ in which case (6.28) clearly holds. If $\alpha(\mathbb{G}_{p_1}, v) \neq \mathcal{V}$, then let $i > 1$ be the least positive integer not exceeding $n-1$ for which $\alpha(\mathbb{G}_{p_{i-1}} \circ \cdots \circ \mathbb{G}_{p_1}, v)$ is a strictly proper subset of $\mathcal{V}$. If $i < n-1$, set $k = i$ in which case (6.28) is clearly true. Therefore suppose $i = n-1$ in which case $\alpha(\mathbb{G}_{p_{j-1}} \circ \cdots \circ \mathbb{G}_{p_1}, v)$ is a strictly proper subset of $\mathcal{V}$ for $j \in \{2, 3, \ldots, n-1\}$. Hence by Lemma 6.5, $\alpha(\mathbb{G}_{p_{j-1}} \circ \cdots \circ \mathbb{G}_{p_1}, v)$ is also a strictly proper subset of $\alpha(\mathbb{G}_{p_j} \circ \cdots \circ \mathbb{G}_{p_1}, v)$ for $j \in \{2, 3, \ldots, n-1\}$. In view of this and (6.26), each containment in the ascending chain

$$\alpha(\mathbb{G}_{p_1}, v) \subset \alpha(\mathbb{G}_{p_2} \circ \mathbb{G}_{p_1}, v) \subset \cdots \subset \alpha(\mathbb{G}_{p_{n-1}} \circ \cdots \circ \mathbb{G}_{p_1}, v)$$

is strict. Since $\alpha(\mathbb{G}_{p_1}, v)$ has at least two vertices in it, and there are n vertices in $\mathcal{V}$, (6.28) must hold with $k = n-1$. $\square$

Proposition 6.3 implies that every sufficiently long composition of graphs from a given subset $\widehat{\mathcal{G}} \subset \mathcal{G}$ will be strongly rooted if each graph in $\widehat{\mathcal{G}}$ is rooted. The converse is also true. To understand why, suppose that it is not in which case there would have to be a graph $\mathbb{G} \in \widehat{\mathcal{G}}$, which is not rooted but for which $\mathbb{G}^m$ is strongly rooted for m sufficiently large where $\mathbb{G}^m$ is the m-fold composition of $\mathbb{G}$ with itself. Thus $\alpha(\mathbb{G}^m, v) = \mathcal{V}$ where v is a root of $\mathbb{G}^m$. But via repeated application of (6.27), $\alpha(\mathbb{G}^m, v) = \alpha^m(\mathbb{G}, v)$ where $\alpha^m(\mathbb{G}, \cdot)$ is the m-fold composition of $\alpha(\mathbb{G}, \cdot)$ with itself. Thus $\alpha^m(\mathbb{G}, v) = \mathcal{V}$. But this can only occur if $\mathbb{G}$ is rooted at v because $\alpha^m(\mathbb{G}, v)$ is the set of vertices reachable from v along paths of length m. Since this is a contradiction, $\mathbb{G}$ cannot be rooted. We summarize.

Proposition 6.5. *Every possible sufficiently long composition of graphs from a given subset $\widehat{\mathcal{G}} \subset \mathcal{G}$ is strongly rooted, if and only if every graph in $\widehat{\mathcal{G}}$ is rooted.*

Sarymsakov Graphs

In the sequel we say that a vertex $v \in \mathcal{V}$ is a *neighbor* of a subset $\mathcal{S} \subset \mathcal{V}$ in a graph $\mathbb{G} \in \bar{\mathcal{G}}$, if v is a neighbor of at least one vertex in $\mathcal{S}$. By a *Sarymsakov Graph* is meant a graph $\mathbb{G} \in \bar{\mathcal{G}}$ with the property that for each pair of non-empty subsets $\mathcal{S}_1$ and $\mathcal{S}_2$ in $\mathcal{V}$ which have no neighbors in common, $\mathcal{S}_1 \cup \mathcal{S}_2$ contains a smaller number of vertices than does the set of neighbors of $\mathcal{S}_1 \cup \mathcal{S}_2$. Such seemingly obscure graphs are so named because they are the graphs of an important class of non-negative matrices studied by Sarymsakov in [29]. In the sequel we will prove that Sarymsakov graphs are in fact rooted graphs. We will also prove that the class of rooted graphs we are primarily interested in, namely those in $\mathcal{G}$, are Sarymsakov graphs.

It is possible to characterize Sarymsakov graph a little more concisely using the following concept. By the *neighbor function* of a graph $\mathbb{G} \in \bar{\mathcal{G}}$, written $\beta(\mathbb{G}, \cdot\,)$, we mean the function $\beta(\mathbb{G}, \cdot\,) : 2^{\mathcal{V}} \rightarrow 2^{\mathcal{V}}$ which assigns to each subset $\mathcal{S} \subset \mathcal{V}$, the subset of vertices in $\mathcal{V}$ which are neighbors of $\mathcal{S}$ in $\mathbb{G}$. Thus in terms of β, a Sarymsakov graph is a graph $\mathbb{G} \in \bar{\mathcal{G}}$ with the property that for each pair of non-empty subsets $\mathcal{S}_1$ and $\mathcal{S}_2$ in $\mathcal{V}$ which have no neighbors in common, $\mathcal{S}_1 \cup \mathcal{S}_2$ contains less vertices than does the set $\beta(\mathbb{G}, \mathcal{S}_1 \cup \mathcal{S}_2)$. Note that if $\mathbb{G} \in \mathcal{G}$, the requirement that $\mathcal{S}_1 \cup \mathcal{S}_2$ contain less vertices than $\beta(\mathbb{G}, \mathcal{S}_1 \cup \mathcal{S}_2)$ simplifies to the equivalent requirement that $\mathcal{S}_1 \cup \mathcal{S}_2$ be a strictly proper subset of $\beta(\mathbb{G}, \mathcal{S}_1 \cup \mathcal{S}_2)$. This is because every vertex in $\mathbb{G}$ is a neighbor of itself if $\mathbb{G} \in \mathcal{G}$.

Proposition 6.6.
1. *Each Sarymsakov graph in $\bar{\mathcal{G}}$ is rooted*
2. *Each rooted graph in $\mathcal{G}$ is a Sarymsakov graph*

It follows that if we restrict attention exclusively to graphs in $\mathcal{G}$, then rooted graphs and Sarymsakov graphs are one and the same.

In the sequel $\beta^m(\mathbb{G}, \cdot)$ denotes the m-fold composition of $\beta(\mathbb{G}, \cdot)$ with itself. The proof of Proposition 6.6 depends on the following ideas.

Lemma 6.6. *Let $\mathbb{G} \in \bar{\mathcal{G}}$ be a Sarymsakov graph. Let $\mathcal{S}$ be a non-empty subset of $\mathcal{V}$ such that $\beta(\mathbb{G}, \mathcal{S}) \subset \mathcal{S}$. Let v be any vertex in $\mathcal{V}$. Then there exists a non-negative integer $m \leq n$ such that $\beta^m(\mathbb{G}, v) \cap \mathcal{S}$ is non-empty.*

Proof: If $v \in \mathcal{S}$, set $m = 0$. Suppose next that $v \notin \mathcal{S}$. Set $\mathcal{T} = \{v\} \cup \beta(\mathbb{G}, v) \cup \cdots \cup \beta^{n-1}(\mathbb{G}, v)$ and note that $\beta^n(\mathbb{G}, v) \subset \mathcal{T}$. Since $\beta(\mathbb{G}, \mathcal{T}) = \beta(\mathbb{G}, v) \cup \beta^2(\mathbb{G}, v) \cup \cdots \cup \beta^n(\mathbb{G}, v)$, it must be true that $\beta(\mathbb{G}, \mathcal{T}) \subset \mathcal{T}$. Therefore

$$\beta(\mathbb{G}, \mathcal{T} \cup \mathcal{S}) \subset \mathcal{T} \cup \mathcal{S}. \tag{6.29}$$

Suppose $\beta(\mathbb{G}, \mathcal{T}) \cap \beta(\mathbb{G}, \mathcal{S})$ is empty. Then because $\mathbb{G}$ is a Sarymsakov graph, $\mathcal{T} \cup \mathcal{S}$ contains fewer vertices than $\beta(\mathbb{G}, \mathcal{T} \cup \mathcal{S})$. This contradicts (6.29) so $\beta(\mathbb{G}, \mathcal{T}) \cap \beta(\mathbb{G}, \mathcal{S})$ is not empty. In view of the fact that $\beta(\mathbb{G}, \mathcal{T}) = \beta(\mathbb{G}, v) \cup \beta^2(\mathbb{G}, v) \cup \cdots \cup \beta^n(\mathbb{G}, v)$ it must therefore be true for some positive integer $m \leq n$, that $\beta^m(\mathbb{G}, v) \cap \mathcal{S}$ is non-empty. $\square$

Lemma 6.7. *Let $\mathbb{G} \in \bar{\mathcal{G}}$ be rooted at r. Each non-empty subset $\mathcal{S} \subset \mathcal{V}$ not containing r is a strictly proper subset of $\mathcal{S} \cup \beta(\mathbb{G}, \mathcal{S})$.*

Proof of Lemma 6.7: Let $\mathcal{S} \subset \mathcal{V}$ be non-empty and not containing r. Pick $v \in \mathcal{S}$. Since $\mathbb{G}$ is rooted at r, there must be a path in $\mathbb{G}$ from r to v. Since $r \notin \mathcal{S}$ there must be a vertex $x \in \mathcal{S}$ which has a neighbor which is not in $\mathcal{S}$. Thus there is a vertex $y \in \beta(\mathbb{G}, \mathcal{S})$ which is not in $\mathcal{S}$. This implies that $\mathcal{S}$ is a strictly proper subset of $\mathcal{S} \cup \beta(\mathbb{G}, \mathcal{S})$. □

By a *maximal rooted subgraph of* $\mathbb{G}$ we mean a subgraph $\mathbb{G}^*$ of $\mathbb{G}$ which is rooted and which is not contained in any rooted subgraph of $\mathbb{G}$ other than itself. Graphs in $\bar{\mathcal{G}}$ may have one or more maximal rooted subgraphs. Clearly $\mathbb{G}^* = \mathbb{G}$ just in case $\mathbb{G}$ is rooted. Note that if $\widehat{\mathcal{R}}$ is the set of all roots of a maximal rooted subgraph $\widehat{\mathbb{G}}$, then $\beta(\mathbb{G}, \widehat{\mathcal{R}}) \subset \widehat{\mathcal{R}}$. For if this were not so, then it would be possible to find a vertex $x \in \beta(\mathbb{G}, \widehat{\mathcal{R}})$ which is not in $\widehat{\mathcal{R}}$. This would imply the existence of a path from x to some root $\widehat{v} \in \widehat{\mathcal{R}}$; consequently the graph induced by the set of vertices along this path together with $\widehat{\mathcal{R}}$ would be rooted at $x \notin \widehat{\mathcal{R}}$ and would contain $\widehat{\mathbb{G}}$ as a strictly proper subgraph. But this contradicts the hypothesis that $\widehat{\mathbb{G}}$ is maximal. Therefore $\beta(\mathbb{G}, \widehat{\mathcal{R}}) \subset \widehat{\mathcal{R}}$. Now suppose that $\widehat{\mathbb{G}}$ is any rooted subgraph in $\mathcal{G}$. Suppose that $\widehat{\mathbb{G}}$'s set of roots $\widehat{\mathcal{R}}$ satisfies $\beta(\mathbb{G}, \widehat{\mathcal{R}}) \subset \widehat{\mathcal{R}}$. We claim that $\widehat{\mathbb{G}}$ must then be maximal. For if this were not so, there would have to be a rooted graph $\mathbb{G}^*$ containing $\widehat{\mathbb{G}}$ as a strictly proper subset. This in turn would imply the existence of a path from a root x^* of $\mathbb{G}^*$ to a root v of $\widehat{\mathbb{G}}$; consequently $x^* \in \beta^i(\mathbb{G}, \widehat{\mathcal{R}})$ for some $i \geq 1$. But this is impossible because $\widehat{\mathcal{R}}$ is $\beta(\mathbb{G}, \cdot\,)$ invariant. Thus $\widehat{\mathbb{G}}$ is maximal. We summarize.

Lemma 6.8. *A rooted subgraph of a graph $\mathbb{G}$ generated by any vertex $v \in \mathcal{V}$ is maximal if and only if its set of roots is $\beta(\mathbb{G}, \cdot\,)$ invariant.*

Proof of Proposition 6.6: Write $\beta(\cdot)$ for $\beta(\mathbb{G}, \cdot)$. To prove the assertion 1, pick $\mathbb{G} \in \bar{\mathcal{G}}$. Let $\mathbb{G}^*$ be any maximal rooted subgraph of $\mathbb{G}$ and write $\mathcal{R}$ for its root set; in view of Lemma 6.8, $\beta(\mathcal{R}) \subset \mathcal{R}$. Pick any $v \in \mathcal{V}$. Then by Lemma 6.6, for some positive integer $m \leq n$, $\beta^m(v) \cap \mathcal{R}$ is non-empty. Pick $z \in \beta^m(v) \cap \mathcal{R}$. Then there is a path from z to v and z is a root of $\mathbb{G}^*$. But $\mathbb{G}^*$ is maximal so v must be a vertex of $\mathbb{G}^*$. Therefore every vertex of $\mathbb{G}$ is a vertex of $\mathbb{G}^*$ which implies that $\mathbb{G}$ is rooted.

To prove assertion 2, let $\mathbb{G} \in \mathcal{G}$ be rooted at r. Pick any two non-empty subsets $\mathcal{S}_1, \mathcal{S}_2$ of $\mathcal{V}$ which have no neighbors in common. If $r \notin \mathcal{S}_1 \cup \mathcal{S}_2$, then $\mathcal{S}_1 \cup \mathcal{S}_2$ must be a strictly proper subset of $\mathcal{S}_1 \cup \mathcal{S}_2 \cup \beta(\mathcal{S}_1 \cup \mathcal{S}_2)$ because of Lemma 6.7.

Suppose next that $r \in \mathcal{S}_1 \cup \mathcal{S}_2$. Since $\mathbb{G} \in \mathcal{G}$, $\mathcal{S}_i \subset \beta(\mathcal{S}_i)$, $i \in \{1, 2\}$. Thus $\mathcal{S}_1$ and $\mathcal{S}_2$ must be disjoint because $\beta(\mathcal{S}_1)$ and $\beta(\mathcal{S}_2)$ are. Therefore r must be in either $\mathcal{S}_1$ or $\mathcal{S}_2$ but not both. Suppose that $r \notin \mathcal{S}_1$. Then $\mathcal{S}_1$ must be a strictly proper subset of $\beta(\mathcal{S}_1)$ because of Lemma 6.7. Since $\beta(\mathcal{S}_1)$ and $\beta(\mathcal{S}_2)$ are disjoint, $\mathcal{S}_1 \cup \mathcal{S}_2$ must be a strictly proper subset of $\beta(\mathcal{S}_1 \cup \mathcal{S}_2)$. By the same reasoning, $\mathcal{S}_1 \cup \mathcal{S}_2$ must be a strictly proper subset of $\beta(\mathcal{S}_1 \cup \mathcal{S}_2)$ if $r \notin \mathcal{S}_2$. Thus in conclusion $\mathcal{S}_1 \cup \mathcal{S}_2$ must be a strictly proper subset of $\beta(\mathcal{S}_1 \cup \mathcal{S}_2)$ whether r is in $\mathcal{S}_1 \cup \mathcal{S}_2$ or not. Since this conclusion holds for all such $\mathcal{S}_1$ and $\mathcal{S}_2$ and $\mathbb{G} \in \mathcal{G}$, $\mathbb{G}$ must be a Sarymsakov graph. $\square$

Neighbor-Shared Graphs

There is a different assumption which one can make about a sequence of graphs from $\bar{\mathcal{G}}$ which also insures that the sequence's composition is strongly rooted. For this we need the concept of a "neighbor-shared graph." Let us call $\mathbb{G} \in \bar{\mathcal{G}}$ *neighbor shared* if each set of two distinct vertices share a common neighbor. Suppose that $\mathbb{G}$ is neighbor shared. Then each pair of vertices is clearly reachable from a single vertex. Similarly each three vertices are reachable from paths starting at one of two vertices. Continuing this reasoning it is clear that each of the graph's n vertices is reachable from paths starting at vertices in some set $\mathcal{V}_{n-1}$ of $n-1$ vertices. By the same reasoning, each vertex in $\mathcal{V}_{n-1}$ is reachable from paths starting at vertices in some set $\mathcal{V}_{n-2}$ of $n-2$ vertices. Thus each of the graph's n vertices is reachable from paths starting at vertices in a set of $n-2$ vertices, namely the set $\mathcal{V}_{n-2}$. Continuing this argument we eventually arrive at the conclusion that each of the graph's n vertices is reachable from paths starting at a single vertex, namely the one vertex in the set $\mathcal{V}_1$. We have proved the following.

Lemma 6.9. *Each neighbor-shared graph in $\bar{\mathcal{G}}$ is rooted.*

It is worth noting that although shared neighbor graphs are rooted, the converse is not necessarily true. The reader may wish to construct a three vertex example which illustrates this. Although rooted graphs in $\mathcal{G}$ need not be neighbor shared, it turns out that the composition of any $n-1$ rooted graphs in $\mathcal{G}$ is.

Proposition 6.7. *The composition of any set of $m \geq n-1$ rooted graphs in $\mathcal{G}$ is neighbor shared.*

To prove Proposition 6.7 we need some more ideas. By the *reverse graph* of $\mathbb{G} \in \bar{\mathcal{G}}$, written $\mathbb{G}'$ is meant the graph in $\bar{\mathcal{G}}$ which results when the directions of all arcs in $\mathbb{G}$ are reversed. It is clear that $\mathcal{G}$ is closed under the reverse operation and that if A is the adjacency matrix of $\mathbb{G}$, then A' is the adjacency matrix of $\mathbb{G}'$. It is also clear that $(\mathbb{G}_p \circ \mathbb{G}_q)' = \mathbb{G}'_q \circ \mathbb{G}'_p$, $p, q \in \bar{\mathcal{P}}$, and that

$$\alpha(\mathbb{G}', \mathcal{S}) = \beta(\mathbb{G}, \mathcal{S}), \quad \mathcal{S} \in 2^{\mathcal{V}} \tag{6.30}$$

Lemma 6.10. *For all $\mathbb{G}_p, \mathbb{G}_q \in \bar{\mathcal{G}}$ and any non-empty subset $\mathcal{S} \subset \mathcal{V}$,*

$$\beta(\mathbb{G}_q, \beta(\mathbb{G}_p, \mathcal{S})) = \beta(\mathbb{G}_p \circ \mathbb{G}_q, \mathcal{S}) \tag{6.31}$$

Proof of Lemma 6.10: In view of (6.27), $\alpha(\mathbb{G}'_p, \alpha(\mathbb{G}'_q, \mathcal{S})) = \alpha(\mathbb{G}'_p \circ \mathbb{G}'_q, \mathcal{S})$. But $\mathbb{G}'_p \circ \mathbb{G}'_q = (\mathbb{G}_q \circ \mathbb{G}_p)'$ so $\alpha(\mathbb{G}'_p, \alpha(\mathbb{G}'_q, \mathcal{S})) = \alpha((\mathbb{G}_q \circ \mathbb{G}_p)', \mathcal{S})$. Therefore $\beta(\mathbb{G}_p, \beta(\mathbb{G}_q, \mathcal{S})) = \beta(\mathbb{G}_q \circ \mathbb{G}_p), \mathcal{S})$ because of (6.30). □

Lemma 6.11. *Let $\mathbb{G}_1$ and $\mathbb{G}_2$ be rooted graphs in $\mathcal{G}$. If u and v are distinct vertices in $\mathcal{V}$ for which*

$$\beta(\mathbb{G}_2, \{u, v\}) = \beta(\mathbb{G}_2 \circ \mathbb{G}_1, \{u, v\}) \tag{6.32}$$

then u and v have a common neighbor in $\mathbb{G}_2 \circ \mathbb{G}_1$

Proof: $\beta(\mathbb{G}_2, u)$ and $\beta(\mathbb{G}_2, v)$ are non-empty because u and v are neighbors of themselves. Suppose u and v do not have a common neighbor in $\mathbb{G}_2 \circ \mathbb{G}_1$. Then $\beta(\mathbb{G}_2 \circ \mathbb{G}_1, u)$ and $\beta(\mathbb{G}_2 \circ \mathbb{G}_1, v)$ are disjoint. But $\beta(\mathbb{G}_2 \circ \mathbb{G}_1, u) = \beta(\mathbb{G}_1, \beta(\mathbb{G}_2, u))$ and $\beta(\mathbb{G}_2 \circ \mathbb{G}_1, v) = \beta(\mathbb{G}_1, \beta(\mathbb{G}_2, v))$ because of (6.31). Therefore $\beta(\mathbb{G}_1, \beta(\mathbb{G}_2, u))$ and $\beta(\mathbb{G}_1, \beta(\mathbb{G}_2, v))$ are disjoint. But $\mathbb{G}_1$ is rooted and thus a Sarymsakov graph because of Proposition 6.6. Thus $\beta(\mathbb{G}_2, \{u, v\})$ is a strictly proper subset of $\beta(\mathbb{G}_2, \{u, v\}) \cup \beta(\mathbb{G}_1, \beta(\mathbb{G}_2, \{u, v\})$. But $\beta(\mathbb{G}_2, \{u, v\}) \subset \beta(\mathbb{G}_1, \beta(\mathbb{G}_2, \{u, v\})$ because all vertices in $\mathbb{G}_2$ are neighbors of themselves and $\beta(\mathbb{G}_1, \beta(\mathbb{G}_2, \{u, v\}) = \beta(\mathbb{G}_2 \circ \mathbb{G}_1, \{u, v\})$ because of (6.31). Therefore $\beta(\mathbb{G}_2, \{u, v\})$ is a strictly proper subset of $\beta(\mathbb{G}_2 \circ \mathbb{G}_1, \{u, v\})$. This contradicts (6.32) so u and v have a common neighbor in $\mathbb{G}_2 \circ \mathbb{G}_1$. □

Proof of Proposition 6.7: Let u and v be distinct vertices in $\mathcal{V}$. Let $\mathbb{G}_1, \mathbb{G}_2, \ldots, \mathbb{G}_{n-1}$ be a sequence of rooted graphs in $\mathcal{G}$. Since $\mathcal{A}(\mathbb{G}_{n-1} \circ \cdots \mathbb{G}_{n-i}) \subset \mathcal{A}(\mathbb{G}_{n-1} \circ \cdots \mathbb{G}_{n-(i+1)})$ for $i \in \{1, 2, \ldots, n-2\}$, it must be true that the $\mathbb{G}_i$ yield the ascending chain

$$\beta(\mathbb{G}_{n-1}, \{u, v\}) \subset \beta(\mathbb{G}_{n-1} \circ \mathbb{G}_{n-2}, \{u, v\}) \subset \cdots \subset \beta(\mathbb{G}_{n-1} \circ \cdots \circ \mathbb{G}_2 \circ \mathbb{G}_1, \{u, v\})$$

Because there are n vertices in $\mathcal{V}$, this chain must converge for some $i < n-1$ which means that

$$\beta(\mathbb{G}_{n-1} \circ \cdots \circ \mathbb{G}_{n-i}, \{u, v\}) = \beta(\mathbb{G}_{n-1} \circ \cdots \circ \mathbb{G}_{n-i} \circ \mathbb{G}_{n-(i+1)}, \{u, v\})$$

This and Lemma 6.11 imply that u and v have a common neighbor in $\mathbb{G}_{n-1} \circ \cdots \circ \mathbb{G}_{n-i}$ and thus in $\mathbb{G}_{n-1} \circ \cdots \circ \mathbb{G}_2 \circ \mathbb{G}_1$. Since this is true for all distinct u and v, $\mathbb{G}_{n-1} \circ \cdots \circ \mathbb{G}_2 \circ \mathbb{G}_1$ is a neighbor shared graph. □

If we restrict attention to those rooted graphs in $\mathcal{G}$ which are strongly connected, we can obtain a neighbor-shared graph by composing a smaller number of rooted graphs that claimed in Proposition 6.7.

Proposition 6.8. *Let k be the integer quotient of n divided by 2. The composition of any set of $m \geq k$ strongly connected graphs in $\mathcal{G}$ is neighbor shared.*

Proof of Proposition 6.8: Let $k < n$ be a positive integer and let v be any vertex in $\mathcal{V}$. Let $\mathbb{G}_1, \mathbb{G}_2, \ldots, \mathbb{G}_k$ be a sequence of strongly connected graphs in $\mathcal{G}$. Since $k < n$ and $\mathcal{A}(\mathbb{G}_k \circ \cdots \mathbb{G}_{k-i}) \subset \mathcal{A}(\mathbb{G}_{k-1} \circ \cdots \mathbb{G}_{k-(i+1)})$ for $i \in \{1, 2, \ldots, k-1\}$, it must be true that the $\mathbb{G}_i$ yield the ascending chain

$$\{v\} \subset \beta(\mathbb{G}_k, \{v\}) \subset \beta(\mathbb{G}_k \circ \mathbb{G}_{k-1}, \{v\}) \subset \cdots \subset \beta(\mathbb{G}_k \circ \cdots \circ \mathbb{G}_2 \circ \mathbb{G}_1, \{v\})$$

Moreover, since any strongly connected graph is one in which every vertex is a root, it must also be true that the subsequence

$$\{v\} \subset \beta(\mathbb{G}_k, \{v\}) \subset \beta(\mathbb{G}_k \circ \mathbb{G}_{k-1}, \{v\}) \subset \cdots \subset \beta(\mathbb{G}_k \circ \cdots \circ \mathbb{G}_{(i+1)} \circ \mathbb{G}_i, \{v\})$$

is strictly increasing where either $i = 1$ or $i > 1$ and $\beta(\mathbb{G}_k \circ \cdots \circ \mathbb{G}_{(i+1)} \circ \mathbb{G}_i, \{v\}) = \mathcal{V}$. In either case this implies that $\beta(\mathbb{G}_k \circ \cdots \circ \mathbb{G}_2 \circ \mathbb{G}_1, \{v\})$ contains at least $k+1$ vertices. Fix k as the integer quotient of $n+1$ divided by 2 in which case $2k \geq n-1$. Let v_1 and v_2 be any pair of distinct vertices in $\mathcal{V}$. Then there must be at least $k+1$ vertices in $\beta(\mathbb{G}_k \circ \cdots \circ \mathbb{G}_2 \circ \mathbb{G}_1, \{v_1\})$ and $k+1$ vertices in $\beta(\mathbb{G}_k \circ \cdots \circ \mathbb{G}_2 \circ \mathbb{G}_1, \{v_2\})$. But $2(k+1) > n$ so $\beta(\mathbb{G}_k \circ \cdots \circ \mathbb{G}_2 \circ \mathbb{G}_1, \{v_1\})$ and $\beta(\mathbb{G}_k \circ \cdots \circ \mathbb{G}_2 \circ \mathbb{G}_1, \{v_2\})$ must have at least one vertex in common. Since this is true for each pair of distinct vertices $v_1, v_2 \in \mathcal{V}$,

$\mathbb{G}_k \circ \cdots \circ \mathbb{G}_2 \circ \mathbb{G}_1$ must be neighbor-shared. □

Lemma 6.9 and Proposition 6.3 imply that any composition of n^2 neighbor-shared graphs in $\mathcal{G}$ is strongly rooted. The following proposition asserts that the composition need only consist of $n-1$ neighbor-shared graphs and moreover that the graphs need only be in $\bar{\mathcal{G}}$ and not necessarily in $\mathcal{G}$.

Proposition 6.9. *The composition of any set of $m \geq n-1$ neighbor-shared graphs in $\bar{\mathcal{G}}$ is strongly rooted.*

To prove this proposition we need a few more ideas. For any integer $1 < k \leq n$, we say that a graph $\mathbb{G} \in \bar{\mathcal{G}}$ is *k-neighbor shared* if each set of k distinct vertices share a common neighbor. Thus a neighbor-shared graph and a two neighbor shared graph are one and the same. Clearly a n neighbor shared graph is strongly rooted at the common neighbor of all n vertices.

Lemma 6.12. *If $\mathbb{G}_p \in \bar{\mathcal{G}}$ is a neighbor-shared graph and $\mathbb{G}_q \in \bar{\mathcal{G}}$ is a k neighbor shared graph with $k < n$, then $\mathbb{G}_q \circ \mathbb{G}_p$ is a $(k+1)$ neighbor shared graph.*

Proof: Let $v_1, v_2, \ldots, v_{k+1}$ be any distinct vertices in $\mathcal{V}$. Since $\mathbb{G}_q$ is a k neighbor shared graph, the vertices $v_1, v_2, \ldots, v_k$ share a common neighbor u_1 in $\mathbb{G}_q$ and the vertices $v_2, v_2, \ldots, v_{k+1}$ share a common neighbor u_2 in $\mathbb{G}_q$ as well. Moreover, since $\mathbb{G}_p$ is a neighbor shared graph, u_1 and u_2 share a common neighbor w in $\mathbb{G}_p$. It follows from the definition of composition that $v_1, v_2, \ldots, v_k$ have w as a neighbor in $\mathbb{G}_q \circ \mathbb{G}_p$ as do $v_2, v_3, \ldots, v_{k+1}$. Therefore $v_1, v_2, \ldots, v_{k+1}$ have w as a neighbor in $\mathbb{G}_q \circ \mathbb{G}_p$. Since this must be true for any set of $k+1$ vertices in $\mathbb{G}_q \circ \mathbb{G}_p$, $\mathbb{G}_q \circ \mathbb{G}_p$ must be a $k+1$ neighbor shared graph as claimed. □

Proof of Proposition 6.9: The preceding lemma implies that the composition of any two neighbor shared graphs is three neighbor shared. From this and induction it follows that for $m < n$, the composition of m neighbor shared graphs is $m+1$ neighbor shared. Thus the composition of $n-1$ neighbor shared graphs is n neighbor shared and consequently strongly rooted. □

In view of Proposition 6.9, we have the following slight improvement on Proposition 6.3.

Proposition 6.10. *The composition of any set of $m \geq (n-1)^2$ rooted graphs in $\mathcal{G}$ is strongly rooted.*

Convergence

We are now in a position to significantly relax the conditions under which the conclusion of Theorem 6.1 holds.

Theorem 6.2. *Let $\mathcal{Q}$ denote the subset of $\mathcal{P}$ consisting of those indices q for which $\mathbb{G}_q \in \mathcal{G}$ is rooted. Let $\theta(0)$ be fixed and let $\sigma : \{0, 1, 2, \ldots\} \to \mathcal{P}$ be a switching signal satisfying $\sigma(t) \in \mathcal{Q}$, $t \in \{0, 1, \ldots\}$. Then there is a constant steady state heading θ_{ss} depending only on $\theta(0)$ and σ for which*

$$\lim_{t\to\infty} \theta(t) = \theta_{ss}\mathbf{1} \tag{6.33}$$

where the limit is approached exponentially fast.

The theorem says that a unique heading is achieved asymptotically along any trajectory on which all neighbor graphs are rooted. The proof of Theorem 6.2 relies on the following generalization of Proposition 6.2.

Proposition 6.11. *Let $\mathcal{S}_r$ be any closed set of stochastic matrices in $\mathcal{S}$ whose graphs are all rooted. Then any product $S_j \ldots S_1$ of matrices from $\mathcal{S}_{sr}$ converges to $\mathbf{1}\lfloor \cdots S_j \cdots S_1 \rfloor$ exponentially fast as $j \to \infty$ at a rate no slower than λ, where λ is a non-negative constant depending on $\mathcal{S}_r$ and satisfying $\lambda < 1$.*

Proof of Proposition 6.11: Set $m = n^2$ and write $\mathcal{S}_r^m$ for the closed set of all products of stochastic matrices of the form $S_m S_{m-1} \cdots S_1$ where each $S_i \in \mathcal{S}_r$. By assumption, $\gamma(S)$ is rooted for $S \in \mathcal{S}$. In view of Proposition 6.3, $\gamma(S_m) \circ \cdots \gamma(S_1)$ is strongly rooted for every list of m matrices $\{S_1, S_2, \ldots, S_m\}$ from $\mathcal{S}_r$. But $\gamma(S_m) \circ \cdots \circ \gamma(S_1) = \gamma(S_m \cdots S_1)$ because of Lemma 6.3. Therefore $\gamma(S_m \cdots S_1)$ is strongly rooted for all products $S_m \cdots S_1 \in \mathcal{S}_r^m$.

Now any product $S_j \cdots S_1$ of matrices in $\mathcal{S}_r$ can be written as

$$S_j \cdots S_1 = \bar{S}(j)\bar{S}_k \cdots \bar{S}_1$$

where

$$\bar{S}_i = S_{im} \cdots S_{(i-1)m+1}, \quad 1 \leq i \leq k$$

is a product in $\mathcal{S}_r^m$,

$$\bar{S}(j) = S_j \cdots S_{(km+1)},$$

and k is the integer quotient of j divided by m. In view of Proposition 6.2, $\bar{S}_k \cdots \bar{S}_1$ must converge to $\mathbf{1}\lfloor \cdots \bar{S}_k \cdots \bar{S}_1 \rfloor$ exponentially fast as $k \to \infty$ at a rate no slower than $\bar{\lambda}$, where

$$\bar{\lambda} = \max_{\bar{S} \in \mathcal{S}_r^m} |||\lfloor \bar{S} \rfloor|||$$

But $\bar{S}(j)$ is a product of at most m stochastic matrices, so it is a bounded function of j. It follows that the product $S_j S_{j-1} \cdots S_1$ must converge to $\mathbf{1}\lfloor \cdots S_j \cdot S_1 \rfloor$ exponentially fast at a rate no slower than $\lambda = \bar{\lambda}^{\frac{1}{m}}$. □

The proof of Proposition 6.11 can be applied to any closed subset $\mathcal{S}_{ns} \subset \mathcal{S}$ of stochastic matrices with neighbor shared graphs. In this case, one would define $m = n - 1$ because of Proposition 6.9.

Proof of Theorem 6.2: By definition, the graph $\mathbb{G}_p$ of each matrix F_p in the finite set $\{F_p : p \in \mathcal{Q}\}$ is rooted. By assumption, $F_{\sigma(t)} \in \{F_p : p \in \mathcal{Q}\}$, $t \geq 0$. In view of Proposition 6.11, the product $F_{\sigma(t)} \cdots F_{\sigma(0)}$ converges to $\mathbf{1}\lfloor \cdots F_{\sigma(t)} \cdots F_{\sigma(0)} \rfloor$ exponentially fast at a rate no slower than

$$\lambda = \{ \max_{F \in \mathcal{F}_r^m} |||\lfloor F \rfloor||| \}^{\frac{1}{m}}$$

where $m = n^2$ and $\mathcal{F}_r^m$ is the finite set of all m-term flocking matrix products of the form $F_{p_m} \cdots F_{p_1}$, $p_i \in \mathcal{Q}$. But it is clear from (6.3) that

$$\theta(t) = F_{\sigma(t-1)} \cdots F_{\sigma(1)} F_{\sigma(0)} \theta(0), \quad t \geq 1$$

Therefore (6.33) holds with $\theta_{ss} = \lfloor \cdots F_{\sigma(t)} \cdots F_{\sigma(0)} \rfloor \theta(0)$ and the convergence is exponential. □

The proof of Theorem 6.2 also applies to the case when all of the $\mathbb{G}_{\sigma(t)}$, $t \geq 0$ are neighbor shared. In this case, one would define $m = n - 1$ because of Proposition 6.9.

Convergence Rates

It is possible to deduce an explicit convergence rate for the situation addressed in Theorem 6.2 [42]. To do this we need a few more ideas.

Scrambling Constants

Let S be an $n \times n$ stochastic matrix. Observe that for any non-negative n-vector x, the ith minus the jth entries of Sx can be written as

$$\sum_{k=1}^{n} (s_{ik} - s_{jk}) x_k = \sum_{k \in \mathcal{K}} (s_{ik} - s_{jk}) x_k + \sum_{k \in \bar{\mathcal{K}}} (s_{ik} - s_{jk}) x_k$$

where

$$\mathcal{K} = \{k : s_{ik} - s_{jk} \geq 0,\ k \in \{1, 2, \ldots, n\}\} \quad \text{and}$$
$$\bar{\mathcal{K}} = \{k : s_{ik} - s_{jk} < 0,\ k \in \{1, 2, \ldots, n\}\}$$

Therefore

$$\sum_{k=1}^{n}(s_{ik}-s_{jk})x_k \leq \left(\sum_{k\in\mathcal{K}}(s_{ik}-s_{jk})\right)\lceil x\rceil + \left(\sum_{k\in\bar{\mathcal{K}}}(s_{ik}-s_{jk})\right)\lfloor x\rfloor$$

But

$$\sum_{k\in\mathcal{K}\cup\bar{\mathcal{K}}}(s_{ik}-s_{jk}) = 0$$

so

$$\sum_{k\in\bar{\mathcal{K}}}(s_{ik}-s_{jk}) = -\sum_{k\in\mathcal{K}}(s_{ik}-s_{jk})$$

Thus

$$\sum_{k=1}^{n}(s_{ik}-s_{jk})x_k \leq \left(\sum_{k\in\mathcal{K}}(s_{ik}-s_{jk})\right)(\lceil x\rceil - \lfloor x\rfloor)$$

Now

$$\sum_{k\in\mathcal{K}}(s_{ik}-s_{jk}) = 1 - \sum_{k\in\bar{\mathcal{K}}} s_{ik} - \sum_{k\in\mathcal{K}} s_{jk}$$

because the row sums of S are all one. Moreover

$$\begin{aligned} s_{ik} &= \min\{s_{ik}, s_{jk}\}, \quad \in \bar{\mathcal{K}} \\ s_{jk} &= \min\{s_{ik}, s_{jk}\}, \quad \in \mathcal{K} \end{aligned}$$

so

$$\sum_{k\in\mathcal{K}}(s_{ik}-s_{jk}) = 1 - \sum_{k=1}^{n}\min\{s_{ik}, s_{jk}\}$$

It follows that

$$\sum_{k=1}^{n}(s_{ik}-s_{jk})x_k \leq \left(1 - \sum_{k=1}^{n}\min\{s_{ik}, s_{jk}\}\right)(\lceil x\rceil - \lfloor x\rfloor)$$

Hence if we define

$$\mu(S) = \max_{i,j}\left(1 - \sum_{k=1}^{n}\min\{s_{ik}, s_{jk}\}\right) \tag{6.34}$$

then

$$\sum_{k=1}^{n}(s_{ik}-s_{jk})x_k \leq \mu(S)(\lceil x\rceil - \lfloor x\rfloor)$$

Since this holds for all i, j, it must hold for that i and j for which

$$\sum_{k=1}^{n} s_{ik}x_k = \lceil Sx\rceil \qquad \text{and} \qquad \sum_{k=1}^{n} s_{jk}x_k = \lfloor Sx\rfloor$$

Therefore

$$\lceil Sx \rceil - \lfloor Sx \rfloor \leq \mu(S)(\lceil x \rceil - \lfloor x \rfloor) \tag{6.35}$$

Now let S_1 and S_2 be any two $n \times n$ stochastic matrices and let e_i be the ith unit n-vector. Then from (6.35),

$$\lceil S_2 S_1 e_i \rceil - \lfloor S_2 S_1 e_i \rfloor \leq \mu(S_2)(\lceil S_1 e_i \rceil - \lfloor S_1 e_1 \rfloor) \tag{6.36}$$

Meanwhile, from (6.11),

$$[\![S_2 S_1]\!] e_i = \mathbf{1}(\lceil S_2 S_1 \rceil - \lfloor S_2 S_1 \rfloor) e_i$$

and

$$[\![S_1]\!] e_i = \mathbf{1}(\lceil S_1 \rceil - \lfloor S_1 \rfloor) e_i$$

But for any non-negative matrix M, $\lceil M \rceil e_i = \lceil M e_i \rceil$ and $\lfloor M \rfloor e_i = \lfloor M e_i \rfloor$ so

$$[\![S_2 S_1]\!] e_i = \mathbf{1}(\lceil S_2 S_1 e_i \rceil - \lfloor S_2 S_1 e_i \rfloor)$$

and

$$[\![S_1]\!] e_i = \mathbf{1}(\lceil S_1 e_i \rceil - \lfloor S_1 e_i \rfloor)$$

From these expressions and (6.36) it follows that

$$[\![S_2 S_1]\!] e_i \leq \mu(s_2) [\![S_1]\!] e_i$$

Since this is true for all i, we arrive at the following fact.

Lemma 6.13. *For any two stochastic matrices in $\mathcal{S}$,*

$$[\![S_2 S_1]\!] \leq \mu(S_2) [\![S_1]\!] \tag{6.37}$$

where for any $n \times n$ stochastic matrix S,

$$\mu(S) = \max_{i,j} \left(1 - \sum_{k=1}^{n} \min\{s_{ik}, s_{jk}\} \right) \tag{6.38}$$

The quantity $\mu(S)$ has been widely studied before [29] and is know as the *scrambling constant* of the stochastic matrix S. Note that since the row sums of S all equal 1, $\mu(S)$ is non-negative. It is easy to see that $\mu(S) = 0$ just in case all the rows of S are equal. Let us note that for fixed i and j, the kth term in the sum appearing in (6.38) will be positive just in case both s_{ik} and s_{jk} are positive. It follows that the sum will be positive if and only if for at least one k, s_{ik} and s_{jk} are both positive. Thus $\mu(S) < 1$ if and only if for each distinct i and j, there is at least one k for which s_{ik} and s_{jk} are both positive. Matrices with this property have been widely studied and are called *scrambling matrices*. Thus a stochastic matrix S is a scrambling matrix if and only if $\mu(S) < 1$. It is easy to see that the definition of a scrambling matrix also implies that S is scrambling if and only if its graph $\gamma(S)$ is neighbor-shared.

The statement of Proposition 6.11 applies to the situation when instead of $\mathcal{S}_r$, one considers a closed subset $\mathcal{S}_{ns} \subset \mathcal{S}_r$ of stochastic matrices with neighbor shared graphs. In this case, the proof of Proposition 6.11 gives a worst case convergence rate bound of

$$\lambda = \left\{ \max_{\bar{S} \in \mathcal{S}_{ns}^m} |||\lfloor \bar{S} \rfloor||| \right\}^{\frac{1}{n-1}}$$

where $m = n - 1$ and $\mathcal{S}_{ns}^m$ is the set of m term matrix products of the form $S_m \cdots S_1, \quad S_i \in \mathcal{S}_{ns}$. Armed with Lemma 6.13, one can do better.

Let $\mathcal{S}_{ns} \subset \mathcal{S}$ be a closed subset consisting of matrices whose graphs are all neighbor shared. Then the scrambling constant $\mu(S)$ defined in (6.38) satisfies $\mu(S) < 1, \ S \in \mathcal{S}_{ns}$ because each such S is a scrambling matrix. Let

$$\lambda = \max_{S \in \mathcal{S}_{ns}} \mu(S)$$

The $\lambda < 1$ because $\mathcal{S}_{ns}$ is closed and bounded and because $\mu(\cdot)$ is continuous. In view of Lemma 6.13,

$$||[S_2 S_1]|| \leq \lambda ||[S_1]||, \quad S_1, S_2 \in \mathcal{S}_{ns}$$

Hence by induction, for any sequence of matrices $S_1, S_2, \ldots$ in $\mathcal{S}_{ns}$

$$||[S_j \cdots S_1]|| \leq \lambda^{j-1} ||[S_1]||, \quad S_i \in \mathcal{S}_{ns}$$

But from (6.10), $\lfloor S \rfloor \leq [S], \ S \in \mathcal{S}$, so $|||\lfloor S \rfloor||| \leq ||[S]||, \ S \in \mathcal{S}$ Therefore for any sequence of stochastic matrices $S_1, S_2, \ldots$ with neighbor shared graphs

$$|||\lfloor S_j \cdots S_1 \rfloor||| \leq \lambda^{j-1} ||[S_1]|| \tag{6.39}$$

Therefore from Proposition 6.1, any such product $S_j \cdots S_1$ converges exponentially at a rate no slower than λ as $j \to \infty$.

Note that because of (6.22) and (6.10), the inequality in (6.39) applies to any sequence of stochastic matrices $S_1, S_2, \ldots$ for which $|||\lfloor S_i \rfloor||| \leq \lambda$. Thus for example (6.39) applies to any sequence of stochastic matrices $S_1, S_2, \ldots$ whose graphs are strongly rooted provided the graphs in the sequence come from a compact set; in this case λ would be the maximum value of $|||\lfloor S \rfloor|||$ as S ranges over the set. Of course any such sequence is far more special then a sequence of stochastic matrices with neighbor-shared graphs since every strongly rooted graph is neighbor shared but the converse is generally not true.

Convergence Rates for Neighbor-Shared Graphs

Suppose that F_p is a flocking matrix for which $\mathbb{G}_p$ is neighbor shared. In view of the definition of a flocking matrix, any non-zero entry in F_p must be bounded below by $\frac{1}{n}$. Fix distinct i and j and suppose that k is a neighbor that i and j share. Then f_{ik} and f_{jk} are both non-zero so $\min\{f_{ik}, f_{jk}\} \geq \frac{1}{n}$. This implies that the sum in (6.38) must be bounded below by $\frac{1}{n}$ and consequently that $\mu(F_p) \leq 1 - \frac{1}{n}$.

Now let F_p be that flocking matrix whose graph $\mathbb{G}_p \in \mathcal{G}$ is such that vertex 1 has no neighbors other than itself, vertex 2 has every vertex as a neighbor, and vertices 3 through n have only themselves and agent 1 as neighbors. Since vertex 1 has no neighbors other than itself, $f_{i,k} = 0$ for all i and for $k > 1$. Thus for all i, j, it must be true that $\sum_{k=1}^{n} \min\{f_{ik}, f_{jk}\} = \min\{f_{i1}, f_{j1}\}$. Now vertex 2 has n neighbors, so $f_{2,1} = \frac{1}{n}$. Thus $\min\{f_{i1}, f_{j1}\}$ attains its lower bound of $\frac{1}{n}$ when either $i = 2$ or $j = 2$. It thus follows that with this F_p, $\mu(F_p)$ attains its upper bound of $1 - \frac{1}{n}$. We summarize.

Lemma 6.14. *Let $\mathcal{Q}$ be the set of indices in $\mathcal{P}$ for which $\mathbb{G}_p$ is neighbor shared. Then*

$$\max_{q \in \mathcal{Q}} \mu(F_q) = 1 - \frac{1}{n} \tag{6.40}$$

Lemma 6.14 can be used as follows. Let $\mathcal{Q}$ denote the set of $p \in \mathcal{P}$ for which $\mathbb{G}_p$ is neighbor shared. It is clear from the discussion at the end of the last section, that any product of flocking matrices $F_{p_j} \cdots F_{p_1}$, $p_i \in \mathcal{Q}$ must converge at a rate no slower than

$$\lambda = \max_{q \in \mathcal{Q}} \mu(F_q)$$

Thus, in view of Lemma 6.14, $1 - \frac{1}{n}$ is a worst case bound on the rate of convergence of products of flocking matrices whose graphs are all neighbor shared. By way of comparison, $1 - \frac{1}{n}$ is a worst case bound if the flocking matrices in the product all have strongly rooted graphs (cf. (6.25)). Of course the latter situation is far more special than the former, since strongly rooted graph are neighbor shared but not conversely.

Convergence Rates for Rooted Graphs

It is also possible to derive a worst case convergence rate for products of flocking matrices which have rooted rather than neighbor-shared graphs. As a first step towards this end we exploit the fact that for any $n \times n$ stochastic scrambling matrix S, the scrambling constant of $\mu(S)$ satisfies the inequality

$$\mu(S) \leq 1 - \phi(S), \tag{6.41}$$

where for any non-negative matrix M, $\phi(M)$ denote the smallest non-zero element of M. Assume that S is any scrambling matrix. Note that for any distinct i and j, there must be a k for which $\min\{s_{ik}, s_{jk}\}$ is non-zero and bounded above by $\phi(S)$. Thus

$$\sum_{k=1}^{n} \min\{s_{ik}, s_{jk}\} \geq \phi(S)$$

so

$$1-\sum_{k=1}^{n}\min\{s_{ik},s_{jk}\}\leq 1-\phi(S)$$

But this holds for all distinct i and j. In view of the definition of $\mu(S)$ in (6.38), (6.41) must therefore be true.

We will also make use of the fact that for any two $n\times n$ stochastic matrices S_1 and S_2,

$$\phi(S_2S_1)\geq\phi(S_2)\phi(S_1). \tag{6.42}$$

To prove that this is so note first that any stochastic matrix S can be written at $S=\phi(S)\bar{S}$ where $\bar{S}$ is a non-zero matrix whose non-zero entries are all bounded below by 1; moreover if $S=\widehat{S}\widehat{S}$ where $\widehat{\phi}(S)$ is a number and $\widehat{S}$ is also a non-zero matrix whose non-zero entries are all bounded below by 1, then $\phi(S)\geq\widehat{\phi}(S)$. Accordingly, write $S_i=\phi(S_i)\bar{S}_i$, $i\in\{1,2\}$ where each $\bar{S}_i$ is a non-zero matrix whose non-zero entries are all bounded below by 1. Since $S_2S_1=\phi(S_2)\phi(S_1)\bar{S}_2\bar{S}_1$ and S_2S_1 is non-zero, $\bar{S}_2\bar{S}_1$ must be non-zero as well. Moreover the nonzero entries of $\bar{S}_2\bar{S}_1$ must be bounded below by 1 because the product of any two $n\times n$ matrices with all non-zero entries bounded below by 1 must be a matrix with the same property. Therefore $\phi(S_2S_1)\geq\phi(S_2)\phi(S_1)$ as claimed.

Suppose next that $\mathcal{S}_r$ is the set of all $n\times n$ stochastic matrices S which have rooted graphs $\gamma(S)\in\mathcal{G}$ and which satisfy $\phi(S)\geq b$ where b is a positive number smaller than 1. Thus for any set of $m=n-1$ $S_i\in\mathcal{S}_r$,

$$\phi(S_m\cdots S_1)\geq b^m \tag{6.43}$$

because of (6.42). Now $\gamma(S_m\cdots S_1)=\gamma(S_m)\cdots\gamma(S_1)$. Moreover $\gamma(S_m\cdots S_1)$ will be neighbor shared if $m\geq n-1$ because of Proposition 6.7. Therefore if $m\geq n-1$, $S_m\cdots S_1$ is a scrambling matrix and

$$\mu(S_m\cdots S_1)\leq 1-b^m \tag{6.44}$$

It turns out that this bound is actually attained if all the $S_i=S$, where S is a stochastic matrix of the form

$$S=\begin{pmatrix}1 & 0 & 0 & 0 & \cdots & 0\\ b & 1-b & 0 & 0 & \cdots & 0\\ 0 & b & 1-b & 0 & \cdots & 0\\ \vdots & \vdots & \vdots & \vdots & \vdots & \vdots\\ \vdots & \vdots & \vdots & \vdots & 1-b & 0\\ 0 & 0 & 0 & 0 & b & 1-b\end{pmatrix} \tag{6.45}$$

with a graph which has no closed cycles other than the self-arcs at each of its vertices. To understand why this is so, note that the first row of S^m is $(1\ 0\ \cdots\ 0)$ and the first element in the last row is b^m. This implies that $\min\{s_{11},s_{n1}\}=b^m$, that

$$1 - \sum_{k=1}^{n} \min\{s_{1k}, s_{nk}\} \geq b^m,$$

and consequently that $\mu(S) \geq b^m$. We summarize.

Lemma 6.15. *Let b be a positive number less than 1 and let $m = n - 1$. Let $\mathcal{S}_r^m$ denote the set of all m-term matrix products $S = S_m S_{m-1} \cdots S_1$ where each S_i is an $n \times n$ stochastic matrix with rooted graph $\gamma(S_i) \in \mathcal{G}$ and all-nonzero entries bounded below by b. Then*

$$\max_{S \in \mathcal{S}_r^m} \mu(S) = 1 - b^{(n-1)}$$

It is possible to apply at least part of the preceding to the case when the S_i are flocking matrices. Towards this end, let $m = n - 1$ and let $\mathbb{G}_{p_1}, \mathbb{G}_{p_2}, \ldots, \mathbb{G}_{p_m}$ be any sequence of m rooted graphs in $\mathcal{G}$ and let $F_{p_1}, \ldots, F_{p_m}$ be the sequence of flocking matrices associated with these graphs. Since each F_p is a flocking matrix, it must be true that $\phi(F_{p_i}) \geq \frac{1}{n}$, $i \in \{1, 2, \ldots, m\}$. Since the hypotheses leading up to (6.44) are satisfied,

$$\mu(F_{p_m} \cdots F_{p_1}) \leq 1 - \left(\frac{1}{n}\right)^{(n-1)} \tag{6.46}$$

Unfortunately, we cannot use the preceding reasoning to show that (6.46) holds with equality for some sequence of rooted graphs. This is because the matrix S in (6.45) is not a flocking matrix when $b = \frac{1}{n}$, except in the special case when $n = 2$. Nonetheless (6.46) can be used as follows to develop a convergence rate for products of flocking matrices whose graphs are all rooted. The development is very similar to that used in the proof of Proposition 6.11. Let $\mathcal{Q}$ denote the set of $p \in \mathcal{P}$ for which $\mathbb{G}_p$ is rooted and write $\mathcal{F}_r^m$ for the closed set of all products of flocking matrices of the form $F_{p_m} F_{p_{m-1}} \cdots F_{p_1}$ where each $p_i \in \mathcal{Q}$. In view of Proposition 6.7, $\mathbb{G}_{p_m} \circ \mathbb{G}_{p_{m-1}} \circ \cdots \mathbb{G}_{p_1}$ is neighbor shared for every list of m indices matrices $\{p_1, p_2, \ldots, p_m\}$ from $\mathcal{Q}$. Therefore (6.46) holds for every such list. Now for any sequence $p(1), p(2), \ldots, p(j)$ of indices in $\mathcal{Q}$, the corresponding product $F_{p(j)} \cdots F_{p(1)}$ of flocking matrices can be written as

$$F_{p(j)} \cdots F_{p(1)} = \bar{S}(j) \bar{S}_k \cdots \bar{S}_1$$

where

$$\bar{S}_i = F_{p(im)} \cdots F_{p((i-1)m+1)}, \quad 1 \leq i \leq k,$$
$$\bar{S}(j) = F_{p(j)} \cdots F_{p(km+1)},$$

and k is the integer quotient of j divided by m. In view of (6.46)

$$\mu(\bar{S}_i) \leq \bar{\lambda}, \quad i \in \{1, 2, \ldots, k\},$$

where
$$\bar{\lambda} = 1 - \left(\frac{1}{n}\right)^{(n-1)}$$
From this and the discussion at the end of the section on scrambling constants it is clear that $\bar{S}_k \cdots \bar{S}_1$ must converge to $\mathbf{1}\lfloor \cdots \bar{S}_k \cdots \bar{S}_1 \rfloor$ exponentially fast as $k \to \infty$ at a rate no slower than $\bar{\lambda}$, But $\bar{S}(j)$ is a product of at most m stochastic matrices, so it is a bounded function of j. It follows that the product $F_{p(j)} \cdots F_{p(1)}$ must converge to $\mathbf{1}\lfloor F_{p(j)} \cdots F_{p(1)} \rfloor$ exponentially fast at a rate no slower than $\lambda = \bar{\lambda}^{\frac{1}{m}}$. We have proved the following corollary to Theorem 6.2.

Corollary 6.1. *Under the hypotheses of Theorem 6.2, convergence of $\theta(t)$ to $\theta_{ss}\mathbf{1}$ is exponential at a rate no slower than*

$$\lambda = \left\{1 - \left(\frac{1}{n}\right)^{(n-1)}\right\}^{\frac{1}{n-1}}$$

Convergence Rates for Strongly Connected Graphs

It is possible to develop results analogous to those in the last section for strongly connected graphs. Consider next the case when the $\mathbb{G}_p$ are all strongly connected. Suppose that $\mathcal{S}_{sc}$ is the set of all $n \times n$ stochastic matrices S which have strongly connected graphs $\gamma(S) \in \mathcal{G}$ and which satisfy $\phi(S) \geq b$ where b is a positive number smaller than 1. Let m denote the integer quotient of n divided by 2. Thus for any set of m stochastic matrices $S_i \in \mathcal{S}_{sc}$,

$$\phi(S_m \cdots S_1) \geq b^m \tag{6.47}$$

because of (6.42). Now $\gamma(S_m \cdots S_1) = \gamma(S_m) \cdots \gamma(S_1)$. Moreover $\gamma(S_m \cdots S_1)$ will be neighbor shared because of Proposition 6.8. Therefore $S_m \cdots S_1$ is a scrambling matrix and

$$\mu(S_m \cdots S_1) \leq 1 - b^m \tag{6.48}$$

Just as in the case of rooted graphs, it turns out that this bound is actually attained if all the $S_i = S$, where S is a stochastic matrix of the form

$$S = \begin{pmatrix} 1-b & 0 & 0 & 0 & \cdots & b \\ b & 1-b & 0 & 0 & \cdots & 0 \\ 0 & b & 1-b & 0 & \cdots & 0 \\ \vdots & \vdots & \vdots & \vdots & \vdots & \vdots \\ \vdots & \vdots & \vdots & \vdots & 1-b & 0 \\ 0 & 0 & 0 & 0 & b & 1-b \end{pmatrix} \tag{6.49}$$

with a graph consisting of one closed cycle containing all vertices plus self-arcs at each of its vertices. The reader may wish to verify that this is so by exploiting the structure of S^m.

Just as in the case of rooted graphs, it is possible to apply at least part of the preceding to the case when the S_i are flocking matrices. The development exactly parallels the rooted graph case and one obtains in the end the following corollary to Theorem 6.2.

Corollary 6.2. *Under the hypotheses of Theorem 6.2, and the additional assumption that σ takes values only in the subset of $\mathcal{Q}$ composed of those indices for which $\mathbb{G}_p$ is strongly connected, convergence of $\theta(t)$ to $\theta_{ss}\mathbf{1}$ is exponential at a rate no slower than*

$$\lambda = \left\{1 - \left(\frac{1}{n}\right)^m\right\}^{\frac{1}{m}}$$

where m is the integer quotient of n divided by 2.

Jointly Rooted Graphs

It is possible to relax still further the conditions under which the conclusion of Theorem 6.1 holds. Towards this end, let us agree to say that a finite sequence of directed graphs $\mathbb{G}_{p_1}, \mathbb{G}_{p_2}, \ldots, \mathbb{G}_{p_k}$ in $\mathcal{G}$ is *jointly rooted* if the composition $\mathbb{G}_{p_k} \circ \mathbb{G}_{p_{k-1}} \circ \cdots \circ \mathbb{G}_{p_1}$ is rooted.

Note that since the arc set of any graph $\mathbb{G}_p, \mathbb{G}_q \in \mathcal{G}$ are contained in the arc set of any composed graph $\mathbb{G}_q \circ \mathbb{G} \circ \mathbb{G}_p$, $\mathbb{G} \in\in \mathcal{G}$, it must be true that if $\mathbb{G}_{p_1}, \mathbb{G}_{p_2}, \ldots, \mathbb{G}_{p_k}$ is a jointly rooted sequence, then so is $\mathbb{G}_q \circ \mathbb{G}_{p_1}, \mathbb{G}_{p_2}, \ldots, \mathbb{G}_{p_k}, \mathbb{G}_p$. In other words, a jointly rooted sequence of graphs in $\mathcal{G}$ remain jointly rooted if additional graphs from $\mathcal{G}$ are added to either end of the sequence.

There is an analogous concept for neighbor-shared graphs. We say that a finite sequence of directed graphs $\mathbb{G}_{p_1}, \mathbb{G}_{p_2}, \ldots, \mathbb{G}_{p_k}$ from $\mathcal{G}$ is *jointly neighbor-shared* if the composition $\mathbb{G}_{p_k} \circ \mathbb{G}_{p_{k-1}} \circ \cdots \circ \mathbb{G}_{p_1}$ is a neighbor-shared graph. Jointly neighbor shared sequences of graphs remains jointly neighbor shared if additional graphs from $\mathcal{G}$ are added to either end of the sequence. The reason for this is the same as for the case of jointly rooted sequences. Although the discussion which follows is just for the case of jointly rooted graphs, the material covered extends in the obvious way to the case of jointly neighbor shared graphs.

In the sequel we will say that an infinite sequence of graphs $\mathbb{G}_{p_1}, \mathbb{G}_{p_2}, \ldots,$ in $\mathcal{G}$ is *repeatedly jointly rooted* if there is a positive integer m for which each finite sequence $\mathbb{G}_{p_{m(k-1)+1}}, \ldots, \mathbb{G}_{p_{mk}}$, $k \geq 1$ is jointly rooted. We are now in a position to generalize Proposition 6.11.

Proposition 6.12. *Let $\bar{\mathcal{S}}$ be any closed set of stochastic matrices in $\mathcal{S}$. Suppose that $S_1, S_2, \ldots$ is an infinite sequence of matrices from $\bar{\mathcal{S}}$ whose corresponding sequence of graphs $\gamma(S_1), \gamma(S_2), \ldots$ is repeatedly jointly rooted. Then the product $S_j \ldots S_1$ converges to $\mathbf{1}\lfloor \cdots S_j \cdots S_1 \rfloor$ exponentially fast as $j \rightarrow \infty$ at a rate no slower than λ, where $\lambda < 1$ is a non-negative constant depending on the sequence.*

Proof of Proposition 6.12: Since $\gamma(S_1), \gamma(S_2), \dots$ is repeatedly jointly rooted, there is a finite integer m such that $\gamma(S_{m(k-1)+1}), \ \dots, \gamma(S_{mk}), \quad k \geq 1$ is jointly rooted. For $k \geq 1$ define $\bar{S}_k = S_{mk} \cdots S_{m(k-1)+1}$. By Lemma 6.3, $\gamma(S_{mk} \cdots S_{m(k-1)+1}) = \gamma(S_{mk}) \circ \cdots \gamma(S_{m(k-1)+1}), \ k \geq 1$. Therefore $\gamma(\bar{S}_k)$ is rooted for $k \geq 1$. Note in addition that $\{\bar{S}_k, k \geq 1\}$ is a closed set because $\bar{S}$ is closed and m is finite. Thus by Proposition 6.11, the product $\bar{S}_k \cdots \bar{S}_1$ converges to $\mathbf{1}\lfloor \cdots \bar{S}_k \cdots \bar{S}_1 \rfloor$ exponentially fast as $k \to \infty$ at a rate no slower than $\bar{\lambda}$, where $\bar{\lambda}$ is a non-negative constant depending on $\{\bar{S}_k, k \geq 1\}$ and satisfying $\bar{\lambda} < 1$.

Now the product $S_j \cdots S_1$ can be written as

$$S_j \cdots S_1 = \widehat{S}(j)\bar{S}_i \cdots \bar{S}_1$$

where

$$\widehat{S}(j) = S_j \cdots S_{(im+1)},$$

and i is the integer quotient of j divided by m. But $\widehat{S}(j)$ is a product of at most m stochastic matrices, so it is a bounded function of j. It follows that the product $S_j S_{j-1} \cdots S_1$ must converge to $\mathbf{1}\lfloor \cdots S_j \cdot S_1 \rfloor$ exponentially fast at a rate no slower than $\lambda = \bar{\lambda}^{\frac{1}{m}}$. □

We are now in a position to state our main result on leaderless coordination.

Theorem 6.3. *Let $\theta(0)$ be fixed and let $\sigma : [0, 1, 2, \dots) \to \bar{\mathcal{P}}$ be a switching signal for which the infinite sequence of graphs $\mathbb{G}_{\sigma(0)}, \mathbb{G}_{\sigma(1)}, \dots$ is repeatedly jointly rooted. Then there is a constant steady state heading θ_{ss}, depending only on $\theta(0)$ and σ, for which*

$$\lim_{t \to \infty} \theta(t) = \theta_{ss}\mathbf{1}, \tag{6.50}$$

where the limit is approached exponentially fast.

Proof of Theorem 6.3: The set of flocking matrices $\mathcal{F} = \{F_p : p \in \mathcal{P}\}$ is finite and $\gamma(F_p) = \mathbb{G}_p, \ p \in \mathcal{P}$. Therefore the infinite sequence of matrices $F_{\sigma(0)}, F_{\sigma(1)}, \dots$ come from a closed set and the infinite sequence of graphs $\gamma(F_{\sigma(0)}), \gamma(F_{\sigma(1)}), \dots$ is repeatedly jointly rooted. It follows from Proposition 6.12 that the product $F_{\sigma(t)} \cdots F_{\sigma(1)}F_{\sigma(0)}$ converges to $\mathbf{1}\lfloor \cdots F_{\sigma(t)} \cdots F_{\sigma(1)}F_{\sigma(0)} \rfloor$ exponentially fast as $t \to \infty$ at a rate no slower than λ, where $\lambda < 1$ is a non-negative constant depending on the sequence. But it is clear from (6.3) that

$$\theta(t) = F_{\sigma(t-1)} \cdots F_{\sigma(1)}F_{\sigma(0)}\theta(0), \quad t \geq 1$$

Therefore (6.50) holds with $\theta_{ss} = \lfloor \cdots F_{\sigma(t)} \cdots F_{\sigma(0)} \rfloor \theta(0)$ and the convergence is exponential. □

6.2 Symmetric Neighbor Relations

It is natural to call a graph in $\bar{\mathcal{G}}$ *symmetric* if for each pair of vertices i and j for which j is a neighbor of i, i is also a neighbor of j. Note that $\mathbb{G}$ is symmetric if and only if its adjacency matrix is symmetric. It is worth noting that for symmetric graphs, the properties of rooted and rooted at v are both equivalent to the property that the graph is strongly connected. Within the class of symmetric graphs, neighbor-shared graphs and strongly rooted graphs are also strongly connected graphs but in neither case is the converse true. It is possible to represent a symmetric directed graph $\mathbb{G}$ with a simple undirected graph $\mathbb{G}^s$ in which each self-arc is replaced with an undirected edge and each pair of directed arcs (i, j) and (j, i) for distinct vertices is replaced with an undirected edge between i and j. Notions of strongly rooted and neighbor shared extend in the obvious way to unconnected graphs. An undirected graph is said to be *connected* if there is an undirected path between each pair of vertices. Thus a strongly connected, directed graph which is symmetric is in essence the same as a connected, undirected graph. Undirected graphs are applicable when the sensing radii r_i of all agents are the same. It was the symmetric version of the flocking problem which Vicsek addressed [25] and which was analyzed in [27] using undirected graphs.

Let $\bar{\mathcal{G}}_s$ and $\mathcal{G}_s$ denote the subsets of symmetric graphs in $\bar{\mathcal{G}}$ and $\mathcal{G}$ respectively. Simple examples show that neither $\bar{\mathcal{G}}_s$ nor $\mathcal{G}_s$ is closed under composition. In particular, composition of two symmetric directed graphs in $\bar{\mathcal{G}}$ is not typically symmetric. On the other hand the "union" is where by the *union* of $\mathbb{G}_p \in \bar{\mathcal{G}}$ and $\mathbb{G}_q \in \bar{\mathcal{G}}$ is meant that graph in $\bar{\mathcal{G}}$ whose arc set is the union of the arc sets of $\mathbb{G}_p$ and $\mathbb{G}_q$. It is clear that both $\bar{\mathcal{G}}^s$ and $\mathcal{G}^s$ are closed under the union operation. The union operation extends to undirected graphs in the obvious way. Specifically, the *union* of two undirected graphs with the same vertex set $\mathcal{V}$, is that graph whose vertex set is $\mathcal{V}$ and whose edge set is the union of the edge sets of the two graphs comprising the union. It is worth emphasizing that union and composition are really quite different operations. For example, as we have already seen with Proposition 6.4, the composition of any $n-1$ strongly connected graphs is complete, symmetric or not, is always complete. On the other hand, the union of $n-1$ $n-1$ strongly connected graphs is not necessarily complete. In terms of undirected graphs, it is simply not true that the union of $n-1$ undirected graphs with vertex set $\mathcal{V}$ is complete, even if each graph in the union has self-loops at each vertex. The root cause of the difference between union and composition stems from the fact that the union and composition of two graphs in $\bar{\mathcal{G}}$ have different arc sets – and in the case of graphs from $\mathcal{G}$, the arc set of the union is always contained in the arc set of the composition, but not conversely.

The development in [27] make use of the notion of a "jointly connected set of graphs." Specifically, a set of undirected graphs with vertex set $\mathcal{V}$ is *jointly connected* if the union of the graphs in the collection is a connected graph. The notion of jointly connected also applies to directed graphs in which case the collection is jointly connected if the union is strongly connected. In the sequel we will say that an infinite sequence of graphs $\mathbb{G}_{p_1}, \mathbb{G}_{p_2}, \ldots,$ in $\mathcal{G}$ is *repeatedly jointly connected* if there is a positive integer m for which each finite sequence $\mathbb{G}_{p_{m(k-1)+1}}, \ldots, \mathbb{G}_{p_{mk}}$, $k \geq 1$ is jointly connected. The main result of [27] is in essence as follows.

Theorem 6.4. *Let $\theta(0)$ be fixed and let $\sigma : [0, 1, 2, \ldots) \to \bar{\mathcal{P}}$ be a switching signal for which the infinite sequence of symmetric graphs $\mathbb{G}_{\sigma(0)}, \mathbb{G}_{\sigma(1)}, \ldots$ in $\mathcal{G}$ is repeatedly jointly connected. Then there is a constant steady state heading θ_{ss}, depending only on $\theta(0)$ and σ, for which*

$$\lim_{t\to\infty} \theta(t) = \theta_{ss}\mathbf{1} \tag{6.51}$$

where the limit is approached exponentially fast.

In view of Theorem 6.3, Theorem 6.4 also holds if the word "connected" is replaced with the word "rooted." The latter supposes that composition replaces union and that jointly rooted replaces jointly connected. Examples show that these modifications lead to a more general result because a jointly rooted sequence of graphs is always jointly connected but the converse is not necessarily true.

Generalization

It is possible to interpret the system we have been studying (6.3) as the closed-loop system which results when a suitably defined decentralized feedback law is applied to the n-agent heading model

$$\theta(t + 1) = \theta(t) + u(t) \tag{6.52}$$

with open-loop control u. To end up with (6.3), u would have to be defined as

$$u(t) = -D_{\sigma(t)}^{-1} e(t) \tag{6.53}$$

where e is the *average heading error* vector

$$e(t) \triangleq L_{\sigma(t)}\theta(t) \tag{6.54}$$

and, for each $p \in \mathcal{P}$, L_p is the matrix

$$L_p = D_p - A_p \tag{6.55}$$

known in graph theory as the *Laplacian* of $\mathbb{G}_p$ [28,43]. It is easily verified that (6.52) to (6.55) do indeed define the system modeled by (6.3). We have elected to call e the average heading error because if $e(t) = 0$ at some time t, then the heading of each agent with neighbors at that time will equal the average of the headings of its neighbors.

In the present context, (6.53) can be viewed as a special case of a more general decentralized feedback control of the form

$$u(t) = -G_{\sigma(t)}^{-1} L_{\sigma(t)} \theta(t) \tag{6.56}$$

where for each $p \in \mathcal{P}$, G_p is a suitably defined, nonsingular diagonal matrix with ith diagonal element g_p^i. This, in turn, is an abbreviated description of a system of n individual agent control laws of the form

$$u_i(t) = -\frac{1}{g_i(t)} \left(\sum_{j \in \mathcal{N}_i(t)} \theta_j(t) \right), \quad i \in \{1, 2, \ldots, n\} \tag{6.57}$$

where for $i \in \{1, 2, \ldots, n\}$, $u_i(t)$ is the ith entry of $u(t)$ and $g_i(t) \triangleq g_{\sigma(t)}^i$. Application of this control to (6.52) would result in the closed-loop system

$$\theta(t+1) = \theta(t) - G_{\sigma(t)}^{-1} L_{\sigma(t)} \theta(t) \tag{6.58}$$

Note that the form of (6.58) implies that if θ and σ were to converge to a constant values $\bar{\theta}$, and $\bar{\sigma}$ respectively, then $\bar{\theta}$ would automatically satisfy $L_{\bar{\sigma}} \bar{\theta} = 0$. This means that control (6.56) automatically forces each agent's heading to converge to the average of its neighbors, if agent headings were to converge at all. In other words, the choice of the G_p does not affect the requirement that each agent's heading equal the average of the headings of its neighbors, if there is convergence at all. In the sequel we will deal only with the case when the graphs $\mathbb{G}_p$ are all symmetric in which case L_p is symmetric as well.

The preceding suggests that there might be useful choices for the G_p alternative to those we have considered so far, which also lead to convergence. One such choice turns out to be

$$G_p = gI, \; p \in \mathcal{P} \tag{6.59}$$

where g is any number greater than n. Our aim is to show that with the G_p so defined, Theorem 6.2 continues to be valid. In sharp contrast with the proof technique used in the last section, convergence will be established here using a common quadratic Lyapunov function.

As before, we will use the model

$$\theta(t+1) = F_{\sigma(t)} \theta(t) \tag{6.60}$$

where, in view of the definition of the G_p in (6.59), the F_p are now symmetric matrices of the form

$$F_p = I - \frac{1}{g} L_p, \quad p \in \mathcal{P} \tag{6.61}$$

To proceed we need to review a number of well known and easily verified properties of graph Laplacians relevant to the problem at hand. For this, let

$\mathbb{G}$ be any given symmetric directed graph in $\mathcal{G}$. Let D be a diagonal matrix whose diagonal elements are the in-degrees of $\mathbb{G}$'s vertices and write A for $\mathbb{G}$'s adjacency matrix. Then, as noted before, the Laplacian of $\mathbb{G}$ is the symmetric matrix $L = D - A$. The definition of L clearly implies that $L\mathbf{1} = 0$. Thus L must have an eigenvalue at zero and $\mathbf{1}$ must be an eigenvector for this eigenvalue. Surprisingly L is always a positive semidefinite matrix [28]. Thus L must have a real spectrum consisting of non-negative numbers and at least one of these numbers must be 0. It turns out that the number of connected components of $\mathbb{G}$ is exactly the same as the multiplicity of L's eigenvalue at 0 [28]. Thus $\mathbb{G}$ is a rooted or strongly connected graph just in case L has exactly one eigenvalue at 0. Note that the trace of L is the sum of the in-degrees of all vertices of $\mathbb{G}$. This number can never exceed $(n-1)n$ and can attain this high value only for a complete graph. In any event, this property implies that the maximum eigenvalue of L is never larger than $n(n-1)$. Actually the largest eigenvalue of L can never be larger than n [28]. This means that the eigenvalues of $\frac{1}{g}L$ must be smaller than 1 since $g > n$. From these properties it clearly follows that the eigenvalues of $(I - \frac{1}{g}L)$ must all be between 0 and 1, and that if $\mathbb{G}$ is strongly connected, then all will be strictly less than 1 except for one eigenvalue at 1 with eigenvector $\mathbf{1}$. Since each F_p is of the form $(I - \frac{1}{g}L)$, each F_p possesses all of these properties.

Let σ be a fixed switching signal with value $p_t \in \mathcal{Q}$ at time $t \geq 0$. What we'd like to do is to prove that as $i \to \infty$, the matrix product $F_{p_i}F_{p_{i-1}} \cdots F_{p_0}$ converges to $\mathbf{1}c$ for some row vector c. As noted near the beginning of Sect. 6.1, this matrix product will so converge just in case

$$\lim_{i \to \infty} \tilde{F}_{p_i}\tilde{F}_{p_{i-1}} \cdots \tilde{F}_{p_0} = 0 \tag{6.62}$$

where as in Sect. 6.1, $\tilde{F}_p$ is the unique solution to $PF_p = \tilde{F}_pP$, $p \in \mathcal{P}$ and P is any full rank $(n-1) \times n$ matrix satisfying $P\mathbf{1} = 0$. For simplicity and without loss of generality we shall henceforth assume that the rows of P form a basis for the orthogonal complement of the span of $\mathbf{e}$. This means that PP' equals the $(n-1) \times (n-1)$ identity $\tilde{I}$, that $\tilde{F}_p = PF_pP', \quad p \in \mathcal{P}$, and thus that each $\tilde{F}_p$ is symmetric. Moreover, in view of (6.6) and the spectral properties of the F_p, $p \in \mathcal{Q}$, it is clear that each $\tilde{F}_p$, $p \in \mathcal{Q}$ must have a real spectrum lying strictly inside of the unit circle. This plus symmetry means that for each $p \in \mathcal{Q}$, $\tilde{F}_p - \tilde{I}$ is negative definite, that $\tilde{F}_p'\tilde{F}_p - \tilde{I}$ is negative definite and thus that $\tilde{I}$ is a common discrete-time Lyapunov matrix for all such $\tilde{F}_p$. Using this fact it is straight forward to prove that Theorem 6.2 holds for system (6.58) provided the G_p are defined as in (6.59) with $g > n$.

In general, each $\tilde{F}_p$ is a discrete-time stability matrix for which $\tilde{F}_p'\tilde{F}_p - \tilde{I}$ is negative definite only if $p \in \mathcal{Q}$. To craft a proof of Theorem 6.3 for the system described by (6.58) and (6.59), one needs to show that for each interval $[t_i, t_{i+1})$ on which $\{\mathbb{G}_{\sigma(t_{i+1}-1)}, \ldots \mathbb{G}_{\sigma(t_i+1)}, \mathbb{G}_{\sigma(t_i)}\}$ is a jointly rooted sequence of graphs, the product $\tilde{F}_{\sigma(t_{i+1}-1)} \cdots \tilde{F}_{\sigma(t_i+1)} \tilde{F}_{\sigma(t_i)}$ is a discrete-time stability matrix and

$$(\tilde{F}_{\sigma(t_{i+1}-1)} \cdots \tilde{F}_{\sigma(t_i+1)} \tilde{F}_{\sigma(t_i)})'(\tilde{F}_{\sigma(t_{i+1}-1)} \cdots \tilde{F}_{\sigma(t_i+1)} \tilde{F}_{\sigma(t_i)}) - \tilde{I}$$

is negative definite. This is a direct consequence of the following proposition.

Proposition 6.13. *If* $\{\mathbb{G}_{p_1}, \mathbb{G}_{p_2}, \ldots, \mathbb{G}_{p_m}\}$ *is a jointly rooted sequence of symmetric graphs, then*

$$(\tilde{F}_{p_1} \tilde{F}_{p_2} \cdots \tilde{F}_{p_m})'(\tilde{F}_{p_1} \tilde{F}_{p_2} \cdots \tilde{F}_{p_m}) - \tilde{I}$$

is a negative definite matrix.

In the light of Proposition 6.13, it is clear that the conclusion Theorem 6.3 is also valid for the system described by (6.58) and (6.59). A proof of this version of Theorem 6.3 will not be given.

To summarize, both the control defined by $u = -D_{\sigma(t)}^{-1} e(t)$ and the simplified control given by $u = -\frac{1}{g} e(t)$ achieve the same emergent behavior. While the latter is much easier to analyze than the former, it has the disadvantage of not being a true decentralized control because each agent must know an upper bound (i.e., g) on the total number of agents within the group. Whether or not this is really a disadvantage, of course depends on what the models are to be used for.

The proof of Proposition 6.13 depends on two lemmas. In the sequel, we state the lemmas, use them to prove Proposition 6.13, and then conclude this section with proofs of the lemmas themselves.

Lemma 6.16. *If* $\mathbb{G}_{p_1}, \mathbb{G}_{p_2}, \ldots, \mathbb{G}_{p_m}$ *is a jointly rooted sequence of symmetric graphs in* $\mathcal{G}$ *with Laplacians* $L_{p_1}, L_{p_2}, \ldots, L_{p_m}$, *then*

$$\bigcap_{i=1}^{m} \text{kernel } L_{p_i} = \text{span } \{\mathbf{1}\} \tag{6.63}$$

Lemma 6.17. *Let* $M_1, M_2, \ldots, M_m$ *be a set of* $n \times n$ *real symmetric, matrices whose induced* 2*-norms are all less than or equal to* 1*. If*

$$\bigcap_{i=1}^{m} \text{kernel } (I - M_i) = 0 \tag{6.64}$$

then the induced 2*-norm of* $M_1 M_2 \cdots M_m$ *is less than* 1.

Proof of Proposition 6.13: The definition of the F_p in (6.61) implies that $I - F_p = \frac{1}{g} L_p$. Hence by Lemma 6.16 and the hypothesis that $\{\mathbb{G}_{p_1}, \mathbb{G}_{p_2}, \ldots, \mathbb{G}_{p_m}\}$ is a jointly rooted sequence,

$$\bigcap_{i=1}^{m} \text{kernel } (I - F_{p_i}) = \text{span } \{\mathbf{1}\} \tag{6.65}$$

We claim that

$$\bigcap_{i=1}^{m} \text{kernel } (\tilde{I} - \tilde{F}_{p_i}) = 0 \tag{6.66}$$

To establish this fact, let $\bar{x}$ be any vector such that $(\tilde{I} - \tilde{F}_{p_i})\bar{x} = 0,\ i \in \{1, 2, \dots, m\}$. Since P has independent rows, there is a vector x such that $\bar{x} = Px$. But $P(I - F_{p_i}) = (\tilde{I} - \tilde{F}_{p_i})P$, so $P(I - F_{p_i})x = 0$. Hence $(I - F_{p_i})x = a_i\mathbf{1}$ for some number a_i. But $\mathbf{1}'(I - F_{p_i}) = \frac{1}{g}\mathbf{1}'L_{p_i} = 0$, so $a_i\mathbf{1}'\mathbf{1} = 0$. This implies that $a_i = 0$ and thus that $(I - F_{p_i})x = 0$. But this must be true for all $i \in \{1, 2, \dots, m\}$. It follows from (6.65) that $x \in \text{span } \{\mathbf{1}\}$ and, since $\bar{x} = Px$, that $\bar{x} = 0$. Therefore (6.66) is true.

As defined, the $\tilde{F}_p$ are all symmetric, positive semi-definite matrices with induced 2-norms not exceeding 1. This and (6.66) imply that the family of matrices $\tilde{F}_{p_1}, \tilde{F}_{p_2}, \dots, \tilde{F}_{p_m}$ satisfy the hypotheses of Lemma 6.17. It follows that Proposition 6.13 is true. □

Proof of Lemma 6.16: In the sequel we write $L(\mathbb{G})$ for the Laplacian of a simple graph $\mathbb{G}$. By the *intersection* of a collection of graphs $\{\mathbb{G}_{p_1}, \mathbb{G}_{p_2}, \dots, \mathbb{G}_{p_m}\}$ in $\mathcal{G}$, is meant that graph $\mathbb{G} \in \mathcal{G}$ with edge set equaling the intersection of the edge sets of all of the graphs in the collection. It follows at once from the definition of a Laplacian that

$$L(\mathbb{G}_p) + L(\mathbb{G}_q) = L(\mathbb{G}_p \cap \mathbb{G}_q) + L(\mathbb{G}_p \cup \mathbb{G}_q)$$

for all $p, q \in \mathcal{P}$. Repeated application of this identity to the set $\{\mathbb{G}_{p_1}, \mathbb{G}_{p_2}, \dots, \mathbb{G}_{p_m}\}$ yields the relation

$$\sum_{i=1}^{m} L(\mathbb{G}_{p_i}) = L\left(\bigcup_{i=1}^{m} \mathbb{G}_{p_i}\right) + \sum_{i=1}^{m-1} L\left(\mathbb{G}_{p_{i+1}} \bigcap \left\{\bigcup_{j=1}^{i} \mathbb{G}_{p_j}\right\}\right) \tag{6.67}$$

which is valid for $m > 1$. Since all matrices in (6.67) are positive semi-definite, any vector x which makes the quadratic form $x'\{L(\mathbb{G}_{p_1}) + L(\mathbb{G}_{p_2}) + \cdots + L(\mathbb{G}_{p_m})\}x$ vanish, must also make the quadratic form $x'L(\mathbb{G}_{p_1} \cup \mathbb{G}_{p_2} \cup \cdots \cup \mathbb{G}_{p_m})x$ vanish. Since any vector in the kernel of each matrix $L(\mathbb{G}_{p_i})$ has this property, we can draw the following conclusion.

$$\bigcap_{i=1}^{m} \text{kernel } L(\mathbb{G}_{p_i}) \subset \text{kernel } L\left(\bigcup_{i=1}^{m} \mathbb{G}_{p_i}\right)$$

Suppose now that $\{\mathbb{G}_{p_1}, \mathbb{G}_{p_2}, \dots, \mathbb{G}_{p_m}\}$ is a jointly rooted collection. Then the union $\mathbb{G}_{p_1} \cup \mathbb{G}_{p_2} \cup \cdots \cup \mathbb{G}_{p_m}$ is rooted so its Laplacian must have exactly span $\{\mathbf{1}\}$ for its kernel. Hence the intersection of the kernels of the $L(\mathbb{G}_{p_i})$

must be contained in span $\{\mathbf{1}\}$. But span $\{\mathbf{1}\}$ is contained in the kernel of each matrix $L(\mathbb{G}_{p_i})$ in the intersection and therefore in the intersection of the kernels of these matrices as well. It follows that (6.63) is true. □

Proof of Lemma 6.17: In the sequel we write $|x|$ for the 2-norm of a real n-vector x and $|M|$ for the induced 2-norm of a real $n \times n$ matrix. Let $x \in \mathbb{R}^n$ be any real, non-zero n-vector. It is enough to show that

$$|M_1 M_2 \cdots M_m x| < |x| \tag{6.68}$$

In view of (6.64) and the assumption that $x \neq 0$, there must be a largest integer $k \in \{1, 2, \ldots, m\}$ such that $x \notin$ kernel $(M_k - I)$. We claim that

$$|M_k x| < |x| \tag{6.69}$$

To show that this is so we exploit the symmetry of M_k to write x as $x = \alpha_1 y_1 + \alpha_2 y_2 + \cdots + \alpha_n y_n$ where $\alpha_1, \alpha_2, \ldots, \alpha_n$ are real numbers and $\{y_1, y_2, \ldots, y_n\}$ is an orthonormal set of eigenvectors of M_k with real eigenvalues $\lambda_1, \lambda_2, \ldots \lambda_n$. Note that $|\lambda_i| \leq 1,\ i \in \{1, 2, \ldots, n\}$, because $|M_k| \leq 1$. Next observe that since $M_k x = \alpha_1\lambda_1 y_1 + \alpha_2\lambda_2 y_2 + \cdots + \alpha_n\lambda_n y_n$ and $M_k x \neq x$, there must be at least one integer j such that $\alpha_j\lambda_j \neq \alpha_j$. Hence $|\alpha_j\lambda_j y_j| < |\alpha_j y_j|$. But $|M_k x|^2 = |\alpha_1\lambda_1 y_1|^2 + \cdots + |\alpha_j\lambda_j y_j|^2 + \cdots + |\alpha_n\lambda_n y_n|^2$ so

$$|M_k x|^2 < |\alpha_1\lambda_1 y_1|^2 + \cdots + |\alpha_j y_j|^2 + \cdots + |\alpha_n\lambda_n y_n|^2$$

Moreover

$$|\alpha_1\lambda_1 y_1|^2 + \cdots + |\alpha_j y_j|^2 + \cdots + |\alpha_n\lambda_n y_n|^2 \leq |\alpha_1 y_1|^2 + \cdots + |\alpha_j y_j|^2 + \cdots + |\alpha_n y_n|^2 = |x|^2$$

so $|M_k x|^2 < |x|^2$; therefore (6.69) is true.

In view of the definition of k, $M_j x = x, j \in \{k+1, \ldots, m\}$. From this and (6.69) it follows that $|M_1 \cdots M_m x| = |M_1 \cdots M_k x| \leq |M_1 \cdots M_{k-1}||M_k x| < |M_1 \cdots M_{k-1}||x|$. But $|M_1 \cdots M_{k-1}| \leq 1$ because each M_i has an induced 2 norm not exceeding 1. Therefore (6.68) is true. □

6.3 Measurement Delays

In this section we consider a modified version of the flocking problem in which integer valued delays occur in sensing the values of headings which are available to agents. More precisely we suppose that at each time $t \in \{0, 1, 2, \ldots\}$, the value of neighboring agent j's headings which agent i may sense is $\theta_j(t - d_{ij}(t))$ where $d_{ij}(t)$ is a delay whose value at t is some integer between 0 and $m_j - 1$; here m_j is a pres-specified positive integer. While well established principles of feedback control would suggest that delays should be dealt with using dynamic compensation, in these notes we will consider the situation in which the delayed value of agent j's heading sensed by agent i at time t is the value which will be used in the heading update law for agent i.

Thus

$$\theta_i(t+1) = \frac{1}{n_i(t)}\left(\sum_{j\in\mathcal{N}_i(t)} \theta_j(t-d_{ij}(t))\right) \tag{6.70}$$

where $d_{ij}(t) \in \{0,1,\ldots,(m_j-1)\}$ if $j \neq i$ and $d_{ij}(t) = 0$ if $i = j$.

It is possible to represent this agent system using a state space model similar to the model discussed earlier for the delay-free case. Towards this end, let $\bar{\mathcal{D}}$ denote the set of all directed graphs with vertex set $\bar{\mathcal{V}} = \mathcal{V}_1 \cup \mathcal{V}_2 \cup \cdots \cup \mathcal{V}_n$ where $\mathcal{V}_i = \{v_{i1} \ldots, v_{im_i}\}$. Here vertex v_{i1} represents agent i and $\mathcal{V}_i$ is the set of vertices associated with agent i. We sometimes write i for v_{i1}, $i \in \{1,2,\ldots,n\}$, and $\mathcal{V}$ for the subset of agent vertices $\{v_{11}, v_{21}, \ldots, v_{n1}\}$. Let $\bar{\mathcal{Q}}$ be an index set parameterizing $\bar{\mathcal{D}}$; i.e, $\bar{\mathcal{D}} = \{\mathbb{G}_q : q \in \bar{\mathcal{Q}}\}$

To represent the fact that each agent can use its own current heading in its update formula (6.70), we will utilize those graphs in $\bar{\mathcal{D}}$ which have self arcs at each vertex in $\mathcal{V}$. We will also require the arc set of each such graph to have, for $i \in \{1,2,\ldots,n\}$, an arc from each vertex $v_{ij} \in \mathcal{V}_i$ except the last, to its successor $v_{i(j+1)} \in \mathcal{V}_i$. Finally we stipulate that for each $i \in \{1,2,\ldots,n\}$, each vertex v_{ij} with $j > 1$ has in-degree of exactly 1. In the sequel we write $\mathcal{D}$ for the subset of all such graphs. Thus unlike the class of graphs $\mathcal{G}$ considered before, there are graphs in $\mathcal{D}$ possessing vertices without self-arcs. Nonetheless each vertex of each graph in $\mathcal{D}$ has positive in-degree. In the sequel we use the symbol $\mathcal{Q}$ to denote that subset of $\bar{\mathcal{Q}}$ for which $\mathcal{D} = \{\mathbb{G}_q : q \in \mathcal{Q}\}$.

The specific graph representing the sensed headings the agents use at time t to update their own headings according to (6.70), is that graph $\mathbb{G}_q \in \mathcal{D}$ whose arc set contains an arc from $v_{ik} \in \mathcal{V}_i$ to $j \in \mathcal{V}$ if agent j uses $\theta_i(t+1-k)$ to update. The set of agent heading update rules defined by (6.70) can now be written in state form. Towards this end define $\theta(t)$ to be that $(m_1 + m_2 + \cdots + m_i)$ vector whose first m_1 elements are $\theta_1(t)$ to $\theta_1(t+1-m_1)$, whose next m_2 elements are $\theta_2(t)$ to $\theta_2(t+1-m_2)$ and so on. Order the vertices of $\bar{\mathcal{V}}$ as $v_{11}, \ldots, v_{1m_1}, v_{21}, \ldots, v_{2m_2}, \ldots, v_{n1}, \ldots, v_{nm_n}$ and with respect to this ordering define

$$F_q = D_q^{-1}A_q', \quad q \in \mathcal{Q} \tag{6.71}$$

where A_q' is the transpose of the adjacency matrix the of $\mathbb{G}_q \in \mathcal{D}$ and D_q the diagonal matrix whose ijth diagonal element is the in-degree of vertex v_{ij} within the graph. Then

$$\theta(t+1) = F_{\sigma(t)}\theta(t), \quad t \in \{0,1,2,\ldots\} \tag{6.72}$$

where $\sigma : \{0,1,\ldots\} \rightarrow \mathcal{Q}$ is a switching signal whose value at time t, is the index of the graph representing which headings the agents use at time t to update their own headings according to (6.70). As before our goal is to characterize switching signals for which all entries of $\theta(t)$ converge to a common steady state value.

There are a number of similarities and a number of differences between the situation under consideration here and the delay-free situation considered earlier. For example, the notion of graph composition defined earlier can be defined in the obvious way for graphs in $\bar{\mathcal{D}}$. On the other hand, unlike the situation in the delay-free case, the set of graphs used to model the system under consideration, namely $\mathcal{D}$, is not closed under composition except in the special case when all of the delays are at most 1; i.e., when all of the $m_i \leq 2$. In order to characterize the smallest subset of $\bar{\mathcal{D}}$ containing $\mathcal{D}$ which is closed under composition, we will need several new concepts.

Hierarchical Graphs

As before, let $\bar{\mathcal{G}}$ be the set of all directed graphs with vertex set $\mathcal{V} = \{1, 2, \dots n\}$. Let us agree to say that a rooted graph $\mathbb{G} \in \bar{\mathcal{G}}$ is a *hierarchical graph* with *hierarchy* $\{v_1, v_2, \dots, v_n\}$ if it is possible to re-label the vertices in $\mathcal{V}$ as $v_1, v_2, \dots v_n$ in such a way so that v_1 is a root of $\mathbb{G}$ with a self-arc and for $i > 1$, v_i has a neighbor v_j "lower " in the hierarchy where by *lower* we mean $j < i$. It is clear that any graph in $\bar{\mathcal{G}}$ with a root possessing a self-arc is hierarchial. Note that a graph may have more than one hierarchy and two graphs with the same hierarchy need not be equal. Note also that even though rooted graphs with the same hierarchy share a common root, examples show that the composition of hierarchial graphs in $\bar{\mathcal{G}}$ need not be hierarchial or even rooted. On the other hand the composition of two rooted graphs in $\bar{\mathcal{G}}$ with the same hierarchy is always a graph with the same hierarchy. To understand why this is so, consider two graphs $\mathbb{G}_1$ and $\mathbb{G}_2$ in $\bar{\mathcal{G}}$ with the same hierarchy $\{v_1, v_2, \dots, v_n\}$. Note first that v_1 has a self-arc in $\mathbb{G}_2 \circ \mathbb{G}_1$ because v_1 has self arcs in $\mathbb{G}_1$ and $\mathbb{G}_2$. Next pick any vertex v_i in $\mathcal{V}$ other than v_1. By definition, there must exist vertex v_j lower in the hierarchy than v_i such that (v_j, v_i) is an arc of $\mathbb{G}_2$. If $v_j = v_1$, then (v_1, v_i) is an arc in $\mathbb{G}_2 \circ \mathbb{G}_1$ because v_1 has a self-arc in $\mathbb{G}_1$. On the other hand, if $v_j \neq v_1$, then there must exist a vertex v_k lower in the hierarchy than v_j such that (v_k, v_j) is an arc of $\mathbb{G}_1$. It follows from the definition of composition that in this case (v_k, v_i) is an arc in $\mathbb{G}_2 \circ \mathbb{G}_1$. Thus v_i has a neighbor in $\mathbb{G}_2 \circ \mathbb{G}_1$ which is lower in the hierarchy than v_i. Since this is true for all v_i, $\mathbb{G}_2 \circ \mathbb{G}_1$ must have the same hierarchy as $\mathbb{G}_1$ and $\mathbb{G}_2$. This proves the claim that composition of two rooted graphs with the same hierarchy is a graph with the same hierarchy.

Our objective is to show that the composition of a sufficiently large number of graphs in $\bar{\mathcal{G}}$ with the same hierarchy is strongly rooted. Note that Proposition 6.3 cannot be used to reach this conclusion, because the v_i in the graphs under consideration here do not all necessarily have self-arcs.

As before, let $\mathbb{G}_1$ and $\mathbb{G}_2$ be two graphs in $\bar{\mathcal{G}}$ with the same hierarchy $\{v_1, v_2, \dots, v_n\}$. Let v_i be any vertex in the hierarchy and suppose that v_j is a neighbor vertex of v_i in $\mathbb{G}_2$. If $v_j = v_1$, then v_i retains v_1 as a neighbor in the composition $\mathbb{G}_2 \circ \mathbb{G}_1$ because v_1 has a self-arc in $\mathbb{G}_1$. On the other hand, if $v_j \neq v_1$, then v_j has a neighboring vertex v_k in $\mathbb{G}_1$ which is lower in the hierarchy than v_j. Since v_k is a neighbor of v_i in the composition $\mathbb{G}_2 \circ \mathbb{G}_1$, we see that in this case v_i has acquired a neighbor in $\mathbb{G}_2 \circ \mathbb{G}_1$ lower in the hierarchy than a neighbor it had in $\mathbb{G}_2$. In summary, any vertex $v_i \in \mathcal{V}$ either has v_1 as neighbor in $\mathbb{G}_2 \circ \mathbb{G}_1$ or has a neighbor in $\mathbb{G}_2 \circ \mathbb{G}_1$ which is one vertex lower in the hierarchy than any neighbor it had in $\mathbb{G}_2$.

Now consider three graphs $\mathbb{G}_1, \mathbb{G}_2, \mathbb{G}_3$ in $\bar{\mathcal{G}}$ with the same hierarchy. By the same reasoning as above, any vertex $v_i \in \mathcal{V}$ either has v_1 as neighbor in $\mathbb{G}_3 \circ \mathbb{G}_2 \circ \mathbb{G}_1$ or has a neighbor in $\mathbb{G}_3 \circ \mathbb{G}_2 \circ \mathbb{G}_1$ which is one vertex lower in the hierarchy than any neighbor it had in $\mathbb{G}_3 \circ \mathbb{G}_2$. Similarly v_i either has v_1 as neighbor in $\mathbb{G}_3 \circ \mathbb{G}_2$ or has a neighbor in $\mathbb{G}_3 \circ \mathbb{G}_2$ which is one vertex lower in the hierarchy than any neighbor it had in $\mathbb{G}_3$. Combining these two observations we see that any vertex $v_i \in \mathcal{V}$ either has v_1 as neighbor in $\mathbb{G}_3 \circ \mathbb{G}_2 \circ \mathbb{G}_1$ or has a neighbor in $\mathbb{G}_3 \circ \mathbb{G}_2 \circ \mathbb{G}_1$ which is two vertices lower in the hierarchy than any neighbor it had in $\mathbb{G}_3$. This clearly generalizes and so after the composition of m such graphs $\mathbb{G}_1, \mathbb{G}_2, \ldots \mathbb{G}_m$, v_i either has v_1 as neighbor in $\mathbb{G}_m \circ \cdots \mathbb{G}_2 \circ \mathbb{G}_1$ or has a neighbor in $\mathbb{G}_m \circ \cdots \mathbb{G}_2 \circ \mathbb{G}_1$ which is $m-1$ vertices lower in the hierarchy than any neighbor it had in $\mathbb{G}_m$. It follows that if $m \geq n$, then v_i must be a neighbor of v_1. Since this is true for all vertices, we have proved the following.

Proposition 6.14. *Let $\mathbb{G}_1, \mathbb{G}_2, \ldots \mathbb{G}_m$ denote a set of rooted graphs in $\bar{\mathcal{G}}$ which all have the same hierarchy. If $m \geq n-1$ then $\mathbb{G}_m \circ \cdots \mathbb{G}_2 \circ \mathbb{G}_1$ is strongly rooted.*

As we have already pointed out, Proposition 6.14 is not a consequence of Proposition 6.3 because Proposition 6.3 requires all vertices of all graphs in the composition to have self-arcs whereas Proposition 6.14 does not. On the other hand, Proposition 6.3 is not a consequence of Proposition 6.14 because Proposition 6.14 only applies to graphs with the same hierarchy whereas Proposition 6.3 does not.

Delay Graphs

We now return to the study of the graphs in $\mathcal{D}$. As before $\mathcal{D}$ is the subset of $\bar{\mathcal{D}}$ consisting of those graphs which (i) have self arcs at each vertex in $\mathcal{V} = \{1, 2, \ldots, \}$, (ii) for each $i \in \{1, 2, \ldots, n\}$, have an arc from each vertex $v_{ij} \in \mathcal{V}_i$ except the last, to its successor $v_{i(j+1)} \in \mathcal{V}_i$, and (iii) for each $i \in \{1, 2, \ldots, n\}$, each vertex v_{ij} with $j > 1$ has in-degree of exactly 1. It can easily be shown by example that $\mathcal{D}$ is not closed under composition. We deal with this problem as follows. A graph $\mathbb{G} \in \bar{\mathcal{D}}$ is said to be a *delay graph* if for each $i \in \{1, 2, \ldots, n\}$, (i) every neighbor of $\mathcal{V}_i$ which is not in $\mathcal{V}_i$ is a neighbor of v_{i1} and (ii) the subgraph of $\mathbb{G}$ induced by $\mathcal{V}_i$ has $\{v_{i1} \ldots, v_{im_i}\}$ as a hierarchy. It is easy to see that every graph in $\mathcal{D}$ is a delay graph. More is true.

Proposition 6.15. *The set of delay graphs in $\bar{\mathcal{D}}$ is closed under composition.*

To prove this proposition, we will need the following fact.

Lemma 6.18. *Let $\mathbb{G}_1, \mathbb{G}_2, \ldots, \mathbb{G}_q$ be any sequence of $q > 1$ directed graphs in $\bar{\mathcal{G}}$. For $i \in \{1, 2, \ldots, q\}$, let $\bar{\mathbb{G}}_i$ be the subgraph of $\mathbb{G}_i$ induced by $\mathcal{S} \subset \mathcal{V}$. Then $\bar{\mathbb{G}}_q \circ \cdots \circ \bar{\mathbb{G}}_2 \circ \bar{\mathbb{G}}_1$ is contained in the subgraph of $\mathbb{G}_q \circ \cdots \circ \mathbb{G}_2 \circ \mathbb{G}_1$ induced by $\mathcal{S}$.*

Proof of Lemma 6.18: It will be enough to prove the lemma for $q = 2$, since the proof for $q > 2$ would then directly follow by induction. Suppose $q = 2$. Let (i, j) be in $\mathcal{A}(\bar{\mathbb{G}}_2 \circ \bar{\mathbb{G}}_1)$. Then $i, j \in \mathcal{S}$ and there exists an integer $k \in \mathcal{S}$ such that $(i, k) \in \mathcal{A}(\bar{\mathbb{G}}_1)$ and $(k, j) \in \mathcal{A}(\bar{\mathbb{G}}_2)$. Therefore $(i, k) \in \mathcal{A}(\mathbb{G}_1)$ and $(k, j) \in \mathcal{A}(\mathbb{G}_2)$. Thus $(i, j) \in \mathcal{A}(\mathbb{G}_2 \circ \mathbb{G}_1)$. But $i, j \in \mathcal{S}$ so (i, j) must be an arc in the subgraph of $\mathbb{G}_2 \circ \mathbb{G}_1$ induced by $\mathcal{S}$. Since this clearly is true for all arcs in $\mathcal{A}(\bar{\mathbb{G}}_2 \circ \bar{\mathbb{G}}_1)$, the proof is complete. □

Proof of Proposition 6.15: Let $\mathbb{G}_1$ and $\mathbb{G}_2$ be two delay graphs in $\bar{\mathcal{D}}$. It will first be shown that for each $i \in \{1, 2, \ldots, n\}$, every neighbor of $\mathcal{V}_i$ which is not in $\mathcal{V}_i$ is a neighbor of v_{i1} in $\mathbb{G}_2 \circ \mathbb{G}_1$. Fix $i \in \{1, 2, \ldots, n\}$ and let v be a neighbor of $\mathcal{V}_i$ in $\mathbb{G}_2 \circ \mathbb{G}_1$ which is not in $\mathcal{V}_i$. Then $(v, k) \in \mathcal{A}(\mathbb{G}_2 \circ \mathbb{G}_1)$ for some $k \in \mathcal{V}_i$. Thus there is a $s \in \bar{\mathcal{V}}$ such that $(v, s) \in \mathcal{A}(\mathbb{G}_1)$ and $(s, k) \in \mathcal{A}(\mathbb{G}_2)$. If $s \notin \mathcal{V}_i$, then $(s, v_{i1}) \in \mathcal{A}(\mathbb{G}_2)$ because $\mathbb{G}_2$ is a delay graph. Thus in this case $(v, v_{i1}) \in \mathcal{A}(\mathbb{G}_2 \circ \mathbb{G}_1)$ because of the definition of composition. If, on the other hand, $s \in \mathcal{V}_i$, then $(v, v_{i1}) \in \mathcal{A}(\mathbb{G}_1)$ because $\mathbb{G}_1$ is a delay graph. Thus in this case $(v, v_{i1}) \in \mathcal{A}(\mathbb{G}_2 \circ \mathbb{G}_1)$ because v_{i1} has a self-arc in $\mathbb{G}_2$. This proves that every neighbor of $\mathcal{V}_i$ which is not in $\mathcal{V}_i$ is a neighbor of v_{i1} in $\mathbb{G}_2 \circ \mathbb{G}_1$. Since this must be true for each $i \in \{1, 2, \ldots, n\}$, $\mathbb{G}_2 \circ \mathbb{G}_1$ has the first property defining delay graphs in $\bar{\mathcal{D}}$.

To establish the second property, we exploit the fact that the composition of two graphs with the same hierarchy is a graph with the same hierarchy. Thus for any integer $i \in \{1, 2, \ldots, n\}$, the composition of the subgraphs of $\mathbb{G}_1$ and $\mathbb{G}_2$ respectively induced by $\mathcal{V}_i$ must have the hierarchy $\{v_{i1}, v_{i2}, \ldots, v_{im_i}\}$. But by Lemma 6.18, for any integer $i \in \{1, 2, \ldots, n\}$, the composition of the subgraphs of $\mathbb{G}_1$ and $\mathbb{G}_2$ respectively induced by $\mathcal{V}_i$, is contained in the subgraph of the composition of $\mathbb{G}_1$ and $\mathbb{G}_2$ induced by $\mathcal{V}_i$. This implies that for $i \in \{1, 2, \ldots, n\}$, the subgraph of the composition of $\mathbb{G}_1$ and $\mathbb{G}_2$ induced by $\mathcal{V}_i$ has $\{v_{i1}, v_{i2}, \ldots, v_{im_i}\}$ as a hierarchy. □

In the sequel we will state and prove conditions under which the composition of a sequence of delay graphs is strongly rooted. To do this, we will need to introduce several concepts. By the *quotient graph* of $\mathbb{G} \in \bar{\mathcal{D}}$, is meant that directed graph with vertex set $\mathcal{V}$ whose arc set consists of those arcs (i, j) for which $\mathbb{G}$ has an arc from some vertex in $\mathcal{V}_i$ to some vertex in $\mathcal{V}_j$. The quotient graph of $\mathbb{G}$ models which headings are being used by each agent in updates without describing the specific delayed headings actually being used. Our main result regarding delay graphs is as follows.

Proposition 6.16. *Let m be the largest integer in the set $\{m_1, m_2, \ldots, m_n\}$. The composition of any set of at least $m(n-1)^2 + m - 1$ delay graphs will be strongly rooted if the quotient graph of each of the graphs in the composition is rooted.*

To prove this proposition we will need several more concepts. Let us agree to say that a delay graph $\mathbb{G} \in \bar{\mathcal{D}}$ has *strongly rooted hierarchies* if for each $i \in \mathcal{V}$, the subgraph of $\mathbb{G}$ induced by $\mathcal{V}_i$ is strongly rooted. Proposition 6.14

states that a hierarchial graph on m_i vertices will be strongly rooted if it is the composition of at least $m_i - 1$ rooted graphs with the same hierarchy. This and Lemma 6.18 imply that the subgraph of the composition of at least $m_i - 1$ delay graphs induced by $\mathcal{V}_i$ will be strongly rooted. We are led to the following lemma.

Lemma 6.19. *Any composition of at least $m - 1$ delay graphs in $\bar{\mathcal{D}}$ has strongly rooted hierarchies.*

To proceed we will need one more type of graph which is uniquely determined by a given graph in $\bar{\mathcal{D}}$. By the *agent subgraph* of $\mathbb{G} \in \bar{\mathcal{D}}$ is meant the subgraph of $\mathbb{G}$ induced by $\mathcal{V}$. Note that while the quotient graph of $\mathbb{G}$ describes relations between distinct agent hierarchies, the agent subgraph of $\mathbb{G}$ only captures the relationships between the roots of the hierarchies.

Lemma 6.20. *Let $\mathbb{G}_p$ and $\mathbb{G}_q$ be delay graphs in $\bar{\mathcal{D}}$. If $\mathbb{G}_p$ has a strongly rooted agent subgraph and $\mathbb{G}_q$ has strongly rooted hierarchies, then the composition $\mathbb{G}_q \circ \mathbb{G}_p$ is strongly rooted.*

Proof of Lemma 6.20: Let v_{i1} be a root of the agent subgraph of $\mathbb{G}_p$ and let v_{jk} be any vertex in $\bar{\mathcal{V}}$. Then $(v_{i1}, v_{j1}) \in \mathcal{A}(\mathbb{G}_p)$ because the agent subgraph of $\mathbb{G}_p$ is strongly rooted. Moreover, $(v_{j1}, v_{jk}) \in \mathcal{A}(\mathbb{G}_q)$ because $\mathbb{G}_q$ has strongly rooted hierarchies. Therefore, in view of the definition of graph composition $(v_{i1}, v_{jk}) \in \mathcal{A}(\mathbb{G}_q \circ \mathbb{G}_p)$. Since this must be true for every vertex in $\bar{\mathcal{V}}$, $\mathbb{G}_q \circ \mathbb{G}_p$ is strongly rooted. □

Lemma 6.21. *The agent subgraph of any composition of at least $(n-1)^2$ delay graphs in $\bar{\mathcal{D}}$ will be strongly rooted if the agent subgraph of each of the graphs in the composition is rooted.*

Proof of Lemma 6.21: Let $\mathbb{G}_1, \mathbb{G}_2, \ldots, \mathbb{G}_q$ be any sequence of $q \geq (n-1)^2$ delay graphs in $\bar{\mathcal{D}}$ whose agent subgraphs, $\bar{\mathbb{G}}_i$ $i \in \{1, 2, \ldots, q\}$, are all rooted. By Proposition 6.10, $\bar{\mathbb{G}}_q \circ \cdots \circ \bar{\mathbb{G}}_2 \circ \bar{\mathbb{G}}_1$ is strongly rooted. But $\bar{\mathbb{G}}_q \circ \cdots \circ \bar{\mathbb{G}}_2 \circ \bar{\mathbb{G}}_1$ is contained in the agent subgraph of $\mathbb{G}_q \circ \cdots \circ \mathbb{G}_2 \circ \mathbb{G}_1$ because of Lemma 6.18. Therefore the agent subgraph of $\mathbb{G}_q \circ \cdots \circ \mathbb{G}_2 \circ \mathbb{G}_1$ is strongly rooted. □

Lemma 6.22. *Let $\mathbb{G}_p$ and $\mathbb{G}_q$ be delay graphs in $\bar{\mathcal{D}}$. If $\mathbb{G}_p$ has a strongly rooted hierarchies and $\mathbb{G}_q$ has a rooted quotient graph, then the agent subgraph of the composition $\mathbb{G}_q \circ \mathbb{G}_p$ is rooted.*

Proof of Lemma 6.22: Let (i, j) be any arc in the quotient graph of $\mathbb{G}_q$ with $i \neq j$. This means that $(v_{ik}, v_{js}) \in \mathcal{A}(\mathbb{G}_q)$ for some $v_{ik} \in \mathcal{V}_i$ and $v_{js} \in \mathcal{V}_j$. Clearly $(v_{i1}, v_{ik}) \in \mathcal{A}(\mathbb{G}_p)$ because $\mathbb{G}_p$ has strongly rooted hierarchies. Moreover since $i \neq j$, v_{ik} is a neighbor of $\mathcal{V}_j$ which is not in $\mathcal{V}_j$. From this and the definition of a delay graph, it follows that v_{ik} is a neighbor of v_{j1}. Therefore $(v_{ik}, v_{j1}) \in \mathcal{A}(\mathbb{G}_q)$. Thus $(v_{i1}, v_{j1}) \in \mathcal{A}(\mathbb{G}_q \circ \mathbb{G}_p)$. We have therefore proved that for any path of length one between any two distinct vertices i, j in the quotient graph of $\mathbb{G}_q$, there is a corresponding path between vertices v_{i1} and v_{j1} in the agent subgraph of $\mathbb{G}_q \circ \mathbb{G}_p$. This implies that for any path of any length between any two distinct vertices i, j in the quotient graph of $\mathbb{G}_q$, there is a corresponding path between vertices v_{i1} and v_{j1} in the agent subgraph of $\mathbb{G}_q \circ \mathbb{G}_p$. Since by assumption, the quotient graph of $\mathbb{G}_q$ is rooted, the agent subgraph of $\mathbb{G}_q \circ \mathbb{G}_p$ must be rooted as well. □

Proof of Proposition 6.16: Let $\mathbb{G}_1, \mathbb{G}_2, \ldots \mathbb{G}_s$ be a sequence of at least $m(n-1)^2+m-1$ delay graphs with strongly rooted quotient graphs. The graph $\mathbb{G}_s \circ \cdots \mathbb{G}_{(m(n-1)^2+1)}$ is composed of at least $m-1$ delay graphs. Therefore $\mathbb{G}_s \circ \cdots \mathbb{G}_{(m(n-1)^2+1)}$ must have strongly rooted hierarchies because of Lemma 6.19. In view of Lemma 6.20, to complete the proof it is enough to show that $\mathbb{G}_{(m(n-1)^2} \circ \cdots \circ \mathbb{G}_1$ has a strongly rooted agent subgraph. But $\mathbb{G}_{(m(n-1)^2} \circ \cdots \circ \mathbb{G}_1$ is the composition of $(n-1)^2$ graphs, each itself a composition of m delay graphs with rooted quotient graphs. In view of Lemma 6.21, to complete the proof it is enough to show that the agent subgraph of any composition of m delay graphs is rooted if each of the quotient graph of each delay graph in the composition is rooted. Let $\mathbb{H}_1, \mathbb{H}_2, \ldots, \mathbb{H}_m$ be such a family of delay graphs. By assumption, $\mathbb{H}_m$ has a rooted quotient graph. In view of Lemma 6.22, the agent subgraph of $\mathbb{H}_m \circ \mathbb{H}_{m-1} \circ \cdots \circ \mathbb{H}_1$ will be rooted if $\mathbb{H}_{m-1} \circ \cdots \circ \mathbb{H}_1$ has strongly rooted hierarchies. But $\mathbb{H}_{m-1} \circ \cdots \circ \mathbb{H}_1$ has this property because of Lemma 6.19. □

Convergence

Using the results from the previous section, it is possible to state results for the flocking problem with measurement delays similar to those discussed earlier for the delay free case. Towards this end let us agree to say that a finite sequence of graphs $\mathbb{G}_{p_1}, \mathbb{G}_{p_2}, \ldots, \mathbb{G}_{p_k}$ in $\mathcal{D}$ is *jointly quotient rooted* if the quotient of the composition $\mathbb{G}_{p_k} \circ \mathbb{G}_{p_{(k-1)}} \circ \cdots \circ \mathbb{G}_{p_1}$ is rooted.

In the sequel we will say that an infinite sequence of graphs $\mathbb{G}_{p_1}, \mathbb{G}_{p_2}, \ldots,$ in $\mathcal{D}$ is *repeatedly jointly quotient rooted* if there is a positive integer m for which each finite sequence $\mathbb{G}_{p_{m(k-1)+1}}, \ldots, \mathbb{G}_{p_{mk}}$, $k \geq 1$ is jointly quotient rooted. We are now in a position to state our main result on leaderless coordination with measurement delays.

Theorem 6.5. *Let $\theta(0)$ be fixed and with respect to (6.72), let $\sigma : [0, 1, 2, \ldots) \rightarrow \bar{\mathcal{Q}}$ be a switching signal for which the infinite sequence of graphs $\mathbb{G}_{\sigma(0)}, \mathbb{G}_{\sigma(1)}, \ldots$ in $\mathcal{D}$ is repeatedly jointly rooted. Then there is a constant steady state heading θ_{ss}, depending only on $\theta(0)$ and σ, for which*

$$\lim_{t \to \infty} \theta(t) = \theta_{ss}\mathbf{1} \tag{6.73}$$

where the limit is approached exponentially fast.

The proof of this theorem exploits Proposition 6.16 and parallels exactly the proof of Theorem 6.3. A proof of Theorem 6.5 therefore will not be given.

6.4 Asynchronous Flocking

In this section we consider a modified version of the flocking problem in which each agent independently updates its heading at times determined by its own clock [44]. We do not assume that the groups' clocks are synchronized together or that the times any one agent updates its heading are evenly spaced. Updating of agent i's heading is done as follows. At its *kth sensing event time* t_{ik}, agent i senses the headings $\theta_j(t_{ik}),\ \ j \in \mathcal{N}_i(t_{ik})$ of its current neighbors and from this data computes its kth way-point $w_i(t_{ik})$. In the sequel we will consider way point rules based on averaging. In particular

$$w_i(t_{ik}) = \frac{1}{n_i(t_{ik})}\left(\sum_{i\in\mathcal{N}_i(t_{ik})} \theta_j(t_{ik})\right), \quad i\{1,2,\ldots,n\} \tag{6.74}$$

where $n_i(t_{ik})$ is the number of neighbor of elements in neighbor index set $\mathcal{N}_i(t_{ik})$. Agent i then changes its heading from $\theta_i(t_{ik})$ to $w_i(t_{ik})$ on the interval $[t_{ik}, t_{i(k+1)})$. In these notes we will consider the case each agent updates its headings instantaneously at its own even times, and that it maintains fixed headings between its event times. More precisely, we will assume that agent i reaches its kth way-point at its $(k+1)$st event time and that $\theta_i(t)$ is constant on each continuous-time interval $(t_{i(k-1)}, t_{ik}]$, $k \geq 1$, where $t_{i0} = 0$ is agent i's zeroth event time. In other words for $k \geq 0$, agent i's heading satisfies is

$$\theta_i(t_{i(k+1)}) = \frac{1}{n_i(t_{ik})}\left(\sum_{j\in\mathcal{N}_i(t_{ik})} \theta_j(t_{ik})\right) \tag{6.75}$$

$$\theta_i(t) = \theta_i(t_{ik}), \quad t_{i(k-1)} < t \leq t_{ik} \tag{6.76}$$

To ensure that each agent's neighbors are unambiguously defined at each of its event times, we will further assume that agents move continuously.

Analytic Synchronization

To develop conditions under which all agents eventually move with the same heading requires the analysis of the asymptotic behavior of the *asynchronous* process which the $2n$ heading equations of the form (6.75), (6.76) define. Despite the apparent complexity of this process, it is possible to capture its salient features using a suitably defined *synchronous* discrete-time, hybrid dynamical system $\mathbb{S}$. We call the sequence of steps involved in defining $\mathbb{S}$ *analytic synchronization*. Analytic synchronization is applicable to any finite family of continuous or discrete time dynamical processes $\{\mathbb{P}_1, \mathbb{P}_2, \ldots, \ldots, \mathbb{P}_n\}$ under the following conditions. First, each process $\mathbb{P}_i$ must be a dynamical system whose inputs consist of functions of the states of the other processes as well as signals which are exogenous to the entire family. Second, each process $\mathbb{P}_i$ must have associated with it an ordered sequence of event times $\{t_{i1}, t_{i2}, \dots\}$ defined in such a way so that the state of $\mathbb{P}_i$ at event time $t_{i(k_i+1)}$ is uniquely determined by values of the exogenous signals and states of the $\mathbb{P}_j,\ \ j \in \{1, 2, \dots, n\}$ at event times t_{jk_j} which occur prior to $t_{i(k_i+1)}$ but in the finite past. Event time sequences for different processes need not be synchronized. Analytic synchronization is a procedure for creating a single synchronous process for purposes of analysis which captures the salient features of the original n asynchronously functioning processes. As a first step, all n event time sequences are merged into a single ordered sequence of even times $\mathcal{T}$. The "synchronized" state of $\mathbb{P}_i$ is then defined to be the original of $\mathbb{P}_i$ at $\mathbb{P}_i$'s event times $\{t_{i1}, t_{i2}, \dots\}$ plus possibly some additional variables; at values of $t \in \mathcal{T}$ between event times t_{ik_i} and $t_{i(k_i+1)}$, the synchronized state of $\mathbb{P}_i$ is taken to be the same at the value of its original state at time $t_{i(k+1)}$. Although it is not always possible to carry out all of these steps, when it is what ultimately results is a synchronous dynamical system $\mathbb{S}$ evolving on the index set of $\mathcal{T}$, with state composed of the synchronized states of the n individual processes under consideration. We now use these ideas to develop such a synchronous system $\mathbb{S}$ for the asynchronous process we have been studying.

Definition of $\mathbb{S}$

As a first step, let $\mathcal{T}$ denote the set of all event times of all n agents. Relabel the elements of $\mathcal{T}$ as $t_0, t_1, t_2, \cdots$ in such a way so that $t_j < t_{j+1},\ j \in \{1, 2, \dots\}$. Next define

$$\bar{\theta}_i(\tau) = \theta_i(t_\tau), \quad \tau \geq 0, \quad i \in \{1, 2, \dots, n\}. \tag{6.77}$$

In view of (6.75), it must be true that if t_τ is an event time of agent i, then

$$\bar{\theta}_i(\tau') = \frac{1}{\bar{n}_i(t_\tau)}\left(\sum_{j \in \bar{\mathcal{N}}_i(\tau)} \bar{\theta}_j(\tau)\right)$$

where $\bar{\mathcal{N}}_i(\tau) = \mathcal{N}_i(t_\tau)$, $\bar{n}_i(\tau) = n_i(t_\tau)$ and $t_{\tau'}$ is the next event time of agent i after t_τ. But $\bar{\theta}_i(\tau') = \bar{\theta}_i(\tau+1)$ because $\theta_i(t)$ is constant for $t_\tau < t \leq t_{\tau'}$ {cf., (6.76)}. Therefore

$$\bar{\theta}_i(\tau+1) = \frac{1}{\bar{n}_i(t_\tau)}\left(\sum_{j \in \bar{\mathcal{N}}_i(\tau)} \bar{\theta}_j(\tau)\right) \tag{6.78}$$

if t_τ is an event time of agent i. Meanwhile if t_τ is not an event time of agent i, then

$$\bar{\theta}_i(\tau+1) = \bar{\theta}_i(\tau), \tag{6.79}$$

again because $\theta_i(t)$ is constant between event times. Note that if we *define* $\bar{\mathcal{N}}_i(\tau) = \{i\}$ and $\bar{n}_i(\tau) = 1$ for every value of τ for which t_τ is not an event time of agent i, then (6.79) can be written as

$$\bar{\theta}_i(\tau+1) = \frac{1}{\bar{n}_i(t_\tau)}\left(\sum_{j\in\bar{\mathcal{N}}_i(\tau)} \bar{\theta}_j(\tau)\right) \tag{6.80}$$

Doing this enables us to combine (6.78) and (6.80) into a single formula valid for all $\tau \geq 0$. In other words, agent i's heading satisfies

$$\bar{\theta}_i(\tau+1) = \frac{1}{\bar{n}_i(t_\tau)}\left(\sum_{j\in\bar{\mathcal{N}}_i(\tau)} \bar{\theta}_j(\tau)\right), \quad \tau \geq 0 \tag{6.81}$$

where

$$\bar{\mathcal{N}}_i(\tau) = \begin{cases} \mathcal{N}_i(t_\tau) \text{ if } t_\tau \text{ is an event time of agent } i \\ \{i\} \quad \text{if } t_\tau \text{ is not an event time of agent } i \end{cases} \tag{6.82}$$

and $\bar{n}_i(\tau)$ is the number of indices in $\bar{\mathcal{N}}_i(\tau)$. For purposes of analysis, it is useful to interpret (6.82) as meaning that between agent i's event times, its only neighbor is itself. There are n equations of the form in (6.81) and together they define a synchronous system $\mathbb{S}$ which models the evolutions of the n agents' headings at event times.

State Space Model

As before, we can represent the neighbor relationships associated with (6.82) using a directed graph $\mathbb{G}$ with vertex set $\mathcal{V} = \{1, 2, \dots n\}$ and arc $\mathcal{A}(\mathbb{G}) \subset \mathcal{V}\times\mathcal{V}$ which is defined in such a way so that (i, j) is an arc from i to j just in case agent i is a neighbor of agent j. Thus as before, $\mathbb{G}$ is a directed graph on n vertices with at most one arc from any vertex to another and with exactly one self-arc at each vertex. We continue to write $\mathcal{G}$ for the set of all such graphs and we also continue to use the symbol $\mathcal{P}$ to denote a set indexing $\mathcal{G}$.

For each $p \in \mathcal{P}$, let $F_p = D_p^{-1}A'_p$, where A'_p is the transpose of the adjacency matrix the of graph $\mathbb{G}_p \in \mathcal{G}$ and D_p the diagonal matrix whose jth diagonal element is the in-degree of vertex j within the graph. The set of agent heading update rules defined by (6.82) can be written in state form as

$$\bar{\theta}(\tau+1) = F_{\sigma(\tau)}\bar{\theta}(\tau), \quad \tau \in \{0, 1, 2, \dots\} \tag{6.83}$$

where $\bar{\theta}$ is the heading vector $\bar{\theta} = (\bar{\theta}_1\ \bar{\theta}_2\ \dots\ \bar{\theta}_n)'$, and $\sigma : \{0, 1, \dots\} \to \mathcal{P}$ is a switching signal whose value at time τ, is the index of the graph representing the agents' neighbor relationships at time τ.

Up to this point things are essentially the same as in the basic flocking problem treated in Sect. 6.1. But when one considers the type of graphs in $\mathcal{G}$ which are likely to be encountered along a given trajectory, things are quite different. Note for example, that the only vertices of $\mathbb{G}_{\sigma(\tau)}$ which can have more than one incoming arc, are those of agents for whom τ is an event time. Thus in the most likely situation when distinct agents have only distinct event times, there will be at most one vertex in each graph $\mathbb{G}_{\sigma(\tau)}$ which has more than one incoming arc. It is this situation we want to explore further. Toward this end, let $\mathcal{G}^* \subset \mathcal{G}$ denote the subclass of all graphs which have at most one vertex with more than one incoming arc. Note that for $n > 2$, there is no rooted graph in $\mathcal{G}^*$. Nonetheless, in the light of Theorem 6.3 it is clear that convergence to a common steady state heading will occur if the infinite sequence of graphs $\mathbb{G}_{\sigma(0)}, \mathbb{G}_{\sigma(1)}, \ldots$ is repeatedly jointly rooted. This of course would require that there exist a jointly rooted sequence of graphs from $\mathcal{G}^*$. We will now explain why such sequences do in fact exist.

Let us agree to call a graph $\mathbb{G} \in \mathcal{G}$ an *all neighbor graph centered at* v if every vertex of $\mathbb{G}$ is a neighbor of v. Thus $\mathbb{G}$ is an all neighbor graph centered at v if and only if its reverse $\mathbb{G}'$ is strongly rooted at v. Note that every all neighbor graph in $\mathcal{G}$ is also in $\mathcal{G}^*$. Note also that all neighbor graphs are maximal in $\mathcal{G}^*$ with respect to the partial ordering of $\mathcal{G}^*$ by inclusion. Note also the composition of any all neighbor graph with itself is itself. On the other hand, because of the union of two graphs in $\mathcal{G}$ is always contained in the composition of the two graphs, the composition of n all neighbor graphs with distinct centers must be a graph in which each vertex is a neighbor of every other; i.e., the complete graph. Thus the composition of n all neighbor graphs with distinct centers is strongly rooted. In summary, the hypothesis of Theorem 6.3 is not vacuous for the asynchronous problem under consideration. When that hypothesis is satisfied, convergence to a common steady state heading will occur.

6.5 Leader Following

In this section we consider two modified versions of the flocking problem for the same group n agents as before, but now with one of the group's members (say agent 1) acting as the group's *leader*. In the first version of the problem, the remaining agents, henceforth called *followers* and labeled 2 through n, do not know who the leader is or even if there is a leader. Accordingly they continue to use the same heading update rule (6.1) as before. The leader on the other hand, acting on its own, ignores update rule (6.1) and moves with a constant heading $\theta_1(0)$. Thus

$$\theta_1(t+1) = \theta_1(t) \tag{6.84}$$

The situation just described can be modeled as a state space system

$$\theta(t+1) = F_{\sigma(t)}\theta(t), \quad t \geq 0 \tag{6.85}$$

just as before, except now agent 1 is constrained to have no neighbors other than itself. The graphs $\mathbb{G}_p$ which model neighbor relations accordingly all have a distinguished *leader vertex* which has no incoming arcs other than its own.

Much like before, our goal here is to show for a large class of switching signals and for any initial set of follower agent headings, that the headings of all n followers converge to the heading of the leader. Convergence in the leaderless case under the most general, required the sequence of graphs $\mathbb{G}_{\sigma(0)}, \mathbb{G}_{\sigma(1)}, \ldots$ encountered along a trajectory to be repeatedly jointly rooted. For the leader follower case now under consideration, what is required is exactly the same. However, since the leader vertex has only one incoming arc, the only way $\mathbb{G}_{\sigma(0)}, \mathbb{G}_{\sigma(1)}, \ldots$ can be repeatedly jointly rooted, is that the sequence be "rooted at the leader vertex $v = 1$." More precisely, an infinite sequence of graphs $\mathbb{G}_{p_1}, \mathbb{G}_{p_2}$, in $\mathcal{G}$ is *repeatedly jointly rooted at* v if there is a positive integer m for which each finite sequence $\mathbb{G}_{p_{m(k-1)+1}}, \ldots, \mathbb{G}_{p_{mk}}, \quad k \geq 1$ is "jointly rooted at v"; a finite sequence of directed graphs $\mathbb{G}_{p_1}, \mathbb{G}_{p_2}, \ldots, \mathbb{G}_{p_k}$ is *jointly rooted at* v if the composition $\mathbb{G}_{p_k} \circ \mathbb{G}_{p_{k-1}} \circ \cdots \circ \mathbb{G}_{p_1}$ is rooted at v. Our main result on discrete-time leader following is next.

Theorem 6.6. *Let $\theta(0)$ be fixed and let $\sigma : [0, 1, 2, \ldots) \rightarrow \mathcal{P}$ be a switching signal for which the infinite sequence of graphs $\mathbb{G}_{\sigma(0)}, \mathbb{G}_{\sigma(1)}, \ldots$ is repeatedly jointly rooted. Then*

$$\lim_{t \to \infty} \theta(t) = \theta_1(0)\mathbf{1} \tag{6.86}$$

where the limit is approached exponentially fast.

Proof of Theorem 6.6: Since any sequence which is repeatedly jointly rooted at v is repeatedly jointly rooted, Theorem 6.3 is applicable. Therefore the headings of all n agents converge exponentially fast to a single common steady state heading θ_{ss}. But since the heading of the leader is fixed, $\theta_s s$ must be the leader's heading. □

References

1. A. S. Morse. Control using logic-based switching. In A. Isidori, editor, *Trends in Control*, pages 69–113. Springer-Verlag, 1995.
2. Daniel Liberzon. *Switching in Systems and Control.* Birkhäuser, 2003.
3. V. D. Blondel and J. N. Tsitsiklis. The boundedness of all products of a pair of matrices is undecidable. *Systems and Control Letters*, 41:135–140, 2000.
4. J. P. Hespanha. Root-mean-square gains of switched linear systems. *IEEE Transactions on Automatic Control*, pages 2040–2044, Nov 2003.
5. J. P. Hespahna and A. S. Morse. Switching between stabilizing controllers. *Automatica*, 38(11), nov 2002.
6. I. M. Lie Ying. Adaptive disturbance rejection. *IEEE Transactions on Automatic Control*, 1991. Rejected for publication.
7. W. M. Wonham and A. S. Morse. Decoupling and pole assignment in linear multivariable systems: A geometric approach. *SIAM Journal on Control*, 8(1):1–18, February 1970.
8. M. A. Aizerman and F. R. Gantmacher. *Absolute Stability of Regulator Systems.* Holden-Day, 1964.

9. H. A. Simon. email communication, Feb. 23, 1997.
10. H. A. Simon. Dynamic programming under uncertainty with a quadratic criterion function. *Econometrica*, pages 171–187, feb 1956.
11. J. P. Hespanha and A. S. Morse. Certainty equivalence implies detectability. *Systems and Control Letters*, pages 1–13, 1999.
12. J. P. Hespanha. *Logic - Based Switching Algorithms in Control.* PhD thesis, Yale University, 1998.
13. F. M. Pait and A. S. Morse. A cyclic switching strategy for parameter-adaptive control. *IEEE Transactions on Automatic Control*, 39(6):1172–1183, June 1994.
14. B. D. O. Anderson, T. S. Brinsmead, F. de Bruyne, J. P. Hespanha, D. Liberzon, and A. S. Morse. Multiple model adaptive control, part 1: Finite coverings. *International Journal on Robust and Nonlinear Control*, pages 909–929, September 2000.
15. F. M. Pait. On the topologies of spaces on linear dynamical systems commonly employed as models for adaptive and robust control design. In *Proceedings of the Third SIAM Conference on Control and Its Applications*, 1995.
16. G. Zames and A. K. El-Sakkary. Unstable systems and feedback: the gap metric. In *Proc. Allerton Conf.*, pages 380–385, 1980.
17. G. Vinnicombe. A ν-gap distance for uncertain and nonlinear systems. In *Proc. of the 38th Conf. on Decision and Contr.*, pages 2557–2562, 1999.
18. A. S. Morse. Supervisory control of families of linear set-point controllers - part 1: Exact matching. *IEEE Transactions on Automatic Control*, pages 1413–1431, oct 1996.
19. A. S. Morse. Supervisory control of families of linear set-point controllers - part 2: Robustness. *IEEE Transactions on Automatic Control*, 42:1500–1515, nov 1997. see also Proceedings of 1995 IEEE Conference on Decision and Control pp. 1750–1755.
20. S. R. Kulkarni and P. J. Ramadge. Model and controller selection policies based on output prediction errors. *IEEE Transactions on Automatic Control*, 41:1594–1604, 1996.
21. D. Borrelli, A. S. Morse, and E. Mosca. Discrete-time supervisory control of families of 2-degree of freedom linear set-point controllers. *IEEE Transactions on Automatic Control*, pages 178–181, jan 1999.
22. K. S. Narendra and J. Balakrishnan. Adaptive control using multiple models. *IEEE Transactions on Automatic Control*, pages 171–187, feb 1997.
23. A. S. Morse. A bound for the disturbance-to-tracking error gain of a supervised set-point control system. In D Normand Cyrot, editor, *Perspectives in Control – Theory and Applications*, pages 23–41. Springer-Verlag, 1998.
24. A. S. Morse. Analysis of a supervised set-point control system containing a compact continuum of finite dimensional linear controllers. In *Proc. 2004 MTNS*, 2004.
25. T. Vicsek, A. Czirók, E. Ben-Jacob, I. Cohen, and O. Shochet. Novel type of phase transition in a system of self-driven particles. *Physical Review Letters*, pages 1226–1229, 1995.
26. C. Reynolds. Flocks, birds, and schools: a distributed behavioral model. *Computer Graphics*, 21:25–34, 1987.
27. A. Jadbabaie, J. Lin, and A. S. Morse. Coordination of groups of mobile autonomous agents using nearest neighbor rules. *IEEE Transactions on Automatic Control*, pages 988–1001, june 2003. also in Proc. 2002 IEEE CDC, pages 2953–2958.

28. C. Godsil and G. Royle. *Algebraic Graph Theory*. Springer Graduate Texts in Mathematics # 207, New York, 2001.
29. E. Seneta. *Non-negative Matrices and Markov Chains*. Springer-Verlag, New York, 1981.
30. J. Wolfowitz. Products of indecomposable, aperiodic, stochastic matrices. *Proceedings of the American Mathematical Society*, 15:733–736, 1963.
31. D. J. Hartfiel. *Markov set-chains*. Springer, Berlin; New York, 1998.
32. H. Tanner, A. Jadbabaie, and G. Pappas. Distributed coordination strategies for groups of mobile autonomous agents. Technical report, ESE Department, University of Pennsylvania, December 2002.
33. L. Moreau. Stability of multi-agent systems with time-dependent communication links. *IEEE Transactions on Automatic Control*, pages 169–182, February 2005.
34. W. Ren and R. Beard. Consensus seeking in multiagent systems under dynamically changing interaction topologies. *IEEE Transactions on Automatic Control*, 50:655–661, 2005.
35. D. Angeli and P. A. Bliman. Stability of leaderless multi-agent systems. 2004. technical report.
36. Z. Lin, B. Francis, and M. Brouche. Local control strategies for groups of mobile autonomous agents. *IEEE Trans. Auto. Control*, pages 622–629, april 2004.
37. V. D. Blondel, J. M. Hendrichx, A. Olshevsky, and J. N. Tsitsiklis. Convergence in multiagent coordination, consensus, and flocking. 2005. Submitted to IEEE CDC 2005.
38. R. Horn and C. R. Johnson. *Matrix Analysis*. Cambridge University Press, New York, 1985.
39. J. L. Doob. *Stochastic Processes*, chapter 5: Markov Processes, Discrete Parameter. John Wiley & Sons, Inc., New York, 1953.
40. J. N. Tsisiklis. *Problems in decentralized decision making and computation.* Ph.D dissertation, Department of Electrical Engineering and Computer Science, M.I.T., 1984.
41. C. F. Martin and W. P. Dayawansa. On the existence of a Lyapunov function for a family of switching systems. In *Proc. of the 35th Conf. on Decision and Contr.*, December 1996.
42. M. Cao, D. A. Spielman, and A. S. Morse. A lower bound on convergence of a distributed network consensus algorithm. In *Proc. 2005 IEEE CDC*, 2005.
43. B. Mohar. The Laplacian spectrum of graphs. *in Graph theory, combinatorics and applications (Ed. Y. Alavi G. Chartrand, O. R. Ollerman, and A. J. Schwenk)*, 2:871–898, 1991.
44. M. Cao, A. S. Morse, and B. D. O. Anderson. Coordination of an asynchronous multi-agent system via averaging. In *Proc. 2005 IFAC Congress*, 2005.
45. Miroslav Krstić, Ioannis Kanellakopoulos, and Petar Kokotović. *Nonlinear and Adaptive Control Design*. Adaptive and Learning Systems for Signal Processing, Communications, and Control. John Wiley & Sons, New York, 1995.
46. Z. Lin, M. Brouche, and B. Francis. Local control strategies for groups of mobile autonomous agents. Ece control group report, University of Toronto, 2003.
47. A. Isidori. *Nonlinear Control Systems*. Springer-Verlag, 1989.

Input to State Stability: Basic Concepts and Results

E.D. Sontag

Department of Mathematics, Hill Center, Rutgers University,
110 Frelinghuysen Rd, Piscataway, NJ 08854-8019, USA
sontag@math.rutgers.edu

1 Introduction

The analysis and design of nonlinear feedback systems has recently undergone an exceptionally rich period of progress and maturation, fueled, to a great extent, by (1) the discovery of certain basic conceptual notions, and (2) the identification of classes of systems for which systematic decomposition approaches can result in effective and easily computable control laws. These two aspects are complementary, since the latter approaches are, typically, based upon the inductive verification of the validity of the former system properties under compositions (in the terminology used in [62], the "activation" of theoretical concepts leads to "constructive" control).

This expository presentation addresses the first of these aspects, and in particular the precise formulation of questions of robustness with respect to disturbances, formulated in the paradigm of *input to state stability*. We provide an intuitive and informal presentation of the main concepts. More precise statements, especially about older results, are given in the cited papers, as well as in several previous surveys such as [103, 105] (of which the present paper represents an update), but we provide a little more detail about relatively recent work. Regarding applications and extensions of the basic framework, we give some pointers to the literature, but we do not focus on feedback design and specific engineering problems; for the latter we refer the reader to textbooks such as [27, 43, 44, 58, 60, 66, 96].

2 ISS as a Notion of Stability of Nonlinear I/O Systems

Our subject is the study of *stability-type questions for input/output ("i/o") systems*. We later define more precisely what we mean by "system," but, in an intuitive sense, we have in mind the situation represented in Fig. 1, where the "system" may well represent a component ("module" or "subsystem") of a more complex, larger, system. In typical applications of control theory,

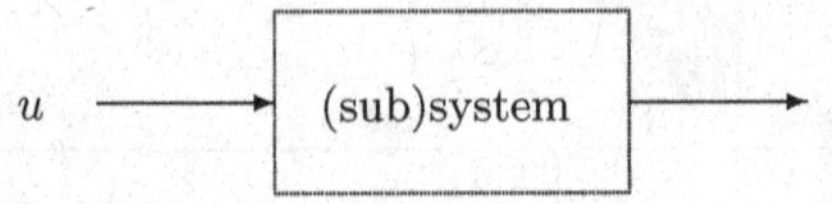

Fig. 1. I/O system

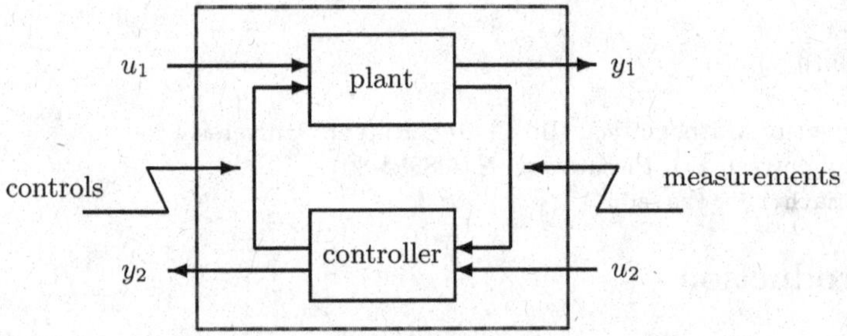

Fig. 2. Plant and controller

our "system" may in turn represent a plant/controller combination (Fig. 2), where the input $u = (u_1, u_2)$ incorporates actuator and measurement noises respectively, as well as disturbances or tracking signals, and where $y = (y_1, y_2)$ might consist respectively of some measure of performance (distance to a set of desired states, tracking error, etc.) and quantities directly available to a controller.

The goals of our work include:

- Helping develop a "toolkit" of concepts for studying systems via decompositions
- The quantification of system response to external signals
- The unification of state-space and input/output stability theories

2.1 Desirable Properties

We wish to formalize the idea of "stability" of the mapping $u(\cdot) \mapsto y(\cdot)$. Intuitively, we look for a concept that encompasses the properties that inputs that are bounded, "eventually small," "integrally small," or convergent, produce outputs with the respective property:

$$u \left\{ \begin{array}{c} \text{bounded} \\ \text{(ev)small} \\ \text{(integ)small} \\ \to 0 \end{array} \right\} \overset{?}{\Rightarrow} y \left\{ \begin{array}{c} \text{bounded} \\ \text{(ev)small} \\ \text{(integ)small} \\ \to 0 \end{array} \right\}$$

and, in addition, we will also want to account appropriately for initial states and transients. A special case is that in which the output y of the system is just the internal state. The key notion in our study will be one regarding such stability from inputs to states; only later do we consider more general outputs. In terms of states, thus, the properties that we would like to encompass in a good stability notion include the the *convergent-input convergent-state* (CICS) and the *bounded-input bounded-state* (BIBS) properties.

We should remark that, for simplicity of exposition, we concentrate here solely on stability notions relative to globally attractive steady states. However, the general theory allows consideration of more arbitrary attractors (so that norms get replaced by, for example, distances to certain compact sets), and one may also consider local versions, as well as robust and/or adaptive concepts associated to the ones that we will define.

2.2 Merging Two Different Views of Stability

Broadly speaking, there are two main competing approaches to system stability: the *state-space approach* usually associated with the name of Lyapunov, and the *operator* approach, of which George Zames was one of the main proponents and developers and which was the subject of major contributions by Sandberg, Willems, Safonov, and others. Our objective is in a sense (Fig. 3) that of merging these "Lyapunov" and "Zames" views of stability. The operator approach studies the i/o mapping

$$\begin{array}{ccc} (x^0, u(\cdot)) & \mapsto & y(\cdot) \\ \mathbb{R}^n \times [\mathcal{L}_q(0,+\infty)]^m & \to & [\mathcal{L}_q(0,+\infty)]^p \end{array}$$

(with, for instance, $q = 2$ or $q = \infty$, and assuming the operator to be well-defined and bounded) and has several advantages, such as allowing the use

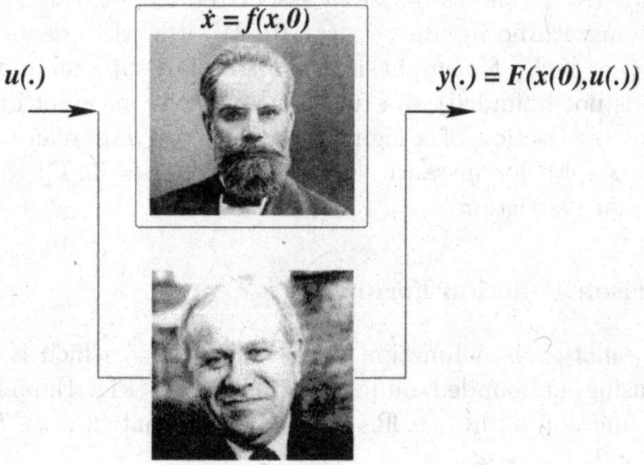

Fig. 3. Lyapunov state-space and Zames-like external stability

of Hilbert or Banach space techniques, and elegantly generalizing many properties of linear systems, especially in the context of robustness analysis, to certain nonlinear situations. The state-space approach, in contrast, is geared to the study of systems without inputs, but is better suited to the study of nonlinear dynamics, and it allows the use of geometric and topological ideas. The ISS conceptual framework is consistent with, and combines several features of, both approaches.

2.3 Technical Assumptions

In order to keep the discussion as informal and simple as possible, we make the assumption from now on that we are dealing with systems with inputs and outputs, in the usual sense of control theory [104]:

$$\dot{x}(t) \;=\; f(x(t), u(t))\,, \quad y(t) = h(x(t))$$

(usually omitting arguments t from now on) with states $x(t)$ taking values in Euclidean space $\mathbb{R}^n$, inputs (also called "controls" or "disturbances" depending on the context) being measurable locally essentially bounded maps $u(\cdot) : [0, \infty) \to \mathbb{R}^m$, and output values $y(t)$ taking values in $\mathbb{R}^p$, for some positive integers n, m, p. The map $f : \mathbb{R}^n \times \mathbb{R}^m \to \mathbb{R}^n$ is assumed to be locally Lipschitz with $f(0,0) = 0$, and $h : \mathbb{R}^n \to \mathbb{R}^p$ is continuous with $h(0) = 0$. Many of these assumptions can be weakened considerably, and the cited references should be consulted for more details. We write $x(t, x^0, u)$ to denote the solution, defined on some maximal interval $[0, t_{\max}(x^0, u))$, for each initial state x^0 and input u. In particular, for systems with no inputs

$$\dot{x}(t) \;=\; f(x(t))\,,$$

we write just $x(t, x^0)$. The *zero-system* associated to $\dot{x} = f(x, u)$ is by definition the system with no inputs $\dot{x} = f(x, 0)$. We use $|x|$ to denote Euclidean norm and $\|u\|$, or $\|u\|_\infty$ for emphasis, the (essential) supremum norm (possibly $+\infty$, if u is not bounded) of a function, typically an input or an output. When only the restriction of a signal to an interval I is relevant, we write $\|u_I\|_\infty$ (or just$\|u_I\|$), for instance $\|u_{[0,T]}\|_\infty$ when $I = [0, T]$, to denote the sup norm of that restriction.

2.4 Comparison Function Formalism

A *class* $\mathcal{K}_\infty$ *function* is a function $\alpha : \mathbb{R}_{\geq 0} \to \mathbb{R}_{\geq 0}$ which is continuous, strictly increasing, unbounded, and satisfies $\alpha(0) = 0$ (Fig. 4), and a *class* $\mathcal{KL}$ *function* is a function $\beta : \mathbb{R}_{\geq 0} \times \mathbb{R}_{\geq 0} \to \mathbb{R}_{\geq 0}$ such that $\beta(\cdot, t) \in \mathcal{K}_\infty$ for each t and $\beta(r, t) \searrow 0$ as $t \to \infty$.

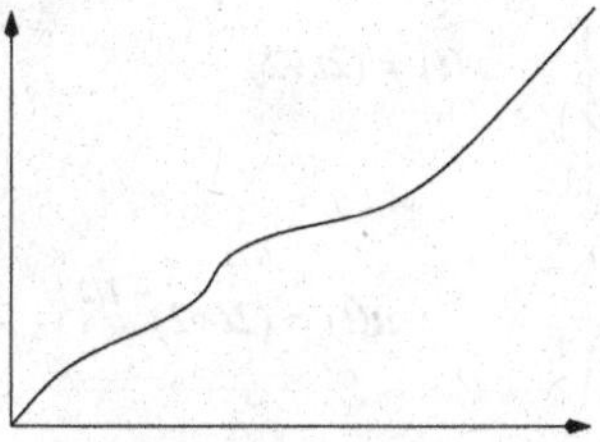

Fig. 4. $\mathcal{K}_\infty$-function

2.5 Global Asymptotic Stability

For a system with no inputs $\dot{x} = f(x)$, there is a well-known notion of global asymptotic stability (for short from now on, *GAS*, or "*0-GAS*" when referring to the zero-system $\dot{x} = f(x, 0)$ associated to a given system with inputs $\dot{x} = f(x, u)$) due to Lyapunov, and usually defined in "ε-δ" terms. It is an easy exercise to show that this standard definition is in fact equivalent to the following statement:

$$(\exists\, \beta \in \mathcal{KL}) \quad |x(t, x^0)| \;\leq\; \beta\left(|x^0|, t\right) \quad \forall\, x^0\,,\, \forall\, t \geq 0\,.$$

Observe that, since β decreases on t, we have, in particular:

$$|x(t, x^0)| \;\leq\; \beta\left(|x^0|, 0\right) \quad \forall\, x^0\,,\, \forall\, t \geq 0\,,$$

which provides the Lyapunov-stability or "small overshoot" part of the GAS definition (because $\beta\left(|x^0|, 0\right)$ is small whenever $|x^0|$ is small, by continuity of $\beta(\cdot, 0)$ and $\beta(0, 0) = 0$), while the fact that $\beta \to 0$ as $t \to \infty$ gives:

$$|x(t, x^0)| \;\leq\; \beta\left(|x^0|, t\right) \underset{t \to \infty}{\longrightarrow} 0 \quad \forall\, x^0\,,$$

which is the attractivity (convergence to steady state) part of the GAS definition.

We also remark a property proved in [102], Proposition 7, namely that for each $\beta \in \mathcal{KL}$ there exist two class $\mathcal{K}_\infty$ functions α_1, α_2 such that:

$$\beta(r, t) \;\leq\; \alpha_2\left(\alpha_1(r) e^{-t}\right) \quad \forall\, s, t \geq 0\,,$$

which means that the GAS estimate can be also written in the form:

$$|x(t, x^0)| \;\leq\; \alpha_2\left(\alpha_1(|x^0|) e^{-t}\right)$$

and thus suggests a close analogy between GAS and an exponential stability estimate $|x(t, x^0)| \leq c\, |x^0|\, e^{-at}$.

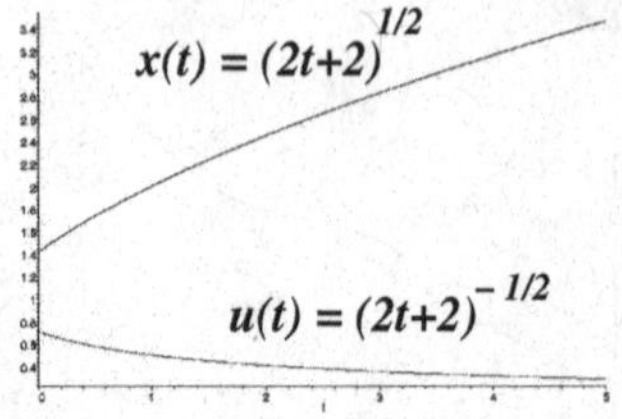

Fig. 5. Diverging state for converging input, for example

2.6 0-GAS Does Not Guarantee Good Behavior with Respect to Inputs

A *linear* system in control theory is one for which both f and h are linear mappings:

$$\dot{x} = Ax + Bu\,, \quad y = Cx$$

with $A \in \mathbb{R}^{n\times n}$, $B \in \mathbb{R}^{n\times m}$, and $C \in \mathbb{R}^{p\times n}$. It is well-known that a linear system is 0-GAS (or "internally stable") if and only if the matrix A is a *Hurwitz* matrix, that is to say, all the eigenvalues of A have negative real parts. Such a 0-GAS linear system automatically satisfies all reasonable input/output stability properties: bounded inputs result in bounded state trajectories as well as outputs, inputs converging to zero imply solutions (and outputs) converging to zero, and so forth; see, e.g., [104]. But *the 0-GAS property is not equivalent*, in general, to input/output, or even input/state, stability of any sort. This is in general false for nonlinear systems. For a simple example, consider the following one-dimensional ($n = 1$) system, with scalar ($m = 1$) inputs:

$$\dot{x} \;=\; -x + (x^2 + 1)u\,.$$

This system is clearly 0-GAS, since it reduces to $\dot{x} = -x$ when $u \equiv 0$. On the other hand, solutions diverge even for some inputs that converge to zero. For example, take the control $u(t) = (2t + 2)^{-1/2}$ and $x^0 = \sqrt{2}$, There results the unbounded trajectory $x(t) = (2t + 2)^{1/2}$ (Fig. 5). This is in spite of the fact that the unforced system is GAS. Thus, the converging-input converging-state property does not hold. Even worse, the bounded input $u \equiv 1$ results in a finite-time explosion. This example is not artificial, as it arises in feedback-linearization design, as we mention below.

2.7 Gains for Linear Systems

For linear systems, the three most typical ways of defining input/output stability in terms of operators

$$\{L^2, L^\infty\} \to \{L^2, L^\infty\}$$

are as follows. (In each case, we mean, more precisely, to ask that there should exist positive c and λ such that the given estimates hold for all $t \geq 0$ and all solutions of $\dot{x} = Ax + Bu$ with $x(0) = x^0$ and arbitrary inputs $u(\cdot)$.)

$$\text{“}L^\infty \to L^\infty\text{”}: \qquad c\,|x(t,x^0,u)| \;\leq\; |x^0|\,e^{-\lambda t} + \sup_{s\in[0,t]} |u(s)|$$

$$\text{“}L^2 \to L^\infty\text{”}: \qquad c\,|x(t,x^0,u)| \;\leq\; |x^0|\,e^{-\lambda t} + \int_0^t |u(s)|^2\,ds$$

$$\text{“}L^2 \to L^2\text{”}: \qquad c\int_0^t |x(s,x^0,u)|^2\,ds \;\leq\; |x^0| + \int_0^t |u(s)|^2\,ds$$

(the missing case $L^\infty \to L^2$ is less interesting, being too restrictive). For linear systems, these are all equivalent in the following sense: if an estimate of one type exists, then the other two estimates exist too. The actual numerical values of the constants c, λ appearing in the different estimates are not necessarily the same: they are associated to the various types of norms on input spaces and spaces of solutions, such as "H_2" and "H_∞" gains, see, e.g., [23]. Here we are discussing only the question of *existence* of estimates of these types. It is easy to see that existence of the above estimates is simply equivalent to the requirement that the A matrix be Hurwitz, that is to say, to 0-GAS, the asymptotic stability of the unforced system $\dot{x} = Ax$.

2.8 Nonlinear Coordinate Changes

A "geometric" view of nonlinear dynamics leads one to adopt the view that

notions of stability should be invariant under (nonlinear) changes of variables

– meaning that if we make a change of variables in a system which is stable in some technical sense, the system in new coordinates should again be stable in the same sense. For example, suppose that we start with the exponentially stable system $\dot{x} = -x$, but we make the change of variables $y = T(x)$ and wish to consider the equation $\dot{y} = f(y)$ satisfied by the new variable y. Suppose that $T(x) \approx \ln x$ for large x. If it were the case that the system $\dot{y} = f(y)$ is globally exponentially stable ($|y(t)| \leq ce^{-\lambda t}\,|y(0)|$ for some positive constants c, λ), then there would exist some time $t_0 > 0$ so that $|y(t_0)| \leq |y(0)|/2$ for all $y(0)$. But $\dot{y} = T'(x)\dot{x} \approx -1$ for large y, so $y(t_0) \approx y(0) - t_0$, contradicting $|y(t_0)| \leq |y(0)|/2$ for large enough $y(0)$. In conclusion, exponential stability is *not* a natural mathematical notion when nonlinear coordinate changes are allowed. This is why the notion of asymptotic stability is important.

Let us now discuss this fact in somewhat more abstract terms, and see how it leads us to GAS and, when adding inputs, to ISS. By a *change of coordinates* we will mean a map

$$T : \mathbb{R}^n \to \mathbb{R}^n$$

such that the following properties hold: $T(0) = 0$ (since we want to preserve the equilibrium at $x = 0$), T is continuous, and it admits an inverse map $T^{-1} : \mathbb{R}^n \to \mathbb{R}^n$ which is well-defined and continuous as well. (In other words, T is a homeomorphism which fixes the origin. We could also add the requirement that T should be differentiable, or that it be differentiable at least for $x \neq 0$, but the discussion to follow does not need this additional condition.) Now suppose that we start with a system $\dot{x} = f(x)$ that is exponentially stable:

$$|x(t, x^{\circ})| \leq c|x^{\circ}|e^{-\lambda t} \quad \forall t \geq 0 \qquad \text{(some } c, \lambda > 0)$$

and we perform a change of variables:

$$x(t) = T(z(t)).$$

We introduce, for this transformation T, the following two functions:

$$\underline{\alpha}(r) := \min_{|x| \geq r} |T(x)| \quad \text{and} \quad \overline{\alpha}(r) := \max_{|x| \leq r} |T(x)|,$$

which are well-defined because T and its inverse are both continuous, and are both functions of class $\mathcal{K}_\infty$ (easy exercise). Then,

$$\underline{\alpha}(|x|) \leq |T(x)| \leq \overline{\alpha}(|x|) \quad \forall x \in \mathbb{R}^n$$

and therefore, substituting $x(t, x^{\circ}) = T(z(t, z^{\circ}))$ in the exponential stability estimate:

$$\underline{\alpha}(|z(t, z^{\circ})|) \leq c\overline{\alpha}(|z^{\circ}|)\, e^{-\lambda t}$$

where $z^{\circ} = T^{-1}(x^{\circ})$. Thus, the estimate in z-coordinates takes the following form:

$$|z(t, z^{\circ})| \leq \beta(|z^{\circ}|, t)$$

where $\beta(r, t) = \underline{\alpha}^{-1}\left(c\overline{\alpha}\left(re^{-\lambda t}\right)\right)$ is a function of class $\mathcal{KL}$. (As remarked earlier, any possible function of class $\mathcal{KL}$ can be written in this factored form, actually.)

In summary, we re-derived the concept of global asymptotic stability simply by making coordinate changes on globally exponentially stable systems. So let us see next where these considerations take us when looking at systems with inputs and starting from the previously reviewed notions of stability for linear systems. Since there are now inputs, in addition to the state transformation $x(t) = T(z(t))$, we must now allow also transformations $u(t) = S(v(t))$, where S is a change of variables in the space of input values $\mathbb{R}^m$. Arguing exactly as for the case of systems without inputs, we arrive to the following three concepts:

$$L^\infty \to L^\infty \rightsquigarrow \alpha(|x(t)|) \leq \beta(|x^{\circ}|, t) + \sup_{s \in [0,t]} \gamma(|u(s)|),$$

$$L^2 \to L^\infty \rightsquigarrow \alpha(|x(t)|) \le \beta(|x^\circ|, t) + \int_0^t \gamma(|u(s)|)\, ds$$

$$L^2 \to L^2 \rightsquigarrow \int_0^t \alpha(|x(s)|)\, ds \le \alpha_0(|x^\circ|) + \int_0^t \gamma(|u(s)|)\, ds\,.$$

From now on, we often write $x(t)$ instead of the more cumbersome $x(t, x^\circ, u)$ and we adopt the convention that, any time that an estimate like the ones above is presented, an unless otherwise stated, we mean that there should exist comparison functions ($\alpha, \alpha_0 \in \mathcal{K}_\infty$, $\beta \in \mathcal{KL}$) such that the estimates hold for all inputs and initial states. We will study these three notions one at a time.

2.9 Input-to-State Stability

The "$L^\infty \to L^\infty$" estimate, under changes of variables, leads us to the first concept, that of *input to state stability* (ISS). That is, there should exist some $\beta \in \mathcal{KL}$ and $\gamma \in \mathcal{K}_\infty$ such that

$$|x(t)| \le \beta(|x^\circ|, t) + \gamma(\|u\|_\infty) \qquad \text{(ISS)}$$

holds for all solutions. By "all solutions" we mean that this estimate is valid for all inputs $u(\cdot)$, all initial conditions x°, and all $t \ge 0$. Note that we did not now include the function "α" in the left-hand side. That is because, redefining β and γ, one can assume that α is the identity: if $\alpha(r) \le \beta(s,t) + \gamma(t)$ holds, then also $r \le \alpha^{-1}(\beta(s,t) + \gamma(t)) \le \alpha^{-1}(2\beta(s,t)) + \alpha^{-1}(2\gamma(t))$; since $\alpha^{-1}(2\beta(\cdot,\cdot)) \in \mathcal{KL}$ and $\alpha^{-1}(2\gamma(\cdot)) \in \mathcal{K}_\infty$, an estimate of the same type, but now with no "α," is obtained. In addition, note that the supremum $\sup_{s\in[0,t]} \gamma(|u(s)|)$ over the interval $[0,t]$ is the same as $\gamma(\|u_{[0,t]}\|_\infty) = \gamma(\sup_{s\in[0,t]}(|u(s)|))$, because the function γ is increasing, and that we may replace this term by $\gamma(\|u\|_\infty)$, where $\|u\|_\infty = \sup_{s\in[0,\infty)} \gamma(|u(s)|)$ is the sup norm of the input, because the solution $x(t)$ depends only on values $u(s), s \le t$ (so, we could equally well consider the input that has values $\equiv 0$ for all $s > t$).

It is important to note that a potentially weaker definition might simply have requested that this condition hold merely for all $t \in [0, t_{\max}(x^\circ, u))$. However, this definition turns out to be equivalent to the one that we gave. Indeed, if the estimate holds a priori only on such a maximal interval of definition, then, since the right-hand is bounded on $[0,T]$, for any $T > 0$ (recall that inputs are by definition assumed to be bounded on any bounded interval), it follows that the maximal solution of $x(t, x^\circ, u)$ is bounded, and therefore that $t_{\max}(x^\circ, u) = +\infty$ (see, e.g., Proposition C.3.6 in [104]). In other words, the ISS estimate holds for all $t \ge 0$ automatically, if it is required to hold merely for maximal solutions.

Since, in general, $\max\{a,b\} \le a + b \le \max\{2a, 2b\}$, one can restate the ISS condition in a slightly different manner, namely, asking for the existence of some $\beta \in \mathcal{KL}$ and $\gamma \in \mathcal{K}_\infty$ (in general different from the ones in the ISS definition) such that

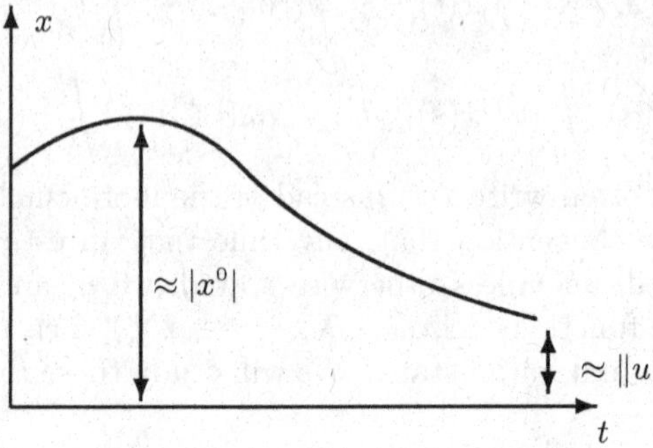

Fig. 6. ISS combines overshoot and asymptotic behavior

$$|x(t)| \leq \max\{\beta(|x^0|, t), \gamma(\|u\|_\infty)\}$$

holds for all solutions. Such redefinitions, using "max" instead of sum, will be possible as well for each of the other concepts to be introduced later; we will use whichever form is more convenient in each context, leaving implicit the equivalence with the alternative formulation.

Intuitively, the definition of ISS requires that, for t large, the size of the state must be bounded by some function of the sup norm – that is to say, the amplitude, – of inputs (because $\beta(|x^0|, t) \to 0$ as $t \to \infty$). On the other hand, the $\beta(|x^0|, 0)$ term may dominate for small t, and this serves to quantify the magnitude of the transient (overshoot) behavior as a function of the size of the initial state x^0 (Fig. 6). The *ISS superposition theorem*, discussed later, shows that ISS is, in a precise mathematical sense, the conjunction of two properties, one of them dealing with asymptotic bounds on $|x^0|$ as a function of the magnitude of the input, and the other one providing a transient term obtained when one ignores inputs.

2.10 Linear Case, for Comparison

For internally stable linear systems $\dot{x} = Ax + Bu$, the variation of parameters formula gives immediately the following inequality:

$$|x(t)| \leq \beta(t)\, |x^0| + \gamma \|u\|_\infty,$$

where

$$\beta(t) = \|e^{tA}\| \to 0 \quad \text{and} \quad \gamma = \|B\| \int_0^\infty \|e^{sA}\| ds < \infty .$$

This is a particular case of the ISS estimate, $|x(t)| \leq \beta(|x^0|, t) + \gamma(\|u\|_\infty)$, with linear comparison functions.

2.11 Feedback Redesign

The notion of ISS arose originally as a way to precisely formulate, and then answer, the following question. Suppose that, as in many problems in control theory, a system $\dot{x} = f(x,u)$ has been stabilized by means of a feedback law $u = k(x)$ (Fig. 7), that is to say, k was chosen such that the origin of the closed-loop system $\dot{x} = f(x,k(x))$ is globally asymptotically stable. (See, e.g., [103] for a discussion of mathematical aspects of state feedback stabilization.) Typically, the design of k was performed by ignoring the effect of possible *input disturbances* $d(\cdot)$ (also called actuator disturbances). These "disturbances" might represent true noise or perhaps errors in the calculation of the value $k(x)$ by a physical controller, or modeling uncertainty in the controller or the system itself. What is the effect of considering disturbances? In order to analyze the problem, we incorporate d into the model, and study the new system $\dot{x} = f(x,k(x)+d)$, where d is seen as an input (Fig. 8). We then ask what is the effect of d on the behavior of the system.

Disturbances d may well destabilize the system, and the problem may arise even when using a routine technique for control design, feedback linearization. To appreciate this issue, we take the following very simple example. We are given the system

$$\dot{x} = f(x,u) = x + (x^2+1)u\,.$$

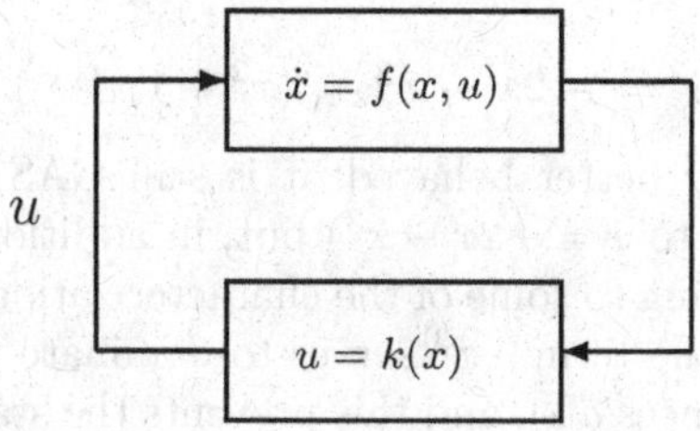

Fig. 7. Feedback stabilization, closed-loop system $\dot{x} = f(x,k(x))$

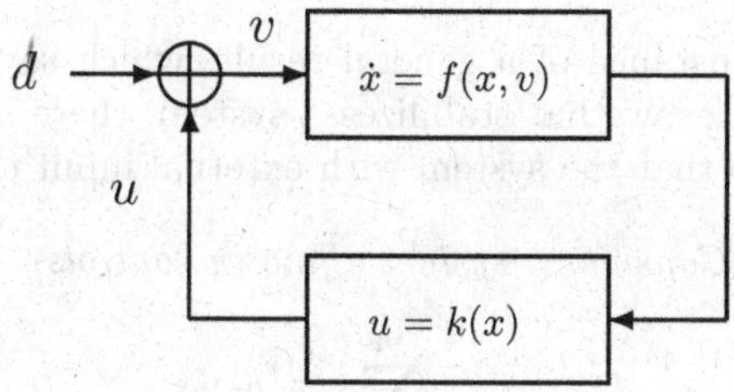

Fig. 8. Actuator disturbances, closed-loop system $\dot{x} = f(x,k(x)+d)$

In order to stabilize it, we first substitute $u = \frac{\tilde{u}}{x^2+1}$ (a preliminary feedback transformation), rendering the system linear with respect to the new input $\tilde{u}$: $\dot{x} = x + \tilde{u}$, and then we use $\tilde{u} = -2x$ in order to obtain the closed-loop system $\dot{x} = -x$. In other words, in terms of the original input u, we use the feedback law:

$$k(x) = \frac{-2x}{x^2+1}$$

so that $f(x, k(x)) = -x$. This is a GAS system. Next, let us analyze the effect of the disturbance input d. The system $\dot{x} = f(x, k(x) + d)$ is:

$$\dot{x} = -x + (x^2 + 1)\, d\,.$$

As seen before, this system has solutions which diverge to infinity even for inputs d that converge to zero; moreover, the constant input $d \equiv 1$ results in solutions that explode in finite time. Thus $k(x) = \frac{-2x}{x^2+1}$ was not a good feedback law, in the sense that its performance degraded drastically once that we took into account actuator disturbances.

The key observation for what follows is that, if we add a correction term "$-x$" to the above formula for $k(x)$, so that we now have:

$$\tilde{k}(x) = \frac{-2x}{x^2+1} - \boldsymbol{x}$$

then the system $\dot{x} = f(x, \tilde{k}(x) + d)$ with disturbance d as input becomes, instead:

$$\dot{x} = -2x - \boldsymbol{x^3} + (x^2 + 1)\, d$$

and this system is much better behaved: it is still GAS when there are no disturbances (it reduces to $\dot{x} = -2x - x^3$) but, in addition, it is ISS (easy to verify directly, or appealing to some of the characterizations mentioned later). Intuitively, for large x, the term $-x^3$ serves to dominate the term $(x^2 + 1)d$, for all bounded disturbances $d(\cdot)$, and this prevents the state from getting too large.

2.12 A Feedback Redesign Theorem for Actuator Disturbances

This example is an instance of a general result, which says that, whenever there is some feedback law that stabilizes a system, there is also a (possibly different) feedback so that the system with external input d (Fig. 9) is ISS.

Theorem 2.1. [99] *Consider a system affine in controls*

$$\dot{x} = f(x, u) = g_0(x) + \sum_{i=1}^{m} u_i g_i(x) \qquad (g_0(0) = 0)$$

and suppose that there is some differentiable feedback law $u = k(x)$ *so that*

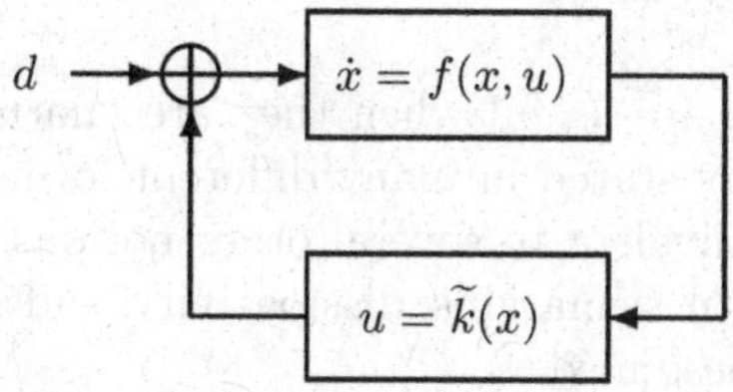

Fig. 9. Different feedback ISS-stabilizes

$$\dot{x} = f(x, k(x))$$

has $x = 0$ as a GAS equilibrium. Then, there is a feedback law $u = \widetilde{k}(x)$ such that

$$\dot{x} = f(x, \widetilde{k}(x) + d)$$

is ISS with input $d(\cdot)$.

The proof is very easy, once that the appropriate technical machinery has been introduced: one starts by considering a smooth Lyapunov function V for global asymptotic stability of the origin in the system $\dot{x} = f(x, k(x))$ (such a V always exists, by classical converse theorems); then $\hat{k}(x) := -(L_G V(x))^T = -(\nabla V(x) G(x))^T$, where G is the matrix function whose columns are the g_i, $i = 1, \ldots, m$ and T indicates transpose, provides the necessary correction term to add to k. This term has the same degree of smoothness as the vector fields making up the original system. Somewhat less than differentiability of the original k is enough for this argument: continuity is enough. However, if no continuous feedback stabilizer exists, then no smooth V can be found. (Continuous stabilization of nonlinear systems is basically equivalent to the existence of what are called smooth control-Lyapunov functions, see, e.g., [103].) In that case, if only discontinuous stabilizers are available, the result can still be generalized, see [79], but the situation is harder to analyze, since even the notion of "solution" of the closed-loop system $\dot{x} = f(x, k(x))$ has to be carefully defined.

There is also a redefinition procedure for systems that are not affine on inputs, but the result as stated above is false in that generality, and is much less interesting; see [101] for a discussion.

The above feedback redesign theorem is merely the beginning of the story. See for instance the book [60], and the references given later, for many further developments on the subjects of recursive feedback design, the "backstepping" approach, and other far-reaching extensions.

3 Equivalences for ISS

Mathematical concepts are useful when they are "natural" in the sense that they can be equivalently stated in many different forms. As it turns out, ISS can be shown to be equivalent to several other notions, including asymptotic gain, existence of robustness margins, dissipativity, and an energy-like stability estimate. We review these next.

3.1 Nonlinear Superposition Principle

Clearly, if a system is ISS, then the system with no inputs $\dot{x} = f(x, 0)$ is GAS: the term $\|u\|_\infty$ vanishes, leaving precisely the GAS property. In particular, then, the system $\dot{x} = f(x, u)$ is *0-stable*, meaning that the origin of the system without inputs $\dot{x} = f(x, 0)$ is stable in the sense of Lyapunov: for each $\varepsilon > 0$, there is some $\delta > 0$ such that $|x^\mathrm{o}| < \delta$ implies $|x(t, x^\mathrm{o})| < \varepsilon$. (In comparison-function language, one can restate 0-stability as: there is some $\gamma \in \mathcal{K}$ such that $|x(t, x^\mathrm{o})| \leq \gamma(|x^\mathrm{o}|)$ holds for all small x^o.)

On the other hand, since $\beta(|x^\mathrm{o}|, t) \to 0$ as $t \to \infty$, for t large one has that the first term in the ISS estimate $|x(t)| \leq \max\{\beta(|x^\mathrm{o}|, t), \gamma(\|u\|_\infty)\}$ vanishes. Thus an ISS system satisfies the following *asymptotic gain property ("AG")*: there is some $\gamma \in \mathcal{K}_\infty$ so that:

$$\overline{\lim_{t \to +\infty}} |x(t, x^\mathrm{o}, u)| \leq \gamma(\|u\|_\infty) \quad \forall\, x^\mathrm{o},\, u(\cdot) \tag{AG}$$

(see Fig. 10). In words, for all large enough t, the trajectory exists, and it gets arbitrarily close to a sphere whose radius is proportional, in a possibly nonlinear way quantified by the function γ, to the amplitude of the input. In the language of robust control, the estimate (AG) would be called an "ultimate boundedness" condition; it is a generalization of attractivity (all trajectories converge to zero, for a system $\dot{x} = f(x)$ with no inputs) to the case of systems with inputs; the "lim sup" is required since the limit of $x(t)$ as $t \to \infty$ may well not exist. From now on (and analogously when defining other properties), we will just say "the system is AG" instead of the more cumbersome "satisfies the AG property."

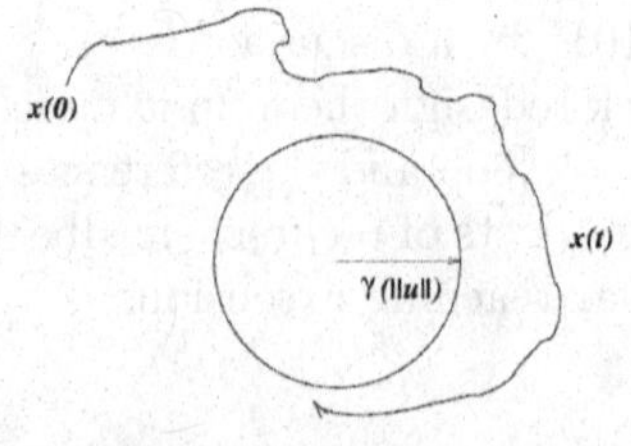

Fig. 10. Asymptotic gain property

Observe that, since only large values of t matter in the limsup, one can equally well consider merely tails of the input u when computing its sup norm. In other words, one may replace $\gamma(\|u\|_\infty)$ by $\gamma(\overline{\lim}_{t\to+\infty}|u(t)|)$, or (since γ is increasing), $\overline{\lim}_{t\to+\infty}\gamma(|u(t)|)$.

The surprising fact is that these two necessary conditions are also sufficient. We call this the *ISS superposition theorem*:

Theorem 3.1. [110] *A system is ISS if and only if it is 0-stable and AG.*

This result is nontrivial. The basic difficulty is in establishing uniform convergence estimates for the states, i.e., in constructing the β function in the ISS estimate, independently of the particular input. As in optimal control theory, one would like to appeal to compactness arguments (using weak topologies on inputs), but there is no convexity to allow this. The proof hinges upon a lemma given in [110], which may be interpreted [41] as a relaxation theorem for differential inclusions, relating global asymptotic stability of an inclusion $\dot{x} \in F(x)$ to global asymptotic stability of its convexification.

A minor variation of the above superposition theorem is as follows. Let us consider the *limit property* (LIM):

$$\inf_{t\geq 0} |x(t,x^\circ,u)| \leq \gamma(\|u\|_\infty) \quad \forall x^\circ,\, u(\cdot) \tag{LIM}$$

(for some $\gamma \in \mathcal{K}_\infty$).

Theorem 3.2. [110] *A system is ISS if and only if it is 0-stable and LIM.*

3.2 Robust Stability

Let us call a system *robustly stable* if it admits a *margin of stability* ρ, by which we mean some smooth function $\rho \in \mathcal{K}_\infty$ which is such that the system

$$\dot{x} = g(x,d) := f(x, d\rho(|x|))$$

is GAS uniformly in this sense: for some $\beta \in \mathcal{KL}$,

$$|x(t,x^\circ,d)| \leq \beta(|x^\circ|\,,t)$$

for all possible $d(\cdot) : [0,\infty) \to [-1,1]^m$. An alternative way to interpret this concept (cf. [109]) is as uniform global asymptotic stability of the origin with respect to all possible time-varying feedback laws Δ bounded by ρ: $|\Delta(t,x)| \leq \rho(|x|)$. In other words, the system

$$\dot{x} \;=\; f(x, \Delta(t,x))$$

(Fig. 11) is stable uniformly over all such perturbations Δ. In contrast to the ISS definition, which deals with all possible "open-loop" inputs, the present notion of robust stability asks about all possible closed-loop interconnections. One may think of Δ as representing uncertainty in the dynamics of the original system, for example.

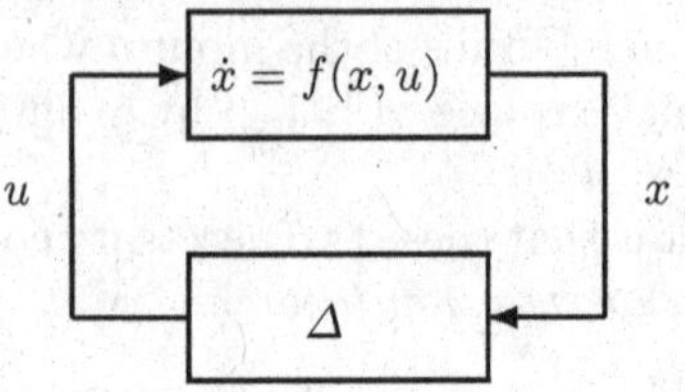

Fig. 11. Margin of robustness

Theorem 3.3. [109] *A system is ISS if and only if it is robustly stable.*

Intuitively, the ISS estimate $|x(t)| \le \max\{\beta(|x^0|,t), \gamma(\|u\|_\infty)\}$ tells us that the β term dominates as long as $|u(t)| \ll |x(t)|$ for all t, but $|u(t)| \ll |x(t)|$ amounts to $u(t) = d(t).\rho(|x(t)|)$ with an appropriate function ρ. This is an instance of a "small gain" argument, about which we will say more later.

One analog for linear systems is as follows: if A is a Hurwitz matrix, then $A + Q$ is also Hurwitz, for all small enough perturbations Q; note that when Q is a nonsingular matrix, $|Qx|$ is a $\mathcal{K}_\infty$ function of $|x|$.

3.3 Dissipation

Another characterization of ISS is as a dissipation notion stated in terms of a Lyapunov-like function.

We will say that a continuous function $V : \mathbb{R}^n \to \mathbb{R}$ is a *storage function* if it is positive definite, that is, $V(0) = 0$ and $V(x) > 0$ for $x \ne 0$, and proper, that is, $V(x) \to \infty$ as $|x| \to \infty$. This last property is equivalent to the requirement that the sets $V^{-1}([0, A])$ should be compact subsets of $\mathbb{R}^n$, for each $A > 0$, and in the engineering literature it is usual to call such functions *radially unbounded*. It is an easy exercise to show that $V : \mathbb{R}^n \to \mathbb{R}$ is a storage function if and only if there exist $\underline{\alpha}, \overline{\alpha} \in \mathcal{K}_\infty$ such that

$$\underline{\alpha}(|x|) \le V(x) \le \overline{\alpha}(|x|) \;\; \forall x \in \mathbb{R}^n$$

(the lower bound amounts to properness and $V(x) > 0$ for $x \ne 0$, while the upper bound guarantees $V(0) = 0$). We also use this notation: $\dot{V} : \mathbb{R}^n \times \mathbb{R}^m \to \mathbb{R}$ is the function:

$$\dot{V}(x,u) := \nabla V(x).f(x,u)$$

which provides, when evaluated at $(x(t), u(t))$, the derivative dV/dt along solutions of $\dot{x} = f(x,u)$.

An *ISS-Lyapunov function* for $\dot{x} = f(x,u)$ is by definition a smooth storage function V for which there exist functions $\gamma, \alpha \in \mathcal{K}_\infty$ so that

$$\dot{V}(x,u) \;\le\; -\alpha(|x|) + \gamma(|u|) \quad \forall x, u\,. \qquad \text{(L-ISS)}$$

In other words, an ISS-Lyapunov function is a smooth (and proper and positive definite) solution of a *partial differential inequality* of this form, for appropriate α, γ. Integrating, an equivalent statement is that, along all trajectories of the system, there holds the following dissipation inequality:

$$V(x(t_2)) - V(x(t_1)) \leq \int_{t_1}^{t_2} w(u(s), x(s))\, ds$$

where, using the terminology of [126], the "supply" function is $w(u,x) = \gamma(|u|) - \alpha(|x|)$. Note that, for systems with no inputs, an ISS-Lyapunov function is precisely the same as a Lyapunov function in the usual sense. Massera's Theorem says that GAS is equivalent to the existence of smooth Lyapunov functions; the following theorem provides a generalization to ISS:

Theorem 3.4. [109] *A system is ISS if and only if it admits a smooth ISS-Lyapunov function.*

Since $-\alpha(|x|) \leq -\alpha(\overline{\alpha}^{-1}(V(x)))$, the ISS-Lyapunov condition can be restated as

$$\dot{V}(x,u) \leq -\tilde{\alpha}(V(x)) + \gamma(|u|) \quad \forall x, u$$

for some $\tilde{\alpha} \in \mathcal{K}_\infty$. In fact, one may strengthen this a bit [93]: for any ISS system, there is a always a smooth ISS-Lyapunov function satisfying the "exponential" estimate $\dot{V}(x,u) \leq -V(x) + \gamma(|u|)$.

The sufficiency of the ISS-Lyapunov condition is easy to show, and was already in the original paper [99]. A sketch of proof is as follows, assuming for simplicity a dissipation estimate in the form $\dot{V}(x,u) \leq -\alpha(V(x)) + \gamma(|u|)$. Given any x and u, either $\alpha(V(x)) \leq 2\gamma(|u|)$ or $\dot{V} \leq -\alpha(V)/2$. From here, one deduces by a comparison theorem that, along all solutions,

$$V(x(t)) \leq \max\left\{\beta(V(x^\circ), t)\, ,\, \alpha^{-1}(2\gamma(\|u\|_\infty))\right\},$$

where we have defined the $\mathcal{KL}$ function $\beta(s,t)$ as the solution $y(t)$ of the initial value problem

$$\dot{y} = -\frac{1}{2}\alpha(y) + \gamma(u)\, , \quad y(0) = s\, .$$

Finally, an ISS estimate is obtained from $V(x^\circ) \leq \overline{\alpha}(x^\circ)$.

The proof of the converse part of the theorem is much harder. It is based upon first showing that ISS implies robust stability in the sense already discussed, and then obtaining a converse Lyapunov theorem for robust stability for the system $\dot{x} = f(x, d\rho(|x|)) = g(x,d)$, which is asymptotically stable uniformly on all Lebesgue-measurable functions $d(\cdot) : \mathbb{R}_{\geq 0} \to B(0,1)$. This last theorem was given in [73], and is basically a theorem on Lyapunov functions for differential inclusions. A classical result of Massera [84] for differential equations becomes a special case.

3.4 Using "Energy" Estimates Instead of Amplitudes

In linear control theory, H_∞ theory studies $L^2 \to L^2$ induced norms. We already saw that, under coordinate changes, we are led to the following type of estimate:

$$\int_0^t \alpha\left(|x(s)|\right) ds \;\leq\; \alpha_0(|x^0|) + \int_0^t \gamma(|u(s)|)\, ds$$

along all solutions, and for some $\alpha, \alpha_0, \gamma \in \mathcal{K}_\infty$. More precisely, let us say, just for the purposes of the next theorem, that a system *satisfies an integral–integral estimate* if for every initial state x^0 and input u, the solution $x(t, x^0, u)$ is defined for all $t > 0$ and an estimate as above holds. (In contrast to ISS, we now have to explicitly demand that $t_{\max} = \infty$.)

Theorem 3.5. [102] *A system is ISS if and only if it satisfies an integral–integral estimate.*

This theorem is quite easy, in view of previous results. A sketch of proof is as follows. If the system is ISS, then there is an ISS-Lyapunov function satisfying $\dot{V}(x, u) \leq -V(x) + \gamma(|u|)$, so, integrating along any solution we obtain

$$\int_0^t V(x(s))\, ds \leq \int_0^t V(x(s))\, ds + V(x(t)) \leq V(x(0)) + \int_0^t \gamma(|u(s)|)\, ds$$

and thus an integral–integral estimate holds. Conversely, if such an estimate holds, one can prove that $\dot{x} = f(x, 0)$ is stable and that an asymptotic gain exists. We show here just the "limit property" $\inf_{t\geq 0} |x(t)| \leq \theta(\|u\|_\infty)$. Indeed, let $\theta := \alpha^{-1} \circ \gamma$. Pick any x^0 and u, and suppose that $\inf_{t\geq 0} |x(t)| > (\alpha^{-1} \circ \gamma)(\|u\|)$, so that there is some $\varepsilon > 0$ so that $\alpha(x(t)) \geq \varepsilon + \gamma(|u(t)|)$ for all $t \geq 0$. Then, $\int_0^t \alpha(x(s))ds \geq \varepsilon t + \int_0^t \gamma(|u(s)|)ds$, which implies $\alpha_0(|x^0|) > \varepsilon t$ for all t, a contradiction. Therefore, the LIM property holds with this choice of θ.

4 Cascade Interconnections

One of the main features of the ISS property is that it behaves well under composition: a cascade (Fig. 12) of ISS systems is again ISS, see [99]. In this section, we will sketch how the cascade result can also be seen as a consequence of the dissipation characterization of ISS, and how this suggests a more general feedback result. We will not provide any details of the rich theory of ISS small-gain theorems, and their use in nonlinear feedback design, for which the references should be consulted, but we will present a very simple example to illustrate the ideas. We consider a cascade as follows:

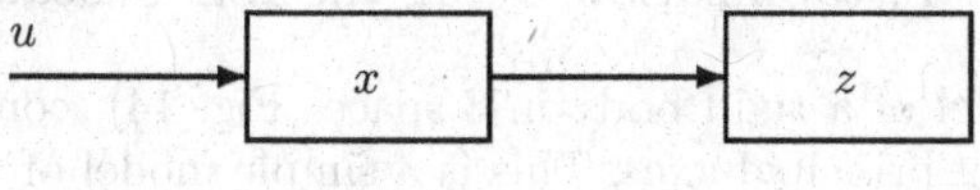

Fig. 12. Cascade

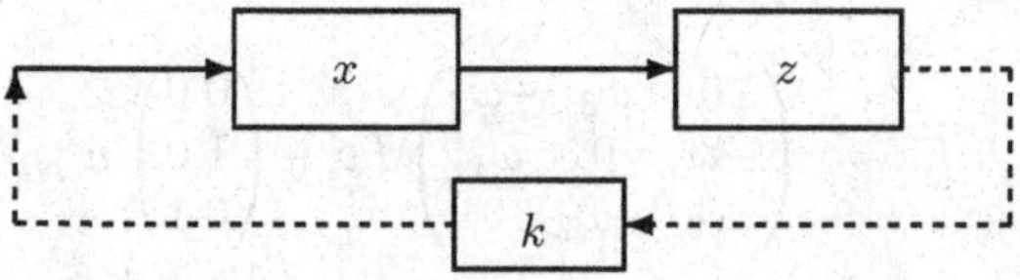

Fig. 13. Adding a feedback to the cascade

$$\begin{aligned}\dot{z} &= f(z, x),\\ \dot{x} &= g(x, u),\end{aligned}$$

where each of the two subsystems is assumed to be ISS. Each system admits an ISS-Lyapunov function V_i. But, moreover, it is always possible (see [106]) to redefine the V_i's so that the comparison functions for both are matched in the following way:

$$\begin{aligned}\dot{V}_1(z, x) &\leq \theta(|x|) - \alpha(|z|),\\ \dot{V}_2(x, u) &\leq \tilde{\theta}(|u|) - 2\theta(|x|)\,.\end{aligned}$$

Now it is obvious why the full system is ISS: we simply use $V := V_1 + V_2$ as an ISS-Lyapunov function for the cascade:

$$\dot{V}((x, z), u) \;\leq\; \tilde{\theta}(|u|) - \theta(|x|) - \alpha(|z|)\,.$$

Of course, in the special case in which the x-subsystem has no inputs, we have also proved that the cascade of a GAS and an ISS system is GAS.

More generally, one may allow a "small gain" feedback as well (Fig. 13). That is, we allow inputs $u = k(z)$ as long as they are small enough:

$$|k(z)| \;\leq\; \tilde{\theta}^{-1}((1-\varepsilon)\alpha(|z|))\,.$$

The claim is that the closed-loop system

$$\begin{aligned}\dot{z} &= f(z, x)\\ \dot{x} &= g(x, k(x))\end{aligned}$$

is GAS. This follows because the same V is a Lyapunov function for the closed-loop system; for $(x, z) \neq 0$:

$$\tilde{\theta}(|u|) \leq (1-\varepsilon)\alpha(|z|) \;\rightsquigarrow\; \dot{V}(x, z) \leq -\theta(|x|) - \varepsilon\alpha(|z|) \;<0\,. \qquad \checkmark$$

A much more interesting version of this result, resulting in a composite system with inputs being itself ISS, is the *ISS small-gain theorem* due to Jiang, Teel, and Praly [53].

4.1 An Example of Stabilization Using the ISS Cascade Approach

We consider a model of a rigid body in 3-space (Fig. 14), controlled by two torques acting along principal axes. This is a simple model of a satellite controlled by an opposing jet pair. If we denote by $\omega = (\omega_1, \omega_2, \omega_3)$ the angular velocity of a body-attached frame with respect to inertial coordinates, and let $I = \text{diag}(I_1, I_2, I_3)$ be the principal moments of inertia, the equations are:

$$I\dot{\omega} = \begin{pmatrix} 0 & \omega_3 & -\omega_2 \\ -\omega_3 & 0 & \omega_1 \\ \omega_2 & -\omega_1 & 0 \end{pmatrix} I\omega + \begin{pmatrix} 0\,0 \\ 1\,0 \\ 0\,1 \end{pmatrix} u$$

Ignoring kinematics, we just look at angular momenta, and we look for a feedback law to globally stabilize this system to $\omega = 0$. Under feedback and coordinate transformations, one can bring this into the following form of a system in $\mathbb{R}^3$ with controls in $\mathbb{R}^2$:

$$\begin{aligned} \dot{x}_1 &= x_2 x_3, \\ \dot{x}_2 &= u_1, \\ \dot{x}_3 &= u_2 . \end{aligned}$$

(We assume that $I_2 \neq I_3$, and use these transformations: $(I_2 - I_3)x_1 = I_1\omega_1$, $x_2 = \omega_2$, $x_3 = \omega_3$, $I_2\widetilde{u}_1 = (I_3 - I_1)\omega_1\omega_2 + u_1$, $I_3\widetilde{u}_2 = (I_1 - I_2)\omega_1\omega_3 + u_2$.) Our claim is that the following feedback law globally stabilizes the system:

$$\begin{aligned} u_1 &= -x_1 - x_2 - x_2 x_3 \\ u_2 &= -x_3 + x_1^2 + 2x_1 x_2 x_3 . \end{aligned}$$

Indeed, as done in [18] for the corresponding local problem, we make the following transformations: $z_2 := x_1 + x_2$, $z_3 := x_3 - x_1^2$, so the system becomes:

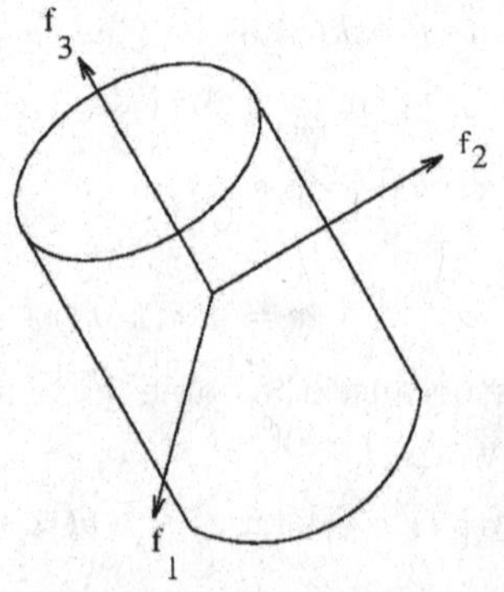

Fig. 14. Rigid body

$$\begin{aligned}\dot{x}_1 &= -x_1^3 + \alpha(x_1, z_2, z_3) \quad (\deg_{x_1} \alpha \leq 2),\\ \dot{z}_2 &= -z_2,\\ \dot{z}_3 &= -z_3\,.\end{aligned}$$

Now, the x_1-subsystem is easily seen to be ISS, and the z_1, z_2 subsystem is clearly GAS, so the cascade is GAS. Moreover, a similar construction produces a feedback robust with respect to input disturbances.

5 Integral Input-to-State Stability

We have seen that several different properties, including "integral to integral" stability, dissipation, robust stability margins, and asymptotic gain properties, all turned out to be exactly equivalent to input to state stability. Thus, it would appear to be difficult to find a general and interesting concept of nonlinear stability that is truly distinct from ISS. One such concept, however, does arise when considering a mixed notion which combines the "energy" of the input with the amplitude of the state. It is obtained from the "$L^2 \to L^\infty$" gain estimate, under coordinate changes, and it provides a genuinely new concept [102].

A system is said to be *integral-input to state stable* (iISS) provided that there exist $\alpha, \gamma \in \mathcal{K}_\infty$ and $\beta \in \mathcal{KL}$ such that the estimate

$$\alpha\left(|x(t)|\right) \;\leq\; \beta(|x^{\mathrm{o}}|\,, t) \,+\, \int_0^t \gamma(|u(s)|)\, ds \qquad \text{(iISS)}$$

holds along all solutions. Just as with ISS, we could state this property merely for all times $t \in t_{\max}(x^{\mathrm{o}}, u)$, but, since the right-hand side is bounded on each interval $[0, t]$ (because, recall, inputs are by definition assumed to be bounded on each finite interval), it is automatically true that $t_{\max}(x^{\mathrm{o}}, u) = +\infty$ if such an estimate holds along maximal solutions. So forward-completeness can be assumed with no loss of generality.

5.1 Other Mixed Notions

We argued that changes of variables transformed linear "finite L^2 gain" estimates into an "integral to integral" property, which we then found to be equivalent to the ISS property. On the other hand, finite operator gain from L^p to L^q, with $p \neq q$ both finite, lead one naturally to the following type of "weak integral to integral" mixed estimate:

$$\int_0^t \underline{\alpha}(|x(s)|)\, ds \;\leq\; \kappa(|x^{\mathrm{o}}|) \,+\, \alpha\left(\int_0^t \gamma(|u(s)|)\, ds\right)$$

for appropriate $\mathcal{K}_\infty$ functions (note the additional "α"). See [12] for more discussion on how this estimate is reached, as well as the following result:

Theorem 5.1. [12] *A system satisfies a weak integral to integral estimate if and only if it is iISS.*

Another interesting variant is found when considering mixed *integral/supremum* estimates:

$$\underline{\alpha}(|x(t)| \leq \beta(|x^0|,t) + \int_0^t \gamma_1(|u(s)|)\,ds + \gamma_2(\|u\|_\infty)$$

for suitable $\beta \in \mathcal{KL}$ and $\underline{\alpha}, \gamma_i \in \mathcal{K}_\infty$. One then has:

Theorem 5.2. [12] *A system satisfies a mixed estimate if and only if it is iISS.*

5.2 Dissipation Characterization of iISS

There is an amazingly elegant characterization of iISS, as follows. Recall that by a storage function we mean a continuous $V : \mathbb{R}^n \to \mathbb{R}$ which is positive definite and proper. Following [11], we will say that a smooth storage function V is an *iISS-Lyapunov function* for the system $\dot{x} = f(x,u)$ if there are a $\gamma \in \mathcal{K}_\infty$ and an $\alpha : [0,+\infty) \to [0,+\infty)$ which is merely *positive definite* (that is, $\alpha(0) = 0$ and $\alpha(r) > 0$ for $r > 0$) such that the inequality:

$$\dot{V}(x,u) \leq -\alpha(|x|) + \gamma(|u|) \qquad \text{(L-iISS)}$$

holds for all $(x,u) \in \mathbb{R}^n \times \mathbb{R}^m$. By contrast, recall that an ISS-Lyapunov function is required to satisfy an estimate of the same form but where α is required to be of class $\mathcal{K}_\infty$; since every $\mathcal{K}_\infty$ function is positive definite, an ISS-Lyapunov function is also an iISS-Lyapunov function.

Theorem 5.3. [11] *A system is iISS if and only if it admits a smooth iISS-Lyapunov function.*

Since an ISS-Lyapunov function is also an iISS one, ISS implies iISS. However, iISS is a strictly weaker property than ISS, because α may be bounded in the iISS-Lyapunov estimate, which means that V may increase, and the state become unbounded, even under bounded inputs, so long as $\gamma(|u(t)|)$ is larger than the range of α. This is also clear from the iISS definition, since a constant input with $|u(t)| = r$ results in a term in the right-hand side that grows like rt. As a concrete example using a nontrivial V, consider the system

$$\dot{x} = -\tan^{-1}x + u,$$

which is not ISS, since $u(t) \equiv \pi/2$ results in unbounded trajectories. This system is nonetheless iISS: if we pick $V(x) = x\tan^{-1}x$, then

$$\dot{V} \leq -(\tan^{-1}|x|)^2 + 2\,|u|$$

so V is an iISS-Lyapunov function. An interesting general class of examples is given by *bilinear* systems

$$\dot{x} = \left(A + \sum_{i=1}^{m} u_i A_i\right) x + Bu$$

for which the matrix A is Hurwitz. Such systems are always iISS (see [102]), but they are not in general ISS. For instance, in the case when $B = 0$, boundedness of trajectories for all constant inputs already implies that $A + \sum_{i=1}^{m} u_i A_i$ must have all eigenvalues with nonpositive real part, for all $u \in \mathbb{R}^m$, which is a condition involving the matrices A_i (for example, $\dot{x} = -x + ux$ is iISS but it is not ISS).

The notion of iISS is useful in situations where an appropriate notion of detectability can be verified using LaSalle-type arguments. There follow two examples of theorems along these lines.

Theorem 5.4. [11] *A system is iISS if and only if it is 0-GAS and there is a smooth storage function V such that, for some $\sigma \in \mathcal{K}_\infty$:*

$$\dot{V}(x,u) \leq \sigma(|u|)$$

for all (x,u).

The sufficiency part of this result follows from the observation that the 0-GAS property by itself already implies the existence of a smooth and positive definite, but not necessarily proper, function V_0 such that $\dot{V}_0 \leq \gamma_0(|u|) - \alpha_0(|x|)$ for all (x,u), for some $\gamma_0 \in \mathcal{K}_\infty$ and positive definite α_0 (if V_0 were proper, then it would be an iISS-Lyapunov function). Now one uses $V_0 + V$ as an iISS-Lyapunov function (V provides properness).

Theorem 5.5. [11] *A system is iISS if and only if there exists an output function $y = h(x)$ (continuous and with $h(0) = 0$) which provides zero-detectability ($u \equiv 0$ and $y \equiv 0 \Rightarrow x(t) \to 0$) and dissipativity in the following sense: there exists a storage function V and $\sigma \in \mathcal{K}_\infty$, α positive definite, so that:*

$$\dot{V}(x,u) \leq \sigma(|u|) - \alpha(h(x))$$

holds for all (x,u).

The paper [12] contains several additional characterizations of iISS.

5.3 Superposition Principles for iISS

We now discuss asymptotic gain characterizations for iISS.

We will say that a system is *bounded energy weakly converging state* (BEWCS) if there exists some $\sigma \in \mathcal{K}_\infty$ so that the following implication holds:

$$\int_0^{+\infty} \sigma(|u(s)|)\,ds < +\infty \;\Rightarrow\; \liminf_{t\to+\infty} |x(t,x^{\circ},u)| = 0 \qquad \text{(BEWCS)}$$

(more precisely: if the integral is finite, then $t_{\max}(x^{\circ},u) = +\infty$ and the liminf is zero), and that it is *bounded energy frequently bounded state* (BEFBS) if there exists some $\sigma \in \mathcal{K}_\infty$ so that the following implication holds:

$$\int_0^{+\infty} \sigma(|u(s)|)\,ds < +\infty \;\Rightarrow\; \liminf_{t\to+\infty} |x(t,x^{\circ},u)| < +\infty \qquad \text{(BEFBS)}$$

(again, meaning that $t_{\max}(x^{\circ},u) = +\infty$ and the liminf is finite).

Theorem 5.6. [6] *The following three properties are equivalent for any given system $\dot{x} = f(x,u)$:*

- *The system is iISS*
- *The system is BEWCS and 0-stable*
- *The system is BEFBS and 0-GAS*

These characterizations can be obtained as consequences of characterizations of input/output to state stability (IOSS), cf. Sect. 8.4. The key observation is that a system is iISS with input gain (the function appearing in the integral) σ if and only if the following auxiliary system is IOSS with respect to the "error" output $y = e$:

$$\begin{aligned}\dot{x} &= f(x,u)\\ \dot{e} &= \sigma(|u|)\,.\end{aligned}$$

The proof of this equivalence is trivial, so we include it here. If the system is iISS, then:

$$\begin{aligned}\alpha(|x(t,x^{\circ},u)|) &\le \beta(|x^{\circ}|,t) + \int_0^t \sigma(|u(s)|)\,ds \;=\; \beta(|x^{\circ}|,t) + e(t) - e(0)\\ &\le \beta(|x^{\circ}|,t) + 2\|y_{[0,t]}\|_\infty\end{aligned}$$

and, conversely, if it is IOSS, then for any x° and picking $e(0) = 0$, we have:

$$\begin{aligned}|x(t,x^{\circ},u)| &\le \beta(|(x^{\circ},0)|,t) + \gamma(\|u\|_\infty) + \|y_{[0,t]}\|_\infty\\ &\le \beta(|x^{\circ}|,t) + \gamma(\|u\|_\infty) + \int_0^t \sigma(|u|)\,ds\,.\end{aligned}$$

5.4 Cascades Involving iISS Systems

We have seen that cascades of ISS systems are ISS, and, in particular, any system of the form:

$$\begin{aligned}\dot{x} &= f(x,z)\\ \dot{z} &= g(z)\end{aligned}$$

for which the x-subsystem is ISS when z is viewed as an input and the g-subsystem is GAS, is necessarily GAS. This is one of the most useful properties of the ISS notion, as it allows proving stability of complex systems by a decomposition approach. The iISS property on the first subsystem, in contrast, is not strong enough to guarantee that the cascade is GAS. As an illustration, consider the following system:

$$\dot{x} = -\text{sat}(x) + xz,$$
$$\dot{z} = -z^3,$$

where $\text{sat}(x) := \text{sgn}(x)\min\{1, |x|\}$. It is easy to see that the x-subsystem with input z is iISS, and the z-subsystem is clearly GAS. On the other hand [13], if we pick $z(0) = 1$ and any $x(0) \geq 3$, then $x(t) \geq e^{(\sqrt{1+2t}-1)}$, so $x(t) \to \infty$ as $t \to \infty$; so the complete system is not GAS. However, under additional conditions, it is possible to obtain a cascade result for a system of the above form. One such result is as follows.

Theorem 5.7. [13] *Suppose that the x-subsystem is iISS and affine in z, and that the z-subsystem is GAS and locally exponentially stable. Then, the cascade is GAS.*

Note that the counterexample shown above is so that the x-subsystem is indeed affine in z, but the *exponential* stability property fails. This theorem is a consequence of a more general result, which is a bit technical to state. We first need to introduce two concepts. The first one qualifies the speed of convergence in the GAS property, and serves to relax exponential stability: we say that the system $\dot{z} = g(z)$ is GAS(α), for a given $\alpha \in \mathcal{K}_\infty$, if there exists a class-$\mathcal{K}_\infty$ function $\theta(\cdot)$ and a positive constant $k > 0$ so that

$$|z(t)| \leq \alpha\left(e^{-kt}\theta(|z^0|)\right)$$

holds for all z^0. (Recall that GAS is always equivalent to the existence of *some* α and θ like this.) The second concept is used to characterize the function γ appearing in the integral in the right-hand side of the iISS estimate, which we call the "iISS gain" of the system: given any $\alpha \in \mathcal{K}_\infty$, we say that the function γ is "class-$\mathcal{H}_\alpha$" if it is of class $\mathcal{K}$ and it also satisfies:

$$\int_0^1 \frac{(\gamma(\alpha(s))}{s}\, ds < \infty\,.$$

The main result says that if the same α can be used in both definitions, then the cascade is GAS:

Theorem 5.8. [13] *Suppose that the x-subsystem is iISS with a class-$\mathcal{H}_\alpha$ iISS gain, and that the z-subsystem is GAS(α). Then, the cascade is GAS.*

See [13] for various corollaries of this general fact, which are based upon checking that the hypotheses are always satisfied, for example for the above-mentioned case of x- subsystem affine in z and exponentially stable z- subsystem.

5.5 An iISS Example

As an example of the application of iISS ideas, we consider as in [11] a robotic device studied by Angeli in [3]. This is a manipulator with one rotational and one linear actuator (Fig. 15). A simple model is obtained considering the arm as a segment with mass M and length L, and the hand as a material point with mass m. The equations for such a system are four-dimensional, using as state variables angular position and velocity $\theta, \dot{\theta}$ and linear extension and velocity $r, \dot{r}$, and they follow from the second order equation

$$\begin{aligned}(mr^2 + ML^2/3)\,\ddot{\theta} + 2mr\dot{r}\dot{\theta} &= \tau \\ m\ddot{r} - mr\dot{\theta}^2 &= F\,,\end{aligned}$$

where the controls are the torque τ and linear force F. We write the state as $(q, \dot{q})$, with $q = (\theta, r)$, We wish to study the standard tracking feedback controller with equations

$$\tau = -k_1\dot{\theta} - k_2(\theta - \theta_d)\,,\ F = -k_3\dot{r} - k_4(r - r_d)$$

where q_d, r_d are the desired trajectories. It is well-known that, for *constant* tracking signals q_d, r_d, one obtains convergence: $\dot{q} \to 0$ and $q \to q_d$ as $t \to \infty$. In the spirit of the ISS approach, however, it is natural to ask what is the sensitivity of the design to *additive measurement noise*, or equivalently, since these errors are potentially arbitrary functions, what is the effect of *time-varying* tracking signals. One could ask if the system is ISS, and indeed the paper [83] proposed the reformulation of tracking problems using ISS as a way to characterize performance.

It turns out that, for this example, even bounded signals may destabilize the system, by a sort of "nonlinear resonance" effect, so the system cannot be ISS (not even bounded-input bounded-state) with respect to q_d and r_d. Fig. 16 plots a numerical example of a de-stabilizing input; the corresponding $r(t)$-component is in Fig. 17. To be precise, the figures show the "r" component of the state of a certain solution which corresponds to the shown input; see [11]

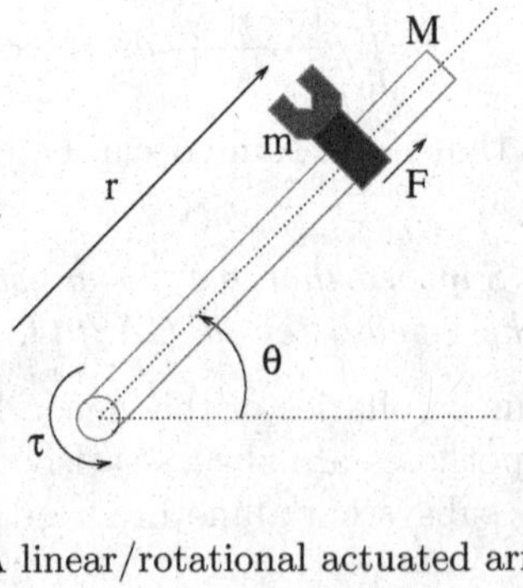

Fig. 15. A linear/rotational actuated arm

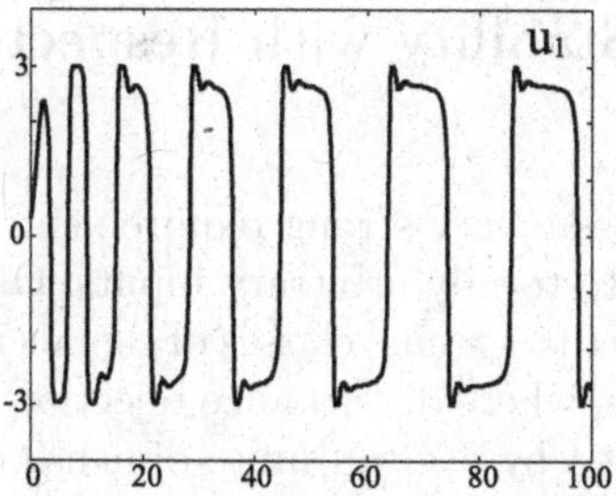

Fig. 16. Destabilizing input

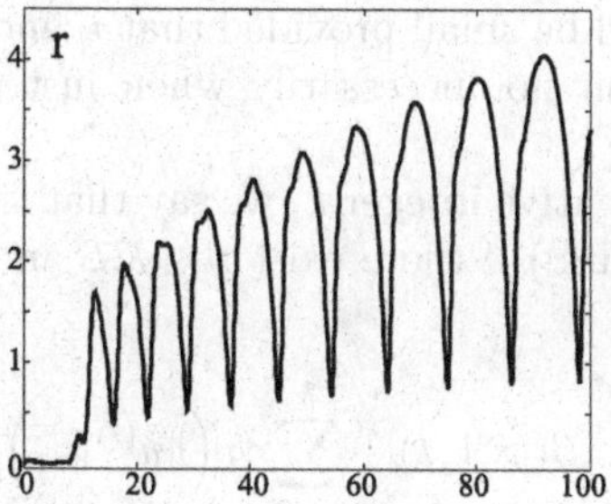

Fig. 17. Corresponding $r(\cdot)$

for details on how this input and trajectory were calculated. Thus, the question arises of how to qualitatively formulate the fact that some other inputs are not destabilizing. We now show that iISS provides one answer to this question,

In summary, we wish to show that the closed loop system

$$\begin{aligned}(mr^2 + ML^2/3)\ddot{\theta} + 2mr\dot{r}\dot{\theta} &= u_1 - k_1\dot{\theta} - k_2\theta, \\ m\ddot{r} - mr\dot{\theta}^2 &= u_2 - k_3\dot{r} - k_4 r,\end{aligned}$$

with states $(q, \dot{q})$, $q = (\theta, r)$, and $u = (k_2\theta_d, k_4 r_d)$ is iISS.

In order to do so, we consider the mechanical energy of the system:

$$V(q, z) := \frac{1}{2}\dot{q}^T H(q)\dot{q} + \frac{1}{2}q^T K q$$

and note [11] the following passivity-type estimate:

$$\frac{d}{dt}V(q(t), \dot{q}(t)) \leq -c_1|\dot{q}(t)|^2 + c_2\,|u(t)|^2$$

for sufficiently small $c_1 > 0$ and large $c_2 > 0$. Taking $\dot{q}$ as an output, the system is zero-detectable and dissipative, since $u \equiv 0$ and $\dot{q} \equiv 0$ imply $q \equiv 0$, and hence, appealing to the given dissipation characterizations, we know that it is indeed iISS.

6 Input to State Stability with Respect to Input Derivatives

The ISS property imposes a very strong requirement, in that stable behavior must hold with respect to totally arbitrary inputs. Often, on the other hand, only stability with respect to specific classes of signals is expected. An example is in regulation problems, where disturbance rejection is usually formulated in terms of signals generated by a given finite-dimensional exosystem. Another example is that of parameter drift in adaptive control systems, where bounds on rates of change of parameters (which we may see as inputs) are imposed. This question motivated the work in [10] on ISS notions in which one asks, roughly, that $x(t)$ should be small provided that u and its derivatives of some fixed order be small, but not necessarily when just u is small. The precise definition is as follows.

For any given nonnegative integer k, we say that the system $\dot{x} = f(x, u)$ is *differentiably k-ISS* (D^kISS) if there exist $\beta \in \mathcal{KL}$ and $\gamma_i \in \mathcal{K}_\infty$, $i = 0, \ldots, k$, such that the estimate:

$$|x(t, x^{\mathrm{o}}, u)| \;\leq\; \beta(|x^{\mathrm{o}}|\,, t) + \sum_{i=0}^{k} \gamma_i \left(\|u^{(i)}\|_\infty \right) \qquad (\mathrm{D}^k\mathrm{ISS})$$

holds for all x^{o}, all inputs $u \in W^{k,\infty}$, and all $t \in t_{\max}(x^{\mathrm{o}}, u)$. (By $W^{k,\infty}$ we are denoting the Sobolev space of functions u: $[0, \infty) \to \mathbb{R}^m$ for which the $(k-1)$st derivative $u^{(k-1)}$ exists and is locally Lipschitz, which means in particular that $u^{(k)}$ exists almost everywhere and is locally essentially bounded.) As with the ISS property, forward completeness is automatic, so one can simply say "for all t" in the definition. Notice that D^0ISS is the same as plain ISS, and that, for every k, D^kISS implies D^{k+1}ISS.

6.1 Cascades Involving the D^kISS Property

Consider any cascade as follows:

$$\begin{aligned} \dot{x} &= f(x, z) \\ \dot{z} &= g(z, u) \end{aligned}$$

where we assume that g is smooth. The following result generalizes the fact that cascading ISS and GAS subsystems gives a GAS system.

Theorem 6.1. [10] *If each subsystem is D^kISS, then the cascade is D^kISS. In particular the cascade of a D^kISS and a GAS system is GAS.*

Actually, somewhat less is enough: the x-subsystem need only be D^{k+1}ISS, and we may allow the input to appear in this subsystem.

It is not difficult to see that a system is D^1ISS if and only if the following system:

$$\begin{aligned}\dot{x} &= f(x, z)\\ \dot{z} &= -z + u\end{aligned}$$

is ISS, and recursively one can obtain a similar characterization of D^kISS. More generally, a system is D^kISS if and only if it is D^{k-1}ISS when cascaded with any ISS "smoothly invertible filter" as defined in [10]. Also very useful is a close relationship with the IOSS concept studied in Sect. 8.2. Consider the following auxiliary system with input u and output y:

$$\begin{aligned}\dot{x} &= f(x, u_0)\\ \dot{u}_0 &= u_1\\ \vdots\ &\ \ \vdots\\ \dot{u}_{k-1} &= u_k\\ y &= [u_0, u_1, \ldots, u_{k-1}].\end{aligned}$$

Theorem 6.2. [10] *The auxiliary system is IOSS if and only if the original system is D^kISS.*

The paper [10] also discusses some relations between the notion of D^1ISS and ISS, for systems of the special form $\dot{x} = f(x + u)$, which are of interest when studying observation uncertainty.

6.2 Dissipation Characterization of D^kISS

Theorem 6.3. [10] *A system is D^1ISS if and only if there exists a smooth function $V(x, u)$ such that, for some $\alpha, \delta_0, \delta_1, \alpha_1, \alpha_2 \in \mathcal{K}_\infty$,*

$$\alpha_1(|x| + |u|) \ \le\ V(x, u) \ \le\ \alpha_2(|x| + |u|)$$

and

$$D_x V(x, u)\, f(x, u) + D_u V(x, u)\dot{u} \ \le\ -\alpha(|x|) + \delta_0(|u|) + \delta_1(|\dot{u}|)$$

for all $(x, u, \dot{u}) \in \mathbb{R}^n \times \mathbb{R}^m \times \mathbb{R}^m$.

Notice that "$\dot{u}$" is just a dummy variable in the above expression. Analogous characterizations hold for D^kISS.

6.3 Superposition Principle for D^kISS

We will say that a forward-complete system satisfies the *k-asymptotic gain (k-AG) property* if there are some $\gamma_0, \gamma_1, \ldots, \gamma_k \in \mathcal{K}$ so that, for all $u \in W^{k,\infty}$, and all x^0, the estimate

$$\overline{\lim_{t\to\infty}}\, |x(t, \xi, u)| \ \le\ \gamma_0(\|u\|_\infty) + \gamma_1(\|\dot{u}\|_\infty) + \ldots + \gamma_k(\|u^{(k)}\|_\infty)$$

holds.

Theorem 6.4. [10] *A system is D^kISS if and only if it is 0-stable and k-AG.*

6.4 A Counter-Example Showing that D^1ISS $\neq$ ISS

Consider the following system:

$$\dot{x} = \|x\|^2 \, U(\theta)' \, \Phi \, U(\theta) \, x,$$

where $x \in \mathbb{R}^2$, and $u = \theta(\cdot)$ is the input,

$$U(\theta) = \begin{bmatrix} \sin(\theta) & \cos(\theta) \\ -\cos(\theta) & \sin(\theta) \end{bmatrix},$$

and where Φ is any 2×2 Hurwitz matrix such that $\Phi' + \Phi$ has a strictly positive real eigenvalue. It is shown in [10] that this system is not forward complete, and in particular it is not ISS, but that it is D^1ISS. This latter fact is shown by proving, through the construction of an explicit ISS-Lyapunov function, that the cascaded system

$$\dot{x} = \|x\|^2 \, U(\theta)' \, \Phi \, U(\theta) \, x, \qquad \dot{\theta} = -\theta + u$$

is ISS.

It is still an open question if D^2ISS is strictly weaker than D^1ISS, and more generally D^{k+1}ISS than D^kISS for each k.

7 Input-to-Output Stability

Until now, we only discussed stability of states with respect to inputs. For systems with outputs $\dot{x} = f(x,u)$, $y = h(x)$, if we simply replace states by outputs in the left-hand side of the estimate defining ISS, we then arrive to the notion of *input-to-output stability* (IOS): there exist some $\beta \in \mathcal{KL}$ and $\gamma \in \mathcal{K}_\infty$ such that

$$|y(t)| \leq \beta(|x^\circ|, t) + \gamma(\|u\|_\infty) \qquad \text{(IOS)}$$

holds for all solutions, where $y(t) = h(x(t, x^\circ, u))$. By "all solutions" we mean that this estimate is valid for all inputs $u(\cdot)$, all initial conditions x°, and all $t \geq 0$, and we are imposing as a requirement that the system be forward complete, i.e., $t_{\max}(x^\circ, u) = \infty$ for all initial states x° and inputs u. As earlier, $x(t)$, and hence $y(t) = h(x(t))$, depend only on past inputs ("causality"), so we could have used just as well simply the supremum of $|u(s)|$ for $s \leq t$ in the estimate.

We will say that a system is *bounded-input bounded-state stable (BIBS)* if, for some $\sigma \in \mathcal{K}_\infty$, the following estimate:

$$|x(t)| \leq \max\{\sigma(|x^\circ|), \sigma(\|u\|_\infty)\}$$

holds along all solutions. (Note that forward completeness is a consequence of this inequality, even if it is only required on maximal intervals, since the state is upper bounded by the right-hand side expression.)

We define an *IOS-Lyapunov function* as any smooth function $V : \mathbb{R}^n \to \mathbb{R}_{\geq 0}$ so that, for some $\alpha_i \in \mathcal{K}_\infty$:

$$\alpha_1(|h(x)|) \leq V(x) \leq \alpha_2(|x|) \qquad \forall\, x \in \mathbb{R}^n,\, u \in \mathbb{R}^m$$

and, for all x, u:

$$V(x) > \alpha_3(|u|) \;\Rightarrow\; \nabla V(x)\, f(x,u) < 0\,.$$

Theorem 7.1. [113] *A BIBS system is IOS if and only if it admits an IOS-Lyapunov function.*

A concept related to IOS is as follows. We call a system *robustly output stable* (ROS) if it is BIBS and there is some smooth $\lambda \in \mathcal{K}_\infty$ such that

$$\dot{x} = g(x,d) := f(x, d\lambda(|y|))\,, \quad y = h(x)$$

is globally output-asymptotically stable uniformly with respect to all $d(\cdot) : [0,\infty) \to [-1,1]^m$: for some $\beta \in \mathcal{KL}$,

$$|y(t,x^{\rm o},d)| \leq \beta(|x^{\rm o}|\,,t)$$

for all solutions. Then, IOS implies ROS, but the converse does not hold in general [112]. We have the following dissipation characterization of ROS:

Theorem 7.2. [113] *A system is ROS if and only if it is BIBS and there are $\alpha_1, \alpha_2 \in \mathcal{K}_\infty$, $\chi \in \mathcal{K}$, and $\alpha_3 \in \mathcal{KL}$, and a smooth function $V : \mathbb{R}^n \to \mathbb{R}$, so that*

$$\alpha_1(|h(x)|) \leq V(x) \leq \alpha_2(|x|)$$

and

$$|h(x)| \geq \chi(|u|) \;\Rightarrow\; \dot{V} \leq -\alpha_3(V(x), |x|)$$

for all (x,u).

The area of *partial stability* studies stability of a subset of variables in a system $\dot{x} = f(x)$. Letting $y = h(x)$ select the variables of interest, one may view partial stability as a special case of output stability, for systems with no inputs. Note that, for systems with no inputs, the partial differential inequality for IOS reduces to $\nabla V(x)\, f(x) < 0$ for all nonzero x, and that for ROS to $\dot{V} \leq -\alpha_3(V(x), |x|)$. In this way, the results in [113] provide a far-reaching generalization of, and converse theorems to, sufficient conditions [125] for partial stability.

There is also a superposition principle for IOS. We will say that a forward-complete system satisfies the *output asymptotic gain* (OAG) property if

$$\overline{\lim_{t\to\infty}}\, |y(t)| \;\leq\; \gamma(\|u\|_\infty) \qquad \text{(OAG)}$$

for some $\gamma \in \mathcal{K}_\infty$ and all solutions. One would like to have a characterization of IOS in terms of OAG, which is an analog of the AG gain property in the state case, and a stability property. Let us define a system to be *output-Lagrange stable (OL)* if it satisfies an estimate, for some $\sigma \in \mathcal{K}_\infty$:

$$|y(t)| \le \sigma(|y(0)|) + \sigma(\|u\|_\infty)$$

along all solutions. Under this assumption, we recover a separation principle:

Theorem 7.3. [6] *An OL system is OAG if and only if it is IOS.*

Observe that the OL property asks that the output be uniformly bounded in terms of the amplitudes of the input and of the initial output (not of the initial state), which makes this property a very strong constraint. If we weaken the assumption to an estimate of the type

$$|y(t)| \le \sigma(|x^0|) + \sigma(\|u\|_\infty)$$

then IOS implies the conjunction of OL and this property, but the converse fails, as shown by the following counter-example, a system with no inputs:

$$\dot{x}_1 = -x_2\,|x_2|\,, \quad \dot{x}_2 = x_1\,|x_2|\,, \qquad y = x_2\,.$$

The set of equilibria is $\{x_2 = 0\}$, and trajectories are half circles traveled counterclockwise. We have that $|y(t)| \le |x(0)|$ for all solutions, and $y(t) \to 0$ as $t \to \infty$, so both properties hold. However, there is no possible IOS estimate $|y(t)| \le \beta(|x^0|, t)$, since, in particular, for a state of the form $x(0) = (1, \varepsilon)$, the time it takes for $y(\cdot)$ to enter an ε-neighborhood of 0 goes to ∞ as $\varepsilon \to 0$; see [6] for more discussion.

8 Detectability and Observability Notions

Recall (see [104] for precise definitions) that an *observer* for a given system with inputs and outputs $\dot{x} = f(x,u)$, $y = h(x)$ is another system which, using only information provided by past input and output signals, provides an asymptotic (i.e., valid as $t \to \infty$) estimate $\hat{x}(t)$ of the state $x(t)$ of the system of interest (Fig. 18). One may think of the observer as a physical system or as an algorithm implemented by a digital computer. The problem of state estimation is one of the most important and central topics in control theory, and it arises

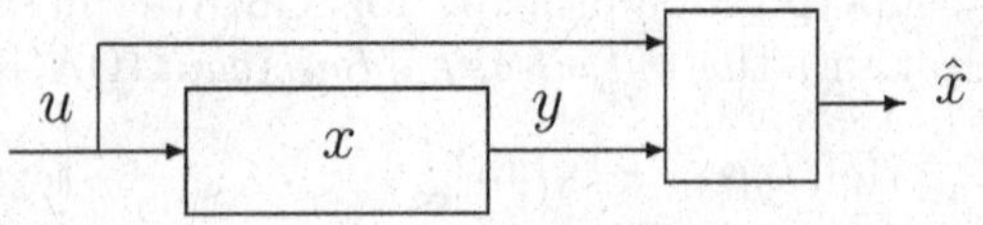

Fig. 18. Observer provides estimate $\hat{x}$ of state x; $\hat{x}(t) - x(t) \to 0$ as $t \to \infty$

in signal processing applications (Kalman filters) as well as when solving the problem of stabilization based on partial information. It is well understood for linear systems, but, a huge amount of research notwithstanding, the theory of observers is not as developed in general.

We will not say much about the general problem of building observers, which is closely related to "incremental" ISS-like notions, a subject not yet studied enough, but will focus on an associated but easier question. When the ultimate goal is that of stabilization to an equilibrium, let us say $x = 0$ in Euclidean space, sometimes a weaker type of estimate suffices: it may be enough to obtain a *norm-estimator* which provides merely an *upper bound* on the norm $|x(t)|$ of the state $x(t)$; see [50, 57, 93]. Before defining norm-estimators, and studying their existence, we need to introduce an appropriate notion of detectability.

8.1 Detectability

Suppose that an observer exists, for a given system. Since $x^0 = 0$ is an equilibrium for $\dot{x} = f(x, 0)$, and also $h(0) = 0$, the solution $x(t) \equiv 0$ is consistent with $u \equiv 0$ and $y \equiv 0$. Thus, the estimation property $\hat{x}(t) - x(t) \to 0$ implies that $\hat{x}(t) \to 0$. Now consider *any* state x^0 for which $u \equiv 0$ and $y \equiv 0$, that is, so that $h(x(t, x^0, 0)) \equiv 0$. The observer output, which can only depend on u and y, must be the same $\hat{x}$ as when $x^0 = 0$, so $\hat{x}(t) \to 0$; then, using once again the definition of observer $\hat{x}(t) - x(t, x^0, 0) \to 0$, we conclude that $x(t, x^0, 0) \to 0$. In summary, a *necessary* condition for the existence of an observer is that the "subsystem" of $\dot{x} = f(x, u)$, $y = h(x)$ consisting of those states for which $u \equiv 0$ produces the output $y \equiv 0$ must have $x = 0$ as a GAS state (Fig. 19); one says in that case that the system is *zero-detectable*. (For *linear* systems, zero-detectability is equivalent to detectability or "asymptotic observability" [104]: two trajectories which produce the same output must approach each other. But this equivalence need not hold for nonlinear systems.) In a nonlinear context, zero-detectability is not "well-posed" enough: to get a well-behaved notion, one should add explicit requirements to ask that small inputs and outputs imply that internal states are small too (Fig. 20), and that inputs and outputs converging to zero as $t \to \infty$ implies that states do, too (Fig. 21), These properties are needed so that "small" errors in measurements of inputs and outputs processed by the observer give rise to small errors. Furthermore, one should impose asymptotic bounds on states as a function

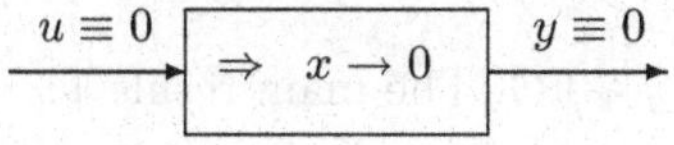

Fig. 19. Zero-detectability

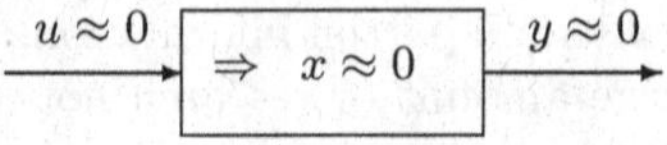

Fig. 20. Small inputs and outputs imply small states

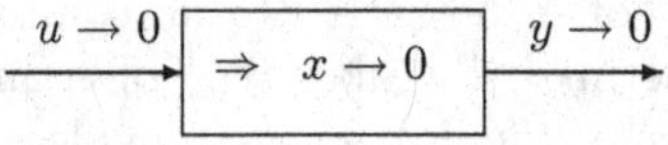

Fig. 21. Converging inputs and outputs imply convergent states

of input/output bounds, and it is desirable to quantify "overshoot" (transient behavior). This leads us to the following notion.

8.2 Dualizing ISS to OSS and IOSS

A system is *input/output to state stable* (IOSS) if, for some $\beta \in \mathcal{KL}$ and $\gamma_u, \gamma_y \in \mathcal{K}_\infty$,

$$x(t) \;\leq\; \beta(|x^0|, t) + \gamma_1\left(\|u_{[0,t]}\|_\infty\right) + \gamma_2\left(\|y_{[0,t]}\|_\infty\right) \qquad \text{(IOSS)}$$

for all initial states and inputs, and all $t \in [0, T_{\xi,u})$. Just as ISS is stronger than 0-GAS, IOSS is stronger than zero-detectability. A special case is when one has no inputs, *output to state stability:*

$$|x(t, x^0)| \;\leq\; \beta(|x^0|, t) + \gamma\left(\|y|_{[0,t]}\|_\infty\right)$$

and this is formally "dual" to ISS, simply replacing inputs u by outputs in the ISS definition. This duality is only superficial, however, as there seems to be no useful way to obtain theorems for OSS by dualizing ISS results. (Note that the outputs y depend on the state, not vice versa.)

8.3 Lyapunov-Like Characterization of IOSS

To formulate a dissipation characterization, we define an *IOSS-Lyapunov function* as a smooth storage function so that

$$\nabla V(x)\, f(x, u) \leq -\alpha_1(|x|) + \alpha_2(|u|) + \alpha_3(|y|)$$

for all $x \in \mathbb{R}^n, u \in \mathbb{R}^m, y \in \mathbb{R}^p$. The main result is:

Theorem 8.1. [65] *A system is IOSS if and only if it admits an IOSS-Lyapunov function.*

8.4 Superposition Principles for IOSS

Just as for ISS and IOS, there are asymptotic gain characterizations of input/output to state stability.

We say that a system satisfies the *IO-asymptotic gain (IO-AG)* property if:

$$\overline{\lim_{t\nearrow t_{\max}(x^0,u)}} |x(t,x^0,u)| \leq \gamma_u(\|u\|_\infty) + \gamma_y(\|y\|_\infty) \qquad \forall\, x^0,\, u(\cdot) \qquad \text{(IO-AG)}$$

(for some γ_u, γ_y), and the *IO-limit* (IO-LIM) property if:

$$\inf_{t\geq 0} |x(t,x^0,u)| \leq \gamma_u(\|u\|_\infty) + \gamma_y(\|y\|_\infty) \qquad \forall\, x^0,\, u(\cdot) \qquad \text{(IO-LIM)}$$

(for some γ_u, γ_y), where sup norms and inf are taken over $[0, t_{\max}(x^0, u))$. We also define the notion of *zero-input local stability modulo outputs* (0-LS) as follows:

$$(\forall\, \varepsilon > 0)\,(\exists\, \delta_\varepsilon) \quad \max\{|x^0|, \|y_{[0,t]}\|_\infty\} \leq \delta_\varepsilon \Rightarrow |x(t,x^0,0)| \leq \varepsilon\,. \qquad \text{(0-LS)}$$

This is a notion of marginal local detectability; for linear systems, it amounts to marginal stability of the unobservable eigenvalues. We have the following result.

Theorem 8.2. [6] *The following three properties are equivalent for any given system* $\dot{x} = f(x,u)$:

- *The system is IOSS*
- *The system is IO-AG and zero-input O-LS*
- *The system is IO-LIM and zero-input O-LS*

Several other characterizations can also be found in [6].

8.5 Norm-Estimators

We define a *state-norm-estimator* (or *state-norm-observer*) for a given system as another system

$$\dot{z} = g(z,u,y)\,, \qquad \text{with output } \; k : \mathbb{R}^\ell \times \mathbb{R}^p \to \mathbb{R}_{\geq 0}$$

evolving in some Euclidean space $\mathbb{R}^\ell$, and driven by the inputs and outputs of the original system. We ask that the output k should be IOS with respect to the inputs u and y, and the true state should be asymptotically bounded in norm by some function of the norm of the estimator output, with a transient (overshoot) which depends on both initial states. Formally:

- There are $\hat{\gamma}_1, \hat{\gamma}_2 \in \mathcal{K}$ and $\hat{\beta} \in \mathcal{KL}$ so that, for each initial state $z^0 \in \mathbb{R}^\ell$, and inputs $\mathbf{u}$ and $\mathbf{y}$, and every t in the interval of definition of the solution $z(\cdot, z^0, \mathbf{u}, \mathbf{y})$

$$k\left(z(t,z^0,\mathbf{u},\mathbf{y}), \mathbf{y}(t)\right) \leq \hat{\beta}(|z^0|, t) + \hat{\gamma}_1\left(\|\mathbf{u}|_{[0,t]}\|\right) + \hat{\gamma}_2\left(\|\mathbf{y}|_{[0,t]}\|\right)$$

- There are $\rho \in \mathcal{K}$, $\beta \in \mathcal{KL}$ so that, for all initial states x° and z° of the system and observer, and every input $\mathbf{u}$

$$|x(t,x^\circ,\mathbf{u})| \leq \beta(|x^\circ|+|z^\circ|,t) + \rho\left(k\left(z(t,z^\circ,\mathbf{u},\mathbf{y}_{x^\circ,\mathbf{u}}),\mathbf{y}_{x^\circ,\mathbf{u}}(t)\right)\right)$$

for all $t \in [0, t_{\max}(x^\circ,\mathbf{u}))$, where $\mathbf{y}_{x^\circ,\mathbf{u}}(t) = y(t,x^\circ,\mathbf{u})$

Theorem 8.3. [65] *A system admits a state-norm-estimator if and only if it is IOSS.*

8.6 A Remark on Observers and Incremental IOSS

As mentioned earlier, for linear systems, "zero-detectability" and detectability coincide, where the latter is the property that *every pair* of distinct states is asymptotically distinguishable. The following is an ISS-type definition of detectability: we say that a system is *incrementally (or Lipschitz) input/output-to-state stable (i-IOSS)* if there exist $\gamma_1, \gamma_2 \in \mathcal{K}$ and $\beta \in \mathcal{KL}$ such that, for any two initial states x° and z°, and any two inputs u_1 and u_2,

$$|x(t,x^\circ,u_1)-x(t,z^\circ,u_2)| \leq \max\left\{\beta(|x^\circ-z^\circ|,t),\gamma_1(\|\Delta u\|),\gamma_2(\|\Delta y\|)\right\} \quad \text{(i-IOSS)}$$

where $\Delta u = (u_1-u_2)$, $\Delta y = (y_{x^\circ,u_1}-y_{z^\circ,u_2})_{[0,t]}$, for all t in the common domain of definition. It is easy to see that i-IOSS implies IOSS, but the converse does not hold in general. The notion of incremental-IOSS was introduced in [111]. A particular case is that in which $h(x) \equiv 0$, in which case we have the following notion: a system is *incrementally ISS* (i-ISS) if there holds an estimate of the following form:

$$|x(t,x^\circ,u_1)-x(t,z^\circ,u_2)| \leq \max\left\{\beta(|x^\circ-z^\circ|,t),\ \gamma_1(\|\Delta u\|)\right\} \qquad \text{(i-ISS)}$$

where $\Delta u = u_1 - u_2$, for all t in the common domain of definition. Several properties of the i-ISS notion were explored in [4], including the fact that i-ISS is preserved under cascades. Specializing even more, when there are no inputs one obtains the property *incremental GAS* (i-GAS). This last property can be characterized in Lyapunov terms using the converse Lyapunov result given in [73] for stability with respect to (not necessarily compact) sets, since it coincides with stability with respect to the diagonal of the system consisting of two parallel copies of the same system. Indeed, i-GAS is equivalent to asking that the system:

$$\dot{x} = f(x)$$
$$\dot{z} = f(z)$$

be asymptotically stable with respect to $\{(x,z) \mid x = z\}$. A sufficient condition for i-ISS in dissipation terms, using a similar idea, was given in [4].

As recalled earlier, an observer is another dynamical system, which processes inputs and outputs of the original system, and produces an estimate $\widehat{x}(t)$ of the state $x(t)$: $x(t)-\widehat{x}(t) \to 0$ as $t \to \infty$, and this difference (the

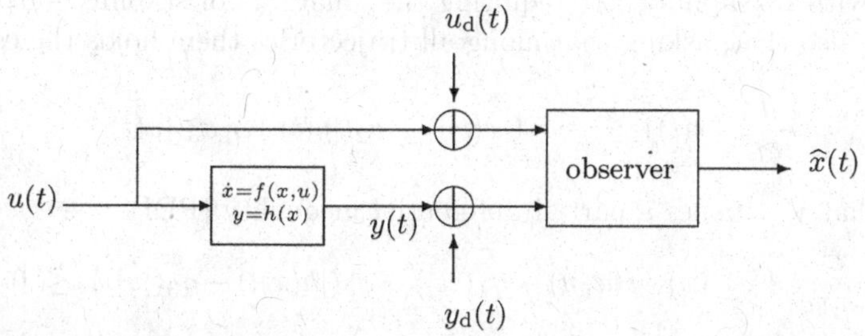

Fig. 22. Observer with perturbations in measurements

estimation error) should be small if it starts small (see [104], Chap. 6). As with zero-detectability, it is more natural in the ISS paradigm to ask that the estimation error $x(t)-\widehat{x}(t)$ should be small even if the measurements of inputs and outputs received by the observer are corrupted by noise. Writing u_d and y_d for the input and output measurement noise respectively, we have the situation shown pictorially in Fig. 22 (see [111] for a precise definition). Existence of an observer implies that the system is i-IOSS [111]. The converse and more interesting problem of building observers under IOSS assumptions is still a largely unsolved, although much progress has been made for systems with special structures, cf. [16, 57].

8.7 Variations of IOSS

The terminology IOSS was introduced in [111], and the name arises from the view of IOSS as "stability from the i/o data to the state." It combines the "strong" observability from [99] with ISS; and was called simply "detectability" in [98], where it was formulated in an input/output language and applied to controller parameterization, and it was called "strong unboundedness observability" in [53] (more precisely, this paper allowed for an additive nonnegative constant in the estimate). IOSS is related to other ISS-like formalisms for observers, see, e.g., [37, 75, 77, 92]. Both IOSS and its incremental variant are very closely related to the OSS-type detectability notions pursued in [59]; see also the emphasis on ISS guarantees for observers in [82].

The dissipation characterization amounts to a Willems'-type dissipation inequality $(d/dt)V(x(t)) \leq -\sigma_1(|x(t)|) + \sigma_2(|y(t)|) + \sigma_3(|u(t)|)$ holding along all trajectories. There have been other suggestions that one should *define* "detectability" in dissipation terms; see, e.g., [76], where detectability was defined by the requirement that there exist a differentiable storage function V as here, but with the special choice $\sigma_2(r) := r^2$ (and no inputs), or as in [85], which asked for a weaker the dissipation inequality:

$$x \neq 0 \;\Rightarrow\; \frac{d}{dt}V(x(t)) < \sigma_2(|y(t)|)$$

(again, with no inputs), not requiring the "margin" of stability $-\sigma_1(|x(t)|)$. Observe also that, asking that along all trajectories there holds the estimate

$$\frac{d}{dt}V(x(t)) \leq -\sigma_1(|x(t)|) + \sigma_2(|y(t)|) + \sigma_3(|u(t)|)$$

means that V satisfies a partial differential inequality (PDI):

$$\max_{u \in \mathbb{R}^m} \{\nabla V(x) \cdot f(x,u) + \sigma_1(|x|) - \sigma_2(|h(x)|) - \sigma_3(|u|)\} \leq 0$$

which is same as the Hamilton–Jacobi inequality

$$g_0(x) + \frac{1}{4}\sum_{i=1}^{m}(\nabla V(x) \cdot g_i(x))^2 + \sigma_1(|x|) - \sigma_2(|h(x)|) \leq 0$$

in the special case of quadratic input "cost" $\sigma_3(r) = r^2$ and systems $\dot{x} = f(x,u)$ affine in controls

$$\dot{x} = g_0(x) + \sum_{i=1}^{m} u_i\, g_i(x)$$

(just replace the right-hand side in the PDI by the maximum value, obtained at $u_i = (1/2)\nabla V(x) \cdot g_i(x)$). Thus the converse result amounts to providing necessary and sufficient conditions for existence of a smooth (and proper and positive definite) solution V to the PDI. In this context, it is worth remarking that the mere existence of a lower semicontinuous V (interpreted in an appropriate weak sense) implies the existence of a C^∞ solution (possibly with different comparison functions); see [64].

8.8 Norm-Observability

There are many notions of observability for nonlinear systems (see, e.g., [104], Chap. 6); here we briefly mention one such notion given in an ISS style, which was presented in [34]. More precisely, we define "norm-observability", which concerns the ability to determine an upper bound on norms, rather than the precise value of the state (an "incremental" version would correspond to true observability). We do so imposing a bound on the norm of the state in terms of the norms of the output and the input, and imposing an additional requirement which says, loosely speaking, that the term describing the effects of initial conditions can be chosen to decay arbitrarily fast.

A system $\dot{x} = f(x,u)$, $y = h(x)$ is *small-time initial-state norm-observable* if:

$$\forall \tau > 0\ \exists \gamma, \chi \in \mathcal{K}_\infty \text{ such that } |x^{\mathrm{o}}| \leq \gamma(\|y_{[0,\tau]}\|_\infty) + \chi(\|u_{[0,\tau]}\|_\infty)\ \forall x^{\mathrm{o}}, u\,,$$

it is *small-time final-state norm-observable* if:

$$\forall \tau > 0 \; \exists \gamma, \chi \in \mathcal{K}_\infty \text{ such that } |x(\tau)| \le \gamma(\|y_{[0,\tau]}\|_\infty) + \chi(\|u_{[0,\tau]}\|_\infty) \; \forall x^{\mathrm{o}},\, u\,,$$

and is *small-time-$\mathcal{KL}$ norm-observable* if for every $\varepsilon > 0$ and every $\nu \in \mathcal{K}$, there exist $\gamma, \chi \in \mathcal{K}_\infty$ and a $\beta \in \mathcal{KL}$ so that $\beta(r, \varepsilon) \le \nu(r)$ for all $r \ge 0$ (i.e., β can be chosen to decay arbitrarily fast in the second argument) such that the IOSS estimate:

$$|x(t)| \;\le\; \beta(|x^{\mathrm{o}}|\,, t) + \gamma(\|y_{[0,t]}\|_\infty) + \chi(\|u_{[0,t]}\|_\infty) \; \forall x^{\mathrm{o}},\, u,\, t \ge 0$$

holds along all solutions.

Theorem 8.4. [34] *The following notions are equivalent:*

- *Small-time initial-state norm-observability*
- *Small-time final-state norm-observability*
- *Small-time-$\mathcal{KL}$ norm-observability*

To be precise, the equivalences assume *unboundedness observability* (UO), which means that for each trajectory defined on some maximal interval $t_{\max} < \infty$, the output becomes unbounded as $t \nearrow t_{\max}$, as well as a similar property for the reversed-time system. The unboundedness observability property is strictly weaker than forward completeness, which is the property that each trajectory is defined for all $t \ge 0$; see [9, 53], the latter especially for complete Lyapunov characterizations of the UO property. Similarly, one can prove equivalences among other definitions, such as asking "$\exists\, \tau$" instead of "$\forall\, \tau$," and one may obtain Lyapunov-like characterizations; the results are used in the derivation of LaSalle-like theorems for verifying stability of switched systems in [34].

9 The Fundamental Relationship Among ISS, IOS, and IOSS

The definitions of the basic ISS-like concepts are consistent and related in an elegant conceptual manner, as follows:

A system is ISS if and only if it is both IOS and IOSS.

In informal terms, we can say that:

external stability and detectability $\iff$ *internal stability*

as it is the case for linear systems. Intuitively, we have the three possible signals in Fig. 23. The basic idea of the proof is as follows. Suppose that external stability and detectability hold, and take an input so that $u \to 0$. Then $y \to 0$ (by external stability), and this then implies that $x \to 0$ (by detectability). Conversely, if the system is internally stable, then we prove i/o stability and detectability. Suppose that $u \to 0$. By internal stability, $x \to 0$,

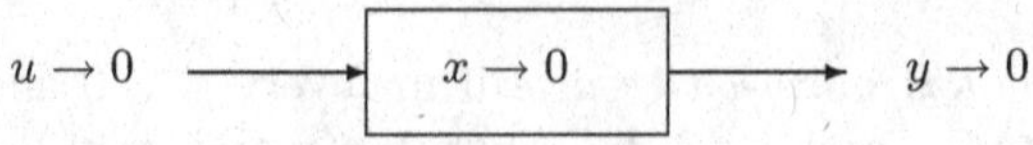

Fig. 23. Convergent input, state, and/or output

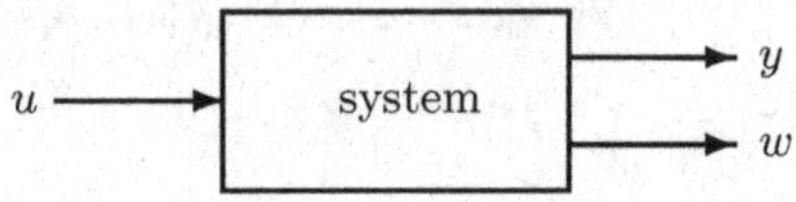

Fig. 24. System with error and measurement outputs $\dot{x} = f(x, u)$, $y = h(x)$, $w = g(x)$

and this gives $y(t) \to 0$ (i/o stability). Detectability is even easier: if both $u(t) \to 0$ and $y(t) \to 0$, then in particular $u \to 0$, so $x \to 0$ by internal stability. The proof that ISS is equivalent to the conjunction of IOS and IOSS must keep careful track of the estimates, but the idea is similar.

10 Systems with Separate Error and Measurement Outputs

We next turn to a topic which was mentioned in [105] as a suggestion for further work, but for which still only incomplete results are available. We will assume that there are *two* types of outputs (Fig. 24), which we think of, respectively as an "error" $y = h(x)$ to be kept small, as in the IOS notion, and a "measurement" $w = g(x)$ which provides information about the state, as in the IOSS notion.

Several ISS-type formulations of the central concept in regulator theory, namely the idea of using the size of w in order to bound y, were given in [38], and are as follows.

10.1 Input-Measurement-to-Error Stability

We will say that a system is *input-measurement-to-error stable* (IMES) if there are $\beta \in \mathcal{KL}$, $\sigma \in \mathcal{K}$, and $\gamma \in \mathcal{K}$ such that the following estimate holds:

$$|y(t)| \le \beta(|x^{\circ}|, t) + \sigma\left(\|w_{[0,t]}\|_{\infty}\right) + \gamma(\|u\|_{\infty}) \qquad \text{(IMES)}$$

for all $t \in t_{\max}(x^{\circ}, u)$, for all solutions, where we are writing $y(t) = h(x(t, x^{\circ}, u))$ and $w(t) = g(x(t, x^{\circ}, u))$. Special cases are all the previous concepts:

- When $h(x) = x$, so $y = x$, and we view w as the output, we recover IOSS
- When the output $g = 0$, that is $w \equiv 0$, we recover IOS
- If both $y = x$ and $g = 0$, we recover ISS

The goal of obtaining general theorems for IMES which will specialize to all known theorems for ISS, IOS, and IOSS is so far unattained. We only know of several partial results, to be discussed next.

For simplicity from now on, we restrict to the case of systems with no inputs $\dot{x} = f(x)$, $y = h(x)$, $w = g(x)$, and we say that a system is measurement to error stable (MES) if an IMES estimate holds, i.e., for suitable $\beta \in \mathcal{KL}$ and $\gamma \in \mathcal{K}$:

$$|y(t)| \leq \beta(|x^{\text{o}}|, t) + \gamma\left(\|w_{[0,t]}\|_\infty\right)$$

for all $t \in [0, T_{\max})$ and for all solutions.

In order to present a dissipation-like version of MES, it is convenient to introduce the following concept. We will say that a system is *relatively error-stable* (RES) if the following property holds, for some $\rho \in \mathcal{K}$ and $\beta \in \mathcal{KL}$:

$$|y(t)| > \rho(|w(t)|) \text{ on } [0,T] \;\Rightarrow\; |y(t)| \leq \beta(|x^{\text{o}}|, t) \text{ on } [0,T] \qquad \text{(RES)}$$

along all solutions and for all $T < t_{\max}(x^{\text{o}}, u)$. In words: while the error is much larger than the estimate provided by the measurement, the error must decrease asymptotically, with an overshoot controlled by the magnitude of the initial state. This property, together with the closely related notion of *stability in three measures* (SIT), was introduced and studied in [38]. It is easy to see that MES implies RES, but that the converse is false. In order to obtain a converse, one requires an additional concept: we say a system satisfies the *relative measurement to error boundedness* (RMEB) property if it admits an estimate of the following form, for some $\sigma_i \in \mathcal{K}$:

$$|y(t)| \leq \max\left\{\sigma_1(|h(x^{\text{o}})|),\ \sigma_2\left(\|w_{[0,t]}\|_\infty\right)\right\} \qquad \text{(RMEB)}$$

along all solutions. For forward complete systems, and assuming RMEB, RES is equivalent to MES [38].

10.2 Review: Viscosity Subdifferentials

So far, no smooth dissipation characterization of any of these properties is available. In order to state a nonsmooth characterization, we first review a notion of weak differential. For any function $V : \mathbb{R}^n \to \mathbb{R}$ and any point $p \in \mathbb{R}^n$ in its domain, one says that a vector ζ is a *viscosity subgradient* of V at p if the following property holds: there is some function $\varphi : \mathbb{R}^n \to \mathbb{R}$, differentiable at zero and with $\nabla\varphi(0) = \zeta$ (that is, $\varphi(h) = \zeta \cdot h + o(h)$), such that

$$V(p+h) \geq V(p) + \varphi(h)$$

for each h in a neighborhood of $0 \in \mathbb{R}^n$. In other words, a viscosity subgradient is a gradient (tangent slopes) of any supporting $\mathcal{C}^1$ function. One denotes then

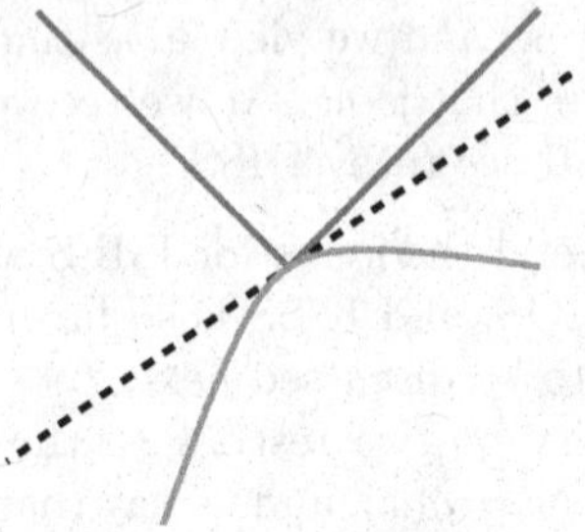

Fig. 25. $\partial_D V(0) = [-1, 1]$

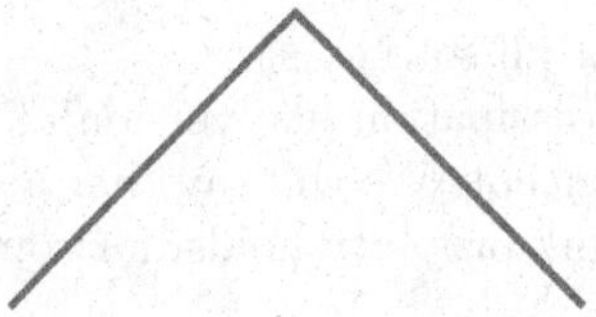

Fig. 26. $\partial_D V(0) = \emptyset$

$\partial_D V(p) := \{$all viscosity subgradients of V at $p\}$. As an illustration, Fig. 25 shows a case where $\partial_D V(0) = [-1, 1]$, for the function $V(x) = |x|$, and Fig. 26 an example where $\partial_D V(0) = \emptyset$, for $V(x) = -|x|$. In particular, if V is differentiable at p, then $\partial_D V(p) = \{\nabla V(p)\}$.

10.3 RES-Lyapunov Functions

The lower semicontinuous V is an *RES-Lyapunov function* if:

- There exist α_1 and $\alpha_2 \in \mathcal{K}_\infty$ so, on the set $C := \{p : |h(p)| > \rho(|g(p)|)\}$, it holds that
$$\alpha_1(|h(p)|) \le V(p) \le \alpha_2(|p|)$$
- For some continuous positive definite $\alpha_3 : \mathbb{R}_{\ge 0} \to \mathbb{R}_{\ge 0}$, on the set C there holds the estimate
$$\zeta \cdot f(x) \;\le\; -\alpha_3(V(p)) \qquad \forall \zeta \in \partial_D V(p)$$
(when V is differentiable, this is just $\nabla V \cdot f(x) \le -\alpha_3(V(p))$)

One can show (cf. [38]) that this estimate is equivalent to the existence of a locally Lipschitz, positive definite $\tilde{\alpha}_3$ such that, for all trajectories:

$$x(t) \in C \;\text{ on }\; [0, t_1] \;\Rightarrow\; V(x(t)) - V(x(0)) \le -\int_0^t \tilde{\alpha}_3(V(x(s)))\,ds\,.$$

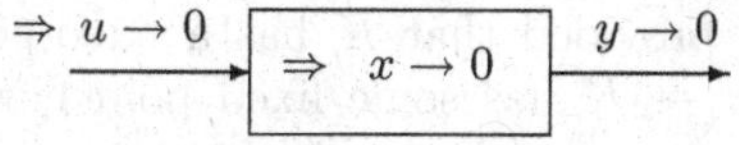

Fig. 27. Inverse of IOS property: small output implies input (and state) small

Theorem 10.1. [38] *A forward-complete system is RES if and only if it admits an RES-Lyapunov function.*

As a corollary, we have that, for RMEB systems, MES is equivalent to the existence of such a lower semicontinuous RES Lyapunov function.

11 Output to Input Stability and Minimum-Phase

We now mention a nonlinear "well-posed" version of Bode's minimum phase systems, which relates to the usual notion (cf. [42]) in the same manner as ISS relates to zero-GAS. We need to say, roughly, what it means for the "inverse system" to be ISS (Fig. 27). The paper [71] defines a smooth system as *output to input stable* (OIS) if there exists an integer $N > 0$, and functions $\beta \in \mathcal{KL}$ and $\gamma \in \mathcal{K}_\infty$, so that, for every initial state x^o and every $(N-1)$-times continuously differentiable input u, the inequality:

$$|u(t)| + |x(t)| \;\leq\; \beta(|x^\text{o}|, t) \,+\, \gamma\left(\|y^N_{[0,t]}\|_\infty\right)$$

holds for all $t \in t_{\max}(x^\text{o}, u)$, where "$y^N$" lists y as well as its first N derivatives (and we use supremum norm, as usual). See [71] for relationships to OSS, an interpretation in terms of an ISS property imposed on the "zero dynamics" of the system, and connections to relative degree, as well as an application to adaptive control.

12 Response to Constant and Periodic Inputs

Systems $\dot{x} = f(x, u)$ that are ISS have certain noteworthy properties when subject to constant or, more generally periodic, inputs, which we now discuss. Let V be an ISS-Lyapunov function which satisfies the inequality $\dot{V}(x, u) \leq -V(x) + \gamma(|u|)$ for all x, u, for some $\gamma \in \mathcal{K}_\infty$.

To start with, suppose that $\bar{u}$ is any fixed bounded input, and let $a := \gamma(\|\bar{u}\|_\infty)$, pick any initial state x^o, and consider the solution $x(t) = x(t, x^\text{o}, \bar{u})$ for this input. Letting $v(t) := V(x(t))$, we have that $\dot{v}(t) + v(t) \leq a$ so, using e^t as an integrating factor, we have that $v(t) \leq a + e^{-t}(v(0) - a)$ for all $t \geq 0$. In particular, if $v(0) \leq a$ it will follow that $v(t) \leq a$ for all $t \geq 0$, that is to say, the sublevel set $K := \{x \mid V(x) \leq a\}$ is a forward-invariant set for this input: if $x^\text{o} \in K$ then $x(t) = x(t, x^\text{o}, \bar{u}) \in K$ for all $t \geq 0$. Therefore

$M_T : x^o \mapsto x(T, x^o, \bar{u})$ is a continuous mapping from K into K, for each fixed $T > 0$, and thus, provided that K has a fixed-point property (every continuous map $M : K \to K$ has some fixed point), we conclude that for each $T > 0$ there exists some state x^o such that $x(T, x^o, \bar{u}) = x^o$. The set K indeed has the fixed-point property, as does any sublevel set of a Lyapunov function. To see this, we note that V is a Lyapunov function for the zero-input system $\dot{x} = f(x, 0)$, and thus, if B is any ball which includes K in its interior, then the map $Q : B \to K$ which sends any $\xi \in B$ into $x(t_\xi, \xi)$, where t_ξ is the first time such that $x(t, \xi) \in K$, is continuous (because the vector field is transversal to the boundary of K since $\nabla V(x).f(x, 0) < 0$), and is the identity on K (that is, Q is a topological retraction). A fixed point of the composition $M \circ Q : B \to B$ is a fixed point of M.

Now suppose that $\bar{u}$ is periodic of period T, $\bar{u}(t+T) = \bar{u}(t)$ for all $t \geq 0$, and pick any x^o which is a fixed point for M_T. Then the solution $x(t, x^o, \bar{u})$ is periodic of period T as well. In other words, *for each periodic input, there is a solution of the same period*. In particular, if $\bar{u}$ is constant, we may pick for each $h > 0$ a state x_h so that $x(h, x_h, \bar{u}) = x_h$, and therefore, picking a convergent subsequence $x_h \to \bar{x}$ gives that $0 = (1/h)(x(h, x_h, \bar{u}) - x_h) \to f(\bar{x}, \bar{u})$, so $f(\bar{x}, \bar{u}) = 0$. Thus we also have the conclusion that *for each constant input, there is a steady state.*

13 A Remark Concerning ISS and H_∞ Gains

We derived the "integral to integral" version of ISS when starting from H_∞-gains, that is, L^2-induced operator norms. In an abstract manner, one can reverse the argument, as this result shows:

Theorem 13.1. [33] *Assume $n \neq 4, 5$. If the system $\dot{x} = f(x, u)$ is ISS, then, under a coordinate change, for all solutions one has:*

$$\int_0^t |x(s)|^2 \, ds \;\leq\; |x^o|^2 + \int_0^t |u(s)|^2 \, ds \,.$$

(A particular case of this is that global exponential stability is equivalent to global asymptotic stability, under such nonsmooth coordinate changes. This would seem to contradict Center Manifold theory, but recall that our "coordinate changes" are not necessarily smooth at the origin, so dimensions of stable and unstable manifolds need not be preserved.) It is still an open question if the theorem generalizes to $n = 4$ or 5. A sketch of proof is as follows.

Let us suppose that the system $\dot{x} = f(x, u)$ is ISS. We choose a "robustness margin" $\rho \in \mathcal{K}_\infty$, i.e., a $\mathcal{K}_\infty$ function with the property that the closed-loop system $\dot{x} = f(x, d\rho(|x|))$ is GAS uniformly with respect to all disturbances such that $\|d\|_\infty \leq 1$. We next pick a smooth, proper, positive definite storage function V so that

$$\nabla V(x) \cdot f(x, d\rho(|x|)) \leq -V(x) \quad \forall \, x, d$$

(such a function always exists, by the results already mentioned). Now suppose that we have been able to find a coordinate change so that $V(x) = |x|^2$, that is, a T so that $W(z) := V(T^{-1}(z)) = |z|^2$ with $z = T(x)$. Then, whenever $|u| \leq \rho(|x|)$, we have

$$d\,|z|^2\,/dt = \dot{W}(z) = \dot{V}(x) \leq -V(x) = -\,|z|^2\,.$$

It follows that, if $\chi \in \mathcal{K}_\infty$ is so that $|T(x)| \leq \chi(\rho(|x|))$, and

$$\alpha(r) := \max_{|u|\leq r, |z|\leq \chi(r)} d\,|z|^2\,/dt$$

then:

$$\frac{d\,|z|^2}{dt} \leq -\,|z|^2 + \alpha(|u|) = -\,|z|^2 + v$$

(where we denote by v the input in new coordinates).

Integrating, one obtains $\int |z|^2 \leq |z^o|^2 + \int |v|^2$, and this gives the L^2 estimate as wanted. The critical technical step, thus, is to show that, up to coordinate changes, every Lyapunov function V is quadratic. That fact is shown as follows. First notice that the level set $S := \{V(x) = 1\}$ is homotopically equivalent to $\mathbb{S}^{n-1}$ (this is well-known: $S \times \mathbb{R} \simeq S$ because $\mathbb{R}$ is contractible, and $S \times \mathbb{R}$ is homeomorphic to $\mathbb{R}^n \setminus \{0\} \simeq \mathbb{S}^{n-1}$ via the flow of $\dot{x} = f(x, 0)$). Thus, $\{V(x) = 1\}$ is diffeomorphic to $\mathbb{S}^{n-1}$, provided $n \neq 4, 5$. (In dimensions $n = 1, 2, 3$ this is proved directly; for $n \geq 6$ the sublevel set $\{V(x) < 1\}$ is a compact, connected smooth manifold with a simply connected boundary, and results on h-cobordism theory due to Smale and Milnor show the diffeomorphism to a ball. Observe that results on the generalized Poincaré conjecture would give a homeomorphism, for $n \neq 4$.) Finally, we consider the normed gradient flow:

$$\dot{x} = \frac{\nabla V(x)'}{|\nabla V(x)|^2}$$

and take the new variable

$$z := \sqrt{V(x)}\,\theta(x')$$

where x' is the translate via the flow back into the level set, and $\theta : \{V = 1\} \simeq \{|z| = 1\}$ is the given diffeomorphism, see Fig. 28. (Actually, this sketch is not quite correct: one needs to make a slight adjustment in order to obtain also continuity and differentiability at the origin; the actual coordinate change is $z = \gamma(V(x))\theta(x')$, so $W(z) = \gamma(|z|)$, for a suitable γ.)

14 Two Sample Applications

For applications of ISS notions, the reader is encouraged to consult textbooks such as [27,43,44,58,60,66,96], as well as articles in journals as well as Proceedings of the various IEEE Conferences on Decision and Control. We highlight

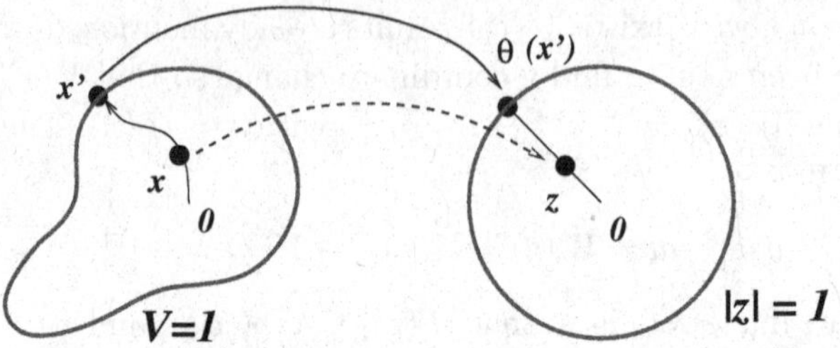

Fig. 28. Making level sets into spheres

next a couple of applications picked quite arbitrarily from the literature. They are chosen as illustrations of the range of possibilities afforded by the ISS viewpoint.

The paper [63] provides a new observer suitable for output-feedback stabilization, and applies the design to the stabilization of surge and the magnitude of the first stall mode, in the single-mode approximation of the Moore–Greitzer PDE axial compressor model used in jet engine studies. The equations are as follows:

$$
\begin{aligned}
\dot{\phi} &= -\psi + \tfrac{3}{2}\phi + \tfrac{1}{2} - \tfrac{1}{2}(\phi+1)^3 - 3(\phi+1)R \\
\dot{\psi} &= \frac{1}{\beta^2}(\phi + 1 - u) \\
\dot{R} &= \sigma R(-2\phi - \phi^2 - R) \qquad (R \geq 0)
\end{aligned}
$$

where ϕ denotes the mass flow relative to a setpoint, ψ the pressure rise relative to the setpoint, and R the magnitude of the first stall mode. The objective is to stabilize this system, using only $y = \psi$.

The systematic use of ISS-type properties is central to the analysis: taking the magnitude of the first stall mode as evolving through uncertain dynamics, the authors require that their estimator have an error that is ISS with respect to this unmodeled dynamics, and that the first mode be IOS with respect to mass flow deviation from its setpoint; an ISS small-gain theorem is then used to complete the design. Abstractly, their general framework in [63] is roughly as follows. One is given a system with the block structure:

$$
\begin{aligned}
\dot{x} &= f(x, z, u) \\
\dot{z} &= g(x, z)
\end{aligned}
$$

and only an output $y = h(x)$ is available for stabilization. The z-subsystem (R in the application) is unknown (robust design). The authors construct a state-feedback $u = k(x)$ and a reduced-order observer that produces an estimate $\widehat{x}$ so that:

- The error $e = x - \widehat{x}$ is ISS with respect to z

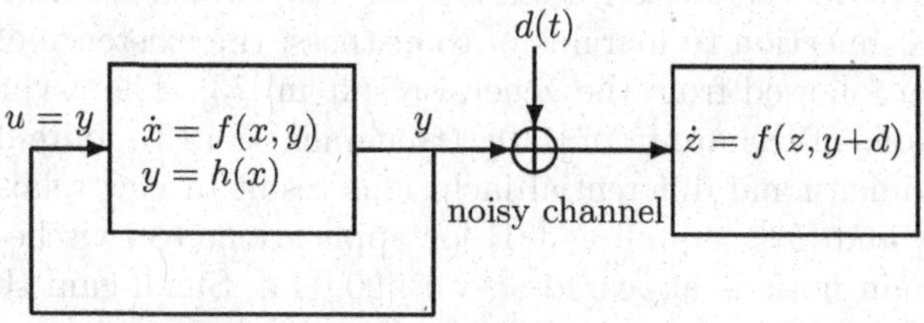

Fig. 29. Synchronized systems

- The system $\dot{x} = f(x, z, k(\widehat{x})) = F(x, z, e)$ is ISS with respect to both e and z
- The system $\dot{z} = g(x, z)$ is ISS with respect to x

Combining with a small-gain condition, the stability of the entire system is guaranteed.

A completely different application, in signal processing, can be found in [4], dealing with the topic of synchronized chaotic systems, which arises in the study of secure communications. A "master-slave" configuration is studied, where a second copy of the system (receiver) is driven by an output from the first (Fig. 29). The main objective is to show that states synchronize:

$$|x(t) - z(t)| \leq \max\{\beta(|x^{\circ} - z^{\circ}|, t), \|d\|\}$$

This can be shown, provided that the system is *incrementally* ISS, in the sense discussed in Sect. 8.6.

One particular example is given by the Lorentz attractor:

$$\begin{aligned}
\dot{x}_1 &= -\beta x_1 + \mathrm{sat}(x_2)\mathrm{sat}(x_3), \\
\dot{x}_2 &= \sigma(x_3 - x_2), \\
\dot{x}_3 &= -x_3 + u, \\
y &= \rho x_2 - x_1 x_2,
\end{aligned}$$

where $\beta = 8/3$, $\sigma = 10$, $\rho = 28$ (the saturation function, $\mathrm{sat}(r) = r$ for $|r| < 1$ and $\mathrm{sat}(r) = \mathrm{sign}(r)$ otherwise, is inserted for technical reasons and does not affect the application). Preservation of the i-ISS property under cascades implies that this system (easily to be seen a cascade of i-ISS subsystems) is i-ISS. The paper [4] provides simulations of the impressive behavior of this algorithm.

15 Additional Discussion and References

The paper [99] presented the definition of ISS, established the result on feedback redefinition to obtain ISS with respect to actuator errors, and provided the sufficiency test in terms of ISS-Lyapunov functions. The necessity of this Lyapunov-like characterization is from [109], which also introduced the "small gain" connection to margins of robustness; the existence of Lyapunov functions then followed from the general result in [73]. The asymptotic gain characterizations of ISS are from [110]. (Generalizations to finite-dimensional and infinite-dimensional differential inclusions result in new relaxation theorems, see [41] and [39], as well as [81] for applications to switched systems.) Asymptotic gain notions appeared also in [20, 114]. Small-gain theorems for ISS and IOS notions originated with [53]. See [40] for an abstract version of such results.

The notion of ISS for time-varying systems appears in the context of asymptotic tracking problems, see, e.g., [124]. In [24], one can find further results on Lyapunov characterizations of the ISS property for time-varying (and in particular periodic) systems, as well as a small-gain theorem based on these ideas. See also [78].

Coprime factorizations are the basis of the parameterization of controllers in the Youla approach. As a matter of fact, as the paper's title indicates, their study was the original motivation for the introduction of the notion of ISS in [99]. Some further work can be found in [98], see also [28], but much remains to be done.

One may of course also study the notion of ISS for discrete-time systems. Many ISS results for continuous time systems, and in particular the Lyapunov characterization and ISS small gain theorems, can be extended to the discrete time case; see [52, 55, 56, 61, 67]. Discrete-time iISS systems are the subject of [2], who proves the very surprising result that, in the discrete-time case, iISS is actually no different than global asymptotic stability of the unforced system (this is very far from true in the continuous-time case, of course); see also [74].

Questions of sampling, relating ISS properties of continuous and discrete-time systems, have been also studied, see [119] which shows that ISS is recovered under sufficiently fast sampling, as well as the papers [86, 87, 90].

The paper [5] introduces a notion of ISS where one merely requires good behavior on a generic (open dense) subset of the state space. Properties of this type are of interest in "almost-global" stabilization problems, where there are topological constraints on what may be achieved by controllers. The area is still largely undeveloped, and there are several open problems mentioned in that reference.

More generally than the question of actuator ISS, one can ask when, given a system $\dot{x} = f(x, d, u)$, is there a feedback law $u = k(x)$ such that the system $\dot{x} = f(x, d, k(x))$ becomes ISS (or iISS, etc) with respect to d. One approach to this problem is in terms of control-Lyapunov function ("cLf") methods, and concerns necessary and sufficient cLf conditions, for the existence of such (possibly dynamic) feedback laws. See for example [120], which deals primarily with systems of the form $\dot{x} = f(x, d) + g(x)u$ (affine in control, and control vector fields are independent of disturbances) and with assigning precise upper bounds to the "nonlinear gain" obtained in terms of d.

A problem of decentralized robust output-feedback control with disturbance attenuation for a class of large-scale dynamic systems, achieving ISS and iISS properties, is studied in [49].

Partial asymptotic stability for differential equations is a particular case of output stability (IOS when there are no inputs) in our sense; see [125] for a survey of the area, as well as the book [94], which contains a converse theorem for a restricted type of output stability. The subject of IOS is also related to the topic of "stability in two measures" (see, e.g., [68]), in the sense that one asks for stability of one "measure" of the state ($h(x)$) relative to initial conditions measured in another one (the norm of the state).

A useful variation of the notion of ISS is obtained when one studies stability with respect to a closed subset K of the state space $\mathbb{R}^n$, but not necessarily $K = \{0\}$. One may generalize the various definitions of ISS, IOS, IOSS, etc. For instance, the definition of ISS becomes

$$|x(t, x^0, u)|_K \leq \beta(|x^0|_K, t) + \gamma(\|u\|_\infty),$$

where $|x|_K$ denotes the distance from x to the set K. (The special case when $u \equiv 0$ implies in particular that the set K must be invariant for the unforced system.) The equivalence of various alternative definitions can be given in much the same way as the equivalence for the particular case $K = \{0\}$ (at least for compact K), since the general results in [73] are already formulated for set stability; see [108] for details. The interest in ISS with respect to sets arises in various contexts, such as the design of robust control laws, where the set K might correspond to equilibria for different parameter values, or problems of so-called "practical stabilization," concerned with stabilization to a prescribed neighborhood of the origin. See [107] for a theorem relating practical stabilization and ISS with respect to compact attractors.

Perhaps the most interesting set of open problems concerns the construction of feedback laws that provide ISS stability with respect to observation errors. Actuator errors are far better understood (cf. [99]), but save for the case of special structures studied in [27], the one-dimensional case (see, e.g., [25]) and the counterexample [26], little is known of this fundamental question. Recent work analyzing the effect of small observation errors (see [103]) might provide good pointers to useful directions of research (indeed, see [69] for some preliminary remarks in that direction). For special classes of systems, even output feedback ISS with respect to observation errors is possible, cf. [88].

A stochastic counterpart of the problem of ISS stabilization is proposed and solved in [22], formulated as a question of stochastic disturbance attenuation with respect to noise covariance. The paper [21], for a class of systems that can be put in output-feedback form (controller canonical form with an added stochastic output injection term), produces, via appropriate clf's, stochastic ISS behavior ("NSS" = noise to state stability, meaning that solutions converge in probability to a residual set whose radius is proportional to bounds on covariances). Stochastic ISS properties are treated in [123].

For a class of block strict-feedback systems including output feedback form systems, the paper [48] provided a global regulation result via nonlinear output feedback, assuming that the zero dynamics are iISS, thus generalizing the ISS-like minimum-phase condition in the previous [47], which in turn had removed the more restrictive assumption that system nonlinearities depend only on the output. See also [45] for iISS and ISS-stabilizing state and output feedback controllers for systems on strict-feedback or output-feedback forms.

For a class of systems including "Euler-Lagrange" models, the paper [17] provides a general result on global output feedback stabilization with a disturbance attenuation property. The notion of OSS and the results on unbounded observability both play a key role in the proof of correctness of the design.

An ISS result for the feedback interconnection of a linear block and a nonlinear element ("Lurie systems") is provided in [14], and an example is worked out concerning boundedness for negative resistance oscillators, such as the van der Pol oscillator.

The authors of [15] obtain robust tracking controllers with disturbance attenuation for a class of systems in strict-feedback form with structurally (non-parametric) unknown dynamics, using neural-network based approximations. One of the key assumptions is an ISS minimum phase condition, when external disturbances are included as inputs to the zero dynamics.

Output-feedback robust stabilization both in the ISS and iISS sense is studied, for large-scale systems with strongly nonlinear interconnections, in [51], using decentralized controllers.

Both ISS and iISS properties have been featured in the analysis of the performance of switching controllers, cf. [36]. The paper [35] dealt with hybrid control strategies for nonlinear systems with large-scale uncertainty, using a logic-based switching among a family of lower-level controllers, each of which is designed by finding an iISS-stabilizing control law for an appropriate system with disturbance inputs. The authors provide a result on stability and finite time switching termination for their controllers. The dissipation characterizations of ISS and of iISS were extended to a class of hybrid switched systems in [80].

A nonstandard application of IOSS, or more precisely of an MES property for turbulent kinetic energy and dissipation, was the method for *de*stabilization of pipe flows (to enhance mixing) studied in [1]. The authors used the wall velocity as inputs (blowing/suction actuators are assumed distributed on the pipe wall) and pressure differences across the pipe as outputs (using pressure sensors to measure). Detectability in the sense of IOSS provided a useful way to express the energy estimates required by the controller.

The papers [115, 116] introduced the notion of "formation input-to-state stability" in order to characterize the internal stability of leader-follower vehicle formations. There, and in related papers by other authors (e.g., [91]), ISS is used as a framework in which to systematically quantify the performance of swarm formations under time-varying signals (leader or enemy to be followed, noise in observation, actuator errors); in this context, the state x in the ISS estimate is in reality a measure of formation error. Thus, in terms of the original data of the problem, this *formation ISS* is an instance not of ISS itself, but rather of input/output stability (IOS), in which a function of state variables is used in estimates.

For results concerning averaging for ISS systems, see [89], and see [19] for singular perturbation issues in this context. See [97] for a notion which is in some sense close to IMES. Neural-net control techniques using ISS are mentioned in [95]. There are ISS-small gain theorems for certain infinite dimensional classes of systems such as delay systems, see [117].

Acknowledgments

The author wishes to thank Daniel Liberzon, Lars Grüne, and others, for very useful comments on a draft of this paper.

References

1. Aamo, O.M., A. Balogh, M. Krstić, "Optimal mixing by feedback in pipe flow," in *Proceedings of the 15th IFAC World Congress on Automatic Control*, Barcelona, Spain, 2002.
2. Angeli, D., "Intrinsic robustness of global asymptotic stability," *Systems & Control Letters* **38**(1999): 297–307.
3. Angeli, D., "Input-to-state stability of PD-controlled robotic systems," *Automatica*, **35**(1999): 1285–1290.
4. Angeli, D., "A Lyapunov approach to incremental stability properties" *IEEE Transactions on Automatic Control* **47**(2002): 410–422.
5. Angeli, D., "An almost global notion of input-to-state stability" *IEEE Transactions on Automatic Control* **49**(2004): 866–874.
6. Angeli, D., B. Ingalls, E.D. Sontag, Y. Wang, "Separation principles for input-output and integral-input to state stability," SIAM J. Control and Opt. **43**(2004): 256–276.
7. Angeli, D., B. Ingalls, E.D. Sontag, Y. Wang, "Uniform global asymptotic stability of differential inclusions," *Journal of Dynamical and Control Systems* **10**(2004): 391–412.
8. Angeli, D., D. Nesic, "Power formulations of input to state stability notions," in *Proc. IEEE Conf. Decision and Control, Orlando, Dec. 2001*, IEEE Publications, 2001, pp. 894–898.
9. Angeli, D., E.D. Sontag, "Forward completeness, unboundedness observability, and their Lyapunov characterizations," *Systems and Control Letters* **38**(1999): 209–217.
10. Angeli, D., E.D. Sontag, Y. Wang, "Input-to-state stability with respect to inputs and their derivatives," *Intern. J. Robust and Nonlinear Control* **13**(2003): 1035–1056.
11. Angeli, D., E.D. Sontag, Y. Wang, "A characterization of integral input to state stability," *IEEE Trans. Autom. Control* **45**(2000): 1082–1097.

12. Angeli, D., E.D. Sontag, Y. Wang, "Further equivalences and semiglobal versions of integral input to state stability," *Dynamics and Control* **10**(2000): 127–149.
13. Arcak, M., D. Angeli, E.D. Sontag, "A unifying integral ISS framework for stability of nonlinear cascades," *SIAM J. Control and Opt.* **40**(2002): 1888–1904.
14. Arcak, M., A. Teel, "Input-to-state stability for a class of Lurie systems," *Automatica* **38**(2002): 1945–1949.
15. Arslan, G., T. Başar, "Disturbance attenuating controller design for strict-feedback systems with structurally unknown dynamics," *Automatica* **37**(2001): 1175–1188.
16. Astolfi, A., L. Praly, "Global complete observability and output-to-state stability imply the existence of a globally convergent observer," in *Proc. IEEE Conf. Decision and Control, Maui, Dec. 2003*, IEEE Publications, 2003, pp. 1562–1567.
17. Besanọn, G., S. Battilotti, L. Lanari, "A new separation result for a class of quadratic-like systems with application to Euler-Lagrange models," *Automatica* **39**(2003): 1085–1093.
18. Brockett, R.W., "Asymptotic stability and feedback stabilization," in *Differential Geometric Control theory* (R.W. Brockett, R.S. Millman, and H.J. Sussmann, eds.), Birkhauser, Boston, 1983, pp. 181–191.
19. Christofides, P.D., A.R. Teel, "Singular perturbations and input-to-state stability," *IEEE Trans. Automat. Control* **41**(1996): 1645–1650.
20. Coron, J.M., L. Praly, A. Teel, "Feedback stabilization of nonlinear systems: sufficient conditions and Lyapunov and input-output techniques," in *Trends in Control*, A. Isidori, ed., Springer-Verlag, London, 1995.
21. Deng, H., and M. Krstić, "Output-feedback stabilization of stochastic nonlinear systems driven by noise of unknown covariance," *Systems Control Lett.* **39**(2000): 173–182.
22. Deng, H., M. Krstic, R.J. Williams, "Stabilization of stochastic nonlinear systems driven by noise of unknown covariance," *IEEE Trans. Automat. Control* **46**(2001): 1237–1253.
23. Doyle, J., B. Francis, A. Tannenbaum, *Feedback Control Systems*, MacMillan Publishing Co, 1992. Also available on the web at: http://www.control.utoronto.capeople/profs/francis/dft.html.
24. Edwards, H., Y. Lin, Y. Wang, "On input-to-state stability for time varying nonlinear systems," in *Proc. 39th IEEE Conf. Decision and Control*, Sydney, Australia, 2000, pp. 3501–3506.
25. Fah, N.C.S., "Input-to-state stability with respect to measurement disturbances for one-dimensional systems," *Control, Optimisation and Calculus of Variations* **4**(1999): 99–121.
26. Freeman, R.A., "Global internal stabilizability does not imply global external stabilizability for small sensor disturbances," *IEEE Trans. Automat. Control* **40**(1996): 2119–2122.
27. Freeman, R.A., P.V. Kokotović, *Robust Nonlinear Control Design, State-Space and Lyapunov Techniques*, Birkhauser, Boston, 1996.
28. Fujimoto, K., T. Sugie, "State-space characterization of Youla parametrization for nonlinear systems based on input-to-state stability, *Proc. 37th IEEE Conf. Decision and Control*, Tampa, Dec. 1998, pp. 2479–2484.

29. Grüne, L., "Input-to-state stability of exponentially stabilized semilinear control systems with inhomogeneous perturbations," *System & Control Letters* **38**(1999): 27–35.
30. Grüne, L., *Asymptotic Behavior of Dynamical and Control Systems under Perturbation and Discretization.* Lecture Notes in Mathematics, Vol. 1783. Springer–Verlag, Heidelberg, 2002.
31. Grüne, L., "Input-to-state dynamical stability and its Lyapunov function characterization," *IEEE Trans. Autom. Control* **47**(2002): 1499–1504.
32. Grüne, L., "Attraction rates, robustness and discretization of attractors," *SIAM J. Numer. Anal.* **41**(2003): 2096–2113.
33. Grune, L., E.D. Sontag, F.R. Wirth, "Asymptotic stability equals exponential stability, and ISS equals finite energy gain - if you twist your eyes," *Systems and Control Letters* **38**(1999): 127–134.
34. Hespanha, J.P., D. Liberzon, D. Angeli, E.D. Sontag, "Nonlinear observability notions and stability of switched systems," *IEEE Trans. Autom. Control*, 2005, to appear.
35. Hespanha, J.P., D. Liberzon, A.S. Morse, "Supervision of integral-input-to-state stabilizing controllers," *Automatica* **38**(2002): 1327–1335.
36. Hespanha, J.P, A.S. Morse, "Certainty equivalence implies detectability," *Systems and Control Letters* 36(1999): 1–13.
37. Hu, X.M., "On state observers for nonlinear systems," *Systems & Control Letters* **17**(1991), pp. 645–473.
38. Ingalls, B., E.D. Sontag, Y. Wang, "Measurement to error stability: a notion of partial detectability for nonlinear systems," in *Proc. IEEE Conf. Decision and Control, Las Vegas, Dec. 2002*, IEEE Publications, 2002, pp. 3946–3951.
39. Ingalls, B., E.D. Sontag, Y. Wang, "A relaxation theorem for differential inclusions with applications to stability properties," in *Proc. 15th Int. Symp. Mathematical Theory of Networks and Systems (MTNS 2002)*, CD-ROM, FM5.3.
40. Ingalls, B., E.D. Sontag, "A small-gain lemma with applications to input/output systems, incremental stability, detectability, and interconnections," *J. Franklin Institute* **339**(2002): 211–229.
41. Ingalls, B., E.D. Sontag, Y. Wang, "An infinite-time relaxation theorem for differential inclusions," *Proc. Amer. Math. Soc.* **131**(2003): 487–499.
42. Isidori, A., *Nonlinear Control Systems, Third Edition*, Springer-Verlag, London, 1995.
43. Isidori, A., *Nonlinear Control Systems II*, Springer-Verlag, London, 1999.
44. Isidori, A., L. Marconi, A. Serrani, *Robust Autonomous Guidance: An Internal Model-Based Approach*, Springer-Verlag, London, 2003.
45. Ito, H., "New characterization and solution of input-to-state stabilization: a state-dependent scaling approach," *Automatica* **37**(2001): 1663–1670.
46. Ito, H., "Scaling supply rates of ISS systems for stability of feedback interconnected nonlinear systems," in *Proc. IEEE Conf. Decision and Control, Maui, Dec. 2003*, IEEE Publications, 2003, pp. 5074–5079.
47. Jiang, Z.-P., "A combined backstepping and small-gain approach to adaptive output feedback control," *Automatica* **35**(1999): 1131–1139.
48. Jiang, Z.-P., I.M. Mareels, D.J. Hill, Jie Huang, "A unifying framework for global regulation via nonlinear output feedback: from ISS to integral ISS," *IEEE Trans. Automat. Control* **49**(2004): 549–562.

49. Jiang, Z.-P., F. Khorrami, D.J. Hill, "Decentralized output-feedback control with disturbance attenuation for large-scale nonlinear systems," *Proc. 38th IEEE Conf. Decision and Control*, Phoenix, Dec. 1999, pp. 3271–3276.
50. Jiang, Z.-P., L. Praly, "Preliminary results about robust Lagrange stability in adaptive nonlinear regulation." *Intern. J. Control* **6**(1992): 285–307.
51. Jiang, Z.-P., D.W. Repperger, D.J. Hill, "Decentralized nonlinear output-feedback stabilization with disturbance attenuation," *IEEE Trans. Automat. Control* **46**(2001): 1623–1629.
52. Jiang, Z.-P., E.D. Sontag, Y. Wang, "Input-to-state stability for discrete-time nonlinear systems," in *Proc. 14th IFAC World Congress* (Beijing), Vol E, pp. 277–282, 1999.
53. Jiang, Z.-P., A. Teel, L. Praly, "Small-gain theorem for ISS systems and applications," *Mathematics of Control, Signals, and Systems* **7**(1994): 95–120.
54. Jiang, Z.-P., Y. Wang, "Small gain theorems on input-to-output stability", in *Proc. of the Third International DCDIS Conference on Engineering Applications and Computational Algorithms*, pp. 220–224, 2003.
55. Jiang, Z.-P., Y. Wang, "A converse Lyapunov theorem for discrete time systems with disturbances", *Systems & Control Letters*, **45**(2002): 49–58.
56. Jiang, Z.P., Y. Wang, "Input-to-state stability for discrete-time nonlinear systems", *Automatica*, **37**(2001): 857–869.
57. Kaliora, G., A. Astolfi, L. Praly, "Norm estimators and global output feedback stabilization of nonlinear systems with ISS inverse dynamics," Proc. 43rd IEEE Conference on Decision and Control, Paradise Island, Bahamas, Dec. 2004, paper ThC07.2, IEEE Publications, Piscataway.
58. Khalil, H.K., *Nonlinear Systems, Second Edition*, Prentice-Hall, Upper Saddle River, NJ, 1996.
59. Krener, A.J., "A Lyapunov theory of nonlinear observers," in *Stochastic analysis, control, optimization and applications*, Birkhäuser Boston, Boston, MA, 1999, pp. 409–420.
60. Krstić, M., I. Kanellakopoulos, P.V. Kokotović, *Nonlinear and Adaptive Control Design*, John Wiley & Sons, New York, 1995.
61. Kazakos, D., J. Tsinias, "The input-to-state stability condition and global stabilization of discrete-time systems," *IEEE Trans. Automat. Control* **39**(1994): 2111–2113.
62. Kokotović, P., M. Arcak, "Constructive nonlinear control: a historical perspective," *Automatica* **37**(2001): 637–662.
63. Kokotović, P., M. Arcak, "Nonlinear observers: a circle criterion design and robustness analysis," *Automatica* **37**(2001): 1923–1930.
64. Krichman, M., E.D. Sontag, "Characterizations of detectability notions in terms of discontinuous dissipation functions," *Intern. J. Control* **75**(2002): 882–900.
65. Krichman, M., E.D. Sontag, Y. Wang, "Input-output-to-state stability," *SIAM J. Control and Optimization* **39**(2001): 1874–1928.
66. Krstić, M., H. Deng, *Stabilization of Uncertain Nonlinear Systems*, Springer-Verlag, London, 1998.
67. Laila, D.S., D. Nesic, "Changing supply rates for input-output to state stable discrete-time nonlinear systems with applications," *Automatica* **39**(2003): 821–835.
68. Lakshmikantham, V., S. Leela, A.A. Martyuk, *Practical Stability of Nonlinear Systems*, World Scientific, New Jersey, 1990.

69. Liberzon, D., "Nonlinear stabilization by hybrid quantized feedback," in *Proc. 3rd International Workshop on Hybrid Systems: Computation and Control, Lecture Notes in Computer Science vol. 1790* (N. Lynch and B. H. Krogh, Eds.), pp/ 243–257, Springer-Verlag, 2000.
70. Liberzon, D., "Output-input stability and feedback stabilization of multivariable nonlinear control systems," in *Proc. IEEE Conf. Decision and Control, Maui, Dec. 2003*, IEEE Publications, 2003, pp. 1550–1555.
71. Liberzon, D., A.S. Morse, E.D. Sontag, "Output-input stability and minimum-phase nonlinear systems," *IEEE Trans. Autom. Control* **47**(2002): 422–436.
72. Liberzon, D., E.D. Sontag, Y. Wang, "Universal construction of feedback laws achieving ISS and integral-ISS disturbance attenuation," *Systems and Control Letters* **46**(2002): 111–127.
73. Lin, Y., E.D. Sontag, Y. Wang, "A smooth converse Lyapunov theorem for robust stability," *SIAM J. Control and Optimization* **34**(1996): 124–160.
74. Loria, A., D. Nesic, "Stability of time-varying discrete-time cascades," *in Proc. 15th. IFAC World Congress, (Barcelona, Spain), 2002*, paper no. 652.
75. Lu, W.M., "A class of globally stabilizing controllers for nonlinear systems," *Systems & Control Letters* **25**(1995), pp. 13–19.
76. Lu, W.M., "A state-space approach to parameterization of stabilizing controllers for nonlinear systems, *IEEE Trans. Automat. Control* **40**(1995): 1576–1588.
77. Lu, W.M., "A class of globally stabilizing controllers for nonlinear systems," *Systems & Control Letters* **25**(1995): 13–19.
78. Malisoff, M., F. Mazenc, "Further remarks on strict input-to-state stable lyapunov functions for time-varying systems," arxiv.org/math/0411150, June 2004.
79. Malisoff, M., L. Rifford, E.D. Sontag, "Asymptotic controllability implies input to state stabilization," *SIAM J. Control and Optimization* **42**(2004): 2221–2238.
80. Mancilla-Aguilar, J.L., R.A. García, "On converse Lyapunov theorems for ISS and iISS switched nonlinear systems," *Systems & Control Letters* **42**(2001): 47–53.
81. Mancilla-Aguilar, J.L., R. García, E. Sontag, Y. Wang, "On the representation of switched systems with inputs by perturbed control systems," *Nonlinear Analysis: Theory, Methods & Applications* **60**(2005): 1111–1150.
82. Marino, R., G. Santosuosso, P. Tomei, "Robust adaptive observers for nonlinear systems with bounded disturbances," *Proc. 38th IEEE Conf. Decision and Control*, Phoenix, Dec. 1999, pp. 5200–5205.
83. Marino, R., P. Tomei, "Nonlinear output feedback tracking with almost disturbance decoupling," *IEEE Trans. Automat. Control* **44**(1999): 18–28.
84. Massera, J.L., "Contributions to stability theory," *Annals of Mathematics* **64**(1956): 182–206.
85. Morse, A.S., "Control using logic-based switching," in Trends in Control: A European Perspective, A. Isidori, ed., Springer-Verlag, London, 1995, pp. 69–114.
86. Nesic, D., D. Angeli, "Integral versions of iss for sampled-data nonlinear systems via their approximate discrete-time models," *IEEE Transactions on Automatic Control* **47**(2002): 2033–2037.

87. Nesic, D., D.S. Laila, "A note on input-to-state stabilization for nonlinear sampled-data systems", *IEEE Transactions on Automatic Control* **47**(2002): 1153–1158.
88. Nešić, D., E.D. Sontag, "Input-to-state stabilization of linear systems with positive outputs," *Systems and Control Letters* **35**(1998): 245–255.
89. Nesic, D., A.R. Teel, "Input-to-state stability for nonlinear time-varying systems via averaging," *Mathematics of Control, Signals and Systems* **14**(2001): 257–280.
90. Nesic, D., A.R. Teel, E.D. Sontag, "Formulas relating KL stability estimates of discrete-time and sampled-data nonlinear systems," *Syst. Contr. Lett.* **38**(1999): 49–60.
91. Ogren, P., N.E. Leonard, "Obstacle avoidance in formation," in *Proc. IEEE Conf. Robotics and Automation,* Taipei, Taiwan, Sept. 2003, pp. 2492–2497.
92. Pan, D.J., Z.Z. Han, Z.J. Zhang, "Bounded-input-bounded-output stabilization of nonlinear systems using state detectors," SCL **21**(1993): 189–198.
93. Praly, L., Y. Wang "Stabilization in spiteof matched unmodelled dynamics and an equivalent definition of input-to-state stability," *Mathematics of Control, Signals, and Systems* **9**(1996): 1–33.
94. Rumyantsev, V.V., A.S. Oziraner, *Stability and Stabilization of Motion with Respect to Part of the Variables* (in Russian), Nauka, Moscow, 1987.
95. Sanchez, E.N., J.P. Perez, "Input-to-state stability (ISS) analysis for dynamic neural networks," *IEEE Trans. Circuits and Systems I: Fundamental Theory and Applications* **46**(1999): 1395–1398.
96. Sepulchre, R., M. Jankovic, P.V. Kokotović, *Constructive Nonlinear Control,* Springer-Verlag, New York, 1997.
97. Shiriaev, A.S., "The notion of V-detectability and stabilization of invariant sets of nonlinear systems," *Proc. 37th IEEE Conf. Decision and Control,* Tampa, Dec. 1998. pp. 2509–2514.
98. Sontag, E.D., "Some connections between stabilization and factorization," *Proc. IEEE Conf. Decision and Control, Tampa, Dec. 1989,* IEEE Publications, 1989, pp. 990–995.
99. Sontag, E.D., "Smooth stabilization implies coprime factorization," *IEEE Trans. Automatic Control,* **34**(1989): 435–443.
100. Sontag, E.D., "Remarks on stabilization and input-to-state stability," *Proc. IEEE Conf. Decision and Control, Tampa, Dec. 1989,* IEEE Publications, 1989, pp. 1376–1378.
101. Sontag, E.D., "Further facts about input to state stabilization", *IEEE Trans. Automatic Control* **35**(1990): 473–476.
102. Sontag, E.D., "Comments on integral variants of ISS," *Systems and Control Letters* **34**(1998): 93–100.
103. Sontag, E.D., "Stability and stabilization: Discontinuities and the effect of disturbances," in *Nonlinear Analysis, Differential Equations, and Control* (Proc. NATO Advanced Study Institute, Montreal, Jul/Aug 1998; F.H. Clarke and R.J. Stern, eds.), Kluwer, Dordrecht, 1999, pp. 551–598.
104. Sontag, E.D., *Mathematical Control Theory: Deterministic Finite Dimensional Systems*, Springer, New York, 1990. Second Edition, 1998.
105. Sontag, E.D., "The ISS philosophy as a unifying framework for stability-like behavior," in *Nonlinear Control in the Year 2000* (Volume 2) (Lecture Notes in Control and Information Sciences, A. Isidori, F. Lamnabhi-Lagarrigue, and W. Respondek, eds.), Springer-Verlag, Berlin, 2000, pp. 443–468.

106. Sontag, E.D., A.R. Teel, "Changing supply functions in input/state stable systems," *IEEE Trans. Autom. Control* **40**(1995): 1476–1478.
107. Sontag, E.D., Y. Wang, "Various results concerning set input-to-state stability," *Proc. IEEE Conf. Decision and Control, New Orleans, Dec. 1995*, IEEE Publications, 1995, pp. 1330–1335.
108. Sontag, E.D., Y. Wang, 'On characterizations of input-to-state stability with respect to compact sets," in *Proceedings of IFAC Non-Linear Control Systems Design Symposium, (NOLCOS '95)*, Tahoe City, CA, June 1995, pp. 226–231.
109. Sontag, E.D., Y. Wang, "On characterizations of the input-to-state stability property," *Systems and Control Letters* **24**(1995): 351–359.
110. Sontag, E.D., Y. Wang, "New characterizations of input to state stability," *IEEE Trans. Autom. Control* **41**(1996): 1283–1294.
111. Sontag, E.D., Y. Wang, "Output-to-state stability and detectability of nonlinear systems," *Systems and Control Letters* **29**(1997): 279–290.
112. Sontag, E.D., Y. Wang, "Notions of input to output stability," *Systems and Control Letters* **38**(1999): 235–248.
113. Sontag, E.D., Y. Wang, "Lyapunov characterizations of input to output stability," *SIAM J. Control and Opt.* **39**(2001): 226–249.
114. Sussmann, H.J., E.D. Sontag, Y. Yang, "A general result on the stabilization of linear systems using bounded controls," *IEEE Trans. Autom. Control* **39**(1994): 2411–2425.
115. Tanner, H.G., G.J. Pappas, V. Kumar, "Input-to-state stability on formation graphs," in *Proc. IEEE Conf. on Decision and Control,* Las Vegas, December 2002.
116. Tanner, H.G., G.J. Pappas, V. Kumar, "Leader-to-formation stability," *IEEE Trans. on Robotics and Automation* **20**(2004): 443–455.
117. Teel, A.R., "Connections between Razumikhin-type theorems and the ISS nonlinear small gain theorem," *IEEE Trans. Automat. Control* **43**(1998): 960–964.
118. Teel, A.R., L. Moreau, D. Nesic, "A note on the robustness of input-to-state stability," in *Proc. IEEE Conf. Decision and Control, Orlando, Dec. 2001*, IEEE Publications, 2001, pp. 875–880.
119. Teel, A.R., D. Nešić, P.V. Kokotović, "A note on input-to-state stability of sampled-data nonlinear systems," *Proc. 37th IEEE Conf. Decision and Control*, Tampa, Dec. 1998, pp. 2473–2479.
120. Teel, A.R., L. Praly, "On assigning the derivative of a disturbance attenuation control," *Mathematics of Control, Signals, and Systems* **3**(2000): 95–124.
121. Su, W., L. Xie, Z. Gong, "Robust input to state stabilization for minimum-phase nonlinear systems," *Int J. Control* **66**(1997): 825–842.
122. Tsinias, J., "Input to state stability properties of nonlinear systems and applications to bounded feedback stabilization using saturation," *Control, Optimisation and Calculus of Variations* **2**(1997): 57–85.
123. Tsinias, J., "Stochastic input-to-state stability and applications to global feedback stabilization," *Int. J. Control* **71**(1998): 907–930.
124. Tsinias, J., I. Karafyllis, "ISS property for time-varying systems and application to partial-static feedback stabilization and asymptotic tracking," *IEEE Trans. Automat. Control* **44**(1999): 2173–2184.
125. Vorotnikov, V.I., "Problems of stability with respect to part of the variables," *Journal of Applied Mathematics and Mechanics* **63**(1999): 695–703.
126. Willems, J.C., "Mechanisms for the stability and instability in feedback systems," *Proc. IEEE* **64**(1976): 24–35.

127. Xie, W., C. Wen, Z. Li, "Controller design for the integral-input-to-state stabilization of switched nonlinear systems: a Cycle analysis method," *Proc. IEEE Conf. Decision and Control, Las Vegas, Dec. 2002*, IEEE Publications, 2002, pp. 3628–3633.
128. Xie, W., C. Wen, Z. Li, "Input-to-state stabilization of switched nonlinear systems," *IEEE Trans. Automatic Control* **46**(2001): 1111–1116.

Generalized Differentials, Variational Generators, and the Maximum Principle with State Constraints

H.J. Sussmann*

Department of Mathematics, Hill Center, Rutgers University,
110 Frelinghuysen Rd, Piscataway, NJ 08854-8019, USA
sussmann@math.rutgers.edu

1 Introduction

In a series of previous papers (cf. [20–23]), we have developed a "primal" approach to the non-smooth Pontryagin Maximum Principle, based on generalized differentials, flows, and general variations. The method used is essentially the one of classical proofs of the Maximum Principle such as that of Pontryagin and his coauthors (cf. Pontryagin et al. [15], Berkovitz [1]), based on the construction of packets of needle variations, but with a refinement of the "topological argument," and with concepts of differential more general than the classical one, and usually set-valued.

In this article we apply this approach to optimal control problems with state space constraints, and at the same time we state the result in a more concrete form, dealing with a specific class of generalized derivatives (the "generalized differential quotients"), rather than in the abstract form used in some of the previous work.

The paper is organized as follows. In Sect. 2 we introduce some of our notations, and review some background material, especially the basic concepts about finitely additive vector-valued measures on an interval. In Sect. 3 we review the theory of "Cellina continuously approximable" (CCA) set-valued maps, and prove the CCA version – due to A. Cellina – of some classical fixed point theorems due to Leray-Schauder, Kakutani, Glicksberg and Fan. In Sect. 4 we define the notions of generalized differential quotient (GDQ), and approximate generalized differential quotient (AGDQ), and prove their basic properties, especially the chain rule, the directional open mapping theorem, and the transversal intersection property. In Sect. 5 we define the two types of variational generators that will occur in the maximum principle, and state and prove theorems asserting that various classical generalized derivatives – such

*Research supported in part by NSF Grants DMS01-03901 and DMS-05-09930.

as classical differentials, Clarke generalized Jacobians, subdifferentials in the sense of Michel–Penot, and (for functions defining state space constraints) the object often referred to as $\partial_x^{>} g$ in the literature – are special cases of our variational generators. In Sect. 6 we discuss the classes of discontinuous vector fields studied in detail in [24]. In Sect. 7 we state the main theorem. The rather lengthy proof will be given in a subsequent paper.

Acknowledgments. The author is grateful to Paolo Nistri and Gianna Stefani, the organizers of the Cetraro 2004 C.I.M.E. Summer School, for inviting him to present the ideas that have led to this paper, to Arrigo Cellina for his useful comments on the condition that is here called "Cellina continuous approximability," and to the audience of the Cetraro lectures for their numerous contributions. He also thanks G. Stefani for the invitation to give a series of lectures in Firenze in the Summer of 2005, and the participants of the Firenze lectures, especially Andrei Sarychev, who explained to the author why it was important to use finitely additive measures.

2 Preliminaries and Background

2.1 Review of Some Notational Conventions and Definitions

Integers and Real Numbers

We use $\mathbb{Z}$, $\mathbb{R}$ to denote, respectively, the set of all integers and the set of all real numbers, and write $\mathbb{N} \stackrel{\text{def}}{=} \{n \in \mathbb{Z} : n > 0\}$, $\mathbb{Z}_+ \stackrel{\text{def}}{=} \mathbb{N} \cup \{0\}$. Also, $\bar{\mathbb{R}}$, $\mathbb{R}_+$, $\bar{\mathbb{R}}_+$, denote, respectively, the extended real line $\mathbb{R} \cup \{-\infty, +\infty\}$, the half-line $[0, +\infty[$, and the extended half-line $[0, +\infty]$ (i.e., $[0, +\infty[\cup \{+\infty\})$.

Intervals

An *interval* is an arbitrary connected subset of $\mathbb{R}$. If $a, b \in \mathbb{R}$ and $a \leq b$, then $\text{INT}([a, b])$ is the set of all intervals J such that $J \subseteq [a, b]$. Hence $\text{INT}([a, b])$ consists of the intervals $[\alpha, \beta]$, $[\alpha, \beta[$, $]\alpha, \beta]$ and $]\alpha, \beta[$, with $a \leq \alpha < \beta \leq b$, as well as the singletons $\{\alpha\}$, for $a \leq \alpha \leq b$), and the empty set. A *nontrivial interval* is one whose length is strictly positive, that is, one that contains at least two distinct points.

Euclidean Spaces and Matrices

The expressions $\mathbb{R}^n$, $\mathbb{R}_n$ will be used to denote, respectively, the set of all real column vectors $x = (x_1, \ldots, x_n)^\dagger$ (where "$\dagger$" stands for "transpose") and the set of all real row vectors $p = (p_1, \ldots, p_n)$. We refer to the members of $\mathbb{R}_n$ as *covectors*. Also, $\mathbb{R}^{m \times n}$ is the space of all real matrices with m rows and n columns.

If $n \in \mathbb{Z}_+$, $x \in \mathbb{R}^n$, $r \in \mathbb{R}$, and $r > 0$, we use $\bar{\mathbb{B}}^n(x,r)$, $\mathbb{B}^n(x,r)$ to denote, respectively, the closed and open balls in $\mathbb{R}^n$ with center x and radius r. We write $\bar{\mathbb{B}}^n(r)$, $\mathbb{B}^n(r)$ for $\bar{\mathbb{B}}^n(0,r)$, $\mathbb{B}^n 0,(r)$, and $\bar{\mathbb{B}}^n$, $\mathbb{B}^n$ for $\bar{\mathbb{B}}^n(1)$, $\mathbb{B}^n(1)$. Also, we will use $\mathbb{S}^n$ to denote the n-dimensional unit sphere, so $\mathbb{S}^n = \{(x_1,\ldots,x_{n+1})^\dagger \in \mathbb{R}^{n+1} : \sum_{j=1}^{n+1} x_j^2 = 1\}$.

Topological Spaces, Metric Spaces, Metric Balls

We will use throughout the standard terminology of point-set topology: a *neighborhood* of a point x in a topological space X is any subset S of X that contains an open set U such that $x \in U$. In the special case of a metric space X, we use $\mathbb{B}_X(x,r)$, $\bar{\mathbb{B}}_X(x,r)$, to denote, respectively, the open ball and the closed ball with center x and radius r.

Quasidistance and Hausdorff Distance

If X is a topological space, then $Comp^0(X)$ will denote the set of all compact subsets of X (including the empty set), and $Comp(X)$ will be the set of all nonempty members of $Comp^0(X)$.

If X is a metric space, with distance function d_X, then we can define the "quasidistance" $\Delta_X^{qua}(A,B)$ from a set $A \in Comp^0(X)$ to another set $B \in Comp^0(X)$ by letting

$$\Delta_X^{qua}(A,B) = \sup\Big\{\inf\{d_X(x,x') : x' \in B\} : x \in A\Big\}. \tag{2.1.1}$$

(This function is not a distance because, for example, it is not symmetric, since $\Delta_X^{qua}(A,B) = 0$ but $\Delta_X^{qua}(B,A) \neq 0$ if $A \subseteq B$ and $A \neq B$. Furthermore, Δ_X^{qua} can take the value $+\infty$, since $\Delta_X^{qua}(A,B) = +\infty$ if $A \neq \emptyset$ but $B = \emptyset$.)

Definition 2.1 *Suppose that X is a metric space. The* ***Hausdorff distance*** *$\Delta_X(K,L)$ between two nonempty subsets K, L of X is the number*

$$\Delta_X(K,L) = \max\Big(\Delta_X^{qua}(K,L), \Delta_X^{qua}(L,K)\Big). \qquad \square$$

It is then clear that the function Δ_X, restricted to $Comp(X) \times Comp(X)$, is a metric.

Linear Spaces and Linear Maps

The abbreviations "FDRLS" and "FDNRLS" will stand for the expressions "finite-dimensional real linear space," and "finite-dimensional normed real linear space," respectively. If X and Y are real linear spaces, then $Lin(X,Y)$ will denote the set of all linear maps from X to Y. We use $X^\dagger$ to denote $Lin(X,\mathbb{R})$, i.e., the dual space of X. If X is a FDNRLS, then $X^{\dagger\dagger}$ is identified with X in the usual way.

If X and Y are FDNRLSs, then $Lin(X,Y)$ is a FDNRLS, endowed with the operator norm $\|\cdot\|_{op}$ given by

$$\|L\|_{op} = \sup\{\|L\cdot x\| : x \in X, \|x\| \leq 1\}\,. \tag{2.1.2}$$

Also, we write $\boldsymbol{L}(X)$ for $Lin(X,X)$, the space of all linear maps $L : X \mapsto X$.

We identify $Lin(\mathbb{R}^n, \mathbb{R}^m)$ with $\mathbb{R}^{m\times n}$ in the usual way, by assigning to each matrix $M \in \mathbb{R}^{m\times n}$ the linear map $\mathbb{R}^n \ni x \mapsto M\cdot x \in \mathbb{R}^m$. In particular, $\boldsymbol{L}(X)$ is identified with $\mathbb{R}^{n\times n}$. Also, we identify $\mathbb{R}_n$ with the dual $(\mathbb{R}^n)^\dagger$ of $\mathbb{R}^n$, by assigning to a $y \in \mathbb{R}_n$ the linear functional $\mathbb{R}^n \ni x \mapsto y\cdot x \in \mathbb{R}$.

If X, Y are FDRLSs, and $L \in Lin(X,Y)$, then the *adjoint* of L is the map $L^\dagger : Y^\dagger \mapsto X^\dagger$ such that $L^\dagger(y) = y \circ L$ for $y \in Y^\dagger$. In the special case when $X = \mathbb{R}^n$ and $Y = \mathbb{R}^m$, so $L \in \mathbb{R}^{m\times n}$, the map $L^\dagger$ goes from $\mathbb{R}_m$ to $\mathbb{R}_n$, and is given by $L^\dagger(y) = y\cdot L$ for $y \in \mathbb{R}_m$.

Manifolds, Tangent Spaces, Differentials

If M is a manifold of class C^1, and $x \in M$, then T_xM will denote the tangent space of M at x. It follows that if M, N are manifolds of class C^1, $x \in M$, F is an N-valued map defined on a neighborhood U of x in M, and F is classically differentiable at x, then the differential $DF(x)$ belongs to $Lin(T_xM, T_{F(x)}N)$.

Single- and Set-Valued Maps

Throughout this paper, the word "map" always stands for "set-valued map." The expression "ppd map" refers to a "possibly partially defined (that is, not necessarily everywhere defined) ordinary (that is, single-valued) map." The precise definitions are as follows. A *set-valued map* is a triple $F = (A,B,G)$ such that A and B are sets and G is a subset of $A\times B$. If $F = (A,B,G)$ is a set-valued map, we say that F *is a set-valued map from A to B*. In that case, we refer to the sets A, B, G as the *source*, *target*, and *graph* of F, respectively, and we write $A = \mathrm{So}(F)$, $B = \mathrm{Ta}(F)$, $G = \mathrm{Gr}(F)$. If $x \in \mathrm{So}(F)$, we write $F(x) = \{y : (x,y) \in \mathrm{Gr}(F)\}$. The set $\mathrm{Do}(F) = \{x \in \mathrm{So}(F) : F(x) \neq \emptyset\}$ is the *domain* of F. If A, B are sets, we use $SVM(A,B)$ to denote the set of all set-valued maps from A to B, and write $F : A \mapsto\!\!\!\rightarrow B$ to indicate that $F \in SVM(A,B)$. A *ppd map from A to B* is an $F \in SVM(A,B)$ such that $F(x)$ has cardinality zero or one for every $x \in A$. We write $F : A \hookrightarrow B$ to indicate that F is a ppd map from A to B. If $F : A \mapsto\!\!\!\rightarrow B$, and $C \subseteq A$, then the *restriction* of F to C is the set-valued map $F\lceil C$ defined by $F\lceil C \stackrel{\mathrm{def}}{=} (C, B, \mathrm{Gr}(F)\cap(C\times B))$.

If F_1 and F_2 are set-valued maps, then the *composite* $F_2\circ F_1$ is defined if and only if $\mathrm{Ta}(F_1) = \mathrm{So}(F_2)$ and, in that case, $\mathrm{So}(F_2\circ F_1) \stackrel{\mathrm{def}}{=} \mathrm{So}(F_1)$, $\mathrm{Ta}(F_2\circ F_1) \stackrel{\mathrm{def}}{=} \mathrm{Ta}(F_2)$, and

$$\mathrm{Gr}(F_2\circ F_1) \stackrel{\mathrm{def}}{=} \Big\{(x,z) : (\exists y)\Big((x,y)\in\mathrm{Gr}(F_1) \text{ and } (y,z)\in\mathrm{Gr}(F_2)\Big)\Big\}\,.$$

If A is a set, then $\mathbb{I}_A$ denotes the *identity map* of A, that is, the triple (A, A, Δ_A), where Δ_A is the set of all pairs (x, x), for all $x \in A$.

Epimaps and Constraint Indicator Maps

If $f : S \hookrightarrow \mathbb{R}$ is a ppd function, then:

- The *epimap* of f is the set-valued map $\check{f} : S \mapsto\!\!\!\rightarrow \mathbb{R}$ whose graph is the epigraph of f, so that $\check{f}(s) = \{f(s)+v : v \in \mathbb{R}, v \geq 0\}$ whenever $s \in \mathrm{Do}(f)$, and $\check{f}(s) = \emptyset$ if $s \in S \backslash \mathrm{Do}(f)$.
- The *constraint indicator map* of f is the set-valued map $\chi_f^{co} : S \mapsto\!\!\!\rightarrow \mathbb{R}$ such that $\chi_f^{co}(s) = \emptyset$ if $f(s) \leq 0$ or $s \in S \backslash \mathrm{Do}(f)$, and $\chi_f^{co}(s) = [0, +\infty[$ if $f(s) > 0$.

Cones and Multicones

A *cone* in a FDRLS X is a nonempty subset C of X such that $r \cdot c \in C$ whenever $c \in C, r \in \mathbb{R}$ and $r \geq 0$. If X is a FDRLS, a *multicone* in X is a nonempty set of convex cones in X. A multicone $\mathcal{C}$ is *convex* if every member C of $\mathcal{C}$ is convex.

Polars

Let X be a FDNRLS. The *polar* of a cone $C \subseteq X$ is the closed convex cone $C^\dagger = \{\lambda \in X^\dagger : \lambda(c) \leq 0 \text{ for all } c \in C\}$. If $\mathcal{C}$ is a multicone in X, the *polar* of $\mathcal{C}$ is the set $\mathcal{C}^\dagger = \mathrm{Clos}\Big(\bigcup\{C^\dagger : C \in \mathcal{C}\}\Big)$, so $\mathcal{C}^\dagger$ is a (not necessarily convex) closed cone in $X^\dagger$.

Boltyanskii Approximating Cones

If X is a FDNRLS, $S \subseteq X$, and $x \in S$, a *Boltyanskii approximating cone to S at x* is a convex cone C in X such that there exist an $n \in \mathbb{Z}_+$, a closed convex cone D in $\mathbb{R}^n$, a neighborhood U of 0 in $\mathbb{R}^n$, a continuous map $F : U \cap D \mapsto S$, and a linear map $L : \mathbb{R}^n \mapsto X$, such that $F(h) = x + L \cdot h + o(\|h\|)$ as $h \to 0$ via values in D, and $C = L \cdot D$. A *limiting Boltyanskii approximating cone to S at x* is a closed convex cone C which is the closure of an increasing union $\bigcup_{j=1}^{\infty} C_j$ such that each C_j is a Boltyanskii approximating cone to S at x.

Some Function Spaces

If A, B are sets, we use $fn(A, B)$ to denote the set of all functions from A to B. If X is a real normed space and A is a set, then $\mathcal{B}dfn(A, X)$ will denote the set of all bounded functions from A to X. The space $\mathcal{B}dfn(A, X)$

is endowed with the norm $\|\cdot\|_{sup}$ given by $\|f\|_{sup} = \sup\{\|f(t)\| : t \in A\}$. Then $\mathcal{B}dfn(A, X)$ is a Banach space if X is a Banach space.

If, in addition, A is a topological space, then $C^0(A, X)$ denotes the space of all bounded continuous functions from A to B, endowed with the norm $\|\cdot\|_{sup}$. It is clear that $C^0(A, X)$ is a closed subspace of $\mathcal{B}dfn(A, X)$, so in particular $C^0(A, X)$ is a Banach space if X is a Banach space.

Tubes

If X is a FDNRLS, $a, b \in \mathbb{R}$, $a \leq b$, $\xi \in C^0([a, b], X)$ and $\delta > 0$, we use $\mathcal{T}^X(\xi, \delta)$ to denote the *δ-tube about ξ in X*, defined by

$$\mathcal{T}^X(\xi, \delta) \stackrel{\text{def}}{=} \{(x, t) : x \in X, a \leq t \leq b, \|x - \xi(t)\| \leq \delta\}. \tag{2.1.3}$$

Vector Fields, Trajectories, and Flow Maps

If X is a FDNRLS, a *ppd time-varying vector field* on X is a ppd map $X \times \mathbb{R} \ni (x, t) \hookrightarrow f(x, t) \in X$. A *trajectory*, or *integral curve*, of a ppd time-varying vector field f on X is a locally absolutely continuous map $\xi : I \mapsto X$, defined on a nonempty real interval I, such that for almost all $t \in I$ the following two conditions hold: (i) $(\xi(t), t) \in \mathrm{Do}(f)$, and (ii) $\dot{\xi}(t) = f(\xi(t), t)$. If f is a ppd time-varying vector field on X, then $\mathrm{Traj}\,(f)$ will denote the set of all integral curves $\xi : I_\xi \mapsto X$ of f. If S is a subset of $X \times \mathbb{R}$, then $\mathrm{Traj}\,(f, S)$ will denote the set of $\xi \in \mathrm{Traj}\,(f)$ such that $(\xi(t), t) \in S$ for all $t \in I_\xi$, and $\mathrm{Traj}_{\,c}(f, S)$ will denote the set of $\xi \in \mathrm{Traj}\,(f, S)$ whose domain I_ξ is a compact interval.

The *flow map* of a ppd time-varying vector field $X \times \mathbb{R} \ni (x, t) \hookrightarrow f(x, t) \in X$ is the set-valued map $\Phi^f : \mathbb{R} \times \mathbb{R} \times X \mapsto\!\!\!\rightarrow X$ that assigns to each triple $(t, s, x) \in \mathbb{R} \times \mathbb{R} \times X$ the set $\Phi^f(t, s, x) = \{\xi(t) : \xi \in \mathrm{Traj}\,(f), \xi(s) = x\}$.

Functions of Bounded Variation

Assume that X is a real normed space, $a, b \in \mathbb{R}$, and $a < b$.

Definition 2.2 *A function $\varphi \in fn([a, b], X)$ is* **of bounded variation** *if there exists a nonnegative real number C such that $\sum_{j=1}^m \|\varphi(t_j) - \varphi(s_j)\| \leq C$ whenever $m \in \mathbb{N}$ and the finite sequences $\{s_j\}_{j=1}^m$, $\{t_j\}_{j=1}^m$ are such that $a \leq s_1 \leq t_1 \leq s_2 \leq t_2 \leq \cdots \leq s_m \leq t_m \leq b$.* □

We use $bvfn([a, b], X)$ to denote the set of all $\varphi \in fn([a, b], X)$ that are of bounded variation, and define the *total variation norm* $\|\varphi\|_{tv}$ of a function $\varphi \in fn([a, b], X)$ by letting $\|\varphi\|_{tv} = \|\varphi(b)\| + C(\varphi)$, where $C(\varphi)$ is the smallest C having the property of Definition 2.2. Also, we let $bvfn^{0,b}([a, b], X)$ denote the set of all $\varphi \in bvfn([a, b], X)$ such that $\varphi(b) = 0$. Then $\|\varphi\|_{tv} = C(\varphi)$ if $\varphi \in bvfn^{0,b}([a, b], X)$. It is then easy to verify that

Fact 2.3 *If X is a Banach space, then the space $bvfn([a,b],X)$, endowed with the total variation norm $\|\cdot\|_{tv}$, is a Banach space, and $bvfn^{0,b}([a,b],X)$ is a closed linear subspace of $bvfn([a,b],X)$ of codimension one.* □

Fact 2.4 *If X is a Banach space and $f \in bvfn(([a,b],X)$, then $\lim_{s\uparrow t} f(s)$ exists for every $t \in]\,a,b]$, and $\lim_{s\downarrow t} f(s)$ exists for every $t \in [a,b\,[$.* □

Remark 2.5 The set $bvfn([a,b],X)$ is clearly a linear subspace of $\mathcal{B}dfn([a,b],X)$. The sup norm and the total variation norm are related by the inequality $\|\varphi\|_{sup} \leq \|\varphi\|_{tv}$, which holds whenever $\varphi \in bvfn([a,b],X)$. On the other hand, $bvfn([a,b],X)$ is clearly *not* closed in $\mathcal{B}dfn([a,b],X)$. □

Measurable Spaces and Measure Spaces

A *measurable space* is a pair $(S,\mathcal{A})$ such that S is a set and $\mathcal{A}$ is a σ-algebra of subsets of S.

If $(S,\mathcal{A})$ is a measurable space, then a *nonnegative measure* on $(S,\mathcal{A})$ is a map $\mu : \mathcal{A} \mapsto [0,+\infty]$ that satisfies $\mu(\emptyset) = 0$ and is countably additive (i.e., such that $\mu(\bigcup_{j=1}^{\infty} A_j) = \sum_{j=1}^{\infty} \mu(A_j)$ whenever $\{A_j\}_{j\in\mathbb{N}}$ is a sequence of pairwise disjoint members of $\mathcal{A}$).

A *nonnegative measure space* is a triple $(S,\mathcal{A},\mu)$ is such that $(S,\mathcal{A})$ is a measurable space and μ is a nonnegative-measure on $(S,\mathcal{A})$. A nonnegative measure space $(S,\mathcal{A},\mu)$ is *finite* if $\mu(A) < \infty$ for all $A \in \mathcal{A}$.

Measurability of Set-Valued Maps; Support Functions

Assume that $(S,\mathcal{A})$ is a measurable space and Y is a FDNRLS.

Definition 2.6 *A set-valued map $\Lambda : S \mapsto\!\!\!\!\to Y$ is said to be* ***measurable*** *if the set $\{s \in S : \Lambda(s) \cap \Omega \neq \emptyset\}$ belongs to $\mathcal{A}$ for every open subset Ω of Y.* □

If Λ has compact values, then we define the *support function* of Λ to be the function $\sigma_\Lambda : S \times Y^\dagger \mapsto \mathbb{R}$ given by

$$\sigma_\Lambda(s,x) = \sup\{\langle x,y\rangle : y \in \Lambda(s)\} \ \text{ for } x \in Y^\dagger, \ s \in S. \tag{2.1.4}$$

(If $\Lambda(s) = \emptyset$ then we define $\sigma_\Lambda(s,y) = -\infty$.) The following fact is well known.

Lemma 2.7 *Assume that $(S,\mathcal{A})$ is a measurable space, Y is a FDNRLS, and $\Lambda : S \mapsto\!\!\!\!\to Y$ is a set-valued map with compact convex values. For each $y \in Y^\dagger$, let $\psi_y(s) = \sigma_\Lambda(s,y)$. Then Λ is measurable if and only if the function $\psi_y : S \mapsto \mathbb{R} \cup \{-\infty\}$ is measurable for every $y \in Y^\dagger$.* □

Integrable Boundedness of Set-Valued Maps

Assume that $(S,\mathcal{A},\nu)$ is a nonnegative measure space.

Definition 2.8 *A* ***ν-integrable bound*** *for a set-valued map $\Lambda : S \mapsto\!\!\!\!\to Y$ is a nonnegative ν-integrable function $k : S \mapsto [0,+\infty]$ having the property that $\Lambda(s) \subseteq \{y \in Y : \|y\| \leq k(s)\}$ for ν-almost all $s \in S$. The map Λ is said to be* ***ν-integrably bounded*** *if there exists a ν-integrable bound for Λ.* □

2.2 Generalized Jacobians, Derivate Containers, and Michel–Penot Subdifferentials

For future use, we will now review the definitions and basic properties of three classical "non-smooth" notions of set-valued derivative, namely, Clarke generalized Jacobians, Warga derivate containers, and Michel–Penot derivatives.

Generalized Jacobians

Assume that X, Y are FDNRLSs, Ω is an open subset of X, $F : \Omega \mapsto Y$ is a Lipschitz-continuous map, and $\bar{x}_* \in \Omega$.

Definition 2.9 *The* ***Clarke generalized Jacobian*** *of* F *at* $\bar{x}_*$ *is the subset* $\partial F(\bar{x}_*)$ *of* $Lin(X,Y)$ *defined as follows:*

- $\partial F(\bar{x}_*)$ *is the convex hull of the set of all limits* $L = \lim_{j\to\infty} DF(x_j)$, *for all sequences* $\{x_j\}_{j\in\mathbb{N}}$ *in* Ω *such that (1)* $\lim_{j\to\infty} x_j = \bar{x}_*$, *(2)* F *is classically differentiable at* x_j *for all* $j \in \mathbb{N}$, *and (3) the limit* L *exists.* □

Warga Derivate Containers

Assume that X, Y are FDNRLSs, Ω is an open subset of X, $F : \Omega \mapsto Y$, and $\bar{x}_* \in \Omega$.

Definition 2.10 *A* ***Warga derivate container*** *of* F *at* $\bar{x}_*$ *is a compact subset* Λ *of* $Lin(X,Y)$ *such that:*

- *For every positive number* δ *there exist (1) an open neighborhood* U_δ *of* $\bar{x}_*$ *such that* $U_\delta \subseteq \Omega$, *and (2) a sequence* $\{F_j\}_{j\in\mathbb{N}}$ *of* Y*-valued functions of class* C^1 *on* U_δ, *such that (i)* $\lim_{j\to\infty} F_j = F$ *uniformly on* U_δ, *(ii)* $\mathrm{dist}(DF_j(x), \Lambda) \leq \delta$ *for every* $(j,x) \in \mathbb{N}\times U_\delta$. □

Michel–Penot Subdifferentials

Assume that X is a FDNRLS, Ω is an open subset of X, $f : \Omega \mapsto \mathbb{R}$ is a Lipschitz-continuous function, and $\bar{x}_* \in \Omega$. For $h \in X$, define

$$d^o f(\bar{x}_*, h) = \sup_{k\in X} \limsup_{t\downarrow 0} t^{-1}\Big(f(\bar{x}_* + t(k+h)) - f(\bar{x}_* + tk)\Big), \qquad (2.2.1)$$

so that $X \ni h \mapsto d^o f(\bar{x}_*, h) \in \bar{\mathbb{R}}$ is a convex positively homogeneous function.

Definition 2.11 *The* ***Michel–Penot subdifferential*** *of* f *at* $\bar{x}_*$ *is the set* $\partial^o f(\bar{x}_*)$ *of all linear functionals* $\omega \in X^\dagger$ *having the property that the inequality* $d^o f(\bar{x}_*, h) \geq \langle \omega, h\rangle$ *holds whenever* $h \in X$. □

2.3 Finitely Additive Measures

If $a, b \in \mathbb{R}$, $a < b$, and X is a FDNRLS, we use $\mathcal{P}c([a,b],X)$ to denote the set of all piecewise constant X-valued functions on $[a,b]$, so that $f \in \mathcal{P}c([a,b],X)$ iff $f : [a,b] \mapsto X$ and there exists a finite partition $\mathcal{P}$ of $[a,b]$ into intervals such that f is constant on each $I \in \mathcal{P}$. We let $\overline{\mathcal{P}c}([a,b],X)$ denote the set of all uniform limits of members of $\mathcal{P}c([a,b],X)$, so $\overline{\mathcal{P}c}([a,b],X)$ is a Banach space, endowed with the sup norm. Furthermore, $\overline{\mathcal{P}c}([a,b],X)$ is exactly the space of all $f : [a,b] \mapsto X$ such that the left limit $f(t-) = \lim_{s \to t, s<t} f(s)$ exists for all $t \in]a,b]$, and the right limit $f(t+) = \lim_{s \to t, s>t} f(s)$ exists for all $t \in [a,b[$.

We define $\overline{\mathcal{P}c}_0([a,b],X)$ to be the set of all $f \in \overline{\mathcal{P}c}([a,b],X)$ that vanish on the complement of a countable (i.e., finite or countably infinite) set. Then $\overline{\mathcal{P}c}_0([a,b],X)$ is the closure in $\overline{\mathcal{P}c}([a,b],X)$ of the space $\mathcal{P}c_0([a,b],X)$ of all $f \in \mathcal{P}c([a,b],X)$ such that f vanishes on the complement of a finite set.

We let $pc([a,b],X)$ be the quotient space $\overline{\mathcal{P}c}([a,b],X)/\overline{\mathcal{P}c}_0([a,b],X)$. Then every equivalence class $F \in pc([a,b],X)$ has a unique left-continuous member F_-, and a unique right-continuous member F_+, and of course $F_- \equiv F_+$ on the complement of a countable set. So $pc([a,b],X)$ can be identified with the set of all pairs (f_-, f_+) of X-valued functions on $[a,b]$ such that f_- is left-continuous, f_+ is right-continuous, and $f_- \equiv f_+$ on the complement of a countable set.

If X is a FDNRLS, then we use $bvadd([a,b],X)$ to denote the dual space $pc([a,b],X^\dagger)^\dagger$ of $pc([a,b],X^\dagger)$. A *reduced additive X-valued interval function of bounded variation* on $[a,b]$ is a member of $bvadd([a,b],X)$. A measure $\mu \in bvadd([a,b],X)$ gives rise to a set function $\hat{\mu} : INT([a,b]) \mapsto X$, defined by $\langle \hat{\mu}(I), y \rangle = \mu(\chi_I^y)$ for $y \in X^\dagger$, where $\chi_I^y(t) = 0$ if $t \notin I$ and $\chi_I^y(t) = y$ if $t \in I$. We then associate to μ its *cumulative distribution* cd_μ, defined by $cd_\mu(t) = -\hat{\mu}([t,b])$ for $t \in [a,b]$. Then cd_μ belongs to the space $bvfn^{0,b}([a,b],X)$ of all functions $\varphi : [a,b] \mapsto X$ that are of bounded variation (cf. Definition 2.2) and such that $\varphi(b) = 0$. The map

$$bvadd([a,b],X) \ni \mu \mapsto cd_\mu \in bvfn^{0,b}([a,b],X)$$

is a bijection. The dual Banach space norm $\|\mu\|$ of a $\mu \in bvadd([a,b],X)$ coincides with $\|cd_\mu\|_{bv}$.

Remark 2.12 The *non-reduced* additive X-valued interval functions of bounded variation on $[a,b]$ are the members of the dual space $\overline{\mathcal{P}c}([a,b],X^\dagger)^\dagger$. Then $bvadd([a,b],X)$ consists of those members of $\overline{\mathcal{P}c}([a,b],X^\dagger)^\dagger$ that vanish on every test function $F \in \overline{\mathcal{P}c}([a,b],X^\dagger)$ such that $F(t) = 0$ for all but finitely many values of t. For a reduced interval function $\mu \in bvadd([a,b],X)$, the measure $\hat{\mu}(\{t\})$ of every singleton is equal to zero, because the function $\chi_{\{t\}}^y$ belongs to $\overline{\mathcal{P}c}_0([a,b],X^\dagger)$ for every $y \in X^\dagger$. □

A $\mu \in bvadd([a,b],X)$ is a *left* (resp.*right*) *delta function* if there exist an $x \in X$ and a $t \in]a,b]$ (resp. a $t \in [a,b[$) such that $\mu(F) = \langle F(t-), x \rangle$

(resp. $\mu(F) = \langle F(t+), x\rangle$) for all $F \in pc([a,b],X)$. We call μ *left-atomic* (resp. *right-atomic*) if it is the sum of a convergent series of left (resp. right) delta functions.

A $\mu \in bvadd([a,b],X)$ is *continuous* if the function cd_μ is continuous. Every $\mu \in bvadd([a,b],X)$ has a unique decomposition into the sum of a continuous part μ_{co}, a left-atomic part $\mu_{at,-}$ and a right-atomic part $\mu_{at,+}$. (This resembles the usual decomposition of a countably additive measure into the sum of a continuous part and an atomic part. The only difference is that in the finitely additive setting there are left and right atoms rather than just atoms.)

If Y is a FDNRLS, a *bounded Y-valued measurable pair on* $[a,b]$ is a pair (γ_-,γ_+) of bounded Borel measurable functions from $[a,b]$ to Y such that $\gamma_- \equiv \gamma_+$ on the complement of a finite or countable set. If X,Y,Z are FDNRLSs, $Y \times X \ni (y,x) \mapsto \langle y,x\rangle \in Z$ is a bilinear map, $\mu \in bvadd([a,b],X)$, and $\gamma = (\gamma_-,\gamma_+)$ is a bounded Y-valued measurable pair on $[a,b]$, then the product measure $\gamma\cdot\mu$ is a member of $bvadd([a,b],Z)$ defined by multiplying the continuous part μ_{co} by γ_- or γ_+, the left-atomic part by γ_-, and the right-atomic part by γ_+. In particular, the product $\gamma\cdot\mu$ is a well defined member of $bvadd([a,b],X)$ whenever $\mu \in bvadd([a,b],\mathbb{R})$ and γ is a bounded X-valued measurable pair on $[a,b]$.

Finally, we briefly discuss the solutions of an "adjoint" Cauchy problem represented formally as

$$dy(t) = -y(t)\cdot L(t)\cdot dt + d\mu(t)\,, \quad y(b) = \bar{y}\,, \tag{2.3.1}$$

where μ belongs to $bvadd([a,b],X^\dagger)$, $L \in L^1([a,b],\boldsymbol{L}(X))$, and we are looking for solutions $y(\cdot) \in bvadd([a,b],X^\dagger)$.

This is done by rewriting our Cauchy problem as the integral equation

$$y(t) - V(t) = \int_t^b y(s)\cdot L(s)\cdot ds\,, \qquad \text{where} \quad V = cd_\mu\,. \tag{2.3.2}$$

Equation (2.3.2) is easily seen to have a unique solution π, given by

$$\pi(t) = \bar{y}\cdot M_L(b,t) - \int_{[t,b]} d\mu(s)\cdot M_L(s,t)\,, \tag{2.3.3}$$

where $M_L : [a,b]\times[a,b] \mapsto \boldsymbol{L}(X)$ is the fundamental solution of $\dot{M} = M\cdot L$, characterized by the identity $M_L(\tau,t) = \mathbb{I}_X + \int_t^\tau L(r)\cdot M_L(r,t)\,dr$.

3 Cellina Continuously Approximable Maps

The CCA maps constitute a class of set-valued maps whose properties are similar to those of single-valued continuous maps. The most important such property is the fixed point theorem that, for single-valued continuous maps, is known as Brouwer's theorem in the finite-dimensional case, and as Schauder's theorem in the infinite-dimensional case. A class of set-valued maps with some of the desired properties was singled out in the celebrated Kakutani fixed point theorem (for the finite-dimensional case), and its infinite-dimensional generalization due to Fan and Glicksberg. This class, whose members are the upper semicontinuous maps with nonempty compact convex values, turns out to be insufficient for our purposes, because is lacks the crucial property that a composite of two maps belonging to the class also belongs to the class. (For example, if $f : \bar{\mathbb{B}}^n(0,1) \mapsto\!\!\!\!\rightarrow \bar{\mathbb{B}}^n(0,1)$ has nonempty convex values and a compact graph, and $g : \bar{\mathbb{B}}^n(0,1) \mapsto \bar{\mathbb{B}}^n(0,1)$ is single-valued and continuous, then g also has a compact graph and nonempty convex values, so g belongs to the class as well, but $g \circ f$ need not belong to the class, because the image of a convex set under a continuous map need not be convex. And yet it is obvious that $g \circ f$ has to have a fixed point, because the same standard argument used to prove the Kakutani theorem applies here as well: we can find a sequence of single-valued continuous maps f_j that converge to f in an appropriate sense, apply Brouwer's theorem to obtain fixed points x_j of the maps $g \circ f_j$, and then pass to the limit.)

The previous example strongly suggests that there ought to exist a class of maps, larger than that of the Kakutani and Fan-Glicksberg theorems, which is closed under composition and such that the usual fixed point theorems hold. This class was introduced by A. Cellina in a series of papers around 1970 (cf. [3–6]). We now study it in detail.

3.1 Definition and Elementary Properties

CCA maps are set-valued maps that are limits of single-valued continuous maps in the sense of an appropriate (non-Hausdorff) notion of convergence. We begin by defining this concept of convergence precisely.

Inward Graph Convergence

If K, Y are metric spaces and K is compact, then $SVM_{comp}(K,Y)$ will denote the subset of $SVM(K,Y)$ whose members are the set-valued maps from K to Y that have a compact graph. We say that a sequence $\{F_j\}_{j\in\mathbb{N}}$ of members of $SVM_{comp}(K,Y)$ *inward graph-converges* to an $F \in SVM_{comp}(K,Y)$ – and write $F_j \xrightarrow{\text{igr}} F$ – if for every open subset Ω of $K \times Y$ such that $\text{Gr}(F) \subseteq \Omega$ there exists a $j_\Omega \in \mathbb{N}$ such that $\text{Gr}(F_j) \subseteq \Omega$ whenever $j \geq j_\Omega$.

The above notion of convergence is a special case of the following more general idea. Recall that $Comp^0(X)$ is the set of all compact subsets of X. Then we can define a topology $\mathcal{T}_{Comp^0(X)}$ on $Comp^0(X)$ by declaring a subset $\mathcal{U}$ of $Comp^0(X)$ to be *open* if for every $K \in \mathcal{U}$ there exists an open subset U of X such that $K \subseteq U$ and $\{J \in Comp^0(X) : J \subseteq U\} \subseteq \mathcal{U}$. (This topology is non-Hausdorff even if X is Hausdorff, because if $J, K \in Comp^0(X)$, $J \subseteq K$, and $J \neq K$, then every neighborhood of K contains J.) Inward graph convergence of a sequence $\{F_j\}_{j\in\mathbb{N}}$ of members of $SVM_{comp}(K,Y)$ to an $F \in SVM_{comp}(K,Y)$ is then equivalent to convergence to $\mathrm{Gr}(F)$ of the sets $\mathrm{Gr}(F_j)$ in the topology $\mathcal{T}_{Comp^0(X)}$.

The convergence of sequences and, more generally, of nets, in the space $\mathcal{T}_{Comp^0(X)}$ can be characterized as follows, in terms of the quasidistance Δ^{qua} defined in (2.1.1).

Fact 3.1 *Let (Z,d_Z) be a metric space, let $\mathbf{K} = \{K_\alpha\}_{\alpha\in A}$ be a net of members of $Comp^0(Z)$, indexed by a directed set $(A,\preceq_A)$, and let $K \in Comp^0(Z)$. Then the net $\mathbf{K}$ converges to K with respect to $\mathcal{T}_{Comp^0(Z)}$ if and only if $\lim_\alpha \Delta_Z^{qua}(K_\alpha,K) = 0$.* □

Fact 3.1 can be applied in the special case when the metric space Z is a product $X \times Y$, equipped with the distance $d_Z : Z \times Z \mapsto \mathbb{R}_+$ given by

$$d_Z\Big((x,y),(x',y')\Big) = d_X(x,x') + d_Y(y,y'). \tag{3.1.1}$$

We then obtain the following equivalent characterization of inward graph convergence.

Fact 3.2 *Let X, Y be metric spaces, with distance functions d_X, d_Y, let $\mathbf{F} = \{F_\alpha\}_{\alpha\in A}$ be a net of members of $SVM_{comp}(X,Y)$, indexed by a directed set $(A,\preceq_A)$, and let $F \in SVM_{comp}(X,Y)$. Then the net $\mathbf{F}$ converges to F in the inward graph convergence sense (that is, the graphs $\mathrm{Gr}(F_\alpha)$ converge to $\mathrm{Gr}(F)$ in $\mathcal{T}_{Comp^0(X\times Y)}$) if and only if $\lim_\alpha \Delta_Z^{qua}\Big(\mathrm{Gr}(F_\alpha),\mathrm{Gr}(F)\Big) = 0$, where $Z = X \times Y$, equipped with the distance d_Z given by (3.1.1).* □

Compactly Graphed Set-Valued Maps

Suppose that X and Y are metric spaces, and $F : X \mapsto\!\!\!\!\rightarrow Y$. Then F is *compactly graphed* if, for every compact subset K of X, the restriction $F\lceil K$ of F to K has a compact graph, i.e., has the property that the set $\mathrm{Gr}(F\lceil K) \stackrel{\mathrm{def}}{=} \{(x,y) : x \in K \wedge y \in F(x)\}$ is compact.

We recall that, if X, Y are topological spaces, then a set-valued map $F : X \mapsto\!\!\!\!\rightarrow Y$ is said to be *upper semicontinuous* if the inverse image of every closed subset U of Y is a closed subset of X. It is then easy to see that

Fact 3.3 *If X and Y are metric spaces and $F : X \mapsto\!\!\!\!\rightarrow Y$ has compact values, then F is upper semicontinuous if and only if it is compactly graphed.* □

CCA Maps

We are now, finally, in a position to define the notion of a "Cellina continuously approximable map."

Definition 3.4 *Assume that X and Y are metric spaces, and $F : X \mapsto\!\!\!\!\rightarrow Y$. We say that F is* **Cellina continuously approximable** *(abbr. "CCA") if F is compactly graphed and*

- *For every compact subset K of X, the restriction $F \lceil K$ is a limit – in the sense of inward graph-convergence – of a sequence of continuous single-valued maps from K to Y.* □

We will use the expression $\mathrm{CCA}(X,Y)$ to denote the set of all CCA set-valued maps from X to Y. It is easy to see that

Fact 3.5 *If $f : X \hookrightarrow Y$ is a ppd map, then the following are equivalent:*

(1) $f \in \mathrm{CCA}(X,Y)$
(2) f is everywhere defined and continuous
(3) f is everywhere defined and compactly graphed □

Composites of CCA Maps

The following simple observation will play a crucial role in the theory of GDQs and AGDQs.

Theorem 3.6 *Assume that X, Y, Z are metric spaces. Let $F \in \mathrm{CCA}(X,Y)$, $G \in \mathrm{CCA}(Y,Z)$. Then the composite map $G \circ F$ belongs to $\mathrm{CCA}(X,Z)$.*

Proof. Let $H = G \circ F$. We prove first that H is compactly graphed. Let K be a compact subset of X, and let $J = \mathrm{Gr}(H \lceil K)$. A pair (x,z) belongs to J if and only if there exists $y \in Y$ such that $(x,y) \in \mathrm{Gr}(F \lceil K)$ and $(y,z) \in \mathrm{Gr}(G)$. Let $Q = \pi(\mathrm{Gr}(F \lceil K))$, where π is the projection $X \times Y \ni (x,y) \mapsto y \in Y$. Then $(x,z) \in J$ iff there exists $y \in Q$ such that $(x,y) \in \mathrm{Gr}(F \lceil K)$ and $(y,z) \in \mathrm{Gr}(G \lceil Q)$. Equivalently, $(x,z) \in J$ iff there exists a point $p = (x,y,\tilde{y},z) \in S$ such that $\Pi(p) = (x,z)$ and $p \in A$, where $A = \{(x,y,\tilde{y},z) \in X \times Y \times Y \times Z : y = \tilde{y}\}$, $S = \mathrm{Gr}(F \lceil K) \times \mathrm{Gr}(G \lceil Q)$, and Π is the projection $X \times Y \times Y \times Z \in (x,y,\tilde{y},z) \mapsto (x,z) \in X \times Z$.

So $J = \Pi(S \cap A)$. Since S is compact and A is closed in $X \times Y \times Y \times Z$, the set $S \cap A$ is compact, so J is compact, since Π is continuous. Hence H is compactly graphed.

We now fix a compact subset K of X, let $h = H \lceil K$, and show that there exists a sequence $\{h_j\}_{j \in \mathbb{N}}$ of continuous maps from K to Z such that $h_j \xrightarrow{\mathrm{igr}} h$. For this purpose, we let $f = F \lceil K$, and use the fact that F is a CCA map to construct a sequence $\{f_j\}_{j \in \mathbb{N}}$ of continuous maps from K to Y such that $f_j \xrightarrow{\mathrm{igr}} f$ as $j \to \infty$. Then the set $B = \mathrm{Gr}(f) \cup \left(\bigcup_{j=1}^{\infty} \mathrm{Gr}(f_j)\right)$ is clearly compact. (*Proof:* Let $\mathcal{O}$ be a set of open subsets of $X \times Y$ such that $B \subseteq \bigcup\{\Omega : \Omega \in \mathcal{O}\}$. Then $\mathrm{Gr}(f) \subseteq \bigcup\{\Omega : \Omega \in \mathcal{O}\}$. Since $\mathrm{Gr}(f)$ is compact, we may pick a finite subset $\mathcal{O}_0$ of $\mathcal{O}$ such that $\mathrm{Gr}(f) \subseteq \bigcup\{\Omega : \Omega \in \mathcal{O}_0\}$. Since the set $\bigcup\{\Omega : \Omega \in \mathcal{O}_0\}$ is open and contains $\mathrm{Gr}(f)$, there exists a $j^* \in \mathbb{N}$ such that $\mathrm{Gr}(f_j) \subseteq \bigcup\{\Omega : \Omega \in \mathcal{O}_0\}$ whenever $j > j^*$. For $j = 1, \ldots, j^*$, use the compactness of $\mathrm{Gr}(f_j)$ to pick a finite subset $\mathcal{O}_j$ of $\mathcal{O}$ such that $\mathrm{Gr}(f_j) \subseteq \bigcup\{\Omega : \Omega \in \mathcal{O}_j\}$. Let $\hat{\mathcal{O}} = \bigcup\{\mathcal{O}_j : j = 0, \dots j^*\}$. Then $\hat{\mathcal{O}}$ is a finite subset of $\mathcal{O}$, and $B \subseteq \bigcup\{\Omega : \Omega \in \hat{\mathcal{O}}\}$.)

Let $C = \pi(B)$, where π is the projection defined above. Then C is a compact subset of Y, and the fact that G is a CCA map implies that there exists a sequence $\{g_j\}_{j\in\mathbb{N}}$ of continuous maps $g_j : C \mapsto Z$ such that $g_j \xrightarrow{\mathrm{igr}} g$, where $g = G\lceil C$.

We now define $h_j = g_j \circ f_j$, and begin by observing that the h_j are well defined continuous maps from K to Z. (The reason that h_j is well defined is that if $x \in K$, then $(x, f_j(x)) \in \mathrm{Gr}(f_j) \subseteq B$, so $(x, f_j(x)) \in B$, and then $f_j(x) \in C$, so $g_j(f_j(x))$ is defined. The continuity of h_j then follows because it is a composite of continuous maps.)

To conclude the proof, we have to establish that $h_j \xrightarrow{\mathrm{igr}} h$. Let us first define $\alpha_j = \sup\{\Xi(x,z) : (x,z) \in \mathrm{Gr}(h_j)\}$, where Ξ is the map given by $\Xi(x,z) = \inf\{d(x,\tilde{x}) + d(z,\tilde{z}) : (\tilde{x},\tilde{z}) \in \mathrm{Gr}(h)\}$. We want to show that $\alpha_j \to 0$ as $j \to \infty$. Suppose not. Then by passing to a subsequence we may assume that $\alpha_j \geq 2\bar{\alpha}$ for all j, for some strictly positive $\bar{\alpha}$. For each j, pick $(x_j, z_j) \in \mathrm{Gr}(h_j)$ such that $\Xi(x_j, z_j) \geq \bar{\alpha}$. Let $y_j = f_j(x_j)$, so $z_j = g_j(y_j)$. The point (x_j, y_j) then belongs to $\mathrm{Gr}(f_j)$, so $\Theta(x_j, y_j) \to 0$, where Θ was defined above. Hence we can find $(\tilde{x}_j, \tilde{y}_j) \in \mathrm{Gr}(f)$ such that $d(x_j, \tilde{x}_j) + d(y_j, \tilde{y}_j) \to 0$. Similarly, we can define $\hat{\Theta}(y,z) = \inf\{d(y,\tilde{y}) + d(z,\tilde{z}) : (\tilde{y},\tilde{z}) \in \mathrm{Gr}(g)\}$, and conclude that $\hat{\Theta}(y_j, z_j) \to 0$, since $g_j \xrightarrow{\mathrm{igr}} g$, so we can find points $(\tilde{y}_j^{\#}, \tilde{z}_j)$, belonging to $\mathrm{Gr}(g)$, such that $d(y_j, \tilde{y}_j^{\#}) + d(z_j, \tilde{z}_j) \to 0$. So all four quantities $d(x_j, \tilde{x}_j)$, $d(y_j, \tilde{y}_j)$, $d(y_j, \tilde{y}_j^{\#})$, and $d(z_j, \tilde{z}_j)$, go to 0. Since $\mathrm{Gr}(f)$ and $\mathrm{Gr}(g)$ are compact we may assume, after passing to a subsequence, that the $(\tilde{x}_j, \tilde{y}_j)$ converge to a limit $(\tilde{x}, \tilde{y}) \in \mathrm{Gr}(f)$, and the $(\tilde{y}_j^{\#}, \tilde{z}_j)$ converge to a limit $d(\tilde{y}^{\#}, \tilde{z}) \in \mathrm{Gr}(g)$. Since $d(y_j, \tilde{y}_j) \to 0$ and $d(y_j, \tilde{y}_j^{\#}) \to 0$, we have $d\tilde{y}_j, \tilde{y}_j^{\#}) \to 0$, so $\tilde{y} = \tilde{y}^{\#}$. So $(\tilde{x}, \tilde{y}) \in \mathrm{Gr}(F)$ and $(\tilde{y}, \tilde{z}) \in \mathrm{Gr}(G)$, from which it follows that $(\tilde{x}, \tilde{z}) \in \mathrm{Gr}(H)$. But $d(x_j, \tilde{x}_j) \to 0$ and $\tilde{x}_j \to \tilde{x}$, so $d(x_j, \tilde{x}) \to 0$. Similarly, $d(z_j, \tilde{z}) \to 0$. Hence $\Xi(x_j, z_j) \to 0$ contradicting the inequalities $\Xi(x_j, z_j) \geq \bar{\alpha} > 0$. So $\alpha_j \to 0$, and our proof is complete. □

3.2 Fixed Point Theorems for CCA Maps

The Space of Compact Connected Subsets of a Compact Metric Space

Recall that, if X is a metric space, then $Comp(X)$ denotes the set of all nonempty compact subsets of X. The Hausdorff distance Δ_X was introduced in Definition 2.1. We write $Comp_c(X)$ to denote the set of all connected members of $Comp(X)$. We will need the following fact about $Comp(X)$.

Proposition 3.7 *Let X be a compact metric space. Then (I) $(Comp(X), \Delta_X)$ is compact, and (II) $Comp_c(X)$ is a closed subset of $Comp(X)$.*

Proof. We first prove (I). Let X be compact, and let D be the diameter of X, that is, $D = \max\{d_X(x,x') : x,x' \in X\}$. Let $\{K_j\}_{j\in\mathbb{N}}$ be a sequence in $Comp(X)$. For each $j \in \mathbb{N}$, let $\varphi_j : X \mapsto \mathbb{R}$ be the function given by $\varphi_j(x) = d_X(x, K_j)$. Then each φ_j is a Lipschitz function on X, with Lipschitz constant 1. Furthermore, the bounds $0 \leq \varphi_j(x) \leq D$ clearly hold. Hence $\{\varphi_j\}_{j\in\mathbb{N}}$ is a uniformly bounded equicontinuous sequence of continuous real-valued functions on the compact space X. Therefore the Ascoli–Arzelà theorem implies that there exist an infinite subset J of $\mathbb{N}$ and a continuous function $\varphi : X \mapsto \mathbb{R}$ such that the φ_j converge uniformly to φ as $j \to \infty$ via values in J. Define $K = \{x : \varphi(x) = 0\}$. Then K is a compact subset of X.

Let us show that $K \neq \emptyset$. For this purpose, use the fact that each K_j is nonempty to find a member x_j of K_j. Since X is compact, there exists an infinite subset J' of J such that the limit $x = \lim_{j\to\infty, j\in J'} x_j$ exists. Since $\varphi_j(x_j) = 0$, and $\varphi_j \to \varphi$ uniformly, it follows that $\varphi(x) = 0$, so $x \in K$, proving that $K \neq \emptyset$, so that $K \in Comp(X)$.

We now show that $K_j \to_J K$ in the Hausdorff metric, where "$\to_J$" means "converges as j goes to ∞ via values in J." First, we prove that $\Delta_X^{qua}(K, K_j) \to_J 0$. By definition, $\Delta_X^{qua}(K, K_j) = \sup\{\varphi_j(x) : x \in K\}$. Since $\varphi_j \to_J \varphi$ uniformly on X, it follows that $\varphi_j \to_J \varphi$ uniformly on K. But $\varphi \equiv 0$ on K, so $\varphi_j \to_J 0$ uniformly on K, and then $\sup\{\varphi_j(x) : x \in K\} \to_J 0$, that is, $\Delta_X^{qua}(K, K_j) \to_J 0$.

Next, we prove that $\Delta_X^{qua}(K_j, K) \to_J 0$. If this was not so, there would exist an infinite subset J' of J and an α such that $\alpha > 0$ and

$$\Delta_X^{qua}(K_j, K) \geq \alpha \quad \text{whenever} \quad j \in J' . \tag{3.2.1}$$

For each $j \in J'$, pick $x_j \in K_j$ such that $\mathrm{dist}_X(x_j, K) = \Delta_X^{qua}(K_j, K)$. Then, using the compactness of X, pick an infinite subset J'' of J' such that the limit $x = \lim_{j\to\infty, j\in J''} x_j$ exists. Clearly, $\varphi_j(x_j) = 0$, because $x_j \in K_j$. Hence $\varphi(x) = 0$, so $x \in K$. But $d_X(x_j, x) \to 0$ as $j \to_{J''} \infty$. Hence $\mathrm{dist}_X(x_j, K) \to 0$ as $j \to_{J''} \infty$, contradicting (3.2.1). This proves (I).

We now prove (II). Let $\{K_j\}_{j\in\mathbb{N}}$ be a sequence in Comp(X) that converges to a $K \in Comp(X)$ and is such that all the K_j are connected. We have to prove that K is connected. Suppose K was not connected. Then there would exist open subsets U_1, U_2 of X such that $K \subseteq U_1 \cup U_2$, $U_1 \cap U_2 = \emptyset$, $K \cap U_1 \neq \emptyset$, and $K \cap U_2 \neq \emptyset$. The fact that $K_j \to K$ clearly implies that there exists a j_* such that, if $j \geq j_*$, then (a) $K_j \subseteq U_1 \cup U_2$, (b) $K_j \cap U_1 \neq \emptyset$, and (c) $K_j \cap U_2 \neq \emptyset$. But then, if we pick any j such that $j \geq j_*$, the set K_j is not connected, and we have reached a contradiction. This completes the proof of (II). □

Connected Sets of Zeros

The following result is a very minor modification of a theorem of Leray and Schauder – stated in [14] and proved by F. Browder in [2] – according to which: *if $K \subseteq \mathbb{R}^n$ is compact convex, $0 \in \mathrm{Int}\, K$, $R > 0$, and $H : K \times [0, R] \mapsto \mathbb{R}^n$*

is a continuous map such that $H(x,0) = x$ whenever $x \in K$ and H never vanishes on $\partial K \times [0,R]$, then there exists a compact connected subset Z of $K \times [0,R]$ such that $H(x,t) = 0$ whenever $(x,t) \in Z$, and the intersections $Z \cap (K \times \{0\})$, $Z \cap (K \times \{R\})$ are nonempty.

Our version allows H to be a set-valued CCA map, and in addition allows 0 to belong to the boundary of K, but requires that 0 be a limit of interior points v_j such that H never takes the value v_j on $\partial K \times [0,R]$.

Theorem 3.8 *Let $n \in \mathbb{Z}_+$, and let K be a compact convex subset of $\mathbb{R}^n$. Assume that $R > 0$ and $H : K \times [0,R] \mapsto\!\!\!\!\rightarrow \mathbb{R}^n$ is a CCA map. Assume, moreover, that:*

(1) $H(x,0) = \{x\}$ whenever $x \in K$
(2) There exists a sequence $\{v_j\}_{j\in\mathbb{N}}$ of interior points of K such that:
  (2.1) $\lim_{j\to\infty} v_j = 0$
  (2.2) $H(x,t) \neq v_j$ whenever $x \in \partial K$, $t \in [0,R]$, $j \in \mathbb{N}$

Then there exists a compact connected subset Z of $K \times [0,R]$ such that:

(a) $0 \in H(x,t)$ whenever $(x,t) \in Z$
(b) $Z \cap (K \times \{0\}) \neq \emptyset$
(c) $Z \cap (K \times \{R\}) \neq \emptyset$

Remark 3.9 If 0 is an interior point of K, and H never takes the value 0 on $\partial K \times [0,R]$, then Hypothesis (2) is automatically satisfied, since in that case we can take $v_j = 0$. If in addition H is single-valued, then Theorem 3.8 specializes to the result of [14] and [2]. □

Remark 3.10 Any point (ξ,τ) of intersection of $Z \cap (K \times \{0\})$ must satisfy $\tau = 0$ and $0 \in H(\xi,0)$. Since $H(\xi,0) = \{\xi\}$, ξ must be 0. So Conclusion (b) is equivalent to the assertion that $(0,0) \in Z$. □

Proof of Theorem 3.8. Pick a sequence $\{H_k^1\}_{k\in\mathbb{N}}$ of ordinary continuous maps $H_k^1 : K \times [0,R] \mapsto \mathbb{R}^n$ such that $H_k^1 \xrightarrow{\text{igr}} H$ as $k \to \infty$. Then, for each k, pick a sequence $\{H_{k,\ell}^2\}_{k\in\mathbb{N}}$ of polynomial maps $H_{k,\ell}^2 : \mathbb{R}^n \times \mathbb{R} \mapsto \mathbb{R}^n$ such that

$$\sup\{\|H_{k,\ell}^2(x,t) - H_k^1(x,t)\| : (x,t) \in K \times [0,R]\} \leq 2^{-\ell}\,.$$

Let $H_k^3 = H_{k,k}^2$, and define $H_k^4(x,t) = H_k^3(x,t) + x - H_k^3(x,0)$. Then the H_k^4 are polynomial maps from $\mathbb{R}^n \times \mathbb{R}$ to $\mathbb{R}^n$ such that $H_k^4(x,0) = x$ for all $x \in \mathbb{R}^n$. We claim that

$$H_k^4 \lceil (K \times [0,R]) \xrightarrow{\text{igr}} H \qquad \text{as} \quad k \to \infty\,. \tag{3.2.2}$$

To prove (3.2.2), we let $\alpha_k = \sup\{\theta_k(\xi,\tau) : (\xi,\tau) \in K \times [0,R]\}$, where

$$\theta_k(\xi,\tau) = \min\Big\{\|\xi - x\| + |\tau - t| + \|H_k^4(\xi,\tau) - y\| : (x,t) \in K \times [0,R],\ y \in H(x,t)\Big\}, \tag{3.2.3}$$

and show that $\alpha_k \to 0$. Assume that α_k does not go to 0. Then assume, after passing to a subsequence if necessary, that $\alpha_k \geq 3\beta$ for a strictly positive β. Then we may pick $(\xi_k, \tau_k) \in K \times [0, R]$ such that $\theta_k(\xi_k, \tau_k) \geq 2\beta$ for all k. After passing once again to a subsequence, we may assume that the limit $(\bar{\xi}, \bar{\tau}) = \lim_{k\to\infty}(\xi_k, \tau_k)$ exists and belongs to $K \times [0, R]$. Then (3.2.3) implies, since $\theta_k(\xi_k, \tau_k) \geq 2\beta$, that $\|\xi_k - \bar{\xi}\| + |\tau_k - \bar{\tau}| + \|H_k^4(\xi_k, \tau_k) - y\| \geq 2\beta$ whenever $y \in H(\bar{\xi}, \bar{\tau})$. If k is large enough then $\|\xi_k - \bar{\xi}\| + |\tau_k - \bar{\tau}| \leq \beta$. So we may assume, after passing to a subsequence, that $\|H_k^4(\xi_k, \tau_k) - y\| \geq \beta$ whenever $y \in H(\bar{\xi}, \bar{\tau})$.

On the other hand, if $y \in H(\bar{\xi}, \bar{\tau})$. then

$$\begin{aligned}
\beta &\leq \|H_k^4(\xi_k, \tau_k) - y\| \\
&\leq \|H_k^4(\xi_k, \tau_k) - H_k^1(\xi_k, \tau_k)\| + \|H_k^1(\xi_k, \tau_k) - y\| \\
&= \|H_k^3(\xi_k, \tau_k) + \xi_k - H_k^3(\xi_k, 0) - H_k^1(\xi_k, \tau_k)\| + \|H_k^1(\xi_k, \tau_k) - y\| \\
&= \|H_k^3(\xi_k, \tau_k) - H_k^1(\xi_k, \tau_k)\| + \|\xi_k - H_k^3(\xi_k, 0)\| + \|H_k^1(\xi_k, \tau_k) - y\| \\
&= \|H_{k,k}^2(\xi_k, \tau_k) - H_k^1(\xi_k, \tau_k)\| + \|\xi_k - H_{k,k}^2(\xi_k, 0)\| + \|H_k^1(\xi_k, \tau_k) - y\| \\
&\leq 2^{-k} + \|\xi_k - H_k^1(\xi_k, 0)\| + \|H_k^1(\xi_k, 0) - H_{k,k}^2(\xi_k, 0)\| + \|H_k^1(\xi_k, \tau_k) - y\| \\
&\leq 2^{1-k} + \|\xi_k - H_k^1(\xi_k, 0)\| + \|H_k^1(\xi_k, \tau_k) - y\| \\
&= 2^{1-k} + \|\xi_k - u_k\| + \|v_k - y\| ,
\end{aligned}$$

where $u_k = H_k^1(\xi_k, 0)$, $v_k = H_k^1(\xi_k, \tau_k)$. Since $(\xi_k, 0, u_k) \in \mathrm{Gr}(H_k^1)$ and $H_k^1 \xrightarrow{\mathrm{igr}} H$, we may pick points $(\tilde{\xi}_k, \tilde{\tau}_k, \tilde{u}_k) \in \mathrm{Gr}(H)$ such that

$$\|\xi_k - \tilde{\xi}_k\| + \tilde{\tau}_k + \|u_k - \tilde{u}_k\| \to 0 \quad \text{as} \quad t \to \infty . \tag{3.2.4}$$

We may then pass to a subsequence and assume that the limit $(\tilde{\xi}_\infty, \tilde{\tau}_\infty, \tilde{u}_\infty)$ of the sequence $\{(\tilde{\xi}_k, \tilde{\tau}_k, \tilde{u}_k)\}_{k\in\mathbb{N}}$ exists and belongs to $\mathrm{Gr}(H)$. Then (3.2.4) implies that $\xi_k \to \tilde{\xi}_\infty$ (from which it follows that $\tilde{\xi}_\infty = \bar{\xi}$), $\tilde{\tau}_\infty = 0$, and, finally $\tilde{u}_\infty = \lim_{k\to\infty} u_k = \lim_{k\to\infty} H_k^1(\xi_k, 0)$.

Since $(\bar{\xi}, 0, \tilde{u}_\infty) = (\tilde{\xi}_\infty, \tilde{\tau}_\infty, \tilde{u}_\infty) \in \mathrm{Gr}(H)$, we conclude that $\tilde{u}_\infty \in H(\bar{\xi}, 0)$, so $\tilde{u}_\infty = \bar{\xi}$. Since $\xi_k \to \bar{\xi}$ and $u_k \to \bar{\xi}$, we see that $\lim_{k\to\infty} \|\xi_k - u_k\| = 0$.

Next, since $(\xi_k, \tau_k, v_k) \in \mathrm{Gr}(H_k^1)$ and $H_k^1 \xrightarrow{\mathrm{igr}} H$, we may pick points $(\hat{\xi}_k, \hat{\tau}_k, \hat{v}_k) \in \mathrm{Gr}(H)$ such that

$$\|\xi_k - \hat{\xi}_k\| + |\tau_k - \hat{\tau}_k| + \|v_k - \hat{v}_k\| \to 0 \quad \text{as} \quad t \to \infty . \tag{3.2.5}$$

It is then possible to pass to a subsequence and assume that the limit $(\hat{\xi}_\infty, \hat{\tau}_\infty, \hat{v}_\infty) = \lim_{k\to\infty}(\hat{\xi}_k, \hat{\tau}_k, \hat{v}_k)$ exists and belongs to $\mathrm{Gr}(H)$. Then (3.2.5) implies that $\xi_k \to \hat{\xi}_\infty$ (so that $\hat{\xi}_\infty = \bar{\xi}$), $\tau_k \to \hat{\tau}_\infty$ (so that $\hat{\tau}_\infty = \bar{\tau}$), and $\hat{v}_\infty = \lim_{k\to\infty} v_k = \lim_{k\to\infty} H_k^1(\xi_k, \tau_k)$ (so that $\|v_k - \hat{v}_\infty\| \to 0$ as $k \to \infty$). Since $(\bar{\xi}, \bar{\tau}, \hat{v}_\infty) = (\hat{\xi}_\infty, \hat{\tau}_\infty, \hat{v}_\infty) \in \mathrm{Gr}(H)$, we conclude that $\hat{v}_\infty \in H(\bar{\xi}, \bar{\tau})$.

Hence we can apply the inequality $\beta \leq 2^{1-k} + \|\xi_k - u_k\| + \|v_k - y\|$ with $y = \hat{v}_\infty$, and conclude that

$$\beta \leq 2^{1-k} + \|\xi_k - u_k\| + \|v_k - \hat{v}_\infty\| . \tag{3.2.6}$$

However, we know that $\lim_{k\to\infty} \|\xi_k - u_k\| = 0$, and $\lim_{k\to\infty} \|v_k - \hat{v}_\infty\| = 0$. So the right-hand side of (3.2.6) goes to zero as $k \to \infty$, contradicting the fact that $\beta > 0$. This contradiction completes the proof of (3.2.2).

The set

$$Q = H(\partial K \times [0, R]) = \{y \in \mathbb{R}^n : (\exists x \in \partial K)(\exists t \in [0, R])(y \in H(x, t))\}$$

is compact, and our hypotheses imply that the points v_j do not belong to Q. Let $Q_k = H_k^4(\partial K \times [0, R])$, so Q_k is also compact. We claim that:

($) *For every $j \in \mathbb{N}$ there exists a $\kappa(j) \in \mathbb{N}$ such that $v_j \notin Q_k$ whenever $k \geq \kappa(j)$.*

To see this, suppose that j is such that $v_j \in Q_k$ for infinitely many values of k. Then we may assume, after passing to a subsequence, that $v_j \in Q_k$ for all k. Let $v_j = H_k^4(x_k, t_k)$, $x_k \in \partial K$, $t_k \in [0, R]$. Since $H_k^4 \lceil (K \times [0, R]) \xrightarrow{\text{igr}} H$, we may pick $(\tilde{x}_k, \tilde{t}_k, \tilde{v}_k) \in \mathrm{Gr}(H)$ such that $\|x_k - \tilde{x}_k\| + \|t_k - \tilde{t}_k\| + \|v_j - \tilde{v}_k\| \to 0$. Since $\mathrm{Gr}(H)$ is compact, we may pass to a subsequence and assume that the limit $(\tilde{x}_\infty, \tilde{t}_\infty, \tilde{v}_\infty) = \lim_{k\to\infty}(\tilde{x}_k, \tilde{t}_k, \tilde{v}_k)$ exists and belongs to $\mathrm{Gr}(H)$. But then $\tilde{x}_\infty = \lim_{k\to\infty} x_k$, so in particular $\tilde{x}_\infty \in \partial K$, because $x_k \in \partial K$, and $\tilde{y}_\infty = \lim_{k\to\infty} t_k$. In addition, $\tilde{v}_k = v_j$. So $v_j \in H(\tilde{x}_\infty, \tilde{t}_\infty)$ and $(\tilde{x}_\infty, \tilde{t}_\infty) \in \partial K \times [0, R]$. Hence $v_j \in Q$, and we have reached a contradiction, proving ($).

We now pick, for each j, an index $k(j)$ such that $k(j) \geq \kappa(j)$ and $k(j) \geq j$, and let $H_j^5 = H_{k(j)}^4$. Then each H_j^5 is a polynomial map such that $H_j^5(x, 0) = x$ whenever $x \in \mathbb{R}^n$, and $H_j^5(x, t) \neq v_j$ whenever (x, t) belongs to $\partial K \times [0, R]$. Furthermore, $H_j^5 \lceil (K \times [0, R]) \xrightarrow{\text{igr}} H$ as $j \to \infty$. Since the set $P_j = H_j^5(\partial K \times [0, R])$ is compact, $v_j \notin P_j$, and $v_j \in \mathrm{Int}(K)$, we may pick for each j an ε_j such that $0 < \varepsilon_j < 2^{-j}$ with the property that the ball $B_j = \{v \in \mathbb{R}^n : \|v - v_j\| < \varepsilon_j\}$ is a subset of $\mathrm{Int}(K)$ and does not intersect P_j. It follows from Sard's theorem that, for any given j, almost every $v \in \mathbb{R}^n$ is a regular value of both maps $\mathbb{R}^n \times \mathbb{R} \ni (x, t) \mapsto H_j^5(x, t) \in \mathbb{R}^n$ and $\mathbb{R}^n \ni x \mapsto H_j^5(x, R) \in \mathbb{R}^n$. So we may pick $w_j \in B_j$ which is a regular value of both maps. Since $v_j \to 0$ as $j \to \infty$ and $\|w_j - v_j\| < \varepsilon_j < 2^{-j}$, we can conclude that $\lim_{j\to\infty} w_j = 0$.

We now fix a j. Let $S = \{(x, t) \in \mathbb{R}^n \times \mathbb{R} : H_j^5(x, t) = w_j\}$. Then S is the set of zeros of the polynomial map

$$\mathbb{R}^n \times \mathbb{R} \ni (x, t) \mapsto H_j^5(x, t) - w_j \in \mathbb{R}^n ,$$

which does not have 0 as a regular value. It follows that S is a closed embedded one-dimensional submanifold of $\mathbb{R}^n \times \mathbb{R}$, so each connected component of

S is a closed embedded one-dimensional submanifold of $\mathbb{R}^n \times \mathbb{R}$ which is diffeomorphic to $\mathbb{R}$ or to the circle $\mathbb{S}^1 = \{(x,y) \in \mathbb{R}^2 : x^2 + y^2 = 1\}$. Since $H_j^5(w_j, 0) = w_j$, the point $(w_j, 0)$ belongs to a connected component C of S. Since C is diffeomorphic to $\mathbb{R}$ or $\mathbb{S}^1$, the set $\mathcal{X}$ of all smooth vector fields X on C such that $\|X(x)\| = 1$ for every $x \in C$ has exactly two members. Fix an $X \in \mathcal{X}$, so the other member of $\mathcal{X}$ is $-X$. The vector $X(w_j, 0)$ is then tangent to C at $(w_j, 0)$, and therefore belongs to the kernel of $DH_j^5(w_j, 0)$. On the other hand, the differential at w_j of the map $\mathbb{R}^n \ni x \mapsto H_j^5(x, 0) \in \mathbb{R}^n$ is the identity map, which is injective. It follows that the vector $X(w_j, 0)$ is not tangent to $\mathbb{R}^n \times \{0\}$. Hence $X(w_j, 0) = (\omega, r)$, with $\omega \in \mathbb{R}^n$, $r \in \mathbb{R}$, and $r \neq 0$. We may then assume, after relabeling $-X$ as X, if necessary, that $r > 0$.

Next, still keeping j fixed, we let γ_q be, for each $q \in C$, the maximal integral curve of X such that $\gamma_q(0) = q$. Then each γ_q is defined, in principle, on an interval $I_q =]\alpha_q, \beta_q[$, where $-\infty \leq \alpha_q < 0 < \beta_q \leq +\infty$. It turns out, however, that the numbers α_q, β_q cannot be finite. (For example, suppose β_q was finite. Then the limit $p = \lim_{t \uparrow \beta_q} \gamma_q(t)$ would exist, as a limit in the ambient space $\mathbb{R}^n \times \mathbb{R}$, because γ_q is Lipschitz. Then p would have to belong to C, since C is closed, and p would also be the limit in C of $\gamma_q(t)$ as $t \uparrow \beta_q$, because C is embedded. Hence we would be able to extend γ_q to a continuous map from the interval $I_q =]\alpha_q, \beta_q]$ to C such that $\gamma_q(\beta_q) = p$, and concatenate this with an integral curve $\tilde{\gamma} : [\beta_q, \beta_q + \varepsilon[\mapsto C$ such that $\tilde{\gamma}(\beta_q) = p$, thereby obtaining an extension of γ_q to a larger interval, and contradicting the maximality. of γ_q. A similar argument works for α_q. So $\alpha_q = -\infty$ and $\beta_q = +\infty$.) Therefore $I_q = \mathbb{R}$ for every $q \in C$. Clearly, the set $A_q = \gamma_q(\mathbb{R})$ is an open submanifold of C. Furthermore, if $q, q' \in C$ then the sets A_q, $A_{q'}$ are either equal or disjoint. Since C is connected, all the sets A_q coincide and are equal to C. In particular, if we let $\bar{q} = (w_j, 0)$, and write $\gamma = \gamma_{\bar{q}}$, then $\gamma(\mathbb{R}) = C$. Write $\gamma(t) = (\xi(t), \tau(t))$, $\xi(t) \in \mathbb{R}^n$, $\tau(t) \in \mathbb{R}$. Then there exists a positive number δ such that $\xi(t) \in \operatorname{Int}(K)$ for $-\delta < t < \delta$ and $t\tau(t) > 0$ for $0 < |t| < \delta$. It follows, after making δ smaller, if necessary, that $\gamma(t)$ is an interior point of $K \times [0, R]$ for $0 < t < \delta$. If C is diffeomorphic to $\mathbb{S}^1$, then γ is periodic, so there exists a smallest time $T > 0$ such that $\gamma_{\bar{q}}(T) = \gamma(0)$. Then $\gamma(T - h) = \gamma(-h)$ for small positive h, so $\tau(T - h) = \tau(-h) < 0$ for such h, implying that $\gamma(t) \notin K \times [0, R]$ when $t < T$ and $T - t$ is small enough. It follows that it is not true that $\gamma(t) \in K \times [0, R]$ for all $t \in [0, T]$. If we let $M = \{t \in [0, T] : \gamma(t) \notin K \times [0, R]\}$, then M is a nonempty relatively open subset of $[0, T]$. Let $T_0 = \inf M$. Then $T_0 > 0$, because $\gamma(t) \in K \times [0, R]$ when $0 \leq t < \delta$. Therefore $T_0 \notin M$, because if $T_0 \in M$ then the facts that M is relatively open in $[0, T]$ and $T_0 > 0$ would imply that $T_0 - h \in M$ for small positive h, contradicting the fact that $T_0 = \inf M$. It follows that:

(&) $T_0 > 0$, $\gamma(t) \in K \times [0, R]$ *for* $0 \leq t \leq T_0$, $\gamma(T_0 + h_\ell) \notin K \times [0, R]$ *for a sequence* $\{h_\ell\}_{\ell \in \mathbb{N}}$ *of positive numbers converging to* 0, *and* γ *is an injective map on* $[0, T_0]$.

So we have proved the existence of a T_0 for which (&) is true, under the hypothesis that C is diffeomorphic to $\mathbb{S}^1$.

We now show, still keeping j fixed, that a T_0 for which (&) holds also exists if C is diffeomorphic to $\mathbb{R}$. To prove this, we define a set M by letting $M = \{t \in [0, +\infty[: \gamma(t) \notin K \times [0, R]\}$. Then M is a relatively open subset of $[0, +\infty[$. Furthermore, $M \neq \emptyset$. (*Proof.* If M was empty, then $\gamma(t)$ would belong to $K \times [0, R]$ for all positive t. So we could pick a sequence $\{t_\ell\}_{\ell\in\mathbb{N}}$ of positive numbers converging to $+\infty$ and such that $\gamma(t_\ell)$ converges to a limit q. But then $q \in C$, because C is closed, and the equality $\lim_{\ell\to\infty} \gamma(t_\ell) = q$ also holds in C, because C is embedded. Since C is embedded, there exists a neighborhood U of q in $\mathbb{R}^n \times \mathbb{R}$ which is diffeomorphic to a product $-]\rho, \rho[^{n+1}$ under a map $\Phi : U \mapsto -]\rho, \rho[^{n+1}$ that sends q to 0 and is such that $\Phi(U \cap C)$ is the arc $A = \{(s, 0, \ldots, 0) : -\rho < s < \rho\}$. Then $\gamma(t_\ell) \in A$ if ℓ is large enough. But A itself, suitably parametrized, is an integral curve $]\alpha, \beta[\ni t \mapsto \zeta(t)$ of X such that $\alpha < 0 < \beta$ and $\zeta(0) = q$. It follows that for large enough ℓ there exist $h_\ell \in]\alpha, \beta[$ such that $h_\ell \to 0$ as $\ell \to \infty$ and $\zeta(h_\ell) = \gamma(t_\ell)$. Let $T \in \mathbb{R}$ be such that $\gamma(T) = q$. Then $\gamma(T + h_\ell) = \zeta(h_\ell) = \gamma(t_\ell)$. Since the t_ℓ go to $+\infty$, but the $T + h_\ell$ are bounded, there must exist at least one ℓ such that $T + h_\ell \neq t_\ell$. Since $\gamma(T + h_\ell) = \zeta(h_\ell) = \gamma(t_\ell)$, it follows that γ is periodic and then $C = \gamma(\mathbb{R})$ is compact, contradicting the assumption that C is diffeomorphic to $\mathbb{R}$.) Let $T_0 = \inf M$. Then $T_0 > 0$, because $\gamma(t) \in K \times [0, R]$ when $0 \le t < \delta$. Therefore $T_0 \notin M$, because if $T_0 \in M$ then the facts that M is relatively open in $[0, +\infty[$ and $T_0 > 0$ would imply that $T_0 - h \in M$ for small positive h, contradicting the fact that $T_0 = \inf M$. Hence (&) holds.

So we have shown that:

(&&) *For every j there exist a positive number T_0^j and a smooth curve $[0, +\infty[\ni s \mapsto \gamma^j(s) = (\xi^j(s), \tau^j(s)) \in \mathbb{R}^n \times \mathbb{R}$ such that:*
- (&&.1) $\gamma^j(0) = (w_j, 0)$
- (&&.2) $\gamma_j(s) \in K \times [0, R]$ *for* $0 \le s \le T_0^j$
- (&&.3) *There exists a sequence $\{h_\ell\}_{\ell\in\mathbb{N}}$ of positive numbers, converging to 0, such that $\gamma^j(T_0^j + h_\ell) \notin K \times [0, R]$ for every ℓ*
- (&&.4) γ^j *is an injective map on* $[0, T_0^j]$
- (&&.5) $H_j^5(\gamma^j(s)) = w_j$ *for every* $s \in [0, T_0^j]$

We now let $Z_j = \gamma^j([0, T_0^j])$ for every $j \in \mathbb{N}$. Then each Z_j is a compact connected subset of $K \times [0, R]$, such that $(w_j, 0) \in Z_j$ and the function $H_j^5 - w_j$ vanishes on Z_j. Furthermore, we claim that $Z_j \cap (K \times \{R\}) \neq \emptyset$. (*Proof.* We show that $\gamma^j(T_0^j) \in K \times \{R\}$. To see this, observe that (&&.2) implies that $\gamma^j(T_0^j) \in K \times [0, R]$, and (&&.3) implies that $\gamma^j(T_0^j)$ is not an interior point of $K \times [0, R]$, so $\gamma^j(T_0^j) \in \partial(K \times [0, R])$. On the other hand, it is clear that $\partial(K \times [0, R]) = (\partial K \times [0, R]) \cup (K \times \{0, R\})$. But $\gamma^j(T_0^j)$ cannot belong to $\partial K \times [0, R]$, because $H_j^5(\gamma^j(T_0^j)) = w_j$ and H_j^5 never takes the value w_j on $\partial K \times [0, R]$ (because $w_j \in B_j$ and $B_j \cap P_j = \emptyset$). So $\gamma^j(T_0^j)$ belongs to $(K \times \{0\}) \cup (K \times \{R\})$. But $\gamma^j(T_0^j)$ cannot belong to $K \times \{0\}$, because

$\gamma^j(T_0^j) \neq \gamma^j(0)$ (thanks to (&&.4)), $\gamma^j(0) = (w_j, 0)$, and $(w_j, 0)$ is the only point of $K \times \{0\}$ where $H_j^5 - w_j$ vanishes (since $H_j^5(x, 0) = x$ for all x). So $\gamma^j(T_0^j) \in K \times \{R\}$, as desired.)

Since $\mathbf{Z} = \{Z_j\}_{j \in \mathbb{N}}$ is a sequence of nonempty compact connected subsets of $K \times [0, R]$, Proposition 3.7 implies that we may assume, after passing to a subsequence, that $\mathbf{Z}$ converges in the Hausdorff metric to a nonempty compact connected subset Z of $K \times [0, R]$. We now show that Z satisfies the three properties of the conclusion of our theorem. First. we prove that $0 \in H(x,t)$ whenever $(x,t) \in Z$. Pick a point (x,t) of Z. Then $\mathrm{dist}((x,t), Z_j)$ goes to 0 as $j \to \infty$. So we may pick $(x_j, t_j) \in Z_j$ such that $x_j \to x$ and $t_j \to t$. Since $(x_j, t_j) \in Z_j$, the point $((x_j, t_j), w_j)$ belongs to $\mathrm{Gr}(H_j^5 \lceil (K \times [0, R]))$. Since $H_j^5 \lceil (K \times [0, R]) \xrightarrow{\mathrm{igr}} H$, we may pick points $((\tilde{x}_j, \tilde{t}_j), \tilde{w}_j)$ in $\mathrm{Gr}(H)$ such that

$$\lim_{j \to \infty} \Big(\|x_j - \tilde{x}_j\| + |t_j - \tilde{t}_j| + \|w_j - \tilde{w}_j\| \Big) = 0 . \tag{3.2.7}$$

Since $(x_j, t_j, w_j) \to (x, t, 0)$, (3.2.7) implies that $((\tilde{x}_j, \tilde{t}_j), \tilde{w}_j) \to ((x,t), 0)$. Since $\mathrm{Gr}(H)$ is compact, $((x,t), 0)$ belongs to $\mathrm{Gr}(H)$, so $0 \in H(x,t)$, as desired. Next we show that $Z \cap (K \times \{0\}) \neq \emptyset$. To see this, it suffices to observe that $(w_j, 0) \in Z_j$ and $w_j \to 0$, so $(0,0) \in Z$. Finally, we prove that $Z_j \cap (K \times \{R\}) \neq \emptyset$. For this purpose, we use the fact that $Z_j \cap (K \times \{R\}) \neq \emptyset$ to pick points $z_j \in K$ such that $(z_j, R) \in Z_j$. Using the compactness of K, pick an infinite subset J of $\mathbb{N}$ such that $z = \lim_{j \to \infty, j \in J} z_j$ exists and belongs to K. Then, since $(z_j, R) \in Z_j$, $(z_j, R) \to (z, R)$, and $Z_j \to Z$ in the Hausdorff metric, it follows that $(z, R) \in Z$, concluding our proof. □

Kakutani–Fan–Glicksberg (KFG) Maps

An important class of examples of CCA maps consists of those that we will call *Kakutani–Fan–Glicksberg* (abbreviated "KFG") *maps*, because they occur in the celebrated finite-dimensional Kakutani fixed point theorem as well as in its infinite-dimensional version due to Fan and Glicksberg.

Definition 3.11 *If X is a metric space and C is a convex subset of a normed space, a* ***KFG map*** *from X to C is a compactly-graphed set-valued map $F : X \mapsto\!\!\!\!\to C$ such that $F(x)$ is convex and nonempty whenever $x \in X$.* □

Remark 3.12 It follows from Fact 3.3 that *a set-valued map $F : X \mapsto\!\!\!\!\to C$ from a metric space X to a convex subset C of a normed space is a KFG map if and only if it is an upper semicontinuous map with nonempty compact convex values.* □

The following result is due to A. Cellina, cf. [3, 4, 6].

Theorem 3.13 *If X is a metric space, C is a convex subset of a normed space Y, $F : X \mapsto\!\!\!\!\to C$, and F is a KFG map, then F is a CCA map.*

Proof. The definition of a KFG map implies that $\mathrm{Gr}(F\lceil K)$ is compact and nonempty whenever K is a nonempty compact subset of X, which is one of the two conditions needed for F to be a CCA map. To prove the other condition, we fix a nonempty compact subset K of X and prove that there exists a sequence $\{F_j\}_{j=1}^{\infty}$ of continuous maps $F_j : K \mapsto C$ such that $F_j \xrightarrow{\text{igr}} F\lceil K$ as $j \to \infty$.

For each positive number ε, select a finite subset S_ε of K such that $K \subseteq \bigcup_{s\in S_\varepsilon} \mathbb{B}_X(s,\varepsilon)$. For $x \in K$, $s \in S_\varepsilon$, let $\psi_{s,\varepsilon}(x) = \max\Big(0, \varepsilon - d_X(x,s)\Big)$, so $\psi_{s,\varepsilon} : K \mapsto \mathbb{R}$ is continuous and nonnegative and $\psi_{s,\varepsilon}(x) > 0$ if and only if $x \in \mathbb{B}_X(s,\varepsilon)$. Define $\varphi_{s,\varepsilon}(x) = \Big(\sum_{s'\in S_\varepsilon} \psi_{s',\varepsilon}(x)\Big)^{-1} \psi_{s,\varepsilon}(x)$, so the $\varphi_{s,\varepsilon}$ are continuous nonnegative real-valued functions on K having the property that $\sum_{s\in S_\varepsilon} \varphi_{s,\varepsilon}(x) = 1$ for all $x \in K$. Using the fact that the sets $F(x)$ are nonempty, pick a $y_{s,\varepsilon} \in F(s)$ for each $s \in S_\varepsilon$. Define $H_\varepsilon : K \mapsto C$ by letting $H_\varepsilon(x) = \sum_{s\in S_\varepsilon} \varphi_{s,\varepsilon}(x) y_{s,\varepsilon}$. Then each H_ε is continuous.

Now let $\{\varepsilon_j\}_{j\in\mathbb{N}}$ be a sequence of positive numbers that converges to zero. We claim that the $H_{\varepsilon_j} \xrightarrow{\text{igr}} F\lceil K$. To see this, we let

$$\alpha_j = \sup\{d_{X\times Y}(q, \mathrm{Gr}(F\lceil K)) : q \in \mathrm{Gr}(H_{\varepsilon_j})\}\,,$$

and prove that $\alpha_j \to 0$. The proof will be by contradiction.

Assume that $\{\alpha_j\}$ does not go to zero. Then we may pass to a subsequence and assume that the α_j are bounded below by a fixed strictly positive number α. Pick a β such that $0 < \beta < \alpha$. Pick $q_j \in \mathrm{Gr}(H_{\varepsilon_j})$ such that

$$d_{X\times Y}(q_j, \mathrm{Gr}(F\lceil K)) \geq \beta\,. \tag{3.2.8}$$

Write $q_j = (x_j, y_j)$. Then the x_j belong to K, so we may assume, after passing to a subsequence, that the limit $\bar{x} = \lim_{j\to\infty} x_j$ exists.

Fix a γ such that $0 < \gamma$ and $2\gamma < \beta$. Pick a positive δ such that $d_Y(z, F(\bar{x})) < \gamma$ whenever $w \in K$, $z \in F(w)$, and $d_X(\bar{x}, w) \leq \delta$. (The existence of such a δ is easily proved: suppose, by contradiction, that there exist sequences $\{w_k\}$, $\{z_k\}$ in K such that $z_k \in F(w_k)$, $w_k \to \bar{x}$ as $k \to \infty$, and $d_Y(z_k, F(\bar{x})) \geq \gamma$; since $\mathrm{Gr}(F\lceil K)$ is compact we may assume, after passing to a subsequence, that the sequence $\{z_k\}$ converges to a limit z; since $z_k \in F(w_k)$, and $w_k \to \bar{x}$, the compactness of $\mathrm{Gr}(F\lceil K)$ also implies that $z \in F(\bar{x})$; since $z_k \to z$, we see that $d_Y(z_k, F(\bar{x})) \to 0$, and we have derived a contradiction.)

Now let $j^* \in \mathbb{N}$ be such that

$$2\,\varepsilon_j \leq \delta \quad \text{and} \quad d_X(x_j, \bar{x}) \leq \min\left(\gamma, \frac{\delta}{2}\right) \tag{3.2.9}$$

whenever $j \geq j^*$. If $j \geq j^*$, $x = x_j$, and $\varepsilon = \varepsilon_j$, then all the terms in the summation defining H_ε for which $d_X(s, \bar{x}) \geq \delta$ vanish, because $d_X(s, \bar{x}) \geq \delta$ implies $d_X(x_j, s) \geq \frac{\delta}{2} \geq \varepsilon_j$ in view of (3.2.9), so $\varphi_{s,\varepsilon_j}(x_j) = 0$. Therefore, if we let $y_j = H_{\varepsilon_j}(x_j)$, we have

$$y_j = H_{\varepsilon_j}(x_j) = \sum_{s\in\hat{S}_{\varepsilon_j,\bar{x}}} \varphi_{s,\varepsilon_j}(x_j) y_{s,\varepsilon_j}, \tag{3.2.10}$$

where $\hat{S}_{\varepsilon_j,\bar{x}} = \left\{s \in S_{\varepsilon_j} : d_X(s,\bar{x}) < \delta\right\}$. For every $s \in \hat{S}_{\varepsilon_j,\bar{x}}$, the point y_{s,ε_j} is in $F(s)$, so $\text{dist}(y_{s,\varepsilon_j}, F(\bar{x})) < \gamma$. Therefore we may pick $\tilde{y}_{s,\varepsilon_j} \in F(\bar{x})$ such that $\|y_{s,\varepsilon_j} - \tilde{y}_{s,\varepsilon_j}\| \le \gamma$. If we let $\tilde{y}_j = \sum_{s\in\hat{S}_{\varepsilon_j,\bar{x}}} \varphi_{s,\varepsilon_j}(x_j)\tilde{y}_{s,\varepsilon_j}$, and compare this with (3.2.10), we find $\|\tilde{y}_j - y_j\| \le \sum_{s\in\hat{S}_{\varepsilon_j,\bar{x}}} \varphi_{s,\varepsilon_j}(x_j)\|\tilde{y}_{s,\varepsilon_j} - y_{s,\varepsilon_j}\| \le \gamma$. On the other hand, $\tilde{y}_j$ clearly is a convex combination of points of $F(\bar{x})$, so $\tilde{y}_j \in F(\bar{x})$, because $F(\bar{x})$ is convex. Since $\|y_j - \tilde{y}_j\| \le \gamma$ and $d_X(x_j,\bar{x}) \le \gamma$ for $j \ge j^*$, and the point $\tilde{q}_j \stackrel{\text{def}}{=} (\bar{x}, \tilde{y}_j)$ belongs to $\text{Gr}(F\lceil K)$, we can conclude that $d_{X\times Y}(q_j, \text{Gr}(F\lceil K)) \le 2\gamma < \beta$ if $j \ge j^*$ This, together with Formula (3.2.8), shows that the assumption that α_j does not go zero leads to a contradiction. So $\alpha_j \to 0$, and the proof is complete. □

The Cellina, Kakutani, and Fan–Glicksberg Fixed Point Theorems

Many fixed point properties of continuous maps are also valid for CCA maps, as we now show. Let us recall that, if A is a set, and $F : A \mapsto\!\!\!\rightarrow A$, then a *fixed point* of F is a point $a \in A$ such that $a \in F(a)$.

Theorem 3.14 *(Cellina, cf. [5]) Let K be a nonempty compact convex subset of a normed space X, and let $F : K \mapsto\!\!\!\rightarrow K$ be a CCA map. Then F has a fixed point.*

Proof. Let $\{F_j\}_{j\in\mathbb{N}}$ be a sequence of continuous maps from K to K such that $F_j \xrightarrow{\text{igr}} F$ as $j \to \infty$. By the Schauder fixed point theorem, there exist x_j such that $F_j(x_j) = x_j$. Since K is compact we may pass to a subsequence, if necessary, and assume that the sequence $\{x_j\}_{j\in\mathbb{N}}$ has a limit $x \in K$. Then $F_j(x_j) \to x$ as well, so $x \in F(x)$. □

Corollary 3.15 *(The Kakutani–Fan–Glicksberg fixed point theorem, cf. Kakutani [13], Fan [10], Glicksberg [11].) Let K be a nonempty compact convex subset of a normed space X. Let $F : K \mapsto\!\!\!\rightarrow K$ be a set-valued map with a compact graph and nonempty convex values. Then F has a fixed point.*

Proof. Theorem 3.13 tells us that F is a CCA map, and then Theorem 3.14 implies that F has a fixed point. □

4 GDQs and AGDQs

We use $\mathbf{\Theta}$ to denote the class of all functions $\theta : [0, +\infty[\mapsto [0, +\infty]$ such that:

- *θ is monotonically nondecreasing (that is, $\theta(s) \le \theta(t)$ whenever s, t are such that $0 \le s \le t < +\infty$)*
- *$\theta(0) = 0$ and $\lim_{s\downarrow 0}\theta(s) = 0$*

If X, Y are FDNRLSs, we endow $Lin(X,Y)$ with the operator norm $\|\cdot\|_{op}$ defined in (2.1.2). If $\Lambda \subseteq Lin(X,Y)$ and $\delta > 0$, we define

$$\Lambda^\delta = \{L \in Lin(X,Y) : \mathrm{dist}(L,\Lambda) \leq \delta\},$$

where $\mathrm{dist}(L,\Lambda) = \inf\{\|L - L'\|_{op} : L' \in \Lambda\}$. Notice that if $L \in Lin(X,Y)$, then $\mathrm{dist}(L,\emptyset) = +\infty$. In particular, if $\Lambda = \emptyset$ then $\Lambda^\delta = \emptyset$. Notice also that Λ^δ is compact if Λ is compact and Λ^δ is convex if Λ is convex.

4.1 The Basic Definitions

Generalized Differential Quotients (GDQs)

We assume that (1) X and Y are FDNRLSs, (2) $F : X \mapsto\!\!\!\!\rightarrow Y$ is a set-valued map; (3) $\bar{x}_* \in X$, (4) $\bar{y}_* \in Y$, and (5) $S \subseteq X$.

Definition 4.1 *A* ***generalized differential quotient*** *(abbreviated "GDQ")* ***of*** F ***at*** $(\bar{x}_*, \bar{y}_*)$ ***in the direction of*** S *is a compact subset Λ of $Lin(X,Y)$ having the property that for every neighborhood $\hat{\Lambda}$ of Λ in $Lin(X,Y)$ there exist U, G such that:*

(I) *U is a neighborhood of $\bar{x}_*$ in X*
(II) *$\bar{y}_* + G(x) \cdot (x - \bar{x}_*) \subseteq F(x)$ for every $x \in U \cap S$*
(III) *G is a CCA set-valued map from $U \cap S$ to $\hat{\Lambda}$* □

We will use $GDQ(F, \bar{x}_*, \bar{y}_*, S)$ to denote the set of all GDQs of F at $(\bar{x}_*, \bar{y}_*)$ in the direction of S.

Remark 4.2 The set Λ can, in principle, be empty. Actually, it is very easy to show that the following three conditions are equivalent:

(1) $\emptyset \in GDQ(F, \bar{x}_*, \bar{y}_*, S)$
(2) *Every compact subset of $Lin(X,Y)$ belongs to $GDQ(F, \bar{x}, \bar{y}, S)$*
(3) *$\bar{x}_*$ does not belong to the closure of S* □

It is not hard to prove the following alternative characterization of GDQs.

Proposition 4.3 *Let X, Y be FDNRLSs, let $F : X \mapsto\!\!\!\!\rightarrow Y$ be a set-valued map, and let Λ be a compact subset of $Lin(X,Y)$. Let $\bar{x}_* \in X$, $\bar{y}_* \in Y$, $S \subseteq X$. Then $\Lambda \in GDQ(F, \bar{x}_*, \bar{y}, S)$ if and only if there exists a function $\theta \in \boldsymbol{\Theta}$ – called a* ***GDQ modulus for*** *$(\Lambda, F, \bar{x}_*, \bar{y}_*, S)$ – having the property that:*

(*) *For every $\varepsilon \in]0, +\infty[$ such that $\theta(\varepsilon) < \infty$ there exists a set-valued map $G^\varepsilon \in CCA(\bar{\mathbb{B}}_X(\bar{x}_*, \varepsilon) \cap S, Lin(X,Y))$ such that for every $x \in \bar{\mathbb{B}}_X(\bar{x}_*, \varepsilon) \cap S$ the inclusions $G^\varepsilon(x) \subseteq \Lambda^{\theta(\varepsilon)}$ and $\bar{y}_* + G^\varepsilon(x) \cdot (x - \bar{x}_*) \subseteq F(x)$ hold*

Proof. Assume that Λ belongs to $GDQ(F, \bar{x}_*, \bar{y}_*, S)$. For each nonnegative real number ε, let $H(\varepsilon)$ be the set of all δ such that (i) $\delta > 0$, and (ii) there exists a $G \in CCA(\bar{\mathbb{B}}_X(\bar{x}_*, \varepsilon) \cap S, Lin(X,Y))$ with the property that

$G(x) \subseteq \Lambda^\delta$ and $\bar{y}_* + G(x) \cdot (x - \bar{x}_*) \subseteq F(x)$ whenever $x \in \bar{\mathbb{B}}_X(\bar{x}_*, \varepsilon) \cap S$. Let $\theta_0(\varepsilon) = \inf H(\varepsilon)$, and then define $\theta(\varepsilon) = \theta_0(\varepsilon) + \varepsilon$. (Notice that the set $H(\varepsilon)$ could be empty, in which case $\theta_0(\varepsilon) = \theta(\varepsilon) = +\infty$.) It is clear that θ is monotonically non-decreasing, since $H(\varepsilon') \subseteq H(\varepsilon)$ whenever $0 \leq \varepsilon < \varepsilon'$. The fact that $\Lambda \in GDQ(F, \bar{x}_*, \bar{y}_*, S)$ implies that, given any positive δ, there exist a neighborhood U of $\bar{x}_*$ and a map $\tilde{G} \in CCA(U \cap S, \Lambda^\delta)$ such that $\bar{y}_* + \tilde{G}(x) \cdot (x - \bar{x}_*) \subseteq F(x)$ whenever $x \in U \cap S$. Find ε such that $\bar{\mathbb{B}}_X(\bar{x}_*, \varepsilon) \subseteq U$, and let $G = \iota_2 \circ \tilde{G} \circ \iota_1$, where $\iota_1 : \bar{\mathbb{B}}_X(\bar{x}_*, \varepsilon) \cap S \mapsto U \cap S$ and $\iota_2 : \Lambda^\delta \mapsto Lin(X, Y)$ are the set inclusions. Then it is clear that G belongs to $CCA(\bar{\mathbb{B}}_X(x_*, \varepsilon) \cap S, Lin(X, Y))$, and also that $G(x) \subseteq \Lambda^\delta$ and $\bar{y}_* + G(x) \cdot (x - \bar{x}_*) \subseteq F(x)$ whenever $x \in \bar{\mathbb{B}}_X(\bar{x}_*, \varepsilon) \cap S$. Therefore $\delta \in H(\varepsilon)$, so $\theta_0(\varepsilon) \leq \delta$. This proves that $\lim_{\varepsilon \downarrow 0} \theta_0(\varepsilon) = 0$, thus establishing that $\theta_0 \in \mathbf{\Theta}$, and then $\theta \in \mathbf{\Theta}$ as well. Finally, if $\theta(\varepsilon) < +\infty$, then we can pick a $\delta \in H(\varepsilon)$ such that $\theta_0(\varepsilon) \leq \delta \leq \theta(\varepsilon)$, and then find a G belonging to $CCA(\bar{\mathbb{B}}_X(\bar{x}_*, \varepsilon) \cap S, Lin(X, Y))$ for which the conditions $G(x) \subseteq \Lambda^\delta$ and $\bar{y}_* + G(x) \cdot (x - \bar{x}_*) \subseteq F(x)$ hold whenever $x \in \bar{\mathbb{B}}_X(\bar{x}_*, \varepsilon) \cap S$. Since $\delta \leq \theta(\varepsilon)$, the map G takes values in $\Lambda^{\theta(\varepsilon)}$. Hence we can choose G^ε to be G, and the condition of (*) is satisfied.

To prove the converse, let θ be a GDQ modulus for $\Lambda, F, \bar{x}_*, \bar{y}_*, S$. Fix a positive number δ. Pick an ε such that $\theta(\varepsilon) < \delta$. Then pick G^ε such that the conditions of (*) hold. Then the map G^ε satisfies the requirement that $\bar{y}_* + G^\varepsilon \cdot (x - \bar{x}_*) \subseteq F(x)$ whenever $x \in \bar{\mathbb{B}}_X(\bar{x}_*, \varepsilon) \cap S$. Furthermore, $G^\varepsilon \in CCA(\bar{\mathbb{B}}_X(\bar{x}_*, \varepsilon) \cap S, Lin(X, Y))$, and G^ε takes values in $\Lambda^{\theta(\varepsilon)}$. Since $\theta(\varepsilon) < \delta$, if K is a compact subset of $\bar{\mathbb{B}}_X(\bar{x}_*, \varepsilon) \cap S$ and $\{G_j\}_{j \in \mathbb{N}}$ is a sequence of continuous maps from K to $Lin(X, Y)$ such that $G_j \xrightarrow{\text{igr}} G^\varepsilon \lceil K$, it follows that G_j takes values in Λ^δ if j is large enough. Therefore G^ε belongs to $CCA(\bar{\mathbb{B}}^n(\bar{x}_*, \varepsilon) \cap S, \Lambda^\delta)$. This shows that $\Lambda \in GDQ(F, \bar{x}_*, \bar{y}_*, S)$, concluding our proof. □

Approximate Generalized Differential Quotients (AGDQs)

Motivated by the characterization of GDQs given in Proposition 4.3, we now define a slightly larger class of generalized differentials. First, if X, Y are FDNRLSs, we let $Aff(X, Y)$ be the set of all affine maps from X to Y, so the members of $Aff(X, Y)$ are the maps $X \ni x \mapsto A(x) = L \cdot x + h$, $L \in Lin(X, Y)$, $h \in Y$. (For a map A of this form, the linear map $L \in Lin(X, Y)$ and the vector $h \in Y$ are the *linear part* and the *constant part* of A.) We identify $Aff(X, Y)$ with $Lin(X, Y) \times Y$ by identifying each $A \in Aff(X, Y)$ with the pair $(L, h) \in Lin(X, Y) \times Y$, where L, h are, respectively, the linear part and the constant part of A.

Definition 4.4 *Assume that X, Y are FDNRLSs, $F : X \mapsto\!\!\!\to Y$ is a set-valued map, Λ is a compact subset of $Lin(X, Y)$, $\bar{x}_* \in X$, $\bar{y}_* \in Y$, and $S \subseteq X$. We say that Λ is an* ***approximate generalized differential quotient of F at*** $(\bar{x}_*, \bar{y}_*)$ ***in the direction of*** S *– and write $\Lambda \in AGDQ(F, \bar{x}_*, \bar{y}_*, S)$ – if there* *exists a function* $\theta \in \mathbf{\Theta}$ – *called an* ***AGDQ modulus for*** $(\Lambda, F, \bar{x}_*, \bar{y}_*, S)$ – *having the property that:*

(**) *For every* $\varepsilon \in]0, +\infty[$ *such that* $\theta(\varepsilon) < \infty$ *there exists a set-valued map* $A^\varepsilon \in CCA(\bar{\mathbb{B}}_X(\bar{x}_*, \varepsilon) \cap S, Aff(X, Y))$ *such that*

$$L \in \Lambda^{\theta(\varepsilon)}, \quad \|h\| \leq \theta(\varepsilon)\varepsilon, \quad \text{and} \quad \bar{y}_* + L \cdot (x - \bar{x}_*) + h \in F(x)$$

whenever $x \in \bar{\mathbb{B}}_X(\bar{x}_*, \varepsilon) \cap S$ *and* (L, h) *belongs to* $A^\varepsilon(x)$. □

4.2 Properties of GDQs and AGDQs

Retracts, Quasiretracts and Local Quasiretracts

In order to formulate and prove the chain rule, we first need some basic facts about retracts.

Definition 4.5 *Let* T *be a topological space and let* S *be a subset of* T. *A* ***retraction from*** T ***to*** S *is a continuous map* $\rho : T \mapsto S$ *such that* $\rho(s) = s$ *for every* $s \in S$. *We say that* S *is a* ***retract of*** T *if there exists a retraction from* T *to* S. □

Often, the redundant phrase "continuous retraction" will be used for emphasis, instead of just saying "retraction."

It follows easily from the definition that

Fact 4.6 *If* T *is a Hausdorff topological space and* S *is a retract of* T, *then* S *is closed.* □

Also, it is easy to show that every retract is a "local retract" at any point, in the following precise sense:

Fact 4.7 *If* T *is a Hausdorff topological space,* S *is a retract of* T, *and* $s \in S$, *then every neighborhood* U *of* s *contains a neighborhood* V *of* s *such that* $S \cap V$ *is a retract of* V. □

It will be convenient to introduce a weaker concept, namely, that of a "quasiretract," as well as its local version.

Definition 4.8 *Let* T *be a topological space and let* S *be a subset of* T. *We say that* S *is a* ***quasiretract of*** T *if for every compact subset* K *of* S *there exist a neighborhood* U *of* K *and a continuous map* $\rho : U \mapsto S$ *such that* $\rho(s) = s$ *for every* $s \in K$. □

Definition 4.9 *Assume that* T *is a topological space,* $S \subseteq T$, *and* $\bar{s}_* \in T$. *We say that* S *is a* ***local quasiretract of*** T ***at*** $\bar{s}_*$ *if there exists a neighborhood* U *of* $\bar{s}_*$ *such that* $S \cap U$ *is a quasiretract of* U. □

It is then easy to verify the following facts.

Fact 4.10 *If T is a topological space and $S \subseteq T$, then:*

(1) If S is a retract of T then S is a quasiretract of T
(2) If S is a quasiretract of T and Ω is an open subset of T then $S \cap \Omega$ is a quasiretract of Ω □

Fact 4.11 *Assume that T is a topological space, $S \subseteq T$, and $\bar{s}_* \in T$. Then the following are equivalent:*

(a) S is a local quasiretract of T at $\bar{s}_*$
(b) Every neighborhood V of $\bar{s}_*$ contains an open neighborhood U of $\bar{s}_*$ in T such that $S \cap U$ is a quasiretract of U □

Fact 4.11 implies, in particular, that *being a local quasiretract is a local-homeomorphism invariant property of the germ of S at $\bar{s}_*$*. Precisely,

Corollary 4.12 *Assume that T, T' are topological spaces, $S \subseteq T$, $S' \subseteq T'$, $\bar{s}_* \in T$, and $\bar{s}'_* \in T'$. Assume that there exist neighborhoods V, V' of $\bar{s}_*$, $\bar{s}'_*$ in T, T', and a homeomorphism h from V onto V' such that $h(S \cap V) = S' \cap V'$ and $h(\bar{s}_*) = \bar{s}'_*$. Then S is a local quasiretract of T at $\bar{s}_*$ if and only if S' is a local quasiretract of T' at $\bar{s}'_*$.*

Proof. It clearly suffices to prove one of the two implications. Assume that S is a local quasiretract of T at $\bar{s}_*$. Then Fact 4.11 implies that there exists an open subset U of T such that $\bar{s}_* \in U$, $U \subseteq V$, and $S \cap U$ is a quasiretract of U. Let $U' = h(U)$. Since h is a homeomorphism, U' is a relatively open subset of V' such that $\bar{s}'_* \in U'$, and $S' \cap U'$ is a quasiretract of U'. Since V' is a neighborhood of $\bar{s}'_*$ in T', it follows that U' is a neighborhood of $\bar{s}'_*$ in T', so Definition 4.9 tells us that S' is a local quasiretract of T' at $\bar{s}'_*$. □

Remark 4.13 The set $S = \{(x, y) \in \mathbb{R}^2 : y > 0\} \cup \{(0, 0)\}$ is a quasiretract of $\mathbb{R}^2$. (Indeed, if K is a compact subset of S, then the convex hull $\hat{K}$ of K is also compact, and $\hat{K} \subseteq S$ because S is convex. Therefore $\hat{K}$ is a retract of $\mathbb{R}^2$. If $\rho : \mathbb{R}^2 \mapsto \hat{K}$ is a retraction, then ρ maps $\mathbb{R}^2$ into S, and $\rho(s) = s$ for every $s \in K$.)

On the other hand, S is not a retract of $\mathbb{R}^2$, because S is not a closed subset of $\mathbb{R}$. This shows that the notion of quasiretract is strictly more general than that of a retract.

The same is true for the notions of "local quasiretract" and "local retract." For example, the set S of our previous example is a local quasiretract at the origin, but it is not a local retract at $(0, 0)$, because there does not exist a neighborhood V of $(0, 0)$ such that $S \cap V$ is a relatively closed subset of V. □

The Chain Rule

We now prove the *chain rule* for GDQs and AGDQs.

Theorem 4.14 *For $i = 1, 2, 3$, let X_i be a FDNRLS, and let $\bar{x}_{*,i}$ be a point of X_i. Assume that, for $i = 1, 2$, (i) $F_i : X_i \mapsto\!\!\!\rightarrow X_{i+1}$ is a set-valued map, (ii) S_i is a subset of X_i, and (iii) $\Lambda_i \in AGDQ(F_i, \bar{x}_{*,i}, \bar{x}_{*,i+1}, S_i)$. Assume, in addition, that (iv) $F_1(S_1) \subseteq S_2$, and*

(v) Either (v.1) S_2 is a local quasiretract of X_2 at $\bar{x}_{*,2}$ or (v.2) there exists a neighborhood U of $\bar{x}_{*,1}$ in X_1 such that the restriction $F_1 \lceil (U \cap S_1)$ of F_1 to $U \cap S_1$ is single-valued.

Then $\Lambda_2 \circ \Lambda_1 \in AGDQ(F_2 \circ F_1, \bar{x}_{*,1}, \bar{x}_{*,3}, S_1)$. Furthermore, if the sets Λ_1, Λ_2 belong to $GDQ(F_1, \bar{x}_{*,1}, \bar{x}_{*,2}, S_1)$ and $GDQ(F_2, \bar{x}_{*,2}, \bar{x}_{*,3}, S_2)$, respectively, then $\Lambda \in GDQ(F, \bar{x}_{*,1}, \bar{x}_{*,3}, S_1)$.

Proof. We assume, as is clearly possible without loss of generality, that $\bar{x}_{*,i} = 0$ for $i = 1, 2, 3$. We let $F \stackrel{\text{def}}{=} F_2 \circ F_1$, $\Lambda \stackrel{\text{def}}{=} \Lambda_2 \circ \Lambda_1$. We will first prove the conclusion for AGDQs, and then indicate how to make a trivial modification to obtain the GDQ result.

To begin with, let us fix AGDQ moduli θ_1, θ_2 for $(\Lambda_1, F_1, 0, 0, S_1)$ and $(\Lambda_2, F_2, 0, 0, S_2)$, respectively. Also, let $\kappa_i = 1 + \sup\Big\{\|L\| : L \in \Lambda_i\Big\}$, for $i = 1, 2$. (We add 1 to make sure that $\kappa_i > 0$ even if $\Lambda_i = \{0\}$.) It is then easy to see that $\Lambda_2^{\delta_2} \circ \Lambda_1^{\delta_1} \subseteq \Lambda^{\kappa_2\delta_1 + \kappa_1\delta_2 + \delta_1\delta_2}$ if $\delta_1 \geq 0$, $\delta_2 \geq 0$. (Indeed, if $L_1 \in \Lambda_1^{\delta_1}$, $L_2 \in \Lambda_2^{\delta_2}$, we may pick $\tilde{L}_1 \in \Lambda_1$, $\tilde{L}_2 \in \Lambda_2$ such that $\|\tilde{L}_1 - L_1\| \leq \delta_1$ and $\|\tilde{L}_2 - L_2\| \leq \delta_2$. Then $\|\tilde{L}_2\tilde{L}_1 - L_2L_1\| \leq \|\tilde{L}_2\tilde{L}_1 - \tilde{L}_2L_1\| + \|\tilde{L}_2L_1 - L_2L_1\|$, so $\|\tilde{L}_2\tilde{L}_1 - L_2L_1\| \leq \|\tilde{L}_2\|\,\|\tilde{L}_1 - L_1\| + \|\tilde{L}_2 - L_2\|\,\|L_1\| \leq (\kappa_2 + \delta_2)\delta_1 + \kappa_1\delta_2$, showing that $L_2L_1 \in \Lambda^{\kappa_2\delta_1 + \kappa_1\delta_2 + \delta_1\delta_2}$.)

We now use Hypothesis (v). If S_2 is a local quasiretract of X_2 at 0, then we choose a neighborhood U of 0 in X_2 such that $S_2 \cap U$ is a quasiretract of U, and then we choose a positive number $\bar{\sigma}$ such that the open ball $\mathbb{B}_{X_2}(0, \bar{\sigma})$ is contained in U. Then Fact 4.10 implies that $S_2 \cap \mathbb{B}_{X_2}(0, \bar{\sigma})$ is a quasiretract of $\mathbb{B}_{X_2}(0, \bar{\sigma})$. If S_2 is not a local quasiretract of X_2, then Hypothesis (v) guarantees that $F_1 \lceil (U \cap S_1)$ is single-valued for some neighborhood U of 0 in X_1. In this case, we choose a positive $\bar{\varepsilon}$ such that F_1 is single-valued on $\bar{\mathbb{B}}_{X_1}(0, \bar{\varepsilon}) \cap S_1$, and then take $\bar{\sigma}$ to be equal to $\bar{\varepsilon}$.

Then, for $\varepsilon \in]0, +\infty[$, we define $\sigma_\varepsilon^0 = (\kappa_1 + 2\theta_1(\varepsilon))\varepsilon$, $\sigma_\varepsilon = \sigma_\varepsilon^0 + \varepsilon$,

$$\theta^0(\varepsilon) = \kappa_2\theta_1(\varepsilon) + \kappa_1\theta_2(\sigma_\varepsilon) + 3\theta_1(\varepsilon)\theta_2(\sigma_\varepsilon)\,, \qquad \theta(\varepsilon) = \begin{cases} \theta^0(\varepsilon) & \text{if } \sigma_\varepsilon < \bar{\sigma} \\ +\infty & \text{if } \sigma_\varepsilon \geq \bar{\sigma}\,. \end{cases}$$

Let us show that θ is an AGDQ modulus for $(\Lambda, F, 0, 0, S_1)$. For this purpose, we first observe that $\theta \in \boldsymbol{\Theta}$. We next fix a positive ε such that $\theta(\varepsilon)$ is finite, and set out to construct a CCA map $A : \bar{\mathbb{B}}_{X_1}(0, \varepsilon) \cap S_1 \mapsto\!\!\!\rightarrow Lin(X_1, X_3) \times X_3$ such that

$$\Big(x \in \bar{\mathbb{B}}_{X_1}(0,\varepsilon) \cap S_1 \wedge (L,h) \in A(x)\Big) \Rightarrow \\ \Big(L \in \Lambda^{\theta(\varepsilon)} \wedge \|h\| \leq \theta(\varepsilon)\varepsilon \wedge L \cdot x + h \in F(x)\Big). \tag{4.2.1}$$

The fact that $\theta(\varepsilon) < +\infty$ clearly implies that $\sigma_\varepsilon < \bar{\sigma}$, $\theta(\varepsilon) = \theta^0(\varepsilon)$, $\theta_1(\varepsilon) < +\infty$, and $\theta_2(\sigma_\varepsilon) < +\infty$. We may therefore choose set-valued maps

$$A_1 \in CCA(\bar{\mathbb{B}}_{X_1}(0,\varepsilon) \cap S_1, Lin(X_1,X_2) \times X_2),$$
$$A_2 \in CCA(\bar{\mathbb{B}}_{X_2}(0,\sigma_\varepsilon) \cap S_2, Lin(X_2,X_3) \times X_3),$$

such that the conditions

$$L_1 \in \Lambda_1^{\theta_1(\varepsilon)}, \quad \|h_1\| \leq \theta_1(\varepsilon)\varepsilon, \quad L_1 \cdot x + h_1 \in F_1(x), \tag{4.2.2}$$

$$L_2 \in \Lambda_2^{\theta_2(\sigma_\varepsilon)}, \quad \|h_2\| \leq \theta_2(\sigma_\varepsilon)\sigma_\varepsilon, \quad L_2 \cdot y + h_2 \in F_2(y) \tag{4.2.3}$$

hold whenever $x \in \bar{\mathbb{B}}_{X_1}(0,\varepsilon) \cap S_1$, $(L_1,h_1) \in A_1(x)$, $y \in \bar{\mathbb{B}}_{X_2}(0,\sigma_\varepsilon) \cap S_2$, and $(L_2,h_2) \in A_2(y)$.

We then define our desired set-valued map A from $\bar{\mathbb{B}}_{X_1}(0,\varepsilon) \cap S$ to $Lin(X_1,X_3) \times X_3$ as follows. For each $x \in \bar{\mathbb{B}}_{X_1}(0,\varepsilon) \cap S_1$, we let

$$A(x) = \Big\{(L_2 \cdot L_1, L_2 h_1 + h_2) : (L_1 h_1) \in A_1(x),\, (L_2,h_2) \in A_2(L_1 \cdot x + h_1)\Big\}.$$

Assume that $x \in \bar{\mathbb{B}}_{X_1}(0,\varepsilon) \cap S_1$ and $(L,h) \in A(x)$, and let $z = L \cdot x + h$. Then there exist $(L_1,h_1) \in A_1(x)$ and $(L_2,h_2) \in A_2(L_1 \cdot x + h_1)$ such that $L = L_2 \cdot L_1$ and $h = L_2 h_1 + h_2$. The fact that $(L_1,h_1) \in A_1(x)$ implies that $L_1 \in \Lambda_1^{\theta(\varepsilon)}$, $\|h_1\| \leq \theta_1(\varepsilon)\varepsilon$, and $y \stackrel{\text{def}}{=} L_1 \cdot x + h_1 \in F_1(x)$. Then $y \in S_2$ (because $F_1(S_1) \subseteq S_2$), and $\|y\| \leq (\kappa_1 + \theta_1(\varepsilon))\varepsilon + \theta_1(\varepsilon)\varepsilon = \sigma_\varepsilon^0 < \sigma_\varepsilon$, so

$$y \in \bar{\mathbb{B}}_{X_2}(0,\sigma_\varepsilon^0) \cap S_2 \subseteq \mathbb{B}_{X_2}(0,\sigma_\varepsilon) \cap S_2 \tag{4.2.4}$$

and then $L_2 \in \Lambda_2^{\theta(\sigma_\varepsilon)}$, $\|h_2\| \leq \theta_2(\sigma_\varepsilon)\sigma_\varepsilon$, and $L_2 \cdot y + h_2 \in F_2(y)$. It follows that $L = L_2 L_1 \in \Lambda^{\kappa_1\theta_2(\sigma_\varepsilon)+\kappa_2\theta_1(\varepsilon)+\theta_2(\sigma_\varepsilon)\theta_1(\varepsilon)} \subseteq \Lambda^{\theta(\varepsilon)}$. Also,

$$\begin{aligned}
\|h\| &\leq \|L_2\|\,\|h_1\| + \|h_2\| \leq (\kappa_2 + \theta_2(\sigma_\varepsilon))\theta_1(\varepsilon)\varepsilon + \theta_2(\sigma_\varepsilon)\sigma_\varepsilon \\
&= (\kappa_2 + \theta_2(\sigma_\varepsilon))\theta_1(\varepsilon)\varepsilon + \theta_2(\sigma_\varepsilon)(\kappa_1 + 2\theta_1(\varepsilon))\varepsilon \\
&= \Big(\kappa_2\theta_1(\varepsilon) + \theta_2(\sigma_\varepsilon)\theta_1(\varepsilon) + \theta_2(\sigma_\varepsilon)\kappa_1 + 2\theta_2(\sigma_\varepsilon)\theta_1(\varepsilon)\Big)\varepsilon \\
&= \Big(\kappa_2\theta_1(\varepsilon) + \theta_2(\sigma_\varepsilon)\kappa_1 + 3\theta_2(\sigma_\varepsilon)\theta_1(\varepsilon)\Big)\varepsilon \\
&= \theta(\varepsilon)\varepsilon.
\end{aligned}$$

Finally,

$$z = L\cdot x + h = L_2 L_1 \cdot x + L_2 \cdot h_1 + h_2 = L_2(L_1 \cdot x + h_1) + h_2 = L_2 \cdot y + h_2 \in F_2(y).$$

Since $y \in F_1(x)$, we conclude that $z \in F(x)$. Hence A satisfies (4.2.1).

To conclude our proof, we have to show that

$$A \in CCA(\bar{\mathbb{B}}^n(0,\varepsilon) \cap S_1, Lin(X_1, X_3) \times X_3)\,. \tag{4.2.5}$$

We let

$$\begin{aligned} \mathcal{Q}_{1,\varepsilon} &= \bar{\mathbb{B}}_{X_1}(0,\varepsilon) \cap S_1\,, & \mathcal{T}_{1,\varepsilon} &= \mathcal{Q}_{1,\varepsilon} \times Lin(X_1, X_2) \times X_2 \times X_2\,, \\ \mathcal{R}_{1,\varepsilon} &= \bar{\mathbb{B}}_{X_2}(0,\sigma_\varepsilon) \cap S_2\,, & \mathcal{T}_{2,\varepsilon} &= \mathcal{Q}_{1,\varepsilon} \times Lin(X_1, X_2) \times X_2 \times \mathcal{R}_{1,\varepsilon}\,, \end{aligned}$$

and let $\Psi_{1,\varepsilon}$ be the set-valued map with source $\mathcal{Q}_{1,\varepsilon}$ and target $\mathcal{T}_{1,\varepsilon}$ that sends each $x \in \mathcal{Q}_{1,\varepsilon}$ to the set $\Psi_{1,\varepsilon}(x)$ of all 4-tuples $(\xi, L_1, h_1, y) \in \mathcal{T}_{1,\varepsilon}$ such that $\xi = x$, $(L_1, h_1) \in A_1(x)$, and $y = L_1 \cdot x + h_1$. We then observe that $\Psi_{1,\varepsilon}$ takes values in $\mathcal{T}_{2,\varepsilon}$. (This is trivial, because we have already established – cf. (4.2.4) – that if $x \in \mathcal{Q}_{1,\varepsilon}$, $(L_1, h_1) \in A_1(x)$, and $y = L_1 \cdot x + h_1$, then $y \in \mathcal{R}_{1,\varepsilon}$.)

Let $\tilde{\Psi}_{1,\varepsilon}$ be "$\Psi_{1,\varepsilon}$ regarded as a set-valued map with target $\mathcal{T}_{2,\varepsilon}$." (Precisely, $\tilde{\Psi}_{1,\varepsilon}$ is the set-valued map with source $\mathcal{Q}_{1,\varepsilon}$, target $\mathcal{T}_{2,\varepsilon}$, and graph $\mathrm{Gr}(\Psi_{1,\varepsilon})$.)

We now show that $\tilde{\Psi}_{1,\varepsilon} \in CCA(\mathcal{Q}_{1,\varepsilon}, \mathcal{T}_{2,\varepsilon})$. To prove this, we pick a compact subset K of $\mathcal{Q}_{1,\varepsilon}$, and show that (a) $\mathrm{Gr}(\tilde{\Psi}_{1,\varepsilon} \lceil K)$ is compact, and (b) there exists a sequence $\mathbf{H} = \{H_j\}_{j\in\mathbb{N}}$ of continuous maps $H_j : K \mapsto \mathcal{T}_{2,\varepsilon}$ such that $H_j \xrightarrow{\text{igr}} \tilde{\Psi}_{1,\varepsilon} \lceil K$ as $j \to \infty$.

The compactness of $\mathrm{Gr}(\tilde{\Psi}_{1,\varepsilon} \lceil K)$ follows from the fact that $\mathrm{Gr}(\tilde{\Psi}_{1,\varepsilon} \lceil K)$ is the image of $\mathrm{Gr}(A_1 \lceil K)$ under the continuous map

$$\mathcal{Q}_{1,\varepsilon} \times Lin(X_1, X_2) \times X_2 \ni (x, L_1, h_1) \mapsto (x, (x, L_1, h_1, L_1 \cdot x + h_1)) \in \mathcal{Q}_{1,\varepsilon} \times \mathcal{T}_{1,\varepsilon}\,.$$

To prove the existence of the sequence $\mathbf{H}$, we use the fact that A_1 belongs to $CCA(\mathcal{Q}_{1,\varepsilon}, Lin(X_1, X_2) \times X_2)$ to produce a sequence $\{A_1^j\}_{j\in\mathbb{N}}$ of ordinary continuous maps from K to $\mathbb{R}^{n_2 \times n_1} \times \mathbb{R}^{n_2}$ such that $A_1^j \xrightarrow{\text{igr}} A_1 \lceil K$ as $j \to \infty$, and we write $A_1^j(x) = (L_1^j(x), h_1^j(x))$ for $x \in K$.

We will construct $\mathbf{H}$ in two different ways, depending on whether *(v.1)* or *(v.2)* holds.

First suppose that *(v.1)* holds. The set

$$\mathcal{K} = \{L_1 \cdot x + h_1 : (x, L_1, h_1) \in \mathrm{Gr}(A_1 \lceil K)\} \tag{4.2.6}$$

is compact, and we know from (4.2.4) that every $y \in \mathcal{K}$ is a member of $\mathbb{B}_{X_2}(0,\sigma_\varepsilon) \cap S_2$. Since $\mathbb{B}_{X_2}(0,\bar{\sigma}) \cap S_2$ is a quasiretract of $\mathbb{B}_{X_2}(0,\bar{\sigma})$, and $\sigma_\varepsilon < \bar{\sigma}$, Fact 4.10 implies that $\mathbb{B}_{X_2}(0,\sigma_\varepsilon) \cap S_2$ is a quasiretract of $\mathbb{B}_{X_2}(0,\sigma_\varepsilon)$. Hence there exist an open subset Ω of the ball $\mathbb{B}_{X_2}(0,\sigma_\varepsilon)$ and a continuous map $\rho : \Omega \mapsto \mathbb{B}_{X_2}(0,\sigma_\varepsilon) \cap S_2$ such that $\rho(y) = y$ whenever $y \in \mathcal{K}$. Since $A_1^j \xrightarrow{\text{igr}} A_1 \lceil K$, the functions A_1^j must satisfy

$$\{L_1^j(x) \cdot x + h_1^j(x) : x \in K\} \subseteq \Omega \tag{4.2.7}$$

for all sufficiently large j. (Otherwise, there would exist an infinite subset J of $\mathbb{N}$ and $x_j \in K$ such that $y_j = L_1^j(x_j) \cdot x_j + h_1^j(x_j) \notin \Omega$. By making J

smaller – but still infinite – if necessary, we may assume that the sequence $\{(x_j, L_1^j, h_1^j)\}_{j\in J}$ converges to a limit $(x, L_1, h_1) \in \mathrm{Gr}(A_1 \lceil K)$. Then if we let $y = L_1 \cdot x + h_1$, we see that $y \in \mathcal{K}$. On the other hand, the y_j are not in Ω, so y is not in Ω either, because Ω is open. Since $\mathcal{K} \subseteq \Omega$, we have reached a contradiction.)

So we may assume, after passing to a subsequence, that (4.2.7) holds for all $j \in \mathbb{N}$. We then define $H_j(x) = (x, L_1^j(x), h_1^j(x), \rho(L_1^j(x)\cdot x + h_1^j(x)))$ for $x \in K$, $j \in \mathbb{N}$. Then the H_j are continuous maps from K to $\mathcal{T}_{2,\varepsilon}$, because ρ takes values in $\mathcal{R}_{1,\varepsilon}$.

We now show that $H_j \xrightarrow{\text{igr}} \tilde{\Psi}_{1,\varepsilon}\lceil K$ as j goes to ∞. To prove this, we let $\nu_j = \sup\{\mathrm{dist}(q, \mathrm{Gr}(\tilde{\Psi}_{1,\varepsilon}\lceil K)) : q \in \mathrm{Gr}(H_j)\}$, and assume that ν_j does not go to zero. We may then assume, after passing to a subsequence, that there exists a $\bar{\nu}$ such that $0 < 2\bar{\nu} \le \nu_j$ for all j. We can then pick $x_j \in K$ such that

$$\|x_j - x\| + \|L_1^j(x_j) - L_1\| + \|h_1^j(x_j) - h_1\| + \|\rho(L_1^j(x_j)\cdot x + h_1^j(x_j)) - y\| \ge \bar{\nu} \quad (4.2.8)$$

whenever $(x, L_1, h_1, y) \in \mathrm{Gr}(\tilde{\Psi}_{1,\varepsilon}\lceil K)$, $j \in \mathbb{N}$. Since $A_1^j \xrightarrow{\text{igr}} A_1\lceil K$, we may clearly assume, after passing to a subsequence if necessary, that the sequence $\{(x_j, L_1^j(x_j), h_1^j(x_j))\}_{j\in\mathbb{N}}$ has a limit $(\bar{x}, \bar{L}_1, \bar{h}_1) \in \mathrm{Gr}(A_1\lceil K)$.

Let $\bar{y}_* = \bar{L}_1 \cdot \bar{x} + \bar{h}_1$. Then $\bar{y}_* \in \mathcal{K}$, because of (4.2.6) and the fact that $(\bar{x}, \bar{L}_1, \bar{h}_1) \in \mathrm{Gr}(A_1\lceil K)$. Therefore $\rho(\bar{y}_*) = \bar{y}_*$. Furthermore, $x_j \to \bar{x}$, $L_1^j(x_j) \to \bar{L}_1$, and $h_1^j(x_j) \to \bar{h}_1$. Hence $L_1^j(x_j)\cdot x_j + h_1^j(x_j)$ converges to $\bar{L}_1(\bar{x})\cdot \bar{x} + \bar{h}_1 = \bar{y}_*$. But then $\lim_{j\to\infty}\left(\rho(L_1^j(x_j)\cdot x + h_1^j(x_j))\right) = \rho(\bar{y}_*)$, since ρ is continuous, so $\lim_{j\to\infty}\left(\rho(L_1^j(x_j)\cdot x + h_1^j(x_j))\right) = \bar{y}_*$, and then $\lim_{j\to\infty}\|\rho(L_1^j(x_j)\cdot x + h_1^j(x_j)) - \bar{y}_*\| = 0$. It follows that

$$\|x_j - \bar{x}\| + \|L_1^j(x_j) - \bar{L}_1\| + \|h_1^j(x_j) - \bar{h}_1\| + \|\rho(L_1^j(x_j)\cdot x_j + h_1^j(x_j)) - \bar{y}_*\| \to 0\,. \quad (4.2.9)$$

Let $\bar{y}_* = (\bar{x}, \bar{L}, \bar{L}\cdot\bar{x})$. Then $\bar{y}_* \in \mathrm{Gr}(\tilde{\Psi}_{1,\varepsilon}\lceil K)$, so (4.2.9) contradicts (4.2.8). This concludes the proof that $H_j \xrightarrow{\text{igr}} \tilde{\Psi}_{1,\varepsilon}\lceil K$ as j goes to ∞. We have thus established that the sequence $\mathbf{H}$ exists, under the assumption that *(v.1)* holds.

Next, we consider the case when *(v.2)* holds. Then $\bar{\sigma} = \bar{\varepsilon}$, so the fact that $\sigma_\varepsilon < \bar{\sigma}$ implies that $\varepsilon < \bar{\varepsilon}$, and then the map F_1 is single-valued on $\mathcal{Q}_{1,\varepsilon}$. Define $\varphi(x) = \{L_1\cdot x + h_1 : (L_1, h_1) \in A_1(x)\}$ for $x \in K$. Since $L_1 \cdot x + h_1 \in F_1(x)$ whenever $x \in K$ and $(L_1, h_1) \in A_1(x)$, the hypothesis that F_1 is single-valued on $\mathcal{Q}_{1,\varepsilon}$ implies that φ is a single-valued CCA map from K to X_2, so φ is an ordinary continuous map from K to X_2. Since $L_1\cdot x + h_1 \in \bar{\mathbb{B}}_{X_2}(0, \sigma_\varepsilon)$ whenever $x \in K$, and $(L_1, h_1) \in A_1(x)$, we conclude that φ is in fact a continuous map from K to $\mathcal{R}_{1,\varepsilon}$. We then define $H_j(x) = (x, L_1^j(x), h_1^j(x), \varphi(x))$ for $x \in K$, $j \in \mathbb{N}$. Then the H_j are continuous maps from K to $\mathcal{T}_{2,\varepsilon}$, and it is easy to see that $H_j \xrightarrow{\text{igr}} \tilde{\Psi}_{1,\varepsilon}\lceil K$ as $j \to \infty$. So the existence of $\mathbf{H}$ has also been proved when *(v.2)* holds.

We are now ready to prove (4.2.5). We do this by expressing A as a composite of CCA maps as follows: $A = \Psi_{3,\varepsilon} \circ \Psi_{2,\varepsilon} \circ \tilde{\Psi}_{1,\varepsilon}$, where:

1. $\mathcal{T}_{3,\varepsilon} = \mathcal{T}_{2,\varepsilon} \times Lin(X_2, X_3) \times X_3$
2. $\Psi_{2,\varepsilon} : \mathcal{T}_{2,\varepsilon} \mapsto\!\!\!\rightarrow \mathcal{T}_{3,\varepsilon}$ is the set-valued map that sends $(x, L_1, h_1, y) \in \mathcal{T}_{2,\varepsilon}$ to the set $\Psi_{2,\varepsilon}(x, L_1, h_1, y) \stackrel{\text{def}}{=} \{x\} \times \{L_1\} \times \{h_1\} \times \{y\} \times A_2(y)$
3. $\mathcal{T}_{4,\varepsilon} = Lin(X_1, X_3) \times X_3$
4. $\Psi_{3,\varepsilon} : \mathcal{T}_{3,\varepsilon} \mapsto \mathcal{T}_{4,\varepsilon}$ is the continuous single-valued map that sends $(x, L_1, h_1, y, L_2, h_2) \in \mathcal{T}_{3,\varepsilon}$ to the pair $(L_2L_1, L_2h_1 + h_2) \in \mathcal{T}_{4,\varepsilon}$

It is clear that $\Psi_{2,\varepsilon}$ and $\Psi_{3,\varepsilon}$ are CCA maps, so A is a CCA map, and our proof for AGDQs is complete.

The proof of the statement for GDQs is exactly the same, except only for the fact in this case all the constant components h of the various pairs (L, h) are always equal to zero. □

GDQs and AGDQs on Manifolds

If M and N are manifolds of class C^1, $\bar{x}_* \in M$, $\bar{y}_* \in N$, $S \subseteq M$, and $F : M \mapsto\!\!\!\rightarrow N$, then it is possible to define sets $GDQ(F, \bar{x}_*, \bar{y}_*, S)$, $AGDQ(F, \bar{x}_*, \bar{y}_*, S)$ of compact subsets of the space $Lin(T_{\bar{x}_*}M, T_{\bar{y}_*}N)$ of linear maps from $T_{\bar{x}_*}M$ to $T_{\bar{y}_*}N$ as follows. We let $m = \dim M$, $n = \dim N$, and pick coordinate charts $M \ni x \hookrightarrow \xi(x) \in \mathbb{R}^m$, $N \ni y \hookrightarrow \eta(y) \in \mathbb{R}^n$, defined near $\bar{x}_*$, $\bar{y}_*$ and such that $\xi(\bar{x}_*) = 0$ and $\eta(\bar{y}_*) = 0$, and declare that a subset Λ of $Lin(T_{\bar{x}_*}M, T_{\bar{y}_*}N)$ belongs to $GDQ(F, \bar{x}_*, \bar{y}_*, S)$ (resp. to $AGDQ(F, \bar{x}_*, \bar{y}_*, S)$) if the composite map $D\eta(\bar{y}_*) \circ \Lambda \circ D\xi(\bar{x}_*)^{-1}$ is in $GDQ(\eta \circ F \circ \xi^{-1}, 0, 0, \xi(S))$ (resp. in $AGDQ(\eta \circ F \circ \xi^{-1}, 0, 0, \xi(S))$). It then follows easily from the chain rule that, with this definition, *the sets $GDQ(F, \bar{x}_*, \bar{y}_*, S)$ and $AGDQ(F, \bar{x}_*, \bar{y}_*, S)$ do not depend on the choice of the charts ξ, η*. In other words, *the notions of GDQ and AGDQ are invariant under C^1 diffeomorphisms and therefore make sense intrinsically on manifolds of class C^1.*

The following facts about GDQs and AGDQs on manifolds are then easily verified.

Proposition 4.15 *If M, N are manifolds of class C^1, $S \subseteq M$, $\bar{x}_* \in M$, $\bar{y}_* \in N$, and $F : M \mapsto\!\!\!\rightarrow N$, then:*

(1) $GDQ(F, \bar{x}_*, \bar{y}_*, S) \subseteq AGDQ(F, \bar{x}_*, \bar{y}_*, S)$.

(2) If (i) U is a neighborhood of $\bar{x}_*$ in M, (ii) the restriction $F \lceil (U \cap S)$ is a continuous everywhere defined map, (iii) $\bar{y}_* = F(\bar{x}_*)$, (iv) F is differentiable at $\bar{x}_*$ in the direction of S, (v) L is a differential of F at $\bar{x}_*$ in the direction of S (that is, L belongs to $Lin(T_{\bar{x}_*}M, T_{\bar{y}_*}N)$ and $\lim_{x \to \bar{x}_*, x \in S} \|x - \bar{x}_*\|^{-1}\Big(F(x) - F(\bar{x}_*) - L \cdot (x - \bar{x}_*)\Big) = 0$ relative to some choice of coordinate charts about $\bar{x}_*$ and $\bar{y}_*$), then $\{L\}$ belongs to $GDQ(F, \bar{x}_*, \bar{y}_*, S)$.

(3) If (i) U is an open neighborhood of $\bar{x}_*$ in M, (ii) the restriction $F\lceil U$ is a Lipschitz-continuous everywhere defined map, (iii) $F(\bar{x}_*) = \bar{y}_*$, and (iv) Λ is the Clarke generalized Jacobian of F at $\bar{x}_*$, then Λ belongs to $GDQ(F, \bar{x}_*, \bar{y}_*, M)$. □

Proposition 4.16 ***(The chain rule.)*** *Assume that (I) for $i = 1, 2, 3$, M_i is a manifold of class C^1 and $\bar{x}_{*,i} \in M_i$, and (II) for $i = 1, 2$, (II.1) $S_i \subseteq M_i$, (II.2) $F_i : M_i \mapsto\!\!\!\!\to M_{i+1}$, and (II.3) $\Lambda_i \in AGDQ(F_i, \bar{x}_{*,i}, \bar{x}_{*,i+1}, S_i)$. Assume, in addition, that either S_2 is a local quasiretract of M_2 or F_1 is single-valued on $U \cap S_1$ for some neighborhood U of $\bar{x}_{*,1}$. Then the composite $\Lambda_2 \circ \Lambda_1$ belongs to $AGDQ(F_2 \circ F_1, \bar{x}_{*,1}, \bar{x}_{*,3}, S_1)$. If in addition $\Lambda_i \in GDQ(F_i, \bar{x}_{*,i}, \bar{x}_{*,i+1}, S_i)$ for $i = 1, 2$, then $\Lambda_2 \circ \Lambda_1 \in GDQ(F_2 \circ F_1, \bar{x}_{*,1}, \bar{x}_{*,3}, S_1)$.* □

Proposition 4.17 ***(The product rule.)*** *Assume that, for $i = 1, 2$, (1) M_i and N_i are manifolds of class C^1, (2) $S_i \subseteq M_i$, (3) $\bar{x}_{*,i} \in M_i$, (4) $\bar{y}_{*,i} \in N_i$, (5) $F_i : M_i \mapsto\!\!\!\!\to N_i$, (6) $\Lambda_i \in AGDQ(F_i, \bar{x}_{*,i}, \bar{y}_{*,i}, S_i)$. Assume also that:*

(7) $\bar{x}_* = (\bar{x}_{*,1}, \bar{x}_{*,2})$, $\bar{y}_* = (\bar{y}_{*,1}, \bar{y}_{*,2})$, and $S = S_1 \times S_2$

(8) $F = F_1 \times F_2$, where $F_1 \times F_2$ is the set-valued map from $M_1 \times M_2$ to $N_1 \times N_2$ that sends each point $(x_1, x_2) \in M_1 \times M_2$ to the subset $F_1(x_1) \times F_2(x_2)$ of $N_1 \times N_2$

(9) $\Lambda = \Lambda_1 \oplus \Lambda_2$, where (i) $\Lambda_1 \oplus \Lambda_2$ is the set of all linear maps $L_1 \oplus L_2$ for all $L_1 \in \Lambda_1$, $L_2 \in \Lambda_2$, (ii) $L_1 \oplus L_2$ is the map

$$T_{\bar{x}_{*,1}} M_1 \oplus T_{\bar{x}_{*,2}} M_2 \ni (v_1, v_2) \mapsto (L_1 v_1, L_2 v_2) \in T_{\bar{y}_{*,1}} N_1 \oplus T_{\bar{y}_{*,2}} N_2 ,$$

and (iii) we are identifying $T_{\bar{x}_{*,1}} M_1 \oplus T_{\bar{x}_{*,2}} M_2$ with $T_{(\bar{x}_{*,1}, \bar{x}_{*,2})}(M_1 \times M_2)$ and $T_{\bar{y}_{*,1}} N_1 \oplus T_{\bar{y}_{*,2}} N_2$ with $T_{(\bar{y}_{*,1}, \bar{y}_{*,2})}(N_1 \times N_2)$

Then $\Lambda \in AGDQ(F, \bar{x}_*, \bar{y}_*, S)$. Furthermore, if $\Lambda_i \in GDQ(F_i, \bar{x}_{*,i}, \bar{y}_{*,i}, S_i)$ for $i = 1, 2$, then $\Lambda \in AGDQ(F, \bar{x}_*, \bar{y}_*, S)$. □

Proposition 4.18 ***(Locality.)*** *Assume that (1) M, N, are manifolds of class C^1, (2) $\bar{x}_* \in M$, (3) $\bar{y}_* \in N$, (4) $S_i \subseteq M$, (5) $F_i : M \mapsto\!\!\!\!\to N$ for $i = 1, 2$, and (6) there exist neighborhoods U, V of $\bar{x}_*$, $\bar{y}_*$, in M, N, respectively, such that $U \cap S_1 = U \cap S_2$ and $(U \times V) \cap \mathrm{Gr}(F_1) = (U \times V) \cap \mathrm{Gr}(F_2)$. Then (a) $AGDQ(F_1, \bar{x}_*, \bar{y}_*, S_1) = AGDQ(F_2, \bar{x}_*, \bar{y}_*, S_2)$, and in addition (b) $GDQ(F_1, \bar{x}_*, \bar{y}_*, S_1) = GDQ(F_2, \bar{x}_*, \bar{y}_*, S_2)$.* □

Remark 4.19 It is easy to exhibit maps that have GDQs at a point $\bar{x}_*$ but are not classically differentiable at $\bar{x}_*$ and do not have differentials at $\bar{x}_*$ in the sense of other theories such as Clarke's generalized Jacobians, Warga's derivate containers, or our "semidifferentials" and "multidifferentials". (A simple example is provided by the function $f : \mathbb{R} \mapsto \mathbb{R}$ given by $f(x) = x \sin 1/x$ if $x \neq 0$, and $f(0) = 0$. The set $[-1, 1]$ belongs to $GDQ(f, 0, 0, \mathbb{R})$, but is not a differential of f at 0 in the sense of any of the other theories.) □

Closedness and Monotonicity

GDQs and AGDQs have an important *closedness property.* In order to state it, we first recall that, if Z is a metric space, then (i) $Comp^0(Z)$ is the set of all compact subsets of Z, (ii) $Comp^0(Z)$ has a natural non-Hausdorff topology $\mathcal{T}_{Comp^0(Z)}$, defined in Sect. 3.1. In particular, if X and Y are FDRLSs, then $Comp^0(Lin(X,Y))$ is the set of all compact subsets of $Lin(X,Y)$. Clearly, a subset $\mathcal{O}$ of $Comp^0(Lin(X,Y))$ is open in the topology $\mathcal{T}_{Comp^0(Lin(X,Y))}$ if and only if for every $\bar{\Lambda} \in \mathcal{O}$ there exists an open subset Ω of $Lin(X,Y)$ such that (i) $\bar{\Lambda} \subseteq \Omega$ and (ii) $\{\Lambda \in Comp^0(Lin(X,Y)) : \Lambda \subseteq \Omega\} \subseteq \mathcal{O}$. It is clear that the topology $\mathcal{T}_{Comp^0(Lin(X,Y))}$ can be entirely characterized by its convergent sequences. (That is, a subset $\mathcal{C}$ of $Comp^0(Lin(X,Y))$ is closed if and only if it is sequentially closed, i.e., such that, whenever $\{\Lambda_k\}_{k\in\mathbb{N}}$ is a sequence of members of $\mathcal{C}$ and $\Lambda \in Comp^0(Lin(X,Y))$ is such that $\Lambda_k \to \Lambda$ in the topology $\mathcal{T}_{Comp^0(Lin(X,Y))}$ as $k \to \infty$, it follows that $\Lambda \in \mathcal{C}$.)

Furthermore, convergence of sequences is easily characterized as follows.

Fact 4.20 *Assume that X and Y are FDRLSs, $\{\Lambda_k\}_{k\in\mathbb{N}}$ is a sequence of members of $Comp^0(Lin(X,Y))$, and Λ belongs to $Comp^0(Lin(X,Y))$. Then $\Lambda_k \to \Lambda$ as $k \to \infty$ in the topology $\mathcal{T}_{Comp^0(Lin(X,Y))}$ if and only if* $\lim_{k\to\infty} \sup\Big\{\mathrm{dist}(L,\Lambda) : L \in \Lambda_k\Big\} = 0.$ □

The following result is then an easy consequence of the definitions of GDQ and AGDQ.

Fact 4.21 *If M, N are manifolds of class C^1, $F : M \mapsto\!\!\!\!\to N$, $(\bar{x}_*, \bar{y}_*) \in M \times N$, $S \subseteq M$, $X = T_{\bar{x}_*}M$, and $Y = T_{\bar{y}_*}N$, then the sets $GDQ(F, \bar{x}_*, \bar{y}_*, S)$ and $AGDQ(F, \bar{x}_*, \bar{y}_*, S)$ are closed relative to the topology $\mathcal{T}_{Comp^0(Lin(X,Y))}$.* □

Fact 4.21 then implies that GDQs and AGDQs also have the following *monotonicity property.*

Fact 4.22 *If M, N are manifolds of class C^1, $F : M \mapsto\!\!\!\!\to N$, $(\bar{x}_*, \bar{y}_*) \in M \times N$, $S \subseteq M$, $\Lambda \in AGDQ(F, \bar{x}_*, \bar{y}_*, S)$, $\tilde{\Lambda} \in Comp^0(Lin(T_{\bar{x}_*}M, T_{\bar{y}_*}N))$, and $\Lambda \subseteq \tilde{\Lambda}$, then $\tilde{\Lambda} \in AGDQ(F, \bar{x}_*, \bar{y}_*, S)$. Furthermore, if $\Lambda \in GDQ(F, \bar{x}_*, \bar{y}_*, S)$ then $\tilde{\Lambda}$ belongs to $GDQ(F, \bar{x}_*, \bar{y}_*, S)$.*

Proof. It suffices to use Fact 4.21 and observe that, under our hypotheses, $\tilde{\Lambda}$ belongs to the closure of the set $\{\Lambda\}$ relative to $\mathcal{T}_{Comp^0(Lin(X,Y))}$. □

In addition, GDQs and AGDQs also have a monotonicity property with respect to F and S. Precisely, the following is a trivial corollary of the definitions of GDQ and AGDQ.

Fact 4.23 *Suppose that M, N are manifolds of class C^1, $(\bar{x}_*, \bar{y}_*) \in M \times N$, $\tilde{S} \subseteq S \subseteq M$, $F : M \mapsto\!\!\!\!\to N$, $\tilde{F} : M \mapsto\!\!\!\!\to N$, and $\mathrm{Gr}(F) \subseteq \mathrm{Gr}(\tilde{F})$. Then*

$$GDQ(F, \bar{x}_*, \bar{y}_*, S) \subseteq GDQ(\tilde{F}, \bar{x}_*, \bar{y}_*, \tilde{S})$$

and

$$AGDQ(F, \bar{x}_*, \bar{y}_*, S) \subseteq AGDQ(\tilde{F}, \bar{x}_*, \bar{y}_*, \tilde{S})\,.$$ □

Fact 4.21 says in particular that every GDQ of a map is also a GDQ of any "larger" map. On the other hand, it is perfectly possible for the "larger" map to have smaller GDQs. For example, if $f : \mathbb{R} \mapsto \mathbb{R}$ is the function given by $f(x) = |x|$, then the interval $[-1, 1]$ is a GDQ of f at 0 in the direction of $\mathbb{R}$, and no proper subset of $[-1, 1]$ has this property. But if we "enlarge" f and consider the set-valued map $F : \mathbb{R} \mapsto\!\!\!\!\rightarrow \mathbb{R}$ given by $F(x) = [0, |x|]$, then $\{0\} \in GDQ(F, 0, 0, \mathbb{R})$.

4.3 The Directional Open Mapping and Transversality Properties

The crucial fact about GDQs and AGDQs that leads to the maximum principle is the transversal intersection property, which is a very simple consequence of the directional open mapping theorem. We will now prove these results. As a preliminary, we need information on pseudoinverses.

Linear (Moore–Penrose) Pseudoinverses

If X, Y are FDRLSs and $L \in Lin(X, Y)$, a *linear right inverse* of L is a linear map $M \in Lin(Y, X)$ such that $L \cdot M = \mathbb{I}_Y$. It is clear that L has a right inverse if and only if it is surjective. Let $Lin_{onto}(X, Y)$ be the set of all surjective linear maps from X to Y. Since every $L \in Lin_{onto}(X, Y)$ has a right inverse, it is natural to ask if it is possible to choose a right inverse $I(L)$ for each L in a way that depends continuously (or smoothly, or real-analytically) on L. One way to make this choice is to let $I(L)$ be $L^{\#}$, the "Moore-Penrose pseudoinverse" of L (with respect to a particular inner product on X).

To define $L^{\#}$, assume X, Y are FDRLSs and endow both X and Y with Euclidean inner products (although, as will become clear below, only the choice of the inner product on X matters). Then every map $L \in Lin(X, Y)$ has an *adjoint* (or *transpose*) $L^{\dagger} \in Lin(Y, X)$, characterized by the property that $\langle L^{\dagger}y, x\rangle = \langle y, Lx\rangle$ whenever $x \in X$, $y \in Y$. It is then easy to see that

Fact 4.24 *If X and Y are FDRLSs endowed with Euclidean inner products, then $L \in Lin_{onto}(X, Y)$ if and only if $LL^{\dagger}$ is invertible.* □

Definition 4.25 *If X and Y are FDRLSs endowed with Euclidean inner products, and $L \in Lin_{onto}(X, Y)$, the* **Moore-Penrose pseudoinverse** *of L is the linear map $L^{\#} \in Lin(Y, X)$ given by $L^{\#} = L^{\dagger}(LL^{\dagger})^{-1}$, where the symbol "$\dagger$" stands for "adjoint."* □

The following result is then a trivial consequence of the definition.

Fact 4.26 *Suppose that X and Y are FDRLSs endowed with Euclidean inner products. Then $Lin_{onto}(X, Y)$ is an open subset of the space $Lin(X, Y)$, and the map $Lin_{onto}(X, Y) \ni L \mapsto L^{\#} \in Lin(Y, X)$ is real-analytic. Furthermore, the identity $LL^{\#} = \mathbb{I}_X$ holds for all $L \in Lin(X, Y)$.* □

Remark 4.27 If X, Y, L are as in Definition 4.25, $y \in Y$, $x = L^{\#}y$, and ξ is any member of $L^{-1}y$, then

$$\langle \xi, x \rangle = \langle \xi, L^{\#}y \rangle = \langle \xi, L^{\dagger}(LL^{\dagger})^{-1}y \rangle = \langle L\xi, (LL^{\dagger})^{-1}y \rangle = \langle y, (LL^{\dagger})^{-1}y \rangle .$$

In particular, the above equalities are true for x in the role of ξ, so that $\langle x, x \rangle = \langle y, (LL^{\dagger})^{-1}y \rangle$, and then $\langle \xi, x \rangle = \langle x, x \rangle$, so $\langle \xi - x, x \rangle = 0$. Therefore

$$\|\xi\|^2 = \|\xi - x + x\|^2 = \|\xi - x\|^2 + \|x\|^2 + 2\langle \xi - x, x \rangle = \|\xi - x\|^2 + \|x\|^2 \geq \|x\|^2 .$$

It follows that *$L^{\#}y$ is the member of $L^{-1}y$ of minimum norm.* This shows, in particular, that *the map $L^{\#}$ does not depend on the choice of a Euclidean inner product on Y.* □

More generally, we would like to find a pseudoinverse P of a given surjective map $L \in Lin(X, Y)$ that, for a given $v \in X$, has the value v when applied to Lv. This is clearly impossible if $Lv = 0$ but $v \neq 0$, because $P0$ has to be 0. But, as we now show, it can be done as long as $Lv \neq 0$, with a P that depends continuously on L and v.

To see this, we first define $\Omega(X, Y) = \{(L, v) : L \in Lin_{onto}(X, Y),\ Lv \neq 0\}$. We then fix inner products $\langle\cdot,\cdot\rangle_X$, $\langle\cdot,\cdot\rangle_Y$, on X, Y, and use $L^{\#}$ to denote, for $L \in Lin_{onto}(X, Y)$, the Moore-Penrose pseudoinverse of L corresponding to these inner products. Then, for $(L, v) \in \Omega(X, Y)$, we define

$$L^{\#,v}(y) = L^{\#}(y) + \frac{\langle y, Lv \rangle_Y}{\langle Lv, Lv \rangle_Y}(v - L^{\#}Lv) . \tag{4.3.1}$$

Then it is clear that

Fact 4.28 *If (L, v) belongs to $\Omega(X, Y)$, then (1) $L^{\#,v}$ is a linear map from Y to X, (2) $LL^{\#,v} = \mathbb{I}_Y$, and (3) $L^{\#,v}Lv = v$. Furthermore, the map $\Omega(X, Y) \ni (L, v) \mapsto L^{\#,v} \in Lin(Y, X)$ is real-analytic.* □

Pseudoinverses on Cones

If X, Y are FDRLSs and C is a convex cone in X, we define

$$\Sigma(X, Y, C) = \Big\{(L, y) \in Lin(X, Y) \times Y : y \in \mathrm{Int}(LC)\Big\}. \tag{4.3.2}$$

(Here "Int(LC)" denotes the absolute interior of LC, i.e., the largest open subset U of Y such that $U \subseteq LC$.)

Lemma 4.29 *Let X, Y be FDRLSs, let C be a convex cone in X, let S_C be the linear span of C, and let $\mathring{C}$ be the interior of C relative to S_C. Then:*

(1) $\Sigma(X, Y, C)$ is an open subset of $Lin(X, Y) \times Y$
(2) There exists a continuous map $\eta_{X,Y,C} : \Sigma(X, Y, C) \mapsto X$ such that the following are true whenever $(L, y) \in \Sigma(X, Y, C)$ and $r \geq 0$:

$$\eta_{X,Y,C}(L,y) \in \mathring{C} \cup \{0\} \tag{4.3.3}$$

$$L\eta_{X,Y,C}(L,y) = y \tag{4.3.4}$$

$$\eta_{X,Y,C}(L,ry) = r\eta_{X,Y,C}(L,y) \tag{4.3.5}$$

Proof. We assume, as we clearly may, that X and Y are endowed with inner products, and we write $\Sigma = \Sigma(X,Y,C)$, $S = S_C$.

Statement (1) is trivial, because if $(\bar{L},\bar{y}) \in \Sigma$, and $m = \dim(Y)$, then we can find $m+1$ points $q_0,\ldots,q_m$ in $\mathrm{Int}(\bar{L}C)$ such that $\bar{y}$ is an interior point of the convex hull of the set $Q = \{q_0,\ldots,q_m\}$. Then we can write $q_j = \bar{L}p_j$, with $p_j \in C$, for $j = 0,\ldots,m$. If $L \in \mathrm{Lin}(X,Y)$ is close to $\bar{L}$, and $y \in Y$ is close to $\bar{y}$, then the points $q_j^L = Lp_j$ belong to LC, and y is an interior point of their convex hull, so $y \in \mathrm{Int}(LC)$, proving (1).

For each $(\bar{L},\bar{y}) \in \Sigma$, we pick a point $x_{\bar{L},\bar{y}} \in \mathring{C}$ such that $\bar{L}\cdot x_{\bar{L},\bar{y}} = \bar{y}$. (To see that such a point exists, fix a $z \in \mathring{C}$, and observe that $\bar{y} - \varepsilon\bar{L}\cdot z \in \bar{L}C$ if ε is positive and small enough, because $\bar{y} \in \mathrm{Int}(\bar{L}C)$, since $(\bar{L},\bar{y}) \in \Sigma$. Pick one such ε, write $\bar{y}-\varepsilon\bar{L}\cdot z = \bar{L}\cdot x$ for an $x \in C$, and then let $x_{\bar{L},\bar{y}} = x+\varepsilon z$. It is then clear that $\bar{L}\cdot x_{\bar{L},\bar{y}} = \bar{y}$ and $x_{\bar{L},\bar{y}} \in \mathring{C}$.) We then define a map $\mu_{\bar{L},\bar{y}} : \Sigma \mapsto X$ by letting $\mu_{\bar{L},\bar{y}}(L,y) = x_{\bar{L},\bar{y}}+(L_S)^{\#}(y-L_S x_{\bar{L},\bar{y}})$ for $(L,y) \in \Sigma$, where L_S denotes the restriction of L to S (so $L_S \in Lin_{onto}(S,Y)$, because $L_S \in Lin(S,Y)$ and $y \in \mathrm{Int}(LC) = \mathrm{Int}(L_S C) \subseteq \mathrm{Int}(L_S S)$, showing that $\mathrm{Int}(L_S S) \neq \emptyset$, so L_S is surjective).

Then $\mu_{\bar{L},\bar{y}}$ is a continuous map from Σ to S, and satisfies the identity $\mu_{\bar{L},\bar{y}}(\bar{L},\bar{y}) = x_{\bar{L},\bar{y}}$. In addition, if $(L,y) \in \Sigma$, then

$$\begin{aligned} L\cdot\mu_{\bar{L},\bar{y}}(L,y) &= L\cdot x_{\bar{L},\bar{y}} + L\cdot(L_S)^{\#}\cdot(y - L\cdot x_{\bar{L},\bar{y}}) \\ &= L\cdot x_{\bar{L},\bar{y}} + y - L\cdot x_{\bar{L},\bar{y}} \\ &= y\,. \end{aligned}$$

Since $\mu_{\bar{L},\bar{y}}(\bar{L},\bar{y}) = x_{\bar{L},\bar{y}} \in \mathring{C}$, $\mathring{C}$ is a relatively open subset of S, and $\mu_{\bar{L},\bar{y}}$ is a continuous map from Σ to S, we can pick an open neighborhood $V_{\bar{L},\bar{y}}$ of $(\bar{L},\bar{y})$ in Σ such that $\mu_{\bar{L},\bar{y}}(L,y) \in \mathring{C}$ whenever $(L,y) \in V_{\bar{L},\bar{y}}$.

The family $\mathcal{V} = \{V_{\bar{L},\bar{y}}\}_{(\bar{L},\bar{y})\in\Sigma}$ of open sets is an open covering of Σ. So we can find a locally finite set $\mathcal{W}$ of open subsets of Σ which is a covering of Σ and a refinement of $\mathcal{V}$. (That is, (a) every $W \in \mathcal{W}$ is an open subset of Σ, (b) for every $W \in \mathcal{W}$ there exists $(\bar{L},\bar{y}) \in \Sigma$ such that $W \subseteq V_{\bar{L},\bar{y}}$, (c) every $(L,y) \in \Sigma$ belongs to some $W \in \mathcal{W}$, and (d) every compact subset K of Σ intersects only finitely many members of $\mathcal{W}$.)

Let $\{\varphi_W\}_{W\in\mathcal{W}}$ be a continuous partition of unity subordinate to the covering $\mathcal{W}$. (That is, (a) each φ_W is a continuous nonnegative real-valued function on Σ such that $\mathrm{support}(\varphi_W) \subseteq W$, and (b) $\sum_{W\in\mathcal{W}}\varphi_W \equiv 1$. Recall that the *support* of a function $\psi : \Sigma \mapsto \mathbb{R}$ is the closure in Σ of the set $\{\sigma \in \Sigma : \psi(\sigma) \neq 0\}$.) Select, for each $W \in \mathcal{W}$, a point $(\bar{L}_W,\bar{y}_W) \in \Sigma$ such that $W \subseteq V_{\bar{L}_W,\bar{y}_W}$, and define $\tilde{\eta}(L,y) = \sum_{W\in\mathcal{W}}\varphi_W(L,y)\mu_{\bar{L}_W,\bar{y}_W}(L,y)$

for $(L, y) \in \Sigma$. Then $\tilde{\eta}$ is a continuous map from Σ to X. If $(L, y) \in \Sigma$, let $\mathcal{W}(L, y)$ be the set of all $W \in \mathcal{W}$ such that $\varphi_W(L, y) \neq 0$. Then $(L, y) \in W$ for every $W \in \mathcal{W}(L, y)$, so $\mathcal{W}(L, y)$ is a finite set. Clearly, $\tilde{\eta}(L, y) = \sum_{W \in \mathcal{W}(L,y)} \varphi_W(L, y)\mu_{\bar{L}_W, \bar{y}_W}(L, y)$, and $\sum_{W \in \mathcal{W}(L,y)} \varphi_W(L, y) = 1$.

If $W \in \mathcal{W}(L, y)$, then $(L, y) \in W \subseteq V_{\bar{L}_W, \bar{y}_W}$, so $\mu_{\bar{L}_W, \bar{y}_W}(y, L) \in \mathring{C}$ and $L \cdot \mu_{\bar{L}_W, \bar{y}_W}(L, y) = y$. So $\tilde{\eta}(L, y)$ is a convex combination of points belonging to $\mathring{C}$, and then $\tilde{\eta}(L, y) \in \mathring{C}$. Furthermore,

$$L \cdot \tilde{\eta}(L, y) = \sum_{W \in \mathcal{W}(L,L)} \varphi_W(L, y) L \cdot \mu_{\bar{L}_W, \bar{y}_W}(L, y) = \Big(\sum_{W \in \mathcal{W}(L,y)} \varphi_W(L, y) \Big) y = y .$$

Hence, if we took $\eta_{X,Y,C}$ to be $\tilde{\eta}$, we would be satisfying all the required conditions, except only for the homogeneity property (4.3.5). In order to satisfy (4.3.5) as well, we define $\eta_{X,Y,C}(L, y)$, for $(L, y) \in \Sigma$, by letting

$$\eta_{X,Y,C}(L, y) = \begin{cases} \|y\| \tilde{\eta}\Big(L, \frac{y}{\|y\|}\Big) & \text{if } y \neq 0 , \\ 0 & \text{if } y = 0 . \end{cases}$$

(This is justified, because if $(L, y) \in \Sigma$ and $y \neq 0$ then $\Big(L . \frac{y}{\|y\|}\Big) \in \Sigma$ as well.)

Then $\eta_{X,Y,C}$ clearly satisfies (4.3.3), (4.3.4) and (4.3.5), and it is easy to verify that $\eta_{X,Y,C}$ is continuous. (Continuity at a point (L, y) of Σ such that $y \neq 0$ is obvious. To prove continuity at a point $(L, 0)$ of Σ, we pick a sequence $\{(L_j, y_j)\}_{j \in \mathbb{N}}$ of members of Σ such that $L_j \to L$ and $y_j \to 0$, and prove that $\eta_{X,Y,C}(L_j, y_j) \to 0$. If this conclusion was not true, there would exist a positive number ε and an infinite subset J of $\mathbb{N}$ such that

$$\|\eta_{X,Y,C}(L_j, y_j)\| \geq \varepsilon \quad \text{for all} \quad j \in J . \tag{4.3.6}$$

In particular, if $j \in J$ then $y_j \neq 0$, so we can define a unit vector $z_j = \frac{y_j}{\|y_j\|}$ and conclude that $(L_j, z_j) \in \Sigma$ and $\eta_{X,Y,C}(L_j, y_j) = \|y_j\| \tilde{\eta}(L_j, z_j)$. Since the z_j are unit vectors, there exists an infinite subset J' of J such that the limit $z = \lim_{j \to \infty, j \in J'} z_j$ exists. Since $(L, 0) \in \Sigma$, 0 is an interior point of the cone LC, so $LC = Y$ and then $z \in \mathrm{Int}(LC)$ as well. Therefore $(L, z) \in \Sigma$. Since $(L_j, z_j) \to (L, z)$ as $j \to \infty$ via values in J', the continuity of $\tilde{\eta}$ on Σ implies that $\tilde{\eta}(L_j, z_j) \to \tilde{\eta}(L, z)$ as $j \to \infty$ via values in J'. But then $\eta_{X,Y,C}(L_j, y_j) \to 0$ as $j \to_{J'} \infty$, because $\eta_{X,Y,C}(L_j, y_j) = \|y_j\| \tilde{\eta}(L_j, z_j)$ and $y_j \to 0$. This contradicts (4.3.6).) So $\eta_{X,Y,C}$ satisfies all our conditions, and the proof is complete. □

The Open Mapping Theorem

We are now ready to prove the open mapping theorem.

Theorem 4.30 *Let X, Y be FDNRLSs, and let C be a convex cone in X. Let $F : X \mapsto\!\!\!\rightarrow Y$ be a set-valued map, and let $\Lambda \in AGDQ(F, 0, 0, C)$. Let $\bar{y} \in Y$ be such that $\bar{y} \in \mathrm{Int}(LC)$ for every $L \in \Lambda$. Then:*

(I) *There exist a closed convex cone D in X such that $\bar{y} \in \text{Int}(D)$, and positive constants $\bar{\alpha}$, κ, having the property that*

(I.$*$) *For every $y \in D$ such that $0 < \|y\| \le \bar{\alpha}$ there exists an $x \in C$ such that $\|x\| \le \kappa\|y\|$ and $y \in F(x)$.*

(II) *Moreover, $\bar{\alpha}$ and κ can be chosen so that*

(II.$*$) *There exists a function $]0, \bar{\alpha}] \ni \alpha \mapsto \rho(\alpha) \in [0,1[$ such that $\lim_{\alpha \downarrow 0} \rho(\alpha) = 0$, for which, if we write $C(r) = C \cap \bar{\mathbb{B}}_X(0,r)$, then:*

(II.$*$.#) *For every $\alpha \in]0, \bar{\alpha}]$ and every $y \in D$ such that $\|y\| = \alpha$ there exists a compact connected subset Z_y of the product $C(\kappa\alpha) \times [\rho(\alpha), 1]$ having the following properties:*

$$Z_y \cap \Big(C(\kappa\alpha) \times \{\rho(\alpha)\}\Big) \neq \emptyset, \qquad Z_y \cap \Big(C(\kappa\alpha) \times \{1\}\Big) \neq \emptyset, \qquad (4.3.7)$$

$$ry \in F(x) \text{ and } \|x\| \le \kappa r\|y\| \text{ whenever } \rho(\alpha) \le r \le 1 \text{ and } (x,r) \in Z_y. \quad (4.3.8)$$

(III) *Finally, if $\Lambda \in GDQ(F,0,0,C)$ then the cone D and the constants $\bar{\alpha}$, κ can be chosen so that the following stronger conclusion holds:*

(III.$*$) *If $y \in D$ and $\|y\| \le \bar{\alpha}$ then there exists a compact connected subset Z_y of $C(\kappa\|y\|) \times [0,1]$ such that $(0,0) \in Z_y$, $Z_y \cap \Big(C(\kappa\alpha) \times \{1\}\Big) \neq \emptyset$, and $ry \in F(x)$ whenever $(x,r) \in Z_y$.*

Remark 4.31 For $\bar{y} \neq 0$, Conclusion (I) of Theorem 4.30 is the *directional open mapping property with linear rate and fixed angle for the restriction of F to C*, since it asserts that there is a neighborhood $\mathcal{N}$ of the half-line $H_{\bar{y}} = \{r\bar{y} : r \ge 0\}$ in the space $\mathcal{H}_Y$ of all closed half-lines emanating from 0 in Y such that, if $D_{\mathcal{N}}$ is the union of all the members of $\mathcal{N}$, then for every sufficiently small ball $\bar{\mathbb{B}}_Y(0,\alpha)$ the set $\Big(\bar{\mathbb{B}}_Y(0,\alpha) \cap D_{\mathcal{N}}\Big) \backslash \{0\}$ is contained in the image under F of a relative neighborhood $\bar{\mathbb{B}}_X(0,r) \cap C$ of 0 in C, whose radius r can be chosen proportional to α.

For $\bar{y} = 0$, Conclusion (I) is the *punctured open mapping property with linear rate for the restriction of F to C*, because in that case the cone D is necessarily the whole space Y, and Conclusion (I) asserts that for every sufficiently small ball $\bar{\mathbb{B}}_Y(0,\alpha)$ the punctured neighborhood $\bar{\mathbb{B}}_Y(0,\alpha) \backslash \{0\}$ is contained in the image under F of a relative neighborhood $\bar{\mathbb{B}}_X(0,r) \cap C$ of 0 in C, whose radius r can be chosen proportional to α. □

Proof of Theorem 4.30. It is clear that (II) implies (I), so there is no need to prove (I), and we may proceed directly to the proof of (II). Furthermore, Conclusion (III) is exactly the same as Conclusion (II), except only for the fact that in (III) $\rho(\alpha)$ is chosen to be equal to 0. So we will just prove (II), making sure that whenever we show the existence of $\rho(\alpha)$ it also follows that $\rho(\alpha)$ can be chosen to be equal to zero when $\Lambda \in GDQ(F,0,0,C)$.

Next, we observe that, once our conclusion is proved for $\bar{y} \neq 0$, its validity for $\bar{y} = 0$ follows by a trivial compactness argument. So we will assume from now on that $\bar{y} \neq 0$, and in that case it is clear that, without loss of generality, we may assume that $\|\bar{y}\| = 1$.

Let S_C be the linear span of C, and let $\overset{\circ}{C}$ be the interior of C relative to S_C. Write $\Sigma = \Sigma(X, Y, C)$ (cf. (4.3.2)). Then Lemma 4.29 tells us that Σ is open in $Lin(X, Y) \times Y$, and there exists a continuous map $\eta_{X,Y,C} : \Sigma \mapsto X$ such that (4.3.3), (4.3.4) and (4.3.5) hold. We write $\eta = \eta_{X,Y,C}$.

Our hypothesis says that the compact set $\Lambda \times \{\bar{y}\}$ is a subset of Σ. Hence we can find numbers $\hat{\gamma}$, δ, such that $\delta > 0$, $0 < \hat{\gamma} < 1$, and $\Lambda^\delta \times \bar{\mathbb{B}}_Y(\bar{y}, \hat{\gamma}) \subseteq \Sigma$. Let $\hat{D} = \{ry : r \in \mathbb{R},\, r \geq 0,\, y \in Y,\, \|y - \bar{y}\| \leq \hat{\gamma}\}$. Then $\hat{D}$ is a closed convex cone in Y and $\bar{y} \in \mathrm{Int}(\hat{D})$. Furthermore, it is clear that $\Lambda^\delta \times (\hat{D} \backslash \{0\}) \subseteq \Sigma$. So $\eta(L, y)$ is well defined whenever $L \in \Lambda^\delta$ and $y \in \hat{D}\backslash\{0\}$. In particular, $\eta(L, y)$ is defined for $(L, y) \in J$, where $J = \{(L, y) : L \in \Lambda^\delta\,,\, y \in \hat{D},\, \|y\| = 1\}$, so J is compact. Let $H = \{\eta(L, y) : (L, y) \in J\}$. Then H is a compact subset of $\overset{\circ}{C}$. Pick a compact subset $\tilde{H}$ of $\overset{\circ}{C}$ such that H is contained in the interior of $\tilde{H}$.

Since $\tilde{H}$ is a compact subset of the convex set $\overset{\circ}{C}$, the convex hull $\hat{H}$ of $\tilde{H}$ is also a compact subset of $\overset{\circ}{C}$. If $0 \notin \hat{H}$, and we define $\mathcal{C} = \{rx : r \geq 0, x \in \hat{H}\}$, then $\mathcal{C}$ is a closed convex cone in $\mathbb{R}^n$ such that

$$\mathcal{C} \subseteq \overset{\circ}{C} \cup \{0\} \qquad \text{and} \qquad \eta(L, y) \in \mathrm{Int}(\mathcal{C}) \text{ whenever } (L, y) \in J\,. \tag{4.3.9}$$

If $0 \in \hat{H}$, then $0 \in \overset{\circ}{C}$, so $C = S_C$ and then in particular C is closed, so we can define $\mathcal{C} = C$, and then then $\mathcal{C}$ is a closed convex cone in $\mathbb{R}^n$ such that (4.3.9) holds. Let $\hat{\kappa} = \max\{\|\eta(L, y)\| : (y, L) \in J\}$. Then $\|\eta(L, y)\| \leq \hat{\kappa}\|y\|$ whenever $(L, y) \in \Lambda^\delta \times (\hat{D}\backslash\{0\})$. This shows that η can be extended to a continuous map from $\Lambda^\delta \times \hat{D}$ to $\mathcal{C}$ by letting $\eta(L, 0) = 0$ for $L \in \Lambda^\delta$.

Fix a $\gamma \in]0, \hat{\gamma}\,[$, and let $D = \{ry : r \in \mathbb{R},\, r \geq 0,\, y \in Y,\, \|y - \bar{y}\| \leq \gamma\}$. Then D is a closed convex cone in Y, $\bar{y} \in \mathrm{Int}(D)$, and $D \subseteq \mathrm{Int}(\hat{D}) \cup \{0\}$. More precisely, we may pick a $\tilde{\sigma}$ such that $\tilde{\sigma} > 0$ and $\bar{\mathbb{B}}_Y(y, \tilde{\sigma}\|y\|) \subseteq \hat{D}$ whenever $y \in D$. (For example, $\tilde{\sigma} = \hat{\gamma} - \gamma$ will do. A simple calculation shows that the best – i.e., largest – possible choice of $\tilde{\sigma}$ is $\tilde{\sigma} = (\hat{\gamma} - \gamma)(1 - \gamma)^{-1/2}$.) We then let $\sigma = \frac{\tilde{\sigma}}{2}$, $\kappa = \hat{\kappa}(1 + 2\sigma)$.

Fix an AGDQ modulus θ for $(F, 0, 0, C)$. For each ε such that $\theta(\varepsilon)$ is finite, pick a map $A_\varepsilon \in CCA(\mathcal{C}(\varepsilon), Lin(X, Y) \times Y)$ such that

$$\Big(x \in \mathcal{C}(\varepsilon) \wedge (L, h) \in A_\varepsilon(x)\Big) \Rightarrow \Big(L \in \Lambda^{\theta(\varepsilon)} \wedge \|h\| \leq \theta(\varepsilon)\varepsilon \wedge L \cdot x + h \in F(x)\Big).$$

Also, observe that when $\Lambda \in GDQ(F, 0, 0, C)$ then A_ε can be chosen so that all the members (L, h) of $A_\varepsilon(x)$ are such that $h = 0$. In that case, we let $G_\varepsilon(x)$ be such that $A_\varepsilon(x) = G_\varepsilon(x) \times \{0\}$.

Next, fix a positive number $\bar{\varepsilon}$ such that $\theta(\bar{\varepsilon}) < \delta$ and $\theta(\bar{\varepsilon}) < \frac{\sigma}{\kappa}$. Let $\bar{\alpha} = \frac{\bar{\varepsilon}}{\kappa}$.

Fix an α such that $0 < \alpha \leq \bar{\alpha}$, and let $\varepsilon = \kappa\alpha$, so $0 < \varepsilon \leq \bar{\varepsilon}$. Then $\theta(\varepsilon) < \delta$ and $\theta(\varepsilon) < \frac{\sigma}{\kappa}$. Let $\mathcal{C}(\varepsilon) = \mathcal{C} \cap \bar{\mathbb{B}}_X(0, \varepsilon)$. Then $\mathcal{C}(\varepsilon)$ is a nonempty compact convex subset of X.

Now choose $\rho(\alpha)$ – for $\alpha \in]\,0,\bar{\alpha}]$ – as follows:

$$\rho(\alpha) = \begin{cases} 0 & \text{if } \Lambda \in GDQ(F,0,0,C) \\ \frac{\kappa\theta(\kappa\alpha)}{\sigma} & \text{if } \Lambda \notin GDQ(F,0,0,,C)\,. \end{cases}$$

It is then clear that $0 \le \rho(\alpha) < 1$, because $\theta(\kappa\alpha) \le \theta(\kappa\bar{\alpha}) = \theta(\bar{\varepsilon}) < \frac{\sigma}{\kappa}$. Furthermore, $\rho(\alpha)$ clearly goes to 0 as $\alpha \downarrow 0$.

Fix a $y \in D$ such that $\|y\| = \alpha$. Let $Q_\varepsilon = \mathcal{C}(\varepsilon) \times [0,1]$, and define a set-valued map $H_\varepsilon : Q_\varepsilon \mapsto\!\!\!\!\rightarrow X$ by letting $H_\varepsilon(x,t) = x - U_\varepsilon(x,t)$ (that is, $H_\varepsilon(x,t) = \{x - \xi : \xi \in U_{\bar{\varepsilon}}(x,t)\}$) for $x \in \mathcal{C}(\varepsilon)$, $t \in [0,1]$, where, for $(x,t) \in Q_\varepsilon$:

- If $\Lambda \notin GDQ(F,0,0,C)$, then $U_\varepsilon(x,t) = \Big\{\eta(L, ty - \varphi_\varepsilon(t)h) : (L,h) \in A_\varepsilon(x)\Big\}$, where the function $\varphi_\varepsilon : [0,1] \mapsto [0,1]$ is defined by
$$\varphi_\varepsilon(t) = \frac{t}{\rho(\alpha)} \quad \text{if} \quad 0 \le t < \rho(\alpha)\,, \qquad \varphi_\varepsilon(t) = 1 \quad \text{if} \quad \rho(\alpha) \le t \le 1\,;$$
- If $\Lambda \in GDQ(F,0,0,C)$, then $U_\varepsilon(x,t) = \Big\{\eta(L,ty) : L \in G_\varepsilon(x)\Big\}$.

We claim that $H_\varepsilon \in CCA(Q_\varepsilon, X)$. To see this, we first show that

$$ty - \varphi_\varepsilon(t)h \in \hat{D} \qquad \text{whenever} \quad (x,t) \in Q_\varepsilon \text{ and } (L,h) \in A_\varepsilon(x)\,. \tag{4.3.10}$$

This conclusion is trivial if $\Lambda \in GDQ(F,0,0,C)$, because in that case $h = 0$. Now consider the case when $\Lambda \notin GDQ(F,0,0,C)$, and observe that if $x \in \mathcal{C}(\varepsilon)$ and $(L,h) \in A_\varepsilon(x)$ then $L \in \Lambda^{\theta(\varepsilon)}$ and

$$\|\varphi_\varepsilon(t)h\| \le \frac{t}{\rho(\alpha)}\|h\| \le \frac{t}{\rho(\alpha)}\theta(\varepsilon)\varepsilon = \frac{t}{\rho(\alpha)}\theta(\kappa\alpha)\kappa\|y\| = \frac{t}{\rho(\alpha)}\theta(\kappa\alpha)\kappa\|y\|$$

$$= \frac{1}{\rho(\alpha)}\frac{\theta(\kappa\alpha)\kappa}{\sigma}t\sigma\|y\| = \frac{1}{\rho(\alpha)}\rho(\alpha)t\sigma\|y\| = t\sigma\|y\|\,.$$

It follows that $ty - \varphi_\varepsilon(t)h$ belongs to the ball $\bar{\mathbb{B}}_Y(ty, t\sigma\|y\|)$, which is contained in $\hat{D}$. So $ty - \varphi_\varepsilon(t)h \in \hat{D}$, completing the proof of (4.3.10).

Next, let μ be the set-valued map with source Q_ε and target $Lin(X,Y) \times Y$, such that $\mu(x,t) = \{(L, ty - \varphi_\varepsilon(t)h) : (L,h) \in A_\varepsilon(x)\}$. Then μ belongs to $CCA(Q_\varepsilon, Lin(X,Y) \times Y)$, because it is the composite of the maps

$$Q_\varepsilon \ni (x,t) \mapsto\!\!\!\!\rightarrow A_\varepsilon(x) \times \{t\} \subseteq Lin(X,Y) \times Y \times \mathbb{R}\,,$$

and $$Lin(X,Y) \times Y \times \mathbb{R} \ni (L,h,t) \mapsto (L, ty - \varphi_\varepsilon(t)h) \in Lin(X,Y) \times Y\,.$$

On the other hand, μ actually takes values in $\Lambda^{\theta(\varepsilon)} \times \hat{D}$. Therefore, if we let ν be the map having exactly the same graph as μ, but with target $\Lambda^\delta \times \hat{D}$, then $\nu \in CCA(Q_\varepsilon, \Lambda^\delta \times \hat{D})$. (Indeed, if $\{\mu_j\}_{j\in\mathbb{N}}$ is a sequence of continuous maps from Q_ε to $Lin(X,Y) \times Y$ with the property that $\mu_j \xrightarrow{\text{igr}} \mu$, and we write $\mu_j(x,t) = (L_j(x,t), \zeta_j(x,t))$, then L_j will take values in Λ^δ if j is large

enough, because Λ^δ is a neighborhood of $\Lambda^{\theta(\varepsilon)}$. On the other hand, $\hat{D}$ is a closed convex subset of Y, so it is a retract of Y. If $\omega : Y \mapsto \hat{D}$ is a retraction, and $\nu_j(x,t) = (L_j(x,t), \omega(\zeta_j(z,t)))$, then $\{\nu_j\}_{j\in\mathbb{N}, j\geq j_*}$ is – for some j_* – a sequence of continuous maps from Q_ε to $\Lambda^\delta \times \hat{D}$ such that $\nu_j \xrightarrow{\text{igr}} \nu$.)
Now, U_ε is the composite $\eta \circ \nu$, and η is a continuous map on $\Lambda^\delta \times \hat{D}$. So $U_\varepsilon \in CCA(Q_\varepsilon, X)$, and then $H_\varepsilon \in CCA(Q_\varepsilon, X)$ as well, completing the proof that $H_\varepsilon \in CCA(Q_\varepsilon, X)$.

It is clear that

$$\text{if} \quad (x,t) \in Q_\varepsilon \quad \text{then} \quad 0 \in H_\varepsilon(x,t) \Longleftrightarrow x \in U_\varepsilon(x,t)\,.$$

We now analyze the implications of the statement "$x \in U_\varepsilon(x,t)$" in two cases.

First, suppose that $\Lambda \in GDQ(F,0,0,C)$. Then $x \in U_\varepsilon(x,t)$ if and only if $(\exists L \in G_\varepsilon(x))(x = \eta(L,ty))$. If such an L exists, then $L \cdot x = L\eta(L,ty) = ty$, so $ty \in G_\varepsilon(x) \cdot x$, and then $ty \in F(x)$. Furthermore, the fact that $x = \eta(L,ty)$ implies that $\|x\| \leq \hat{\kappa}t\|y\|$, so *a fortiori* $\|x\| \leq \kappa t\|y\|$.

Now suppose that $\Lambda \notin GDQ(F,0,0,C)$. Then $x \in U_\varepsilon(x,t)$ if and only if $\Big(\exists (L,h) \in A_\varepsilon(x)\Big)\Big(x = \eta(L, ty - \varphi_\varepsilon(t)h)\Big)$. If such a pair (L,h) exists, and $t \geq \rho(\alpha)$, then

$$L \cdot x = L\eta(L, ty - \varphi_\varepsilon(t)h) = L\eta(L, ty - h) = ty - h$$

so $L \cdot x + h = ty$, and then $ty \in F(x)$. On the other hand, the fact that $x = \eta(L, ty - \varphi_\varepsilon(t)h)$ implies that $\|x\| \leq \hat{\kappa}(t\|y\| + t\sigma\|y\|)$, since we have already established that $\|\varphi_\varepsilon(t)h\| \leq t\sigma\|y\|$. Hence $\|x\| \leq \kappa t\|y\|$.

So we have shown, in both cases, that:

(A) *If* $(x,t) \in Q_\varepsilon$, $0 \in H_\varepsilon(x,t)$ *and* $\rho(\alpha) \leq t \leq 1$, *then* $ty \in F(x)$ *and* $\|x\| \leq \kappa t\|y\|$.

In addition, H_ε obviously satisfies:

(B) $H_\varepsilon(x,0) = \{x\}$ *whenever* $x \in \mathcal{C}(\varepsilon)$.

Next, choose a sequence $\{v_j\}_{j\in\mathbb{N}}$ of interior points of $\mathcal{C}$ such that $v_j \to 0$ as $j \to \infty$ and $\|v_j\| < \sigma\hat{\kappa}\|y\|$ for all j. We claim that:

(C) $v_j \notin H_\varepsilon(x,t)$ *whenever* $x \in \partial\mathcal{C}(\varepsilon)$, $t \in [0,1]$, *and* $j \in \mathbb{N}$.

To see this, we first observe that the condition $v_j \in H_\varepsilon(x,t)$ is equivalent to $x \in v_j + U_\varepsilon(x,t)$. If $x \in \partial\mathcal{C}(\varepsilon)$, then either $x \in \partial\mathcal{C}$ or $\|x\| = \kappa\|y\|$. If $x \in \partial\mathcal{C}$, then x cannot belong to $v_j + U_\varepsilon(x,t)$, because $U_\varepsilon(x,t) \subseteq \mathcal{C}$ and $v_j \in \text{Int}(\mathcal{C})$, so $v_j + U_\varepsilon(x,t) \subseteq \text{Int}(\mathcal{C})$. If $\|x\| = \kappa\|y\|$, then x cannot belong to $v_j + U_\varepsilon(x,t)$ either, because if $(L,h) \in A_{\bar{\varepsilon}}(x)$ then

$$\begin{aligned}\|\eta(L, ty - \varphi_\varepsilon(t)h)\| &\leq \hat{\kappa}\|ty - \varphi_\varepsilon(t)h\| \leq \hat{\kappa}\|ty\| + \hat{\kappa}\|\varphi_\varepsilon(t)h\| \\ &\leq t\hat{\kappa}\|y\| + t\hat{\kappa}\|\sigma\|y\| = t\hat{\kappa}(1+\sigma)\|y\| \leq \hat{\kappa}(1+\sigma)\|y\|\,,\end{aligned}$$

so $\|v_j + \eta(L, ty)\| \leq \hat{\kappa}(1+\sigma)\|y\| + \|v_j\| < \hat{\kappa}(1+\sigma)\|y\| + \hat{\kappa}\sigma\|y\| = \kappa\|y\|$.

Hence we can apply Theorem 3.8 and conclude that there exists a compact connected subset Z of $\mathcal{C}(\varepsilon) \times [0,1]$ such that (i) the sets $Z \cap (\mathcal{C}(\varepsilon) \times \{0\})$ and $Z \cap (\mathcal{C}(\varepsilon) \times \{1\})$ are nonempty, and (ii) $0 \in H_\varepsilon(x,t)$ whenever $(x,t) \in Z$.

For β such that $0 < \beta < 1-\rho(\alpha)$, let $Z^{(\beta)}$ be the open β-neighborhood of Z in Q_ε, so that $Z^{(\beta)} = \{q \in Q_\varepsilon : \operatorname{dist}(q, Z) < \beta\}$. Then $Z^{(\beta)}$ is a relatively open subset of Q_ε. It is clear that $Z^{(\beta)}$ is connected, so it is path-connected. Since $Z^{(\beta)}$ intersects both sets $\mathcal{C}(\varepsilon) \times \{0\}$ and $\mathcal{C}(\varepsilon) \times \{1\}$, there exists a continuous map $\xi : [0,1] \mapsto Z^{(\beta)}$ such that $\xi(0) \in \mathcal{C}(\varepsilon) \times \{0\}$ and $\xi(1) \in \mathcal{C}(\varepsilon) \times \{1\}$. Let

$$I = \Big\{t \in [0,1] : \xi(t) \in \mathcal{C}(\varepsilon) \times [0, \rho(\alpha)+\beta]\Big\},$$

Then it is clear that I is a nonempty compact subset of $[0,1]$, so I has a largest element τ. Then $\xi(\tau) \in \mathcal{C}(\varepsilon) \times \{\rho(\alpha)+\beta\}$; and

$$\xi(t) \in \mathcal{C}(\varepsilon) \times [\rho(\alpha)+\beta, 1] \quad \text{whenever} \quad \tau \leq t \leq 1. \tag{4.3.11}$$

Hence, if we define $W^\beta = \gamma([\tau,1])$, we see that (i) W^β is compact and connected, (ii) $W^\beta \subseteq \mathcal{C}(\varepsilon) \times [\rho(\alpha)+\beta, 1]$, (iii) $W^\beta \cap \Big(\mathcal{C}(\varepsilon) \times \{\rho(\alpha)+\beta\}\Big) \neq \emptyset$, (iv) $W^\beta \cap \Big(\mathcal{C}(\varepsilon) \times \{1\}\Big) \neq \emptyset$, and (v) $\operatorname{dist}(w, Z) \leq \beta$ whenever $w \in W^\beta$.

Let $\tilde{Z} = Z \cap (\mathcal{C}(\varepsilon) \times [\rho(\alpha), 1])$. Then $\tilde{Z}$ is a compact subset of $\mathcal{C}(\varepsilon) \times [\rho(\alpha), 1]$. If $w \in W^\beta$, then the point $z_w \in Z$ closest to w is at a distance $\leq \beta$ from w, and must therefore belong to $\mathcal{C}(\varepsilon) \times [\rho(\alpha), 1]$, since $w \in \mathcal{C}(\varepsilon) \times [\rho(\alpha)+\beta, 1]$. It follows that $z_w \in \tilde{Z}$. Therefore

$$\operatorname{dist}(w, \tilde{Z}) \leq \beta \text{ whenever } w \in W^\beta. \tag{4.3.12}$$

We now use Theorem 3.7 to pick a sequence $\{\beta_j\}_{j\in\mathbb{N}}$ converging to zero, such that the sets W^{β_j} converge in $Comp(Q_\varepsilon)$ to a compact connected set W. It then follows from (4.3.12) that $W \subseteq \tilde{Z}$. On the other hand, since the sets $W^{\beta_j} \cap \Big(\mathcal{C}(\varepsilon) \times \{\rho(\alpha)+\beta_j\}\Big)$ and $W^{\beta_j} \cap \Big(\mathcal{C}(\varepsilon) \times \{1\}\Big)$ are nonempty for each j, we can easily conclude that $W \cap \Big(\mathcal{C}(\varepsilon) \times \{\rho(\alpha)\}\Big) \neq \emptyset$ and $W \cap \Big(\mathcal{C}(\varepsilon) \times \{1\}\Big) \neq \emptyset$. Hence, if we take Z_y to be the set W, we see that (i) Z_y is compact connected, (ii) $Z_y \subseteq Z \cap \Big(\mathcal{C}(\varepsilon) \times [\rho(\alpha), 1]\Big)$, and (iii) Z_y has a nonempty intersection with both $\mathcal{C}(\varepsilon) \times \{\rho(\alpha)\}$ and $\mathcal{C}(\varepsilon) \times \{1\}$.

Now, if $(x,t) \in Z_y$, we know that $0 \in H_\varepsilon(x,t)$, and then (A) implies that $ty \in F(x)$ and $\|x\| \leq \kappa t\|y\|$, since $\rho(\alpha) \leq t \leq 1$. This shows that Z_y satisfies all the conditions of our statement, and completes our proof. □

Approximating Multicones

Assume that M is a manifold of class C^1, S is a subset of M and $\bar{x}_* \in S$ .Recall that "multicones" were defined on Page 225.

Definition 4.32 *An* ***AGDQ approximating multicone to*** S ***at*** $\bar{x}_*$ *is a convex multicone* $\mathcal{C}$ *in* $T_{\bar{x}_*}M$ *such that there exist an* $m \in \mathbb{Z}_+$, *a set-valued map* $F : \mathbb{R}^m \mapsto\!\!\!\!\to M$, *a convex cone* D *in* $\mathbb{R}^m$, *and a* $\Lambda \in AGDQ(F, 0, \bar{x}_*, D)$, *such that* $F(D) \subseteq S$ *and* $\mathcal{C} = \{LD : L \in \Lambda\}$. *If* Λ *can be chosen so that* $\Lambda \in GDQ(F, 0, \bar{x}_*, D)$, *then* $\mathcal{C}$ *is said to be a* ***GDQ approximating multicone to*** S ***at*** x_*. □

Transversality of Cones and Multicones

If S_1, S_2 are subsets of a linear space X, we define the *sum* $S_1 + S_2$ and the *difference* $S_1 - S_2$ by letting

$$S_1+S_2=\{s_1+s_2 : s_1 \in S_1, s_2 \in S_2\}, \quad S_1-S_2=\{s_1-s_2 : s_1 \in S_1, s_2 \in S_2\}.$$

Definition 4.33 *Let* X *be a FDRLS. We say that two convex cones* C^1, C^2 *in* X *are* ***transversal***, *and write* $C^1 \overline{\pitchfork} C^2$, *if* $C^1 - C^2 = X$. □

Definition 4.34 *Let* X *be a FDRLS. We say that two convex cones* C^1, C^2 *in* X *are* ***strongly transversal***, *and write* $C^1 \overline{\Pitchfork} C^2$, *if* $C^1 \overline{\pitchfork} C^2$ *and in addition* $C^1 \cap C^2 \neq \{0\}$. □

The definition of "transversality" of multicones is a straightforward extension of that of transversality of cones.

Definition 4.35 *Let* X *be a FDRLS. We say that two convex multicones* $\mathcal{C}^1$ *and* $\mathcal{C}^2$ *in* X *are* ***transversal***, *and write* $\mathcal{C}^1 \overline{\pitchfork} \mathcal{C}^2$, *if* $C^1 \overline{\pitchfork} C^2$ *for all pairs* $(C^1, C^2) \in \mathcal{C}^1 \times \mathcal{C}^2$. □

The definition of "strong transversality" for multicones requires more care. It is clear that two convex cones C^1, C^2 are strongly transversal if and only if (i) $C^1 \overline{\pitchfork} C^2$, and (ii) there exists a nontrivial linear functional $\lambda \in X^\dagger$ such that $C^1 \cap C^2 \cap \{x \in X : \lambda(x) > 0\} \neq \emptyset$. It is under this form that the definition generalizes to multicones.

Definition 4.36 *Let* X *be a finite-dimensional real linear space. Let* $\mathcal{C}^1$, $\mathcal{C}^2$ *be convex multicones in* X. *We say that* $\mathcal{C}^1$ *and* $\mathcal{C}^2$ *are* ***strongly transversal***, *and write* $\mathcal{C}^1 \overline{\Pitchfork} \mathcal{C}^2$, *if (i)* $\mathcal{C}^1 \overline{\pitchfork} \mathcal{C}^2$, *and (ii) there exists a nontrivial linear functional* $\lambda \in X^\dagger$ *such that* $C^1 \cap C^2 \cap \{x \in X : \lambda(x) > 0\} \neq \emptyset$ *for every* $(C^1, C^2) \in \mathcal{C}^1 \times \mathcal{C}^2$. □

The Nonseparation Theorem

If S_1, S_2 are subsets of a topological space T, and $\bar{s}_* \in S_1 \cap S_2$, we say that S_1 and S_2 are *locally separated* at $\bar{s}_*$ if there exists a neighborhood U of $\bar{s}_*$ such that $S_1 \cap S_2 \cap U = \{\bar{s}_*\}$. If T is metric, then it is clear that S_1 and S_2 are locally separated at $\bar{s}_*$ if and only if there does not exist a sequence $\{s_j\}_{j \in \mathbb{N}}$ of points of $(S_1 \cap S_2)\backslash\{0\}$ converging to $\bar{s}_*$.

Theorem 4.37 *Let M be a manifold of class C^1, let S_1, S_2 be subsets of M, and let $\bar{s}_* \in S_1 \cap S_2$. Let $\mathcal{C}_1$, $\mathcal{C}_2$ be AGDQ-approximating multicones to S_1, S_2 at $\bar{s}_*$ such that $\mathcal{C}_1 \pitchfork \mathcal{C}_2$. Then S_1 and S_2 are not locally separated at $\bar{s}_*$ (that is, the set $S_1 \cap S_2$ contains a sequence of points s_j converging to $\bar{s}_*$ but not equal to $\bar{s}_*$). Furthermore:*

(1) If $\xi : \Omega \mapsto \mathbb{R}^n$ is a coordinate chart of M, defined on an open set Ω containing $\bar{s}_*$, and such that $\xi(\bar{s}_*) = 0$, then there exist positive numbers $\bar{\alpha}$, κ, σ, and a function $\rho :]0, \bar{\alpha}] \mapsto [0, 1[$ such that $\lim_{\alpha \downarrow 0} \rho(\alpha) = 0$, having the property that, whenever $0 < \alpha \leq \bar{\alpha}$, the set $\xi(S_1 \cap S_2 \cap \Omega)$ contains a nontrivial compact connected set Z_α such that Z_α contains points $x_-(\alpha)$, $x_+(\alpha)$, for which $\|x_-(\alpha)\| \leq \kappa\rho(\alpha)\alpha$ and $\|x_+(\alpha)\| \geq \sigma\alpha$.
(2) If $\mathcal{C}_1$, $\mathcal{C}_2$ are GDQ-approximating multicones to S_1, S_2 at $\bar{s}_*$. then $S_1 \cap S_2$ contains a nontrivial compact connected set Z such that $\bar{s}_* \in Z$.

In view of our definitions, Theorem 4.37 will clearly follow if we prove:

Theorem 4.38 *Let n_1, n_2, m be positive integers. Assume that, for $i = 1, 2$, (1) C_i is a convex cone in $\mathbb{R}^{n_i}$, (2) $F_i : \mathbb{R}^{n_i} \mapsto\!\!\!\!\!\rightarrow \mathbb{R}^m$ is a set-valued map, and (3) $\Lambda_i \in AGDQ(F_i, 0, 0, C_i)$. Assume that the transversality condition*

$$L_1C_1 - L_2C_2 = \mathbb{R}^m \quad \text{for all} \quad (L_1, L_2) \in \Lambda_1 \times \Lambda_2 \tag{4.3.13}$$

holds, and there exists a nontrivial linear functional $\mu : \mathbb{R}^m \mapsto \mathbb{R}$ such that

$$L_1C_1 \cap L_2C_2 \cap \{y \in \mathbb{R}^m : \mu(y) > 0\} \neq \emptyset \quad \text{for all} \quad (L_1, L_2) \in \Lambda_1 \times \Lambda_2 \,. \tag{4.3.14}$$

Let $\mathcal{I} = \Big\{(x_1, x_2, y) \in C_1 \times C_2 \times \mathbb{R}^m : y \in F_1(x_1) \cap F_2(x_2)\Big\}$. Then there exist positive constants $\bar{\alpha}$, κ, σ, and a function $\rho :]0, \bar{\alpha}] \mapsto [0, 1[$ such that $\lim_{\alpha \downarrow 0} \rho(\alpha) = 0$, having the property that:

(*) For every α for which $0 < \alpha \leq \bar{\alpha}$ there exist a compact connected subset Z_α of $\mathcal{I}$, and points $(x_{1,\alpha,-}, x_{2,\alpha,-}, y_{\alpha,-})$, $(x_{1,\alpha,+}, x_{2,\alpha,+}, y_{\alpha,+})$ of Z_α, for which $\|y_{\alpha,+}\| \geq \sigma\alpha$ and $\|y_{\alpha,-}\| \leq \kappa\rho(\alpha)\alpha$.

Furthermore, if $\Lambda_i \in GDQ(F_i, 0, 0, C_i)$ for $i = 1, 2$, then it is possible to choose $\rho(\alpha) \equiv 0$.

Proof. Define a set-valued map $\mathcal{F} : \mathbb{R}^{n_1} \times \mathbb{R}^{n_2} \times \mathbb{R}^m \mapsto\!\!\!\!\!\rightarrow \mathbb{R}^m \times \mathbb{R}^m \times \mathbb{R}$ by letting $\mathcal{F}(x_1, x_2, y) = (y - F_1(x_1), y - F_2(x_2), \mu(y))$ for $x_1 \in \mathbb{R}^{n_1}$, $x_2 \in \mathbb{R}^{n_2}$, $y \in \mathbb{R}^m$. (Precisely, this means that $\mathcal{F}(x_1, x_2, y)$ is the set of all triples $(y - y_1, y - y_2, \mu(y))$, for all $y_1 \in F_1(x_1)$, $y_2 \in F_2(x_2)$.)

Also, define a cone $C \subseteq \mathbb{R}^{n_1} \times \mathbb{R}^{n_2} \times \mathbb{R}^m$ by letting $C = C_1 \times C_2 \times \mathbb{R}^m$, and a subset $\mathcal{L}$ of $Lin(\mathbb{R}^{n_1} \times \mathbb{R}^{n_2} \times \mathbb{R}^m, \mathbb{R}^m \times \mathbb{R}^m \times \mathbb{R})$ by letting $\mathcal{L}$ be the set of all linear maps $\mathcal{L}_{L_1,L_2}$, for all $(L_1, L_2) \in \Lambda_1 \times \Lambda_2$, where $\mathcal{L}_{L_1,L_2}$ is the map from $\mathbb{R}^{n_1} \times \mathbb{R}^{n_2} \times \mathbb{R}^m$ to $\mathbb{R}^m \times \mathbb{R}^m \times \mathbb{R}$ such that

$$\begin{aligned} \mathcal{L}_{L_1,L_2}(x_1, x_2, y) &= (y - L_1x_1, y - L_2x_2, \mu(y)) \\ &\text{if} \quad (x_1, x_2, y) \in \mathbb{R}^{n_1} \times \mathbb{R}^{n_2} \times \mathbb{R}^m. \end{aligned} \tag{4.3.15}$$

It then follows immediately from the definition of AGDQs and GDQs that $\mathcal{L} \in AGDQ(\mathcal{F}, (0,0,0), (0,0), C)$, and also that $\mathcal{L} \in GDQ(\mathcal{F}, (0,0,0), (0,0), C)$ if Λ_i is in $GDQ(F_i, 0, 0, C_i)$ for $i = 1, 2$.

Let $\bar{w}_* = (0,0,1)$. We want to show that the conditions of the directional open mapping theorem are satisfied, that is, that $\bar{w}_* \in \mathrm{Int}(LC)$ whenever $L \in \mathcal{L}$. Let $L \in \mathcal{L}$, and write $L = \mathcal{L}_{L_1,L_2}$, with $L_1 \in \Lambda_1$, $L_2 \in \Lambda_2$. Using (4.3.14), find $\bar{c}_1 \in C_1$, $\bar{c}_2 \in C_2$, such that $L_1\bar{c}_1 = L_2\bar{c}_2$ and $\mu(L_1\bar{c}_1) > 0$. Let $\bar{\alpha} = \mu(L_1\bar{c}_1)$. Let $v_1, v_2 \in \mathbb{R}^m$ be arbitrary vectors. We claim that the equation

$$L(x_1, x_2, y) = (v_1, v_2, r) \tag{4.3.16}$$

has a solution $(x_1, x_2, y) \in C$ provided that r is large enough. To see this, observe first that (4.3.13) implies that we can express $v_2 - v_1$ as a difference

$$v_2 - v_1 = L_1 c_1 - L_2 c_2\,, \quad c_1 \in C_1\,,\ c_2 \in C_2\,. \tag{4.3.17}$$

Then, if we let $\tilde{y} = v_1 + L_1 c_1$ (so that (4.3.17) implies that $\tilde{y} = v_2 + L_2 c_2$ as well), it is clear that $L(c_1, c_2, \tilde{y}) = (\tilde{y} - L_1 c_1, \tilde{y} - L_2 c_2, \mu(\tilde{y})) = (v_1, v_2, \tilde{r})$, if we let $\tilde{r} \stackrel{\mathrm{def}}{=} \mu(\tilde{y})$. If $r \geq \tilde{r}$, then we can choose

$$y = \tilde{y} + \frac{r - \tilde{r}}{\bar{\alpha}} \cdot L_1\bar{c}_1\,, \quad x_1 = c_1 + \frac{r - \tilde{r}}{\bar{\alpha}} \cdot \bar{c}_1\,, \quad x_2 = c_2 + \frac{r - \tilde{r}}{\bar{\alpha}} \cdot \bar{c}_2\,.$$

With this choice, we have

$$\begin{aligned} y - L_1 x_1 &= \tilde{y} - L_1 c_1 + \frac{r - \tilde{r}}{\bar{\alpha}} \cdot L_1\bar{c}_1 - \frac{r - \tilde{r}}{\bar{\alpha}} \cdot L_1\bar{c}_1 = \tilde{y} - L_1 c_1 = v_1\,, \\ y - L_2 x_2 &= \tilde{y} - L_2 c_2 + \frac{r - \tilde{r}}{\bar{\alpha}} \cdot L_1\bar{c}_1 - \frac{r - \tilde{r}}{\bar{\alpha}} \cdot L_2\bar{c}_2 \\ &= \tilde{y} - L_2 c_2 + \frac{r - \tilde{r}}{\bar{\alpha}} \cdot L_1\bar{c}_1 - \frac{r - \tilde{r}}{\bar{\alpha}} \cdot L_1\bar{c}_1 = \tilde{y} - L_2 c_2 = v_2\,, \end{aligned}$$

and $\mu(y) = \mu(\tilde{y}) + \frac{r-\tilde{r}}{\bar{\alpha}}\mu(L_1\bar{c}_1) = \tilde{r} + r - \tilde{r} = r$. It then follows that $L(x_1, x_2, y) = (v_1, v_2, r)$, and we have found our desired solution of (4.3.16).

So we have shown that for every $(v_1, v_2) \in \mathbb{R}^m \times \mathbb{R}^m$ the vector (v_1, v_2, r) belongs to $L \cdot C$ if r is large enough. This easily implies that the point $\bar{w}_* = (0,0,1)$ belongs to the interior of $L \cdot C$. (This can be proved in many ways. For example, let $E = (e_0, \ldots, e_{2m})$ be a sequence of $2m+1$ affinely independent vectors in $\mathbb{R}^m \times \mathbb{R}^m$ such that the origin of $\mathbb{R}^m \times \mathbb{R}^m$ is an interior point of the convex hull of E. Then we can find an $\bar{r}$ such that $\bar{r} > 0$ and $(e_i, \bar{r}) \in LC$ whenever $r \geq \bar{r}$. It then follows that the vectors $(e_i, \bar{r})$ and $(e_i, \bar{r} + 2)$ belong to LC, so the vector $(0, 0, \bar{r} + 1)$ is in $\mathrm{Int}(LC)$, and then $(0,0,1) \in \mathrm{Int}(LC)$ as well.)

We can then apply Theorem 4.30 to the map $\mathcal{F}$ and conclude that there exist positive numbers $\bar{\alpha}$, κ, and a function $\rho :]0, \bar{\alpha}] \mapsto [0, 1[$ such that, if $\alpha \in]0, \bar{\alpha}]$ and we let $\hat{w}_*(\alpha) = \alpha\bar{w}_*$, then there exists a compact connected subset $\hat{Z}_\alpha$ of $C(\kappa\alpha) \times [\rho(\alpha), 1]$ such that Z_α intersects the sets $C(\kappa\alpha) \times \{\rho(\alpha)\}$ and $C(\kappa\alpha) \times \{1\}$, and the conditions

$$r\hat{w}_*(\alpha) \in \mathcal{F}(x_1, x_2, y) \quad \text{and} \quad \|x_1\| + \|x_2\| + \|y\| \leq \kappa r \alpha$$

hold whenever $((x_1, x_2, y), r) \in \hat{Z}_\alpha$ and $\rho(\alpha) \leq r \leq 1$. (Here we are writing $C(r) = \{(x_1, x_2, y) \in C : \|x_1\| + \|x_2\| + \|y\| \leq r\}$.) We let $\sigma = \|\mu\|^{-1}$.

If we now define $Z_\alpha = \Big\{(x_1, x_2, y) : (\exists r \in [\rho(\alpha), 1])\Big(((x_1, x_2, y), r) \in \hat{Z}_\alpha\Big)\Big\}$, then Z_α is a continuous projection of a compact connected set, so Z_α is compact and connected. If $(x_1, x_2, y) \in Z_\alpha$, then there is an $r \in [\rho(\alpha), 1]$ such that $((x_1, x_2, y), r) \in \hat{Z}_\alpha$, and then $(0, 0, r\alpha) \in \mathcal{F}(x_1, x_2, y)$, so in particular $0 = y - y_1 = y - y_2$ for some $y_1 \in F_1(x_1)$ and some $y_2 \in F_2(x_2)$. But then $y_1 = y_2 = y$, so $y \in F_1(x_1) \cap F_2(x_2)$, showing that $(x_1, x_2, y) \in \mathcal{I}$. So $Z_\alpha \subseteq \mathcal{I}$, as desired.

Finally, we must show that Z_α contains points $(x_{1,\alpha,-}, x_{2,\alpha,-}, y_{\alpha,-})$ and $(x_{1,\alpha,+}, x_{2,\alpha,+}, y_{\alpha,+})$ for which $\|y_{\alpha,-}\| \leq \kappa\rho(\alpha)\alpha$ and $\|y_{\alpha,+}\| \geq \sigma\alpha$. Let $((x_{1,\alpha,-}, x_{2,\alpha,-}, y_{\alpha,-}), r_{\alpha,-})$ and $((x_{1,\alpha,+}, x_{2,\alpha,+}, y_{\alpha,+}), r_{\alpha,+})$ be members of $\hat{Z}_\alpha \cap (C(\kappa\alpha) \times \{\rho(\alpha)\})$ and $\hat{Z}_\alpha \cap (C(\kappa\alpha) \times \{1\})$, respectively. Then $r_{\alpha,-} = \rho(\alpha)$, and $(0, 0, \rho(\alpha)\alpha) = (0, 0, r_{\alpha,-}\alpha) = r_{\alpha,-}\hat{w}_*(\alpha) \in \mathcal{F}(x_{1,\alpha,-}, x_{2,\alpha,-}, y_{\alpha,-})$, from which it follows that $\|y_{\alpha,-}\| \leq \kappa r_{\alpha,-}\alpha$. On the other hand, $r_{\alpha,+} = 1$, and then $(0, 0, \alpha) = (0, 0, r_{\alpha,+}\alpha) = r_{\alpha,+}\hat{w}_*(\alpha) \in \mathcal{F}(x_{1,\alpha,+}, x_{2,\alpha,+}, y_{\alpha,+})$, from which it follows that $\mu(y_{\alpha,+}) = \alpha$, so that $\alpha = \mu(y_{\alpha,+}) \leq \|\mu\| \, \|y_{\alpha,+}\|$, and then $\|y_{\alpha,+}\| \geq \sigma\alpha$. □

5 Variational Generators

5.1 Linearization Error and Weak GDQs

Assume that X and Y are FDNRLSs, $S \subseteq X$, $F : S \mapsto Y$, and $\bar{x}_* \in X$. Recall that a linear map $L : X \mapsto Y$ is said to be a *differential* of F at $\bar{x}_*$ in the direction of S if the *linearization error* $E^{lin}_{F,L,\bar{x}_*}(h) = F(\bar{x}_* + h) - F(\bar{x}_*) - L \cdot h$ is $o(\|h\|)$ as $h \to 0$ via values such that $\bar{x}_* + h \in S$.

Remark 5.1 The precise meaning of the sentence "$E^{lin}_{F,L,\bar{x}_*}(h)$ is $o(\|h\|)$ as $h \to 0$ via values such that $\bar{x}_* + h \in S$" is:

- *There exists a function $\theta \in \mathbf{\Theta}$ (cf. Sect. 4, page 243) having the property that $\|E^{lin}_{F,\Lambda,\bar{y}_*}(\bar{x}_*, h)\| \leq \theta(\|h\|)\|h\|$ for every h such that $\bar{x}_* + h \in S$.* □

A natural generalization of that, when Λ is a set of linear maps, F is set-valued, and we have picked a point $\bar{y}_* \in Y$ to play the role of $F(\bar{x}_*)$, is obtained by defining the linearization error via the formula

$$E^{lin}_{F,\Lambda,\bar{x}_*,\bar{y}_*}(h) \stackrel{\text{def}}{=} \inf\Big\{\|y - \bar{y}_* - L \cdot h\| : y \in F(\bar{x}_* + h),\ L \in \Lambda\Big\}. \tag{5.1.1}$$

Definition 5.2 *Assume that X and Y are FDNRLSs, $(\bar{x}_*, \bar{y}_*) \in X \times Y$, $F : X \mapsto\!\!\!\to Y$, and $S \subseteq X$. A **weak GDQ** of F at $(\bar{x}_*, \bar{y}_*)$ in the direction of S is a compact set Λ of linear maps from X to Y such that the linearization error $E^{lin}_{F,\Lambda,\bar{x}_*,\bar{y}_*}(h)$ is $o(\|h\|)$ as $h \to 0$ via values such that $\bar{x}_* + h \in S$.* □

In other words, a weak GDQ is just the same as a classical differential, except for the fact that, since the map F is set-valued and the "differential" Λ is a set, we compute the linearization error by choosing the $y \in F(\bar{x}_* + h)$ and the linear map $L \in \Lambda$ that give the smallest possible error.

We will write $WGDQ(F, \bar{x}_*, \bar{y}_*, S)$ to denote the set of all weak GDQs of F at $(\bar{x}_*, \bar{y}_*)$ in the direction of S.

The following trivial observations will be important, so we state them explicitly. (The second assertion is true because the infimum of the empty subset of $[0, +\infty]$ is $+\infty$.)

Fact 5.3 *Assume that* X *and* Y *are FDNRLSs,* $(\bar{x}_*, \bar{y}_*) \in X \times Y$, $F : X \mapsto\!\!\!\rightarrow Y$, *and* $S \subseteq X$. *Then:*

- *If* $\Lambda \in WGDQ(F, \bar{x}_*, \bar{y}_*, S)$, $\Lambda' \in Comp^0(X, Y)$, *and* $\Lambda \subseteq \Lambda'$, *then* $\Lambda' \in WGDQ(F, \bar{x}_*, \bar{y}_*, S)$.
- $\emptyset \in WGDQ(F, \bar{x}_*, \bar{y}_*, S)$ *if and only if* $\bar{x}_* \notin \text{Closure}(S)$. □

We recall that the *distance* $\text{dist}(S, S')$ between two subsets S, S' of a metric space (M, d_M) is defined by $\text{dist}(S, S') = \inf\{d_M(s, s') : s \in S,\ s' \in S'\}$. It follows that $\text{dist}(S, S') \geq 0$, and also that $\text{dist}(S, S') < +\infty$ if and only if both S and S' are nonempty. Furthermore, the linearization error $E^{lin}_{F,\Lambda,\bar{x}_*,\bar{y}_*}(h)$ defined in (5.1.1) is exactly equal to $\text{dist}(\bar{y}_* + \Lambda \cdot h, F(\bar{x}_* + h))$.

The following two propositions are rather easy to prove, but we find it convenient to state them explicitly, because they will be the key to the notion of "variational generator" in GDQ theory.

Proposition 5.4 *Suppose* X, Y *are FDNRLSs,* $F : X \mapsto\!\!\!\rightarrow Y$, $S \subseteq X$, $(\bar{x}_*, \bar{y}_*)$ *belongs to* $X \times Y$, *and* Λ *is a compact set of linear maps from* X *to* Y. *Then the following three conditions are equivalent:*

(1) $\Lambda \in WGDQ(F, \bar{x}_*, \bar{y}_*, S)$

(2) There exist a positive number $\bar{\delta}_*$ *and a family* $\{\kappa^\delta\}_{0<\delta\leq\bar{\delta}_*}$ *of positive numbers such that* $\lim_{\delta\downarrow 0} \kappa^\delta = 0$, *having the property that*

$$\text{dist}\Big(\bar{y}_* + \Lambda \cdot h, F(\bar{x}_* + h)\Big) \leq \delta\kappa^\delta \quad \text{whenever} \quad \|h\| \leq \delta \leq \bar{\delta}_* \text{ and } \bar{x}_* + h \in S \tag{5.1.2}$$

(3) If $\{h_j\}_{j\in\mathbb{N}}$ *is a sequence in* X *such that* $\lim_{j\to\infty} h_j = 0$ *and* $\bar{x}_* + h_j \in S$ *for all* j, *then there exist (i) a sequence* $\{L_j\}_{j\in\mathbb{N}}$ *of members of* Λ *(ii) a sequence* $\{y_j\}_{j\in\mathbb{N}}$ *for which* $y_j \in F(\bar{x}_* + h_j)$ *for each* j, *(iii) a sequence* $\{r_j\}_{j\in\mathbb{N}}$ *of positive numbers such that* $\|y_j - \bar{y}_* - L_j \cdot h_j\| \leq r_j \|h_j\|$ *for all* $j \in \mathbb{N}$ *and* $\lim_{j\to\infty} r_j = 0$ □

Proposition 5.5 *Let* $X, Y, F, S, \bar{x}_*, \bar{y}_*$ *be as in Proposition 5.4. Then:*

- *If* $\Lambda \in AGQD(F, \bar{x}_*, \bar{y}_*, S)$ *it follows that* $\Lambda \in WGQD(F, \bar{x}_*, \bar{y}_*, S)$.
- *If* Λ *belongs to* $WGQD(F, \bar{x}_*, \bar{y}_*, S)$, Λ *is convex, and the restriction* $F \lceil S$ *is upper semicontinuous with closed convex values, then it follows that* $\Lambda \in GQD(F, \bar{x}_*, \bar{y}_*, S)$. □

5.2 GDQ Variational Generators

For a set-valued map $F : X \times \mathbb{R} \mapsto\!\!\!\!\rightarrow Y$, we write F_x, F^t, if $x \in X$, $t \in \mathbb{R}$, to denote the partial maps $F_x : \mathbb{R} \mapsto\!\!\!\!\rightarrow Y$, $F^t : X \mapsto\!\!\!\!\rightarrow Y$, given by

$$F_x(s) = F(x, s) \text{ and } F^t(u) = F(u, t) \text{ if } s \in \mathbb{R},\ u \in X.$$

For a subset S of $X \times \mathbb{R}$, we write S_x, S^t, if $x \in X$, $t \in \mathbb{R}$, to denote the sections $S_x \subseteq \mathbb{R}$, $S^t \subseteq X$, given by $S_x = \{s \in \mathbb{R} : (x, s) \in S\}$ and $S^t = \{u \in X : (u, t) \in S\}$.

We would like to define the notion of "variational generator" as follows, assuming that:

(VGA1) *X and Y are FDNRLSs, $a, b \in \mathbb{R}$, and $a \leq b$*
(VGA2) *$\xi_* \in C^0([a, b]\,; X)$ and σ_* is a ppd single-valued function from $[a, b]$ to Y*
(VGA3) $S \subseteq X \times \mathbb{R}$
(VGA4) *$F : X \times \mathbb{R} \mapsto\!\!\!\!\rightarrow Y$ is a set-valued map*

Tentative definition: Assume that (VGA1,2,3,4) hold. A *GDQ variational generator of F along (ξ_*, σ_*) in the direction of S* is a set-valued map $\Lambda : [a, b] \mapsto\!\!\!\!\rightarrow Lin(X, Y)$ such that, for every $t \in [a, b]$, the set $\Lambda(t)$ is a weak GDQ of F^t at $(\xi_*(t), \sigma_*(t))$ in the direction of S^t. □

The trouble with this definition is twofold:

- First of all, there at least two natural ways to define the "linearization error" at a particular time t, because we could:
  (1) Use the "fixed time error" $h \mapsto E^{lin}_{F,\Lambda,\xi_*,\sigma_*}(h, t) \stackrel{\text{def}}{=} E^{lin}_{F^t,\Lambda(t),\xi_*(t),\sigma_*(t)}(h)$, where $E^{lin}_{F^t,\Lambda(t),\xi_*(t),\sigma_*(t)}(h)$ is obtained by applying Formula (5.1.1) to the map F^t, so that

$$E^{lin}_{F,\Lambda,\xi_*,\sigma_*}(h, t) = \operatorname{dist}(\sigma_*(t) + \Lambda(t) \cdot h, F(\xi_*(t) + h, t)) \qquad (5.2.1)$$

  (2) Work instead with a "robust" version of the error, in which we try to approximate $F(\xi_*(t + s) + h, t + s) - \sigma_*(t + s)$ by $\Lambda(t) \cdot h$ not just for $s = 0$ but also for s in some neighborhood of 0; this leads to defining

$$E^{lin,rob}_{F,\Lambda,\xi_*,\sigma_*}(h, s, t) = \operatorname{dist}(\sigma_*(t + s) + \Lambda(t) \cdot h, F(\xi_*(t + s) + h, t + s)) \qquad (5.2.2)$$

- Second, once we have settled on which form of the error to use, this will lead to introducing functions $t \mapsto \kappa^\delta(t)$, $t \mapsto \kappa^{\delta,s}(t)$ such that $\|E^{lin}_{F,\Lambda,\xi_*,\sigma_*}(h, t)\| \leq \delta\kappa^\delta(t)$ and $\|E^{lin,rob}_{F,\Lambda,\xi_*,\sigma_*}(h, s, t)\| \leq \delta\kappa^{\delta,s}(t)$ whenever $\|h\| \leq \delta$, and require that these functions "go to zero." However, when functions are involving, "going to zero" can mean many different things, since the convergence could be, for example, pointwise, in L^1, or uniform.

It follows that, in principle, there are at least twice as many reasonable notions of "variational generators" as there are notions of convergence of functions, since for each convergence notion we can require that the convergence take place for the fixed-time error or for the robust one.

It turns out, however, that of all these possible notions of "variational generator," only two will be important to us. So we will define these two notions and ignore all the others.

L^1 Fixed-Time GDQ Variational Generators

Let us assume that X, Y, a, b, ξ_*, σ_*, S, F are such that (VGA1,2,3,4) hold.

Definition 5.6 *An L^1* ***fixed-time GDQ variational generator of the map*** F ***along*** (ξ_*, σ_*) ***in the direction of the set*** S *is a set-valued map* $\Lambda : [a,b] \mapsto\!\!\!\to Lin(X,Y)$ *such that:*

- *There exist a positive number $\bar{\delta}$ and a family $\{\kappa^\delta\}_{0<\delta\le\bar{\delta}}$ of measurable functions $\kappa^\delta : [a,b] \mapsto [0,+\infty]$ such that $\lim_{\delta\downarrow 0} \int_a^b \kappa^\delta(t)\,dt = 0$ and, in addition, $\mathrm{dist}(\sigma_*(t) + \Lambda(t)\cdot h, F(\xi_*(t)+h,t)) \le \delta\kappa^\delta(t)$ whenever $h \in X$, $t \in [a,b]$, $(\xi_*(t)+h,t) \in S$, and $\|h\| \le \delta$.* □

We will write $VG^{L^1,ft}_{GDQ}(F,\xi_*,\sigma_*,S)$ to denote the set of all L^1 fixed-time GDQ variational generators of F along (ξ_*,σ_*) in the direction of S.

Pointwise Robust GDQ Variational Generators

Again, let us assume that X, Y, a, b, ξ_*, σ_*, S, F are such that (VGA1,2,3,4) hold.

Definition 5.7 *A* ***pointwise robust GDQ variational generator of the map*** F ***along*** (ξ_*, σ_*) ***in the direction of the set*** S *is a set-valued map* $\Lambda : [a,b] \mapsto\!\!\!\to Lin(X,Y)$ *such that:*

- *There exist $\bar{\delta} > 0$, $\bar{s} > 0$, and a family $\{\kappa^{\delta,s}\}_{0<\delta\le\bar{\delta},0<s\le\bar{s}}$ of functions $\kappa^{\delta,s} : [a,b] \mapsto [0,+\infty]$, such that (i) $\lim_{\delta\downarrow 0,s\downarrow 0} \kappa^{\delta,s}(t) = 0$ for every $t \in [a,b]$ and (ii) $\mathrm{dist}(\sigma_*(t+s)+\Lambda(t)\cdot h, F(\xi_*(t+s)+h,t+s)) \le \delta\kappa^{\delta,s}(t)$ whenever $h \in X$, $\|h\| \le \delta$, $t \in [a,b]$, $t+s \in [a,b]$, and $(\xi_*(t+s)+h,t+s) \in S$.* □

We write $VG^{pw,rob}_{GDQ}(F,\xi_*,\sigma_*,S)$ to denote the set of all pointwise robust GDQ variational generators of F along (ξ_*,σ_*) in the direction of S.

5.3 Examples of Variational Generators

We now prove four propositions giving important examples of variational generators.

Clarke Generalized Jacobians

Recall that $\partial_x f(q,t)$ denotes the Clarke generalized Jacobian (cf. Definition 2.9) at $x = q$ of the map $x \mapsto f(x,t)$.

Proposition 5.8 *Assume that X, Y are FDNRLSs, and f is a single-valued ppd map from $X \times \mathbb{R}$ to Y, whose domain contains a tube $\mathcal{T}^X(\xi_*, \bar{\delta})$ about a continuous curve $\xi_* : [a,b] \mapsto X$. Assume that each partial map $t \mapsto f(x,t)$ is measurable and each partial map $x \mapsto f(x,t)$ is Lipschitz with a Lipschitz constant $C(t)$ such that the function $C(\cdot)$ is integrable. Let $Z = Lin(X,Y)$, and define $\Lambda(t) = \partial_x f(\xi_*(t),t)$ and $\sigma_*(t) = f(\xi_*(t),t)$ for $t \in [a,b]$. Then Λ is an integrably bounded measurable set-valued function from $[a,b]$ to Z with a.e. nonempty compact convex values, and Λ is an L^1 fixed-time variational GDQ of f along (ξ_*, σ_*) in the direction of $X \times [a,b]$.*

Proof. To begin with, we observe that the bound $\|L\| \leq C(t)$ holds for every $t \in [a,b]$ and every $L \in \Lambda(t)$, so Λ is integrably bounded. Furthermore, Λ clearly has compact convex a.e. nonempty values. A somewhat tedious but elementary argument proves that Λ is measurable.

Now, let $\kappa^\delta(t)$ denote the maximum of the distances $\text{dist}(L, \Lambda(t))$ for all $L \in \Lambda^{(\delta)}(t)$, where $\Lambda^{(\delta)}(t)$ is the closed convex hull of the set of all the differentials $Df^t(x)$ for all $x \in \mathcal{D}^t_\delta$, and $\mathcal{D}^t_\delta$ is the set of all points x in the the open ball $\mathbb{B}_X(\xi_*(t), \delta)$ such that f^t is differentiable at x. Then κ^δ is easily seen to be measurable, and such that $\lim_{\delta \downarrow 0} \kappa^\delta(t) = 0$ for every t. Furthermore, if $\|h\| \leq \delta$, then the equality $f(\xi_*(t)+h,t) - f(\xi_*(t),t) = \tilde{L} \cdot h$ holds for some $\tilde{L} \in \Lambda^{(\delta)}(t)$, and we can pick $L \in \Lambda(t)$ such that $\|\tilde{L} - L\| \leq \kappa^\delta(t)$, and conclude that

$$f(\xi_*(t)+h,t) - f(\xi_*(t),t) - L \cdot h = (\tilde{L} - L) \cdot h\,,$$

from which it follows that $\|f(\xi_*(t)+h,t) - f(\xi_*(t),t) - L \cdot h\| \leq \delta\kappa^\delta(t)$.

On the other hand, it is clear that $\kappa^\delta(t) \leq 2C(t)$. So the functions κ^δ converge pointwise to zero and are bounded by a fixed integrable function. Hence $\lim_{\delta \downarrow 0} \int_a^b \kappa^\delta(t)\,dt = 0$, and our proof is complete. □

Michel–Penot Subdifferentials

Recall that if $f : X \times \mathbb{R} \hookrightarrow \mathbb{R}$ then $\partial^o_x f(q,t)$ is the Michel–Penot subdifferential (cf. Definition 2.11) at $x = q$ of the function $x \mapsto f(x,t)$, and that the notion of *epimap* was defined in Sect. 2.1, page 225.

Proposition 5.9 *Let X be a FDNRLS, and let f be a single-valued ppd map from $X \times \mathbb{R}$ to $\mathbb{R}$, whose domain contains a tube $\mathcal{T}^X(\xi_*, \bar{\delta})$ about a continuous curve $\xi_* : [a,b] \mapsto X$. Assume that each partial map $t \mapsto f(x,t)$ is measurable and each partial map $x \mapsto f(x,t)$ is Lipschitz with a Lipschitz constant $C(t)$ such that the function $C(\cdot)$ is integrable. Let $\Lambda(t) = \partial^o_x f(\xi_*(t),t)$, and let $\sigma_*(t) = f(\xi_*(t),t)$. Let F be the epimap of f. Then Λ is an integrably bounded measurable set-valued function with a.e. nonempty compact convex* *values, and Λ is an L^1 fixed-time variational GDQ of F along (ξ_*, σ_*) in the direction of $X \times [a, b]$.*

Proof. To begin with, we observe, as in the previous proof, that (i) the bound $\|L\| \leq C(t)$ holds for every $t \in [a, b]$ and every $L \in \Lambda(t)$, so Λ is integrably bounded, and (ii) Λ clearly has compact convex a.e. nonempty values.

Next, we prove that Λ is measurable. For this purpose, we need to review how the Michel–Penot subdifferential $\Lambda(t)$ is defined: for each $t \in [a, b]$, let f^t be the function $\bar{\mathbb{B}}_X(\xi_*(t), \bar{\delta}) \ni x \mapsto f(x, t) \in \mathbb{R}$; extend f^t to all of X by defining it in an arbitrary fashion outside $\bar{\mathbb{B}}_X(\xi_*(t), \bar{\delta})$; for $x, h \in X$, define $d^o f^t(x, h) = \sup_{k \in X} \limsup_{t \downarrow 0} t^{-1}\Big(f(x+t(k+h)) - f(x+tk)\Big)$, so that, for each $x \in X$, the function $X \ni h \mapsto df(x, h) \in [-\infty, +\infty]$ is convex and positively homogeneous; then $\Lambda(t)$ is the set of all linear functionals $\omega \in X^\dagger$ such that $d^o f^t(\xi_*(t), h) \geq \langle \omega, h \rangle$ whenever $h \in X$.

We define the support function σ_Λ using (2.1.4), with $\mathbb{R}$ in the role of Y, and $X^\dagger = Lin(X, \mathbb{R})$ in the role of X, so σ_Λ is a function on $[a, b] \times X$. The measurability of Λ will follow if we prove that the function $[a, b] \ni t \mapsto \sigma_\Lambda(t, \bar{h})$ is measurable for each $\bar{h} \in X$.

Fix an $\bar{h} \in X$ and a $t \in [a, b]$. If $\omega \in \Lambda(t)$, then $\langle \omega, \bar{h} \rangle \leq d^o f^t(\xi_*(t), \bar{h})$. Therefore $\sigma_\Lambda(t, \bar{h}) \leq d^o f^t(\xi_*(t), \bar{h})$. We will prove that the opposite inequality is also true. Define $E = \{(h, r) \in X \times \mathbb{R} : r \geq d^o f^t(\xi_*(t), h)\}$, Then E is the epigraph of the function $X \ni h \mapsto d^o f^t(\xi_*(t), h) \in \mathbb{R}$, which is everywhere finite, convex, and positively homogeneous. In particular, E is a closed convex cone in $X \times \mathbb{R}$ with nonempty interior. If we let $\bar{r} = d^o f^t(\xi_*(t), \bar{h})$, then the point $(\bar{h}, \bar{r})$ belongs to the boundary of E. Hence the Hahn–Banach theorem implies that there exists a linear functional $\Omega \in (X \times \mathbb{R})^\dagger \backslash \{0\}$ such that $0 = \Omega(\bar{h}, \bar{r}) \leq \Omega(h, r)$ for all $(h, r) \in E$. Then there exist a linear functional $\omega : X \mapsto \mathbb{R}$ and a real number ω_0 such that $\Omega(h, r) = -\omega(h) + \omega_0 r$ for all $(h, r) \in X \times \mathbb{R}$, and $(\omega, \omega_0) \neq (0, 0)$. Clearly, $\omega_0 \geq 0$, because $0 = -\omega(\bar{h}) + \omega_0 \bar{r} \leq -\omega(\bar{h}) + \omega_0(\bar{r} + 1)$. Furthermore, $\omega_0 \neq 0$, because if $\omega_0 = 0$ then $\omega(\bar{h}) = -\Omega(\bar{h}, \bar{r}) = 0$, and then the inequality $\Omega(\bar{h}, \bar{r}) \leq \Omega(h, r)$ implies $0 = -\omega(\bar{h}) \leq -\omega(h)$ for all $h \in X$, so $\omega = 0$ as well. So we may assume that $\omega_0 = 1$, and then $0 = -\omega(\bar{h}) + \bar{r} \leq -\omega(h) + r$ for all $(h, r) \in E$. Hence $\omega(h) \leq r$ for all $(h, r) \in E$, so in particular $\omega(h) \leq d^o f^t(\xi_*(t), h)$ for all $h \in X$. It follows that $\omega \in \Lambda(t)$. On the other hand, the fact that $-\omega(\bar{h}) + \bar{r} = 0$ tells us that $\omega(\bar{h}) = d^o f^t(\xi_*(t), \bar{h})$. Hence $\sigma_\Lambda(t, \bar{h}) \geq d^o f^t(\xi_*(t), \bar{h})$.

It follows that $\sigma_\Lambda(t, \bar{h}) = d^o f^t(\xi_*(t), \bar{h})$ for all $\bar{h} \in X$. This implies the desired measurability of the function $[a, b] \ni t \mapsto \sigma_\Lambda(t, \bar{h}) \in \mathbb{R}$, because $[a, b] \ni t \mapsto d^o f^t(\xi_*(t), \bar{h})$ is clearly measurable.

Now fix $t \in [a, b]$. For $h \in \mathbb{R}^n$ such that $\|h\| \leq \bar{\delta}$, let

$$\hat{\theta}^t(h) = \min\{f(\xi_*(t) + h, t) - \sigma_*(t) - \omega \cdot h : \omega \in \Lambda(t)\}\,. \tag{5.3.1}$$

If in addition $h \neq 0$, write $\theta^t(h) = \frac{\hat{\theta}^t(h)}{\|h\|}$. We claim that $\limsup_{h \to 0, h \neq 0} \theta^t(h) \leq 0$. Indeed, if this was not so there would exist a positive ε and a sequence $\{h_j\}_{j \in \mathbb{N}}$

converging to zero and such that $h_j \neq 0$ and $\theta^t(h_j) \geq \varepsilon$ for all j. Then $f(\xi_*(t) + h_j, t) - f(\xi_*(t), t) - \omega \cdot h_j \geq \varepsilon \|h_j\|$ for all j and all $\omega \in \Lambda(t)$. Let $\tau_j = \|h_j\|$, $w_j = \frac{h_j}{\tau_j}$, so $\|w_j\| = 1$. By passing to a subsequence, if necessary, assume that the limit $w = \lim_{j\to\infty} w_j$ exists. Let $e_j = w_j - w$, so $e_j \to 0$. Then $h_j = \tau_j w_j = \tau_j(w + e_j)$, so $f(\xi_*(t) + \tau_j(w + e_j), t) - f(\xi_*(t), t) - \omega \cdot h_j \geq \varepsilon\tau_j$ for all $j \in \mathbb{N}$ and all $\omega \in \Lambda(t)$.

It follows that $\limsup_{j\to\infty} \tau_j^{-1}\Big(f(\xi_*(t)+\tau_j(w+e_j), t) - f(\xi_*(t), t) - \omega\cdot h_j\Big) \geq \varepsilon$ if $\omega \in \Lambda(t)$. But $f(\xi_*(t)+\tau_j(w+e_j), t) - f(\xi_*(t)+\tau_j w, t) \leq C(t)\tau_j\|e_j\|$. Hence $\limsup_{j\to\infty} \tau_j^{-1}\Big(f(\xi_*(t) + \tau_j w, t) - f(\xi_*(t), t) - \omega \cdot h_j\Big) \geq \varepsilon$, and then we find that $\limsup_{j\to\infty} \tau_j^{-1}\Big(f(\xi_*(t) + \tau_j w, t) - f(\xi_*(t), t)\Big) \geq \varepsilon + \omega \cdot w$, from which it follows that $\limsup_{\tau\downarrow 0} \tau^{-1}\Big(f(\xi_*(t) + \tau w, t) - f(\xi_*(t), t)\Big) \geq \varepsilon + \omega \cdot w$. So we have shown that $d^o f^t(\xi_*(t), w) \geq \varepsilon + \omega \cdot w$ for all $\omega \in \Lambda(t)$. But this is impossible, because we already know that $d^o f^t(\xi_*(t), w) = \sigma_\Lambda(t, w)$, so $d^o f^t(\xi_*(t), w) = \omega \cdot w$ for some $\omega \in \Lambda(t)$. This proves our claim that $\limsup_{h\to 0} \theta^t(h) \leq 0$.

Now define $\kappa^\delta(t) = \max\Big(0, \sup\{\theta^t(h) : \|h\| \leq \delta\}\Big)$. Then the functions κ^δ are measurable and nonnegative, and converge pointwise to zero. In addition, they clearly satisfy $\kappa^\delta \leq 2C(t)$, since (5.3.1) implies that $\hat{\theta}(h) \leq 2C(t)\|h\|$. Therefore $\lim_{\delta\downarrow 0} \int_a^b \kappa^\delta(t)\, dt = 0$.

Given $t \in [a, b]$ and $h \in X$ such that $\|h\| \leq \delta$, we can pick $\omega \in \Lambda(t)$ such that $f(\xi_*(t) + h, t) - \sigma_*(t) - \omega \cdot h = \hat{\theta}^t(h)$, and then

$$f(\xi_*(t) + h, t) - \sigma_*(t) - \omega \cdot h = \|h\|\theta^t(h) \leq \|h\|\kappa^\delta(t) \leq \delta\kappa^\delta(t).$$

It then follows that we can pick a real number $r \in F(\xi_*(t) + h, t)$ such that $|r - \sigma_*(t) - \omega \cdot h| \leq \delta\kappa^\delta(t)$. (Indeed, if $f(\xi_*(t) + h, t) - \sigma_*(t) - \omega \cdot h \geq 0$, we may pick $r = f(\xi_*(t) + h, t)$, and if $f(\xi_*(t) + h, t) - \sigma_*(t) - \omega \cdot h < 0$ pick $r = \sigma_*(t) + \omega \cdot h$.) But then $\mathrm{dist}\Big(F(\xi_*(t) + h, t), \sigma_*(t) + \Lambda(t) \cdot h\Big) \leq \delta\kappa^\delta(t)$, since $\sigma_*(t) + \omega \cdot h \in \sigma_*(t) + \Lambda(t) \cdot h$ and $r \in F(\xi_*(t) + h, t)$. This completes our proof. □

Classical Differentials

If (M, d_M), (N, d_N) are metric spaces, and $\bar{x}_* \in M$, a map $F : M \hookrightarrow N$ is *calm* at $\bar{x}_*$ if there exist positive constants $C, \bar{\delta}$ such that $x \in \mathrm{Do}(F)$ and $d_N(F(x), F(\bar{x}_*) \leq C d_M(x, \bar{x}_*)$ whenever $d_M(x, \bar{x}_*) \leq \delta$. If $a, b \in \mathbb{R}$, $a < b$, and $\xi_* : [a, b] \mapsto M$ is continuous, then a ppd map $F : M \times [a, b] \hookrightarrow N$ is *integrably calm* along ξ_* if there exist a positive constant $\bar{\delta}$ and an integrable function $C : [a, b] \mapsto [0, +\infty]$ such that, for almost all $t \in [a, b]$, the following two conditions are satisfied whenever $d_M(x, \xi_*(t)) \leq \delta$: (i) $(x, t) \in \mathrm{Do}(F)$, and (ii) $d_N(F(x, t), F(\xi_*(t), t)) \leq C(t) d_M(x, \xi_*(t))$. Then the following is easily proved.

Proposition 5.10 *Assume that X, Y are FDNRLSs, and f is a single-valued ppd map from $X \times \mathbb{R}$ to Y whose domain contains a tube $\mathcal{T}^X(\xi_*, \bar{\delta})$ about a continuous curve $\xi_* : [a,b] \mapsto X$. Assume that each partial map $t \mapsto f(x,t)$ is measurable. Assume in addition that:*

- *For each t the map $x \mapsto f(x,t)$ is differentiable at $\xi_*(t)$*
- *f is integrably calm along ξ_**

Let $\sigma_*(t) = f(\xi_*(t), t)$, and let $\Lambda(t) = \{D_x f(\xi_*(t), t)\}$. Then Λ is an integrable single-valued map. Furthermore, Λ is an L^1 fixed-time variational GDQ of f along (ξ_*, σ_*) in the direction of $X \times [a,b]$. □

The Set-Valued Maps $\partial > g$

We are going to assume that:

(A) *X is a FDNRLS, $\xi_* \in C^0([a,b], X)$, $\bar{\delta} > 0$, and $T = \mathcal{T}^X(\xi_*, \bar{\delta})$.*
(B) *$g : T \mapsto \mathbb{R}$ is a single-valued everywhere defined function such that (i) $g(\xi_*(t), t) \leq 0$ for all $t \in [a,b]$, and (ii) each partial map $x \mapsto g(x,t)$ is Lipschitz on $\{x \in X : \|x - \xi_*(t)\| \leq \bar{\delta}\}$, with a Lipschitz constant C which is independent of t for $t \in [a,b]$.*

We define $Av_g = \{(x,t) \in \mathcal{T}^X(\xi_*, \bar{\delta}) : g(x,t) > 0\}$, so Av_g is the domain of the constraint indicator map χ_g^{co} (cf. Sect. 2.1, page 225).

Remark 5.11 For an optimal control problem with an inequality state space constraint $g(x,t) \leq 0$, Av_g is the *set to be avoided*, that is, the set of points (x,t) such that any trajectory ξ for which $(\xi(t), t)$ is one of these points, for some t, fails to be admissible. □

We define $\partial_x^> g(\bar{x}, t)$ to be the convex hull of the set of all limits $\lim_{j\to\infty} \omega_j$, for all sequences $\{(x_j, t_j, \omega_j)\}_{j\in\mathbb{N}}$ such that $\lim_{j\to\infty}(x_j, t_j) \to (\bar{x}, t)$ and, for all j, (1) $(x_j, t_j) \in Av_g$, (2) the function $x \mapsto g(x, t_j)$ is differentiable at x_j, and (3) $\omega_j = \nabla_x g(x_j, t_j)$.

We let K be the set of all $t \in [a,b]$ such that $(\xi_*(t), t)$ belongs to the closure of Av_g. Then K is compact.

Remark 5.12 The set K could be empty. (This happens if and only if the closure of Av_g does not contain any point of the form $(\xi_*(t), t)$, $t \in [a,b]$.) □

Proposition 5.13 *Assume that X, a, b, ξ_*, $\bar{\delta}$, $T = \mathcal{T}^X(\xi_*, \bar{\delta})$, g, C are such that (A), (B) hold, and Av_g, $\partial_x^> g$, K are defined as above. Let $\sigma_*(t) = 0$ for $t \in [a,b]$, and define $\Lambda(t) = \partial_x^> g(\xi_*(t), t)$ for $t \in [a,b]$. Then:*

(1) Λ is an upper semicontinuous set-valued map with compact convex values
(2) $K = \{t \in [a,b] : \Lambda(t) \neq \emptyset\}$
(3) Λ is a pointwise robust GDQ variational generator of χ_g^{co} along (ξ_*, σ_*) in the direction of Av_g

Proof. The desired conclusions do not depend on the choice of a norm on X, so we will assume that the norm on X is Euclidean. For each $t \in [a,b]$, let g^t denote the function $x \mapsto g(x,t)$, with domain $B^t = \bar{\mathbb{B}}_X(\xi_*(t),\bar{\delta})$, and let D^t be the set of points $x \in B^t$ such that g^t is differentiable at x. Then D^t is a subset of full measure of B^t.

Let us show that Λ is upper semicontinuous and has compact convex values. The convexity of the sets $\Lambda(t)$ is clear from the definition of Λ. We will prove that the graph of Λ is compact, from which it will follow that Λ is upper semicontinuous and has compact values.

First, we observe that every member (t,ω) of $\mathrm{Gr}(\Lambda)$ is the limit of a sequence $\{(t_j,\omega_j)\}_{j\in\mathbb{N}}$ such that $\|\omega_j\| \le C$ for all j. Therefore $\|\omega\| \le C$ whenever $t \in [a,b]$ and $\omega \in \Lambda(t)$.

Now, take a sequence $\{(t_j,\omega_j)\}_{j\in\mathbb{N}}$ of points in $\mathrm{Gr}(\Lambda)$. Then $\|\omega_j\| \le C$ for all j, so we may find an infinite subset J of $\mathbb{N}$ such that the sequence $\{(t_j,\omega_j)\}_{j\in J}$ converges to a limit $(t,\omega) \in [a,b] \times X^\dagger$. We need to show that $\omega \in \Lambda(t)$. For each $j \in J$, the covector ω_j is a convex combination $\sum_{k=0}^n \alpha_{j,k}\omega_{j,k}$, where $\alpha_{j,k} \ge 0$, $\sum_{k=0}^n \alpha_{j,k} = 1$, and $\omega_{j,k} = \lim_{\ell\to\infty}\omega_{j,k,\ell}$, with $x_{j,k,\ell} \in D^{t_{j,k,\ell}}$, $g(x_{j,k,\ell},t_{j,k,\ell}) > 0$, $\omega_{j,k,\ell} = \frac{\partial g}{\partial x}(x_{j,k,\ell},t_{j,k,\ell})$, and $\lim_{\ell\to\infty}(x_{j,k,\ell},t_{j,k,\ell}) = (\xi_*(t_j),t_j)$. Pick an infinite subset J' of J such that the limits $\tilde{\omega}_k = \lim_{j\to\infty, j\in J'}\omega_{j,k}$ and $\tilde{\alpha}_k = \lim_{j\to\infty, j\in J'}\alpha_{j,k}$ exist. Then $\tilde{\alpha}_k \ge 0$, $\sum_{k=0}^n \tilde{\alpha}_k = 1$, and $\sum_{k=0}^n \tilde{\alpha}_k\tilde{\omega}_k = \omega$. Therefore the conclusion that $\omega \in \Lambda(t)$ will follow if we show that $\tilde{\omega}_k \in \Lambda(t)$ for each k. For $j \in J'$, $k \in \{0,\ldots,n\}$, pick $\ell(j,k) \in \mathbb{N}$ such that

$$\|\hat{\omega}_{j,k} - \omega_{j,k}\| + \|\hat{x}_{j,k} - \xi_*(t_j)\| + |\hat{t}_{j,k} - t_j| \le 2^{-j},$$

where $\hat{\omega}_{j,k} = \omega_{j,k,\ell(j,k)}$, $\hat{x}_{j,k} = x_{j,k,\ell(j,k)}$, $\hat{t}_{j,k} = t_{j,k,\ell(j,k)}$. Then $\tilde{\omega}_k = \lim_{j\to\infty, j\in J'}\hat{\omega}_{j,k}$, with $\hat{\omega}_{j,k} \in \partial_x g(\hat{x}_{j,k},\hat{t}_{j,k})$, $g(\hat{x}_{j,k},\hat{t}_{j,k}) > 0$, and $\lim_{j\to\infty}(\hat{x}_{j,k},\hat{t}_{j,k}) = (\xi_*(t),t)$. Therefore $\tilde{\omega}_k \in \Lambda(t)$ for each k, and then $\omega \in \Lambda(t)$, completing the proof that Λ is upper semicontinuous and has compact values. So we have proved (1).

Now let us prove (2). Fix a $t \in K$. Then there exist, for $j \in \mathbb{N}$, pairs $(\tilde{x}_j,t_j) \in S_g$ such that $g(\tilde{x}_j,t_j) > 0$ and $\|\tilde{x}_j - \xi_*(t)\| + |t_j - t| < 2^{-j}$. Pick $x_j \in D^{t_j}$ such that $g(x_j,t_j) > 0$ and $\|x_j - \tilde{x}_j\| < 2^{-j}$. Let $\omega_j = \frac{\partial g}{\partial x}(x_j,t_j)$. Then $\|\omega_j\| \le C$ for all j. Therefore, we can pick an infinite subset J of $\mathbb{N}$ such that $\omega = \lim_{j\to\infty, j\in J}\omega_j$ exists. Then $\omega \in \Lambda(t)$, so $\Lambda(t) \neq \emptyset$. Next, fix a $t \in [a,b]\backslash K$. Then no sequence $\{(x_j,t_j,\omega_j)_{j\in\mathbb{N}}$ of the kind specified in the definition of $\partial_x^> g$ exists, so $\partial_x^> g(\bar{x},t)$ is empty, that is, $\Lambda(t) = \emptyset$. This completes the proof of (2).

We now prove (3). We take a point $\bar{t} \in [a,b]$, a sequence $\{(t_j,h_j)\}_{j\in\mathbb{N}}$ of points of S_g such that $\lim_{j\to\infty} h_j = 0$, and $\lim_{j\to\infty} t_j = \bar{t}$, and show that

$$\lim_{j\to\infty}\mu_j = 0, \text{ where } \mu_j = \frac{\rho_j}{\|h_j\|},\ \rho_j = \mathrm{dist}(\chi_g^{co}(\xi_*(t_j)+h_j,t_j), \Lambda(\bar{t})\cdot h_j). \quad (5.3.2)$$

Write $x_j = \xi_*(t_j) + h_j$, $\bar{x} = \xi_*(\bar{t})$ (so that $\lim_{j\to\infty} x_j = \bar{x}$). Suppose (5.3.2) is not true. Then we can pick an infinite subset J of $\mathbb{N}$ and an $\varepsilon \in \mathbb{R}$ such that $\varepsilon > 0$

and $\mu_j \geq \varepsilon$ for all $j \in J$. Fix a $j \in J$. Then $g(x_j, t_j) > 0$. Let $\gamma_j = g(x_j, t_j)$, and use Σ_j to denote the sphere $\{h \in X : \|h\| = \|h_j\|\}$. (Recall that $h_j \neq 0$, so Σ_j is a true sphere, not reduced to a point.) For $h \in X \backslash \{0\}$, let σ_h denote the segment $\{\xi_*(t_j) + sh : 0 \leq s \leq 1\}$. It then follows from Fubini's theorem and Rademacher's theorem that the function g^{t_j} is differentiable at almost all points of σ_h (that is, $\xi_*(t_j) + sh \in D^{t_j}$ for almost all $s \in [0,1]$) for almost all $h \in \Sigma_j$. Therefore we can pick $\tilde{h}_j \in \Sigma_j$ such that, if we let $\tilde{x}_j = \xi_*(t_j) + \tilde{h}_j$, then $\|\tilde{h}_j - h_j\| \leq (2C)^{-1}\gamma_j$ and $\xi_*(t_j) + s\tilde{h}_j \in D^{t_j}$ for almost all $s \in [0,1]$. Therefore $\|\tilde{x}_j - x_j\| \leq (2C)^{-1}\gamma_j$ and $g(x_j, t_j) - g(\tilde{x}_j, t_j) \leq C\|x_j - \tilde{x}_j\| \leq \frac{\gamma_j}{2}$ from which it follows (since $g(x_j, t_j) = \gamma_j$) that $g(\tilde{x}_j, t_j) \geq \frac{\gamma_j}{2}$. Clearly,

$$g(\tilde{x}_j, t_j) = g(\xi_*(t_j), t_j) + \left(\int_0^1 \frac{\partial g}{\partial x}(\xi_*(t_j) + s\tilde{h}_j, t_j)\, ds \right) \cdot \tilde{h}_j .$$

Since $g(\xi_*(t_j), t_j) \leq 0$, and $g(\tilde{x}_j, t_j) \geq \frac{\gamma_j}{2}$, we conclude that

$$\left(\int_0^1 \frac{\partial g}{\partial x}(\xi_*(t_j) + s\tilde{h}_j, t_j)\, ds \right) \cdot \tilde{h}_j \geq \frac{\gamma_j}{2} .$$

We claim that we can pick $s_j \in [0,1]$ such that the three conditions

$$\xi_*(t_j) + s_j\tilde{h}_j \in D^{t_j}, \quad g(\xi_*(t_j) + s_j\tilde{h}_j, t_j) > 0, \quad \frac{\partial g}{\partial x}(\xi_*(t_j) + s_j\tilde{h}_j, t_j) \cdot \tilde{h}_j \geq \frac{\gamma_j}{2} \tag{5.3.3}$$

hold. To see this, let $\eta(s) = g(\xi_*(t_j) + s\tilde{h}_j, t_j) - g(\xi_*(t_j), t_j)$ for $s \in [0,1]$, so $\eta(0) = 0$, $\eta(1) > 0$, η is Lipschitz, and $\dot{\eta}(s) = \frac{\partial g}{\partial x}(\xi_*(t_j) + s\tilde{h}_j, t_j) \cdot \tilde{h}_j$ for almost all $s \in [0,1]$. Let $\tau = \sup\{s \in [0,1] : \eta(s) \leq 0\}$. Then $\tau < 1$, $\eta(\tau) = 0$, $\eta(1) \geq \frac{\gamma_j}{2}$, and $\eta(s) > 0$ for $s > \tau$. Therefore, there exists an s such that $\tau < s < 1$, $\xi_*(t_j) + s\tilde{h}_j \in D^{t_j}$, and $\dot{\eta}(s) \geq \frac{\gamma_j}{2}$ (because if such an s did not exist it would follow that $\dot{\eta}(s) < \frac{\gamma_j}{2}$ for all $s \in]\tau, 1[$ such that $\xi_*(t_j) + s\tilde{h}_j \in D^{t_j}$, i.e., that $\dot{\eta}(s) < \frac{\gamma_j}{2}$ for almost all $s \in [\tau, 1]$, and then $\int_\tau^1 \dot{\eta}(s)\, ds < \frac{\gamma_j}{2}$, so $\eta(1) - \eta(\tau) < \frac{\gamma_j}{2}$, contradicting the fact that $\eta(\tau) = 0$ and $\eta(1) \geq \frac{\gamma_j}{2}$). This s is our desired s_j, and the claim is proved.

Now let $\hat{h}_j = s_j\tilde{h}_j$,k $\hat{x}_j = \xi_*(t_j) + \hat{h}_j$, $\omega_j = \frac{\partial g}{\partial x}(\hat{x}_j, t_j)$. Then the sequence $\{\omega_j\}_{j \in J_3}$ is bounded (because $\|\omega_j\| \leq C$) so we may find an infinite subset J' of J such that $\omega = \lim_{j \to \infty, j \in J'} \omega_j$ exists. It then follows from the definition of Λ that $\omega \in \Lambda(\bar{t})$. Then, if $j \in J'$, we have

$$\omega \cdot h_j = (\omega - \omega_j) \cdot h_j + \omega_j \cdot (h_j - \tilde{h}_j) + \omega_j \cdot \tilde{h}_j. \tag{5.3.4}$$

It follows from (5.3.3) that $\omega_j \cdot \tilde{h}_j \geq \frac{\gamma_j}{2}$, while on the other hand we also have $|\omega_j \cdot (h_j - \tilde{h}_j)| \leq \frac{\gamma_j}{2}$, since $\|\tilde{h}_j - h_j\| \leq (2C)^{-1}\gamma_j$ and $\|\omega_j\| \leq C$. Then (5.3.4) allows us to conclude that

$$\omega \cdot h_j \geq -\|\omega - \omega_j\| \cdot \|h_j\| . \tag{5.3.5}$$

Since $\chi_g^{co}(\xi_*(t_j) + h_j, t_j) = [0, +\infty[$, and $\omega \cdot h_j$ belongs to $\Lambda(\bar{t}) \cdot h_j$, (5.3.5) implies that the distance ρ_j between the sets $\Lambda(\bar{t}) \cdot h_j$ and $\chi_g^{co}(\xi_*(t_j) + h_j, t_j)$ is not greater than $\|\omega - \omega_j\| \cdot \|h_j\|$. Hence $\mu_j \leq \|\omega - \omega_j\|$. Therefore $\lim_{j\to\infty, j\in J'} \mu_j = 0$. But this contradicts the facts that $J' \subseteq J$ and $\mu_j \geq \varepsilon$ for all $j \in J$. This contradiction concludes our proof. □

6 Discontinuous Vector Fields

In this section we will study classes of discontinuous vector fields f that have good properties, such as local existence of trajectories, local Cellina approximability of flow maps, and differentiability of the flow maps $(t, s, x) \mapsto \Phi^f(t, s, x)$ at points $(\bar{t}, \bar{t}, \bar{x})$. (The *flow map* of a ppd time-varying vector field was defined in Sect. 2.1, p. 226.)

These classes have already been studied in great detail in [24], so here we will just limit ourselves to presenting the relevant definitions, referring the reader to [24] for the proofs.

6.1 Co-Integrably Bounded Integrally Continuous Maps

The goal of this subsection is to define (i) the class of "co-IBIC" time-varying maps $K \ni (x, t) \mapsto f(x, t) \in Y$, where X, Y are FDNRLSs and K is a compact subset of $X \times \mathbb{R}$, and (ii) the lower semicontinuous analogue of the co-IBIC condition – called "co-ILBILSC," – in the case when $Y = \mathbb{R}$. (The two abbreviations "co-IBIC" and "co-ILBILSC" stand, respectively, for "co-integrably bounded and integrally continuous" and "co-integrably lower bounded and integrally lower semicontinuous.")

The co-IBIC class will be interesting when $Y = X$, i.e., when f is a time-varying vector field on X. Roughly speaking the co-IBIC condition is the minimum requirement that has to be satisfied so that local existence of trajectories can be proved using the Schauder fixed point theorem. For a time-varying vector field $f : X \times \mathbb{R} \mapsto X$, and an initial condition $(\bar{t}, \bar{x})$, one would like to prove existence of a trajectory ξ of f, defined on some interval $[\bar{t} - \varepsilon, \bar{t} + \varepsilon]$, by finding a fixed point of the map

$$\Xi_{\bar{t},\varepsilon,\bar{x}} \ni \xi \mapsto \mathcal{I}(\xi) \in \mathcal{Z} \quad \text{such that} \quad \mathcal{I}(\xi)(t) = \bar{x} + \int_{\bar{t}}^{t} f(\xi(s), s)\, ds\,,$$

where $\mathcal{Z} = C^0([\bar{t}-\varepsilon, \bar{t}+\varepsilon], X)$, and $\Xi_{\bar{t},\varepsilon,\bar{x}}$ is the set of all $\xi \in C^0([\bar{t}-\varepsilon, \bar{t}+\varepsilon], X)$ for which $\xi(\bar{t}) = \bar{x}$. To guarantee the existence of a fixed point, one needs $\mathcal{I}$ to map $\Xi_{\bar{t},\varepsilon,\bar{x}}$ continuously into a compact convex subset of $\Xi_{\bar{t},\varepsilon,\bar{x}}$.

Traditionally, this is done – if, for example, f is continuous with respect to x for each t and measurable with respect to t for each x – by assuming that a bound $\|f(x, t)\| \leq k(t)$ is satisfied for all x, t, where the function $k : \mathbb{R} \mapsto [0, +\infty]$ is locally integrable. (Naturally, it suffices to assume that a function k_J exists for every compact subset J of X.) In that case, the functions $\mathcal{I}(\xi)$, for $\xi \in \Xi_{\bar{t},\varepsilon,\bar{x}}$, are absolutely continuous with derivatives $\dot{\xi}(t)$ bounded in norm by $k(t)$, and the Ascoli–Arzelà theorem guarantees the desired compactness, while the continuity of the map follows from the Lebesgue dominated convergence theorem.

Here we will consider a much large class of time-varying vector fields, and in particular we will not require that $f(x,t)$ be continuous with respect to x. The main condition is going to be the continuity of the map $\mathcal{I}$. We will still want to assume the existence of the integral bounds k, and the continuity of the integral map will only be assumed on the set of absolutely continuous arcs ξ whose derivatives are bounded by the same function k. That is, we will single out, for each compact subset S of $X \times \mathbb{R}$, the set $\mathrm{Arc}\,(S)$ of all arcs $\xi : I \mapsto X$, defined on a ξ-dependent compact interval I, and such that $(\xi(t),t) \in S$ for all $t \in I$, and the subset $\mathrm{Arc}_k(S)$ of $\mathrm{Arc}\,(S)$ consisting of all absolutely continuous $\xi \in \mathrm{Arc}\,(S)$ such that $\|\dot{\xi}(t)\| \leq k(t)$ for almost all t. This leads us to the concept of "co-IBIC" time-varying ppd vector fields, that is, maps $f : X \times \mathbb{R} \hookrightarrow X$ such that, on a given compact subset S of $X \times \mathbb{R}$, satisfy a bound $\|f(x,t)\| \leq k(t)$ and also give rise to a continuous integral map $\mathcal{I}$ on $\mathrm{Arc}_{k'}(S)$, *with the integrable functions k and k' equal to each other.*

Finally, we point out that, for the integral map to be continuous, an obvious prerequisite is that it be well defined. If $\xi \in \mathrm{Arc}\,(S)$, and $\mathrm{Do}(\xi) = I$, then of course the map $I \ni t \mapsto f(\xi(t),t)$ will be bounded by an integrable function of t as long as f satisfies a bound $\|f(x,t)\| \leq k(t)$. But in addition we have to make sure that the map is measurable, and this will require that f be measurable with respect to (x,t) in some appropriate sense. This is why our discussion will begin with the definition of "essential Borel×Lebesgue measurability."

Measurability Conditions

If X is a FDNRLS, we use $\mathcal{B}o(X)$, $\mathcal{L}eb(X)$, $\mathcal{BL}eb(X,\mathbb{R})$, to denote, respectively, the Borel and Lebesgue σ-algebras of subsets of X, and the product σ-algebra $\mathcal{B}o(X) \otimes \mathcal{L}eb(\mathbb{R})$. We let $\mathcal{N}(X,\mathbb{R})$ denote the set of all subsets S of $X \times \mathbb{R}$ such that $\Pi_X(S)$ is a Lebesgue-null subset of $\mathbb{R}$, where Π_X is the canonical projection $X \times \mathbb{R} \ni (x,t) \to t \in \mathbb{R}$. Finally, we use $\mathcal{BL}^e(X,\mathbb{R})$ to denote the σ-algebra of subsets of $X \times \mathbb{R}$ generated by $\mathcal{BL}eb(X,\mathbb{R}) \cup \mathcal{N}(X,\mathbb{R})$. It is then clear that the relations $\mathcal{B}o(X \times \mathbb{R}) \subset \mathcal{BL}eb(X,\mathbb{R}) \subset \mathcal{BL}^e(X,\mathbb{R})$ hold, and both inclusions are strict.

Definition 6.1 *Let X, Y be FDNRLSs, let f be a ppd map from $X \times \mathbb{R}$ to Y, and let K be a compact subset of $X \times \mathbb{R}$:*

- *We say that f is* ***essentially Borel×Lebesgue measurable on** K, or* $\mathcal{BL}^e(X,\mathbb{R})$***-measurable on** K, if $K \subseteq \mathrm{Do}(f)$ and $f^{-1}(U) \cap K$ belongs to $\mathcal{BL}^e(X,\mathbb{R})$ for all open subsets U of Y.*

- *We use* $\mathcal{M}_{\mathcal{BL}^e}(X \times \mathbb{R}, K, Y)$ *to denote the set of ppd maps from* $X \times \mathbb{R}$ *to* Y *that are* $\mathcal{BL}^e(X, \mathbb{R})$*-measurable on* K.

Integrable Boundedness

Assume that X, Y are FDNRLSs, f is a ppd map from $X \times \mathbb{R}$ to Y, and K is a compact subset of $X \times \mathbb{R}$:

- *An* ***integrable bound*** *for* f *on the set* K *is an integrable function* $\mathbb{R} \ni t \to \varphi(t) \in [0, +\infty]$ *such that* $\|f(x,t)\| \leq \varphi(t)$ *for all* $(x,t) \in K$.
- *If* $Y = \mathbb{R}$, *an* ***integrable lower bound*** *for* f *on* K *is an integrable function* $\mathbb{R} \ni t \to \varphi(t) \in [0, +\infty]$ *such that* $f(x,t) \geq -\varphi(t)$ *for all* $(x,t) \in K$.
- *We call* f ***integrably bounded*** *(IB) – resp.* ***integrably lower bounded*** *(ILB) – on* K *if* f *is* $\mathcal{BL}^e(X, \mathbb{R})$*-measurable on* K *and there exists an integrable bound – resp. an integrable lower bound – for* f *on* K.
- *We write* $\mathcal{IB}(X \times \mathbb{R}, K, Y)$, $\mathcal{ILB}(X \times \mathbb{R}, K, \mathbb{R})$ *to denote, respectively, the sets of (i) all ppd maps from* $X \times \mathbb{R}$ *to* Y *that are IB on* K, *and (ii) all ppd maps from* $X \times \mathbb{R}$ *to* $\mathbb{R}$ *that are ILB on* K. □

Spaces of Arcs

If $S \subseteq X \times \mathbb{R}$, and I is a nonempty compact interval, we write $\text{Arc}\,(I, S)$ to denote the set of all curves $\xi \in C^0(I; X)$ such that $(\xi(t), t) \in S$ for all $t \in I$. If $k : \mathbb{R} \mapsto \mathbb{R}_+ \cup \{+\infty\}$ is a locally integrable function, then $\text{Arc}_k(I, S)$ will denote the set of all $\xi \in \text{Arc}\,(I, S)$ such that ξ is absolutely continuous and $\|\dot{\xi}(t)\| \leq k(t)$ for almost all $t \in I$. We then write $\text{Arc}\,(S)$, $\text{Arc}_k(S)$ to denote, respectively, the union of the sets $\text{Arc}\,(J, S)$ and the union of the $\text{Arc}_k(J, S)$, taken over all nonempty compact subintervals J of $\mathbb{R}$. It is then easy to show that

Fact 6.2 *If* X, Y *are FDNRLSs,* $K \subseteq X \times \mathbb{R}$ *is compact,* $\xi \in \text{Arc}\,(K)$, *and* f *belongs to* $\mathcal{M}_{\mathcal{BL}^e}(X \times \mathbb{R}, K, Y)$, *then the function* $\text{Do}(\xi) \ni t \mapsto f(\xi(t), t) \in Y$ *is measurable.* □

The sets $\text{Arc}\,(S)$ are metric spaces, with the distance $d(\xi, \xi')$ between two members $\xi : [a, b] \mapsto X$, $\xi' : [a', b'] \mapsto X$ of $\text{Arc}\,(S)$ defined by

$$d(\xi, \xi') = |a - a'| + |b - b'| + \sup\{\|\tilde{\xi}(t) - \tilde{\xi}'(t)\| : t \in \mathbb{R}\}$$

where, for any continuous map $\gamma : [\alpha, \beta] \mapsto X$, $\tilde{\gamma}$ denotes the extension of γ to $\mathbb{R}$ which is identically equal to $\gamma(\alpha)$ on $]-\infty, \alpha]$ and to $\gamma(\beta)$ on $[\beta, +\infty[$. Clearly, then

Fact 6.3 *If* X *is a FDNRLS and* $S \subseteq X \times \mathbb{R}$, *then:*

(1) If $\{\xi_j\}_{j \in \mathbb{N}}$ *is a sequence of members of* $\text{Arc}\,(S)$, *with domains* $[a_j, b_j]$, *and* $\xi \in \text{Arc}\,(S)$ *has domain* $[a, b]$, *then* $\{\xi_j\}_{j \in \mathbb{N}}$ *converges to* ξ *if and only if (a)* $\lim_{j \to \infty} a_j = a$, *(b)* $\lim_{j \to \infty} b_j = b$, *and (c)* $\lim_{j \to \infty} \xi_j(t_j) = \xi(t)$

whenever $\{t_j\}_{j\in\mathbb{N}}$ *is a sequence such that* $t_j \in [a_j, b_j]$ *for each* j *and* $\lim_{j\to\infty} t_j = t \in [a,b]$,

(2) If S *is compact, and* $k : \mathbb{R} \mapsto \mathbb{R}_+ \cup \{+\infty\}$ *is locally integrable, then* $\mathrm{Arc}_k(S)$ *is compact.* □

Integral Continuity

If X, Y are FDNRLSs, $K \subseteq X \times \mathbb{R}$ is compact, and $f \in \mathcal{IB}(X \times \mathbb{R}, K, Y)$, then it is convenient to define a real-valued ***integral map*** $\mathcal{I}_{f,K} : \mathrm{Arc}(K) \mapsto \mathbb{R}$, by letting $\mathcal{I}_{f,K}(\xi) = \int_{\mathrm{Do}(\xi)} f(\xi(s), s)\, ds$ for every $\xi \in \mathrm{Arc}(K)$. If $\mathcal{S} \subseteq \mathrm{Arc}(K)$, we call f ***integrally continuous*** (abbr. IC) ***on*** $\mathcal{S}$ if $\mathcal{I}_{f,K} \lceil \mathcal{S}$ is continuous. If $f \in \mathcal{ILB}(X \times \mathbb{R}, K, \mathbb{R})$, then $\mathcal{I}_{f,K}$ is still well defined as a map into $\mathbb{R} \cup \{+\infty\}$, and we call f ***integrally lower semicontinuous*** (abbr. ILSC) ***on*** $\mathcal{S}$ if $\mathcal{I}_{f,K} \lceil \mathcal{S}$ is lower semicontinuous.

We will be particularly interested in maps f that, for some integrable function k, are both integrably bounded with integral bound k and integrally continuous on $\mathrm{Arc}_k(K)$.

Definition 6.4 *If* X, Y *are FDNRLSs,* K *is a compact subset of* $X \times \mathbb{R}$*, and* $f : X \times \mathbb{R} \hookrightarrow Y$*, we call* f ***co-IBIC*** *("co-integrably bounded and integrally continuous")* ***on*** K *if* $f \in \mathcal{IB}(X \times \mathbb{R}, K, Y)$ *and there exists an integrable bound* $k : \mathbb{R} \mapsto [0, +\infty]$ *for* f *on* K *such that* f *is integrally continuous on* $\mathrm{Arc}_k(K)$*. If* $f : X \times \mathbb{R} \hookrightarrow \mathbb{R}$*, we call* f ***co-ILBILSC*** *("co-integrably bounded and integrally lower semicontinuous")* ***on*** K *if* $f \in \mathcal{ILB}(X \times \mathbb{R}, K, \mathbb{R})$ *and there exists an integrable lower bound* $k : \mathbb{R} \mapsto [0, +\infty]$ *for* f *on* K *such that* f *is integrally lower semicontinuous on* $\mathrm{Arc}_k(K)$. □

6.2 Points of Approximate Continuity

Suppose that X and Y are FDNRLSs, f is a ppd map from $X \times \mathbb{R}$ to Y, and $(\bar{x}_*, \bar{t}_*) \in X \times \mathbb{R}$. A *modulus of approximate continuity* (abbr. MAC) for f near $(\bar{x}_*, \bar{t}_*)$ is a function $]0, +\infty[\times \mathbb{R} \ni (\beta, r) \mapsto \psi(\beta, r) \in]0, +\infty]$ such that:

(MAC.1) *The function* $\mathbb{R} \ni r \mapsto \psi(\beta, r) \in]0, +\infty]$ *is measurable for each* $\beta \in]0, +\infty[$,

(MAC.2) $\lim_{(\beta,\rho)\to(0,0),\beta>0,\rho>0} \frac{1}{\rho} \int_{-\rho}^{\rho} \psi(\beta, r)\, dr = 0$,

(MAC.3) *There exist positive numbers* β_*, ρ_*, *such that:*

- (MAC.3.a) $f(x,t)$ *is defined whenever* $\|x - \bar{x}_*\| \leq \beta_*$ *and* $|t - \bar{t}_*| \leq \rho_*$,
- (MAC.3.b) *The inequality* $\|f(x,t) - f(\bar{x}_*, \bar{t}_*)\| \leq \psi(\beta, t - \bar{t}_*)$ *holds whenever* $\beta \in \mathbb{R}$, $x \in X$, $t \in \mathbb{R}$ *are such that* $\|x - \bar{x}_*\| \leq \beta \leq \beta_*$ *and* $|t - \bar{t}_*| \leq \rho_*$.

Definition 6.5 *A* ***point of approximate continuity*** *(abbr. PAC) for* f *is a point* $(\bar{x}_*, \bar{t}_*) \in X \times \mathbb{R}$ *having the property that there exists a MAC for* f *near* $(\bar{x}_*, \bar{t}_*)$. □

An important example of a class of maps with many points of approximate continuity is given by the following corollary of the well-known Scorza-Dragoni theorem.

Proposition 6.6 *Suppose* X, Y *are FDNRLSs,* Ω *is open in* X, $a, b \in \mathbb{R}$, $a < b$, *and* $f : \Omega \times [a, b] \mapsto Y$ *is such that:*

- *The partial map* $[a, b] \ni t \mapsto f(x, t) \in Y$ *is measurable for every* $x \in \Omega$
- *The partial map* $\Omega \ni x \mapsto f(x, t) \in Y$ *is continuous for every* $t \in [a, b]$
- *There exists an integrable function* $[a, b] \ni t \mapsto k(t) \in [0, +\infty]$ *such that the bound* $\|f(x, t)\| \leq k(t)$ *holds whenever* $(x, t) \in \Omega \times [a, b]$

Then there exists a subset G *of* $[a, b]$ *for which* $\text{meas}([a, b] \backslash G) = 0$, *such that every* $(\bar{x}_*, \bar{t}) \in \Omega \times G$ *is a point of approximate continuity of* f. □

Another important example of maps with many PACs is given by the following result, proved in [24].

Proposition 6.7 *Suppose that* X *and* Y *are FDNRLSs,* $a, b \in \mathbb{R}$, $a < b$, *and* $F : X \times [a, b] \mapsto\!\!\!\!\to Y$ *is an almost lower semicontinuous set-valued map with closed nonempty values such that for every compact subset* K *of* X *the function* $[a, b] \ni t \mapsto \sup\{\min\{\|y\| : y \in F(x, t)\} : x \in K\}$ *is integrable. Then there exists a subset* G *of* $[a, b]$ *such that* $\text{meas}([a, b] \backslash G) = 0$, *having the property that, whenever* $x_* \in X$, $t_* \in G$, $v_* \in F(x_*, t_*)$, *and* $K \subseteq X$ *is compact, there exists a map* $K \times [a, b] \ni (x, t) \mapsto f(x, t) \in F(x, t)$ *which is co-IBIC on* $K \times [a, b]$ *and such that* (x_*, t_*) *is a PAC of* f *and* $f(x_*, t_*) = v_*$. □

7 The Maximum Principle

We consider a *fixed time-interval optimal control problem with state space constraints,* of the form

$$\text{minimize} \quad \varphi(\xi(b)) + \textstyle\int_a^b f_0(\xi(t), \eta(t), t)\, dt$$

$$\text{subject to} \begin{cases} \xi(\cdot) \in W^{1,1}([a, b], X) \quad \text{and} \quad \dot{\xi}(t) = f(\xi(t), \eta(t), t) \ \text{a.e.}, \\ \xi(a) = \bar{x}_* \quad \text{and} \quad \xi(b) \in S\,, \\ g_i(\xi(t), t) \leq 0 \ \text{ for } t \in [a, b]\,, \ i = 1, \ldots, m\,, \\ h_j(\xi(b)) = 0 \ \text{ for } j = 1, \ldots, \tilde{m}\,, \\ \eta(t) \in U \ \text{ for all } t \in [a, b] \quad \text{and} \quad \eta(\cdot) \in \mathcal{U}\,, \end{cases}$$

and a *reference trajectory-control pair* (ξ_*, η_*).

The Technical Hypotheses

We will make the assumption that the data 14-tuple $\mathcal{D} = (X, m, \tilde{m}, U, a, b, \varphi, f_0, f, \bar{x}_*, \mathbf{g}, \mathbf{h}, S, \mathcal{U})$ satisfies:

(H1) *X is a normed finite-dimensional real linear space, $\bar{x}_* \in X$, and $m, \tilde{m}$ are nonnegative integers*
(H2) *U is a set, $a, b \in \mathbb{R}$ and $a < b$*
(H3) *f_0, f are ppd functions from $X \times U \times \mathbb{R}$ to $\mathbb{R}$, X, respectively*
(H4) *$\mathbf{g} = (g_1, \ldots, g_m)$ is an m-tuple of ppd functions from $X \times \mathbb{R}$ to $\mathbb{R}$*
(H5) *$\mathbf{h} = (h_1, \ldots, h_{\tilde{m}})$ is an $\tilde{m}$-tuple of ppd functions from X to $\mathbb{R}$*
(H6) *φ is a ppd function from X to $\mathbb{R}$*
(H7) *S is a subset of X*
(H8) *$\mathcal{U}$ is a set of ppd functions from $\mathbb{R}$ to U such that the domain of every $\eta \in \mathcal{U}$ is a nonempty compact interval*

Given such a $\mathcal{D}$, a *controller* is a ppd function $\eta : \mathbb{R} \hookrightarrow U$ whose domain is a nonempty compact interval. (Hence (H8) says that $\mathcal{U}$ is a set of controllers.) An *admissible controller* is a member of $\mathcal{U}$. If $\alpha, \beta \in \mathbb{R}$ and $\alpha \leq \beta$, then we use $W^{1,1}([\alpha, \beta], X)$ to denote the space of all absolutely continuous maps $\xi : [\alpha, \beta] \mapsto X$. A *trajectory* for a controller $\eta : [\alpha, \beta] \mapsto U$ is a map $\xi \in W^{1,1}([\alpha, \beta], X)$ such that, for almost every $t \in [\alpha, \beta]$, $(\xi(t), \eta(t), t)$ belongs to $\mathrm{Do}(f)$ and $\dot{\xi}(t) = f(\xi(t), \eta(t), t)$. A *trajectory-control pair* (abbr. TCP) is a pair (ξ, η) such that η is a controller and ξ is a trajectory for η. The *domain* of a TCP (ξ, η) is the domain of η, which is, by definition, the same as the domain of ξ. A TCP (ξ, η) is *admissible* if $\eta \in \mathcal{U}$.

A TCP (ξ, η) with domain $[\alpha, \beta]$ is *cost-admissible* if:

- *(ξ, η) is admissible*
- *The function $[\alpha, \beta] \ni t \mapsto f_0(\xi(t), \eta(t), t)$ is a.e. defined, measurable, and such that $\int_\alpha^\beta \min\Big(0, f_0(\xi(t), \eta(t), t)\Big)\, dt > -\infty$*
- *The terminal point $\xi(\beta)$ belongs to the domain of φ*

It follows that if (ξ, η) is cost-admissible then the number

$$J(\xi, \eta) = \varphi(\xi(\beta)) + \int_\alpha^\beta f_0(\xi(t), \eta(t), t)\, dt$$

– called the *cost* of (ξ, η) – is well defined and belongs to $]-\infty, +\infty]$.

A TCP (ξ, η) with domain $[\alpha, \beta]$ is *constraint-admissible* if it satisfies all our state space constraints, that is, if:

(CA1) $\xi(\alpha) = \bar{x}_*$
(CA2) $(\xi(t), t) \in \mathrm{Do}(g_i)$ *and* $g_i(\xi(t), t) \leq 0$ *if* $t \in [\alpha, \beta]$ *and* $i \in \{1, \ldots, m\}$
(CA3) $\xi(\beta) \in S \cap \Big(\cap_{j=1}^{\tilde{m}} \mathrm{Do}(h_j)\Big)$ *and* $h_j(\xi(\beta)) = 0$ *for* $j = 1, \ldots, \tilde{m}$

For the data tuple $\mathcal{D}$, we use $ADM(\mathcal{D})$ to denote the set of all cost-admissible, constraint-admissible TCPs (ξ, η), and $ADM_{[a,b]}(\mathcal{D})$ to denote the set of all $(\xi, \eta) \in ADM(\mathcal{D})$ whose domain is $[a, b]$.

The hypothesis on the reference TCP (ξ_*, η_*) is that it is a cost-minimizer in $ADM_{[a,b]}(\mathcal{D})$, i.e., an admissible, cost admissible, constraint-admissible TCP with domain $[a, b]$ that minimizes the cost in the class of all admissible, cost-admissible, constraint-admissible, TCP's with domain $[a, b]$. That is:

(H9) *The pair* (ξ_*, η_*) *satisfies* $(\xi_*, \eta_*) \in ADM_{[a,b]}(\mathcal{D})$, $J(\xi_*, \eta_*) < +\infty$, *and* $J(\xi_*, \eta_*) \leq J(\xi, \eta)$ *for all pairs* $(\xi, \eta) \in ADM_{[a,b]}(\mathcal{D})$.

To the data $\mathcal{D}$, ξ_*, η_* as above, we associate the *cost-augmented dynamics* $\mathbf{f} : X \times U \times \mathbb{R} \hookrightarrow \mathbb{R} \times X$ defined by

$$\mathrm{Do}(\mathbf{f}) = \mathrm{Do}(f_0) \cap \mathrm{Do}(f), \text{ and } \mathbf{f}(z) = (f_0(z), f(z)) \text{ for } z = (x, u, t) \in \mathrm{Do}(\mathbf{f}).$$

We also define the *epi-augmented dynamics* $\check{\mathbf{f}} : X \times U \times \mathbb{R} \mapsto\!\!\!\rightarrow \mathbb{R} \times X$, given, for each $z = (x, u, t) \in X \times U \times \mathbb{R}$, by

$$\check{\mathbf{f}}(z) = [f_0(z), +\infty[\times\{f(z)\} \text{ if } z \in \mathrm{Do}(\mathbf{f}), \quad \check{\mathbf{f}}(z) = \emptyset \text{ if } z \notin \mathrm{Do}(\mathbf{f}).$$

We will also use the constraint indicator maps $\chi^{co}_{g_i} : X \times \mathbb{R} \mapsto\!\!\!\rightarrow \mathbb{R}$, for $i = 1, \ldots, m$, and the epimap $\check{\varphi} : X \mapsto\!\!\!\rightarrow \mathbb{R}$. (These two notions were defined in Sect. 2.1.)

For $i \in \{1, \ldots, m\}$, we let

$$\begin{array}{l} \sigma^{\mathbf{f}}_*(t) = \mathbf{f}(\xi_*(t), \eta_*(t), t) \text{ and } \sigma^{g_i}_*(t) = g_i(\xi_*(t), t) \quad \text{if } t \in [a, b], \\ Av_{g_i} = \{(x, t) \in X \times [a, b] : g_i(x, t) > 0\} \end{array} .$$

(so the Av_{g_i} are the "sets to be avoided"). We then define K_i to be the set of all $t \in [a, b]$ such that $(\xi_*(t), t)$ belongs to the closure of Av_{g_i}. Then K_i is obviously a compact subset of $[a, b]$.

We now make technical hypotheses on $\mathcal{D}$, ξ_*, η_*, and five new objects called $\Lambda^{\mathbf{f}}$, $\Lambda^{\mathbf{g}}$, $\Lambda^{\mathbf{h}}$, Λ^{φ}, and C. To state these hypotheses, we let $\mathcal{U}_{c,[a,b]}$ denote the set of all constant U-valued functions defined on $[a, b]$, and define $\mathcal{U}_{c,[a,b],*} = \mathcal{U}_{c,[a,b]} \cup \{\eta_*\}$. The technical hypotheses are then as follows:

(H10) *For each* $\eta \in \mathcal{U}_{c,[a,b],*}$. *there exist a positive number* δ_η *such that*
- (H10.a) $\mathbf{f}(x, \eta(t), t)$ *is defined for all* (x, t) *in the tube* $\mathcal{T}^X(\xi_*, \delta_\eta)$
- (H10.b) *The time-varying vector field* $\mathcal{T}^X(\xi_*, \delta_\eta) \ni (x, t) \mapsto f(x, \eta(t), t)$ *is co-IBIC on* $\mathcal{T}^X(\xi_*, \delta_\eta)$
- (H10.c) *The time-varying function* $\mathcal{T}^X(\xi_*, \delta_\eta) \ni (x, t) \mapsto f_0(x, \eta(t), t) \in \mathbb{R}$ *is co-ILBILSC on* $\mathcal{T}^X(\xi_*, \delta_\eta)$

(H11) *The number* δ_{η_*} *can be chosen so that (i) each function* g_i *is defined on* $\mathcal{T}^X(\xi_*, \delta_{\eta_*})$, *and (ii) for each* $i \in \{1, \ldots, m\}$, $t \in [a, b]$, *the set* $\{x \in X : g_i(x, t) > 0, \|x - \xi_*(t)\| \leq \delta_{\eta_*}\}$ *is relatively open in the ball* $\{x \in X : \|x - \xi_*(t)\| \leq \delta_{\eta_*}\}$

(H12) $\Lambda^{\mathbf{f}}$ *is a measurable integrably bounded set-valued map from* $[a, b]$ *to* $X^\dagger \times \boldsymbol{L}(X)$ *with compact convex values such that* $\Lambda^{\mathbf{f}}$ *belongs to* $VG^{L^1,ft}_{GDQ}(\check{\mathbf{f}}, [a, b], \xi_*, \sigma^{\mathbf{f}}_*, X \times \mathbb{R})$

(H13) $\Lambda^{\mathbf{g}}$ *is an* $\hat{m}$*-tuple* $(\Lambda^{g_1}, \ldots, \Lambda^{g_{\hat{m}}})$ *such that,. for each* $i \in \{1, \ldots, \hat{m}\}$, Λ^{g_i} *is an upper semicontinuous set-valued map from* $[a, b]$ *to* $X^\dagger$ *with compact convex values, such that* $\Lambda^{g_i} \in VG^{pw,rob}_{GDQ}(\chi^{co}_{g_i}, \xi_*, \sigma^{g_i}_*, Av_{g_i})$

(H14) $\Lambda^{\mathbf{h}}$ *is a generalized differential quotient of* $\mathbf{h}$ *at* $(\xi_*(b), \mathbf{h}(\xi_*(b)))$ *in the direction of* X

(H15) *Λ^{φ} is a generalized differential quotient of the epifunction $\check{\varphi}$ at the point $(\xi_*(b), \varphi(\xi_*(b)))$ in the direction of X*

(H16) *C is a limiting Boltyanskii approximating cone of S at $\xi_*(b)$*

Our last hypothesis will require the concept of an *equal-time interval-variational neighborhood* (abbr. ETIVN) of a controller η. We say that a set $\mathcal{V}$ of controllers is an ETIVN of a controller η if:

- *For every $n \in \mathbb{Z}_+$ and every n-tuple $\mathbf{u} = (u_1, \ldots, u_n)$ of members of U, there exists a positive number $\varepsilon = \varepsilon(n, \mathbf{u})$ such that whenever $\eta' : \mathrm{Do}(\eta) \mapsto U$ is a map obtained from η by first selecting an n-tuple $\mathbf{I} = (I_1, \ldots, I_n)$ of pairwise disjoint subintervals of $\mathrm{Do}(\eta)$ with the property that $\sum_{j=1}^{n} \mathrm{meas}(I_j) \leq \varepsilon$, and then substituting the constant value u_j for the value $\eta(t)$ for every $t \in I_j$, $j = 1, \ldots, n$, it follows that $\eta' \in \mathcal{U}$.*

We will then assume:

(H17) *The class $\mathcal{U}$ is an equal-time interval-variational neighborhood of η_*.*

We define the *Hamiltonian* to be the function $H_\alpha : X \times U \times X^{\dagger} \times \mathbb{R} \hookrightarrow \mathbb{R}$ given by $H_\alpha(x, u, p, t) = p \cdot f(x, u, t) - \alpha f_0(x, u, t)$, so H_α depends on the real parameter α.

The Main Theorem

The following is our version of the maximum principle.

Theorem 7.1 *Assume that the data $\mathcal{D}$, ξ_*, η_*, $\Lambda^{\mathbf{f}}$, $\Lambda^{\mathbf{g}}$, $\Lambda^{\mathbf{h}}$, Λ^{φ}, C satisfy Hypotheses (H1) to (H17). Let I be the set of those indices $i \in \{1, \ldots, m\}$ such that K_i is nonempty. Then there exist:*

1. *A covector $\bar{\pi} \in X^{\dagger}$, a nonnegative real number π_0, and an $\tilde{m}$-tuple $\boldsymbol{\lambda} = (\lambda_1, \ldots, \lambda_{\tilde{m}})$ of real numbers*
2. *A measurable selection $[a, b] \ni t \mapsto (L_0(t), L(t)) \in X^{\dagger} \times \boldsymbol{L}(X)$ of the set-valued map $\Lambda^{\mathbf{f}}$*
3. *A family $\{\nu_i\}_{i \in I}$ of nonnegative additive measures $\nu_i \in bvadd([a, b], \mathbb{R})$ such that $\mathrm{support}(\nu_i) \subseteq |\Lambda_i|$ for every $i \in I$*
4. *A family $\{\gamma_i\}_{i \in I}$ of pairs $\gamma_i = (\gamma_i^-, \gamma_i^+)$ such that $\gamma_i^- : |\Lambda^{g_i}| \mapsto X^{\dagger}$ and $\gamma_i^+ : |\Lambda^{g_i}| \mapsto X^{\dagger}$ are measurable selections of Λ^{g_i}, and $\gamma_i^-(t) = \gamma_i^+(t)$ for all t in the complement of a finite or countable set*
5. *A member $L^{\mathbf{h}} = (L^{h_1}, \ldots, L^{h_{\tilde{m}}}) \in (X^{\dagger})^{\tilde{m}}$ of $\Lambda^{\mathbf{h}}$ and a member L^{φ} of Λ^{φ}*

having the property that, if we let $\pi : [a, b] \mapsto X^{\dagger}$ be the unique solution of the adjoint Cauchy problem

$$\begin{cases} d\pi(t) = (-\pi(t) \cdot L(t) + \pi_0 L_0(t))dt + \sum_{i \in I} d\mu_i(t) \\ \pi(b) = \bar{\pi} - \sum_{j=1}^{\tilde{m}} \lambda_j L_j^{\mathbf{h}} - \pi_0 L^{\varphi} \end{cases}$$

(where $\mu_i \in bvadd(\Lambda^{g_i})$ is the finitely additive $X^{\dagger}$-valued measure such that $d\mu_i = \gamma_i \cdot d\nu_i$, defined in Page 230), then the following conditions are true:

I. the ***Hamiltonian maximization condition:*** *the inequality*

$$H_{\pi_0}(\xi_*(\bar{t}), \eta_*(\bar{t}), \pi(\bar{t})) \geq H_{\pi_0}(\xi_*(\bar{t}), u, \pi(\bar{t}))$$

holds whenever $u \in U$, $\bar{t} \in [a,b]$ *are such that* $(\xi_*(\bar{t}), \bar{t})$ *is a point of approximate continuity of both augmented vector fields* $(x,t) \mapsto \mathbf{f}(x,u,t)$ *and* $(x,t) \mapsto \mathbf{f}(x, \eta_*(t), t)$,

II. the ***transversality condition:*** $-\bar{\pi} \in C^{\dagger}$,

III. The ***nontriviality condition:*** $\|\bar{\pi}\| + \pi_0 + \sum_{j=1}^{\tilde{m}} |\lambda_j| + \sum_{i \in I} \|\nu_i\| > 0$.

Remark 7.2 The adjoint equation satisfied by π can be written in integral form, incorporating the terminal condition at b. The result is the formula

$$\pi(t) = \bar{\pi} - \sum_{j=1}^{\tilde{m}} \lambda_j L_j^{\mathbf{h}} - \pi_0 L^{\varphi} + \int_t^b \Big(\pi(s) \cdot L(s) - \pi_0 L_0(s)\Big) ds - \sum_{i \in I} \int_{[t,b]} \gamma^i(s) d\nu_i(s),$$

from which it follows, in particular, that $\pi(b) = \bar{\pi} - \sum_{j=1}^{\tilde{m}} \lambda_j L_j^{\mathbf{h}} - \pi_0 L^{\varphi}$. □

Remark 7.3 The adjoint covector π can also be expressed using (2.3.3). This yields $\pi(t) = \pi(b) - \int_t^b M_L(s,t)^{\dagger} \Big(\pi_0 L_0(s)\, ds + \sum_{i \in I} d(\gamma^i \cdot \nu_i)(s)\Big)$, where $\pi(b) = \bar{\pi} - \sum_{j=1}^{\tilde{m}} \lambda_j L_j^{\mathbf{h}} - \pi_0 L^{\varphi}$, and M_L is the fundamental solution of the equation $\dot{M} = L \cdot M$. □

References

1. Berkovitz, L. D., *Optimal Control Theory.* Springer-Verlag, New York, 1974.
2. Browder, F. E., "On the fixed point index for continuous mappings of locally connected spaces." *Summa Brazil. Math.* **4**, 1960, pp. 253–293.
3. Cellina, A., "A theorem on the approximation of compact multivalued mappings." *Atti Accad. Naz. Lincei Rend. Cl. Sci. Fis. Mat. Natur.* (8) **47** (1969), pp. 429–433.
4. Cellina, A., "A further result on the approximation of set-valued mappings." *Atti Accad. Naz. Lincei Rend. Cl. Sci. Fis. Mat. Natur.* (8) **48** (1970), pp. 412–416.
5. Cellina, A., "Fixed points of noncontinuous mappings." *Atti Accad. Naz. Lincei Rend. Cl. Sci. Fis. Mat. Natur.* (8) **49** (1970), pp. 30–33.
6. Cellina, A., "The role of approximation in the theory of multivalued mappings." In *1971 Differential Games and Related Topics (Proc. Internat. Summer School, Varenna, 1970)*, North-Holland, Amsterdam, 1970, pp. 209–220.
7. Clarke, F. H., "The Maximum Principle under minimal hypotheses." *SIAM J. Control Optim.* **14**, 1976, pp. 1078–1091.
8. Clarke, F. H., *Optimization and Nonsmooth Analysis.* Wiley Interscience, New York, 1983.
9. Conway, J. B., *A Course in Functional Analysis.* Grad. Texts in Math. **96**, Springer-Verlag, New York, 1990.
10. Fan, K., "Fixed point and minimax theorems in locally convex topological linear spaces." *Proc. Nat. Acad. Sci. U.S.* **38**, 1952, pp. 121–126.

11. Glicksberg, I. L., "A further generalization of the Kakutani fixed point theorem, with application to Nash equilibrium points." *Proc. Amer. Math. Soc.* **3**, 1952, pp. 170–174.
12. Halkin, H., "Necessary conditions for optimal control problems with differentiable or nondifferentiable data." In *Mathematical Control Theory*, Lect. Notes in Math. **680**, Springer-Verlag, Berlin, 1978, pp. 77–118.
13. Kakutani, S., "A generalization of Brouwer's fixed point theorem." *Duke Math. J.* **7**, 1941, pp. 457–459.
14. Leray, J. and J. Schauder, "Topologie et équations fonctionnelles." *Ann. Sci. École Norm. Sup.* **51**, 1934, pp. 45–78.
15. Pontryagin, L. S., V. G. Boltyanskii, R. V. Gamkrelidze, and E. F. Mischenko, *The Mathematical Theory of Optimal Processes.* Wiley, New York, 1962.
16. Sussmann, H. J., "A strong version of the Łojasiewicz maximum principle." In *Optimal Control of Differential Equations,* N. H. Pavel Ed., Marcel Dekker Inc., New York, 1994, pp. 293–309.
17. Sussmann, H. J., *An introduction to the coordinate-free maximum principle.* In *Geometry of Feedback and Optimal Control,* B. Jakubczyk and W. Respondek Eds., M. Dekker, Inc., New York, 1997, pp. 463–557.
18. Sussmann, H. J., "Some recent results on the maximum principle of optimal control theory." In *Systems and Control in the Twenty-First Century,* C. I. Byrnes, B. N. Datta, D. S. Gilliam and C. F. Martin Eds., Birkhäuser, Boston, 1997, pp. 351–372.
19. Sussmann, H. J., "Multidifferential calculus: chain rule, open mapping and transversal intersection theorems." In *Optimal Control: Theory, Algorithms, and Applications,* W. W. Hager and P. M. Pardalos Eds., Kluwer, 1998, pp. 436–487.
20. Sussmann, H. J., "A maximum principle for hybrid optimal control problems." In *Proc. 38th IEEE Conf. Decision and Control, Phoenix, AZ, Dec. 1999.* IEEE publications, New York, 1999, pp. 425–430.
21. Sussmann, H. J., "Résultats récents sur les courbes optimales." In *15e Journée Annuelle de la Société Mathémathique de France (SMF),* Publications de la SMF, Paris, 2000, pp. 1–52.
22. Sussmann, H. J., "New theories of set-valued differentials and new versions of the maximum principle of optimal control theory." In *Nonlinear Control in the year 2000,* A. Isidori, F. Lamnabhi-Lagarrigue and W. Respondek Eds., Springer-Verlag, London, 2000, pp. 487–526.
23. Sussmann, H. J., "Set-valued differentials and the hybrid maximum principle." In *Proc. 39th IEEE Conf. Decision and Control, Sydney, Australia, Dec. 12–15, 2000,* IEEE publications, New York, 2000.
24. Sussmann, H. J., "Needle variations and almost lower semicontinuous differential inclusions." *Set-valued analysis,* Vol. 10, Issue 2–3, June–September 2002, pp. 233–285.
25. Sussmann, H. J., "Combining high-order necessary conditions for optimality with nonsmoothness." In *Proceedings of the 43rd IEEE 2004 Conference on Decision and Control (Paradise Island, the Bahamas, December 14-17, 2004),* IEEE Publications, New York, 2004.
26. Warga, J., "Fat homeomorphisms and unbounded derivate containers." *J. Math. Anal. Appl.* **81**, 1981, pp. 545–560.

27. Warga, J., "Controllability, extremality and abnormality in nonsmooth optimal control." *J. Optim. Theory Applic.* **41**, 1983, pp. 239–260.
28. Warga, J., "Optimization and controllability without differentiability assumptions." *SIAM J. Control and Optimization* **21**, 1983, pp. 837–855.
29. Warga, J., "Homeomorphisms and local C^1 approximations." *Nonlinear Anal. TMA* **12**, 1988, pp. 593–597.

Sliding Mode Control: Mathematical Tools, Design and Applications

V.I. Utkin

Department of Electrical Engineering, 205 Dreese Laboratory, The Ohio State University, 2015 Neil Avenue, Columbus, OH 43210, USA
utkin@ee.eng.ohio-state.edu

1 Introduction

The sliding mode control approach is recognized as one of the efficient tools to design robust controllers for complex high-order nonlinear dynamic plant operating under uncertainty conditions. The research in this area were initiated in the former Soviet Union about 40 years ago, and then the sliding mode control methodology has been receiving much more attention from the international control community within the last two decades.

The major advantage of sliding mode is low sensitivity to plant parameter variations and disturbances which eliminates the necessity of exact modeling. Sliding mode control enables the decoupling of the overall system motion into independent partial components of lower dimension and, as a result, reduces the complexity of feedback design. Sliding mode control implies that control actions are discontinuous state functions which may easily be implemented by conventional power converters with "on-off" as the only admissible operation mode. Due to these properties the intensity of the research at many scientific centers of industry and universities is maintained at high level, and sliding mode control has been proved to be applicable to a wide range of problems in robotics, electric drives and generators, process control, vehicle and motion control.

2 Examples of Dynamic Systems with Sliding Modes

Sliding modes as a phenomenon may appear in a dynamic system governed by ordinary differential equations with discontinuous right hand sides.

The term "*sliding mode*" first appeared in the context of relay systems. It may happen that the control as a function of the system state switches at high (theoretically infinite) frequency and this motion is referred to as "*sliding*

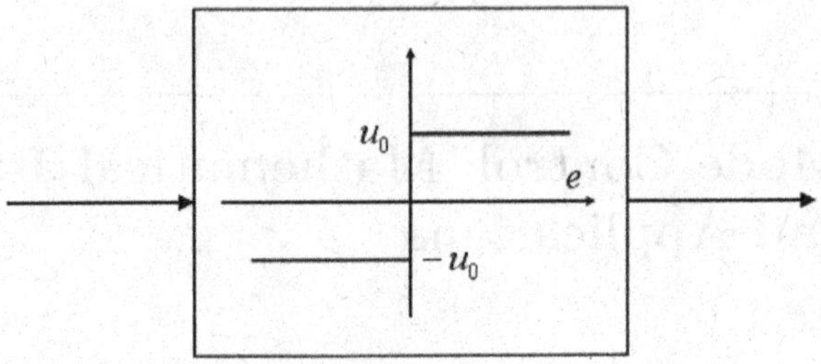

Fig. 1.

mode." This motion may be enforced in the simplest first order tracking relay system with the state variable $x(t)$

$$\dot{x} = f(x) + u$$

with the bounded function

$$f(x)\ , \quad |f(x)| < f_0 = const$$

and the control as a relay function (Fig. 1) of the tracking error $e = r(t) - x$ and $r(t)$ being the reference input,

$$u = \begin{cases} u_0 & \text{if} \quad e > 0 \\ -u_0 & \text{if} \quad e < 0 \end{cases} \qquad \text{or} \qquad u = u_0 \ sign(e), \quad u_0 = const.$$

The values of e and $\frac{de}{dt} = \dot{e} = \dot{r} - f(x) - u_0 \ sign(e)$ have different signs if $u_0 > f_0 + |\dot{r}|$. It means that the magnitude of the tracking error decays at a finite rate and the error is equal to zero identically after a finite time interval T (Fig. 2). The argument of the control function, e is equal to zero which is the discontinuity point. For any real-life implementation due to imperfections in switching device the control switches at high frequency or takes intermediate values for continuous approximation of the relay function. The motion for $t > T$ is called "*sliding mode.*"

The conventional example to demonstrate sliding modes in terms of the state space method is a second-order time-invariant relay system with relay control input

$$\begin{aligned} &\ddot{x} + a_2\dot{x} + a_1 x = u, \\ &u = -M sign(s), \qquad s = cx + \dot{x}, \qquad a_1, a_2, M, c \ \text{-} \ const \end{aligned} \tag{2.1}$$

The system behavior may be analyzed in the state plane $(x, \dot{x})$. The state plane in Fig. 3 is shown for $a_1 = a_2 = 0$. The control u undergoes discontinuities at the switching line $s = 0$ and the state trajectories are constituted

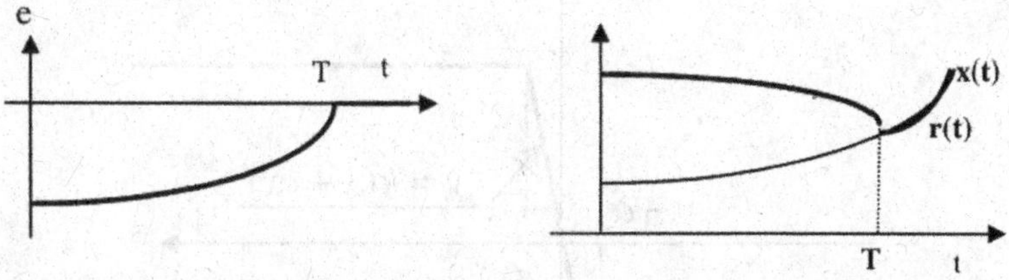

Fig. 2.

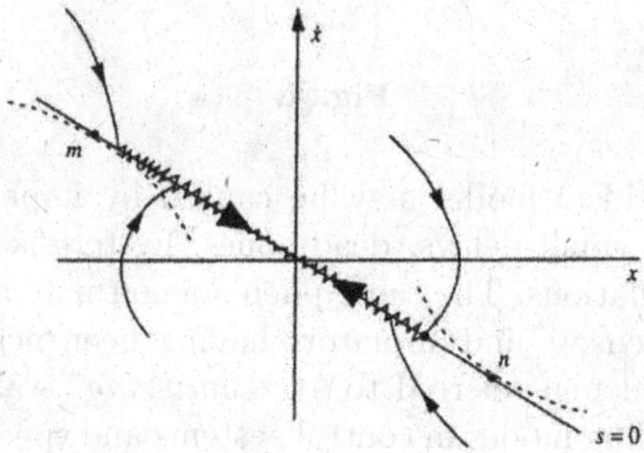

Fig. 3.

by two families: the first family corresponds to $s > 0$ and $u = -M$ (upper semiplane), the second to $s < 0$ and $u = M$ (lower semiplane). Within the sector $m - n$ on the switching line the state trajectories are oriented towards the line. Having reached the sector at some time t_1, the state can not leave the switching line. This means that the state trajectory will belong to the switching line for $t > t_1$. This motion with state trajectories in the switching line is *sliding mode* as well. Since in the course of sliding mode, the state trajectory coincides with the switching line $s = 0$, its equation may be interpreted as the motion equation, i.e.

$$\dot{x} + cx = 0. \tag{2.2}$$

It is important that its solution

$$x(t) = x(t_1)e^{-c(t-t_1)}$$

depends neither on the plant parameters nor the disturbance. This so-called "*invariance*" property looks promising for designing feedback control for the dynamic plants operating under uncertainty conditions.

We have just described an ideal mathematical model. In real implementations, the trajectories are confined to some vicinity of the switching line. The

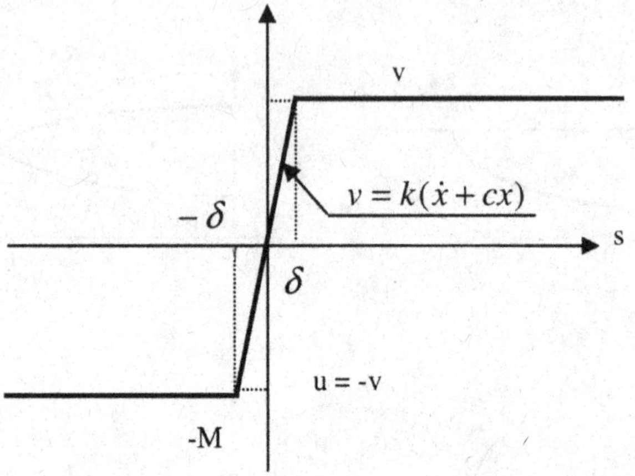

Fig. 4.

deviation from the ideal model may be caused by imperfections of switching devices such as small delays, dead zones, hysteresis which may lead to high frequency oscillations. The same phenomenon may appear due to small time constants of sensors and actuators having been neglected in the ideal model. This phenomenon referred to as "*chattering*" was a serious obstacle for utilization of sliding modes in control systems and special attention will be paid to chattering suppression methods in Chap. 8. It is worth noting that the state trajectories are also confined to some vicinity of the switching line for continuous approximation of a discontinuous relay function (Fig. 4) as well.

In a δ-vicinity of the line $s = 0$, control is the linear state function with a high gain k and the eigenvalues of the linear system are close to $-k$ and $-c$. This means that the motion in the vicinity consists of the fast component decaying rapidly and the slow component coinciding with solution to the ideal sliding mode equation.

Sliding mode became the principle operation mode in so-called variable structures systems. A variable structure system consists of a set of continuous subsystems with a proper switching logic and, as a result, control actions are discontinuous functions of the system state, disturbances (if they are accessible for measurement), and reference inputs. The term "Variable Structure System" (VSS) first appeared in the late fifties. Naturally at the very beginning several specific control tasks for second-order linear and non-linear systems were tackled and advantages of the new approach were demonstrated. Then the main directions of further research were formulated. In the course of further development the first expectations of such systems were modified, their real potential has been revealed. Some research trends proved to be unpromising while the others, being enriched by new achievements of the control theory and technology, have become milestones in VSS theory.

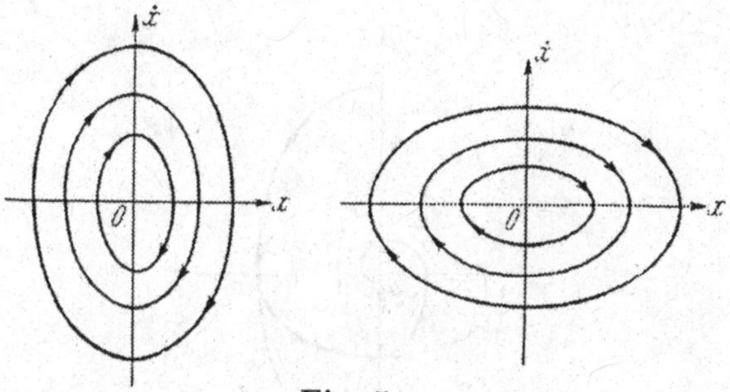

Fig. 5.

Three papers were published by S. Emel'yanov in late fifties on feedback design for the second-order linear systems ([1] was the first of them). The novelty of the approach was that the feedback gains could take several constant values depending on the system state. Although the term "Variable Structure System" was not introduced in the papers, each of the systems consisted of a set linear structures and was supplied with a switching logic and actually was VSS. The author observed that due to altering the structure in the course of control process the properties could be attained which were not inherent in any of the structures. For example the system consisting of two conservative subsystems (Fig. 5)

$$\ddot{x} = -kx,$$

$$k = \begin{cases} k_1 & \text{if} \quad x\dot{x} > 0 \\ k_2 & \text{if} \quad x\dot{x} < 0 \end{cases}, \qquad k_1 > k_2 > 0.$$

becomes asymptotically stable due to varying its structure on coordinate axes (Fig. 6).

Another way for stabilization is to find a trajectory in a state plane of one of the structures with converging motion. Then the switching logic should be found such that the state reaches this trajectory for any initial conditions and moves along it. If in the system $\ddot{x} = a_2\dot{x} - kx$, $a_2 > 0$, there are two structures with $k_2 < 0$ and $k_1 > 0$ (Fig. 7a, b), then such trajectory exists in the first structure (straight line $s = c{*}x_1 + x_2 = 0$ $c{*} = a_2/2 + \sqrt{a_2^2/4 - k_2}$ in Fig. 7a). As it can be seen in Fig. 7c the variable structure system with switching logic

$$k = \begin{cases} k_1 & \text{if} \quad xs > 0 \\ k_2 & \text{if} \quad xs < 0 \end{cases} \tag{2.3}$$

is asymptotically stable with monotonous processes. It is of interest that a similar approach was offered by A. Letov [2], but his approach implied calculation of the switching time as a function of initial conditions. As a result

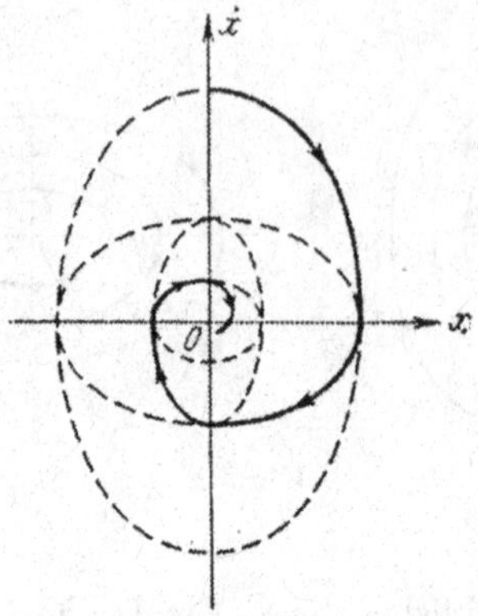

Fig. 6.

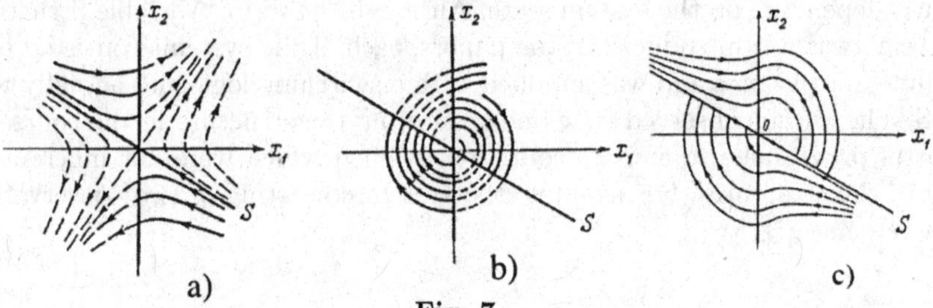

Fig. 7.

any calculation error led to instability, therefore the system was called "conditionally stable."

Starting from early sixties term "Variable Structure Control" appeared in titles of the papers by S. Emel'yanov and his colleagues. Interesting attempts were made to stabilize second-order nonlinear plants [3]. The plants under study were unstable with several equilibrium points and could not be stabilized by any linear control. Such situation was common for many processes of chemical technology. The universal design recipe could hardly be developed, so for any specific case the authors tried to "cut and glue" different pieces of available structures such that the system turned to be globally asymptotically stable. In Fig. 8a–c the state planes ($x_1 = x,\ x_2 = \dot{x}$) of three structures with linear feedback are shown. Equilibrium points of each of them are unstable. Partitioning of the state plane into six parts led to an asymptotically stable variable structure system (Fig. 9).

Note that the state trajectories are oriented towards switching line $s = x_2 + cx_1 = 0$ in the above example (Fig. 9). It means that that having reached this line the state trajectory can not leave it and for further motion the state vector will be on this line. This motion is called *sliding mode*. Sliding mode will play the dominant role in the development of VSS theory. Sliding mode may

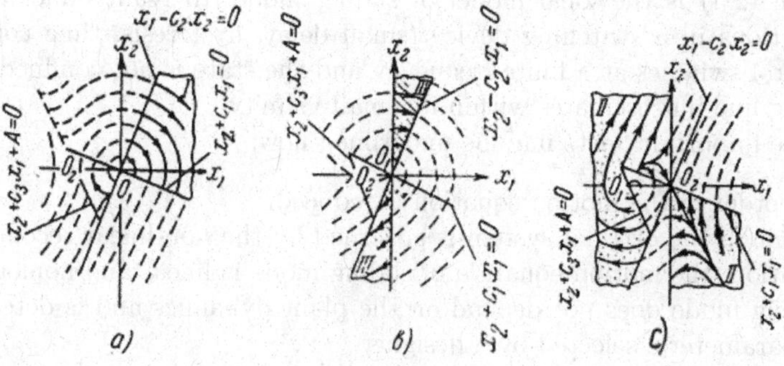

Fig. 8.

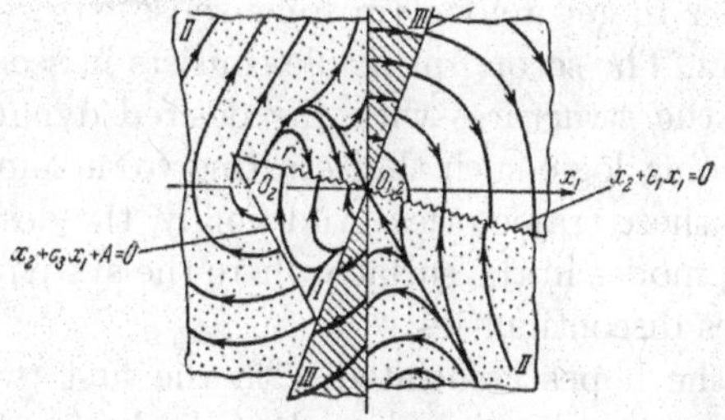

Fig. 9.

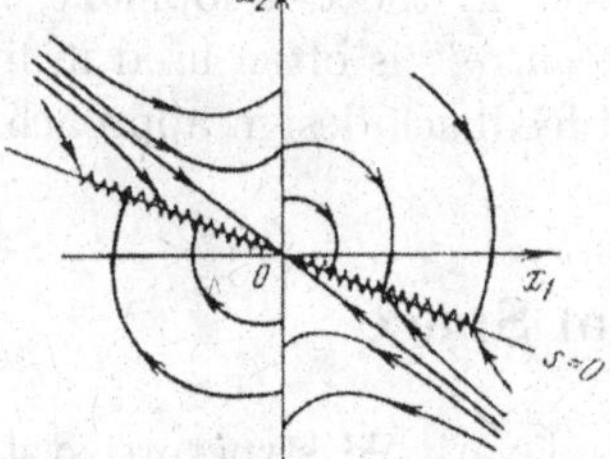

Fig. 10.

appear in our second example on the switching line, if $0 < c < c*$ (Fig. 10). Since the state trajectory coincides with the switching line in sliding mode, similarly to the relay system, its equation may be interpreted as the sliding mode equation

$$\dot{x} + cx = 0. \tag{2.4}$$

Equation (2.4) is the ideal model of sliding mode. In reality due to small imperfections in a switching device (small delay, hysteresis, time constant) the control switches at a finite frequency and the state is not confined to the switching line but oscillates within its small vicinity.

Three important facts may be underlined now:

1. The order of the motion equation is reduced.
2. Although the original system is governed by the non-linear second order equation, the motion equation of sliding mode is linear and homogenous.
3. Sliding mode does not depend on the plant dynamics and is determined by parameter c selected by a designer.

The above example let us outline various design principles offered at the initial stage of VSS theory development. The first most obvious principle implies taking separate pieces of trajectories of the existing structures and combining them together to get a good (in some sense) trajectory of the motion of a feedback system. The second principle consists in seeking individual trajectories in one of the structures with the desired dynamic properties and designing the switching logic such that starting from some instant the state moves along one of these trajectories. And finally, the design principle based on enforcing sliding modes in the surface where the system structure is varied or control undergoes discontinuities.

Unfortunately, the hopes associated with the first two approaches have not been justified; their applications have been limited to the study of several specific systems of low order. Only control design principles based on enforcing sliding modes proved to be promising due to the properties observed in our second-order examples. As a result sliding modes have played, and are still playing, an exceptional role in the development of VSS theory. Therefore the term "*Sliding Mode Control*" is often used in literature on VSS as more adequate to the nature of feedback design approach.

3 VSS in Canonical Space

The examples of relay and variable structure systems demonstrated order reduction and invariance with respect to plant uncertainties of the systems with sliding modes. Utilization of these properties was the key idea of variable structure theory at the first stage when only single-input-single-output systems with motion equations in canonical space were studied [4]. A control variable $x = x_1$ and its time-derivatives $x^{(i-1)} = x_i$, $i = 1, \ldots, n$ are components of a state vector in the canonical space

$$\begin{aligned} \dot{x}_i &= x_{i+1}, \qquad i = 1, \ldots, n-1 \\ \dot{x}_n &= -\sum_{i=1}^{n} a_i(t)x_i + f(t) + b(t)u, \end{aligned} \tag{3.1}$$

where $a_i(t)$ and $b_i(t)$ are unknown parameters and $f(t)$ is an unknown disturbance.

Control undergoes discontinuities on some plane $s(x) = 0$ in the state space

$$u = \begin{cases} u^+(x,t) & \text{if } s(x) > 0 \\ u^-(x,t) & \text{if } s(x) < 0, \end{cases}$$

where $u^+(x,t)$ and $u^-(x,t)$ are continuous state functions, $u^+(x,t) \neq u^-(x,t)$,

$$s(x) = \sum_{i=1}^{n} c_i x_i, \quad c_n = 1 \text{ and } c_1 \dots c_{n-1} \text{ are constant coefficients.}$$

The discontinuous control was selected such that the state trajectories are oriented towards the switching plane $s = 0$ and as a result, sliding mode arises in this plane (Fig. 11).

After the sliding mode starts, the motion trajectories of system (3.1) are in the switching surface

$$x_n = -\sum_{i=1}^{n-1} c_i x_i.$$

Substitution into the $(n-1)$ -st equation yields the sliding mode equations

$$\dot{x}_i = x_{i+1}, \qquad i = 1, ..., n-2$$
$$\dot{x}_{n-1} = -\sum_{i=1}^{n-1} c_i x_i \quad \text{or} \quad x^{(n-1)} + c_{n-1}x^{(n-2)} + \cdots + c_1 = 0.$$

The motion equation is of reduced order and depends neither the plant parameters nor the disturbance. The desired dynamics of the sliding mode may be assigned by a proper choice of the parameters of switching plane c_i.

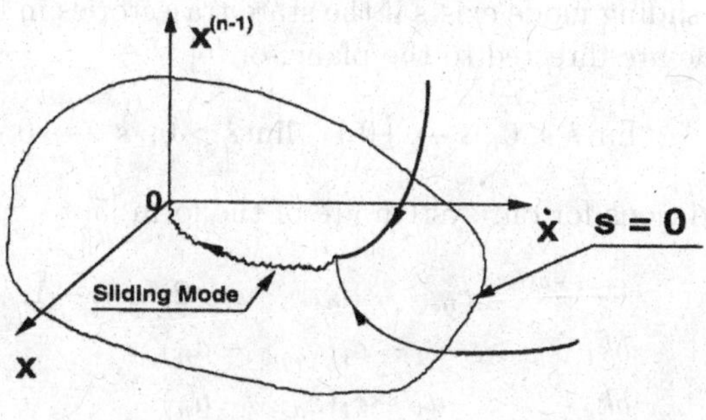

$x^{(n-1)} + c_{n-1}x^{(n-2)} + \cdots + c_1 x = 0$

Fig. 11.

3.1 Control of Free Motion

The sliding mode control methodology, demonstrated for the second-order systems, can be easily generalized for linear SISO dynamic systems of an arbitrary order under strong assumption that their behavior is represented in canonical space–space of output and its time derivatives:

$$\begin{aligned} \dot{x}_i &= x_{i+1} \qquad i = 1, ..., n-1 \\ \dot{x}_n &= -\sum_{i=1}^{n} a_i x_i + bu, \qquad a_i, b \text{ are plant parameters, } u \text{ is control input.} \end{aligned} \tag{3.2}$$

Similar to the second-order systems, control was designed as a piece-wise linear function of system output

$$u = -kx_1, \qquad k = \begin{cases} k_1 & \text{if } x_1 s > 0 \\ k_2 & \text{if } x_1 s < 0 \end{cases} \tag{3.3}$$

with switching plane

$$s = \sum_{i=1}^{n} c_i x_i = 0, \qquad c_i = \text{ const}, \quad c_n = 1.$$

The design method of VSS (3.2), (3.3) was developed after V. Taran joined the research team [5].

The methodology, developed for second-order systems, was preserved:

- Sliding mode should exist at any point of switching plane, then it is called *sliding plane*.
- Sliding mode should be stable.
- The state should reach the plane for any initial conditions.

Unfortunately the first and second requirements may be in conflict. On one hand, sliding mode exists if the state trajectories in the vicinity of the switching plane are directed to the plane, or [6]

$$\lim \dot{s} < 0, \; s \to +0 \qquad \lim \dot{s} > 0, \; s \to -0 \tag{3.4}$$

These conditions for our system are of the form [5]

$$\begin{aligned} \tfrac{c_{i-1} - a_i}{c_i} &= c_{n-1} - a_n, \qquad i = 2, ..., n-1. \\ bk_1 &> -a_1 - c_1(c_{n-1} - a_n) \\ bk_2 &< -a_1 - c_1(c_{n-1} - a_n) \end{aligned} \tag{3.5}$$

On the other hand coefficients c_i in sliding mode equation

$$x_1^{(n-1)} + c_{n-1} x_1^{(n-2)} + \cdots + c_1 x_1 = 0 \tag{3.6}$$

should satisfy Hurwitz conditions.

The result of [5]: a sliding plane with stable motion exists if and only if there exists k_0 , $k_2 < k_0 < k_1$ such that the linear system with control $u = k_0 x_1$ has $(n-1)$ eigenvalues with negative real parts.

The result of [7]: for the state to reach a switching plane from any initial position it is necessary and sufficient that the linear system with $u = -k_1 x_1$ does not have real positive eigenvalues.

The above results mean that for asymptotical stability of VSS system each of the structures may be unstable. For example the third-order VSS system,

$$\dddot{x} = u, \qquad u = -kx, \qquad k = \begin{cases} 1 & \text{if} \quad xs > 0 \\ -1 & \text{if} \quad xs < 0 \end{cases} \qquad s = c_1 x + c_2 \dot{x} + \ddot{x}$$

consisting of two unstable linear structures is asymptotically stable for $c_1 = c_2^2$ (sliding plane existence condition), $c_1 > 0$, $c_2 > 0$ (Hurwitz condition for sliding mode). The reaching condition holds as well since the linear system with $k = 1$ does not have real positive eigenvalues.

Again sliding mode equation (3.6) is of a reduced order, linear, homogenous, does not depend on plant dynamics and is determined by coefficients in switching plane equation. This property looks promising when controlling plants with unknown time-varying parameters. Unfortunately control (3.3) is not applicable for this purpose because the conditions for sliding plane to exist (3.5) need knowledge on the parameters a_i.

For the modified version of VSS control

$$u = -\sum_{i=1}^{n-1} k_i x_i, \tag{3.7}$$

$$k_i = \begin{cases} k_i' & \text{if} \quad x_i s > 0 \\ k_i'' & \text{if} \quad x_i s < 0 \end{cases}$$

$s = 0$ is a sliding plane for any value of c_i if

$$\begin{aligned} bk_i' &> \max_{a_i, a_n}(c_{i-1} - a_i - c_{n-1}c_i + a_n c_i), \\ bk_i'' &> \min_{a_i, a_n}(c_{i-1} - a_i - c_{n-1}c_i + a_n c_i) \end{aligned} \qquad c_0 = 0, \qquad i = 1, ..., n-1. \tag{3.8}$$

The design procedure of VSS (3.7) consisting of 2^{n-1} linear structures implies selection of switching plane or sliding mode equation (3.6) with desired dynamics and then coefficients k_i' and k_i'' (3.8) such that $s = 0$ is a sliding plane (assuming that the ranges of parameter variations are known). Reaching this plane may be provided by increasing coefficient k_1' [7]. Sliding mode with the desired properties starts in the VSS after a finite time interval. The time, preceding the sliding mode, may be decreased by increasing the gains in control (3.7).

Development of special methods is needed if the last equation in (3.1) depends on time derivatives of control, since trajectories in the canonical space become discontinuous. Two approaches were offered in the framework of the VSS theory: first, designing a switching surface with a part of state variables, and, second, using a pre-filter in controller [8]. For the both cases the conventional sliding mode with the desired properties can be enforced. Traces of the approaches may be found in modern publications.

3.2 Disturbance Rejection

The property of insensitivity of sliding modes to plant dynamics may be utilized for control of plant subjected to *unknown* external disturbances. It is obvious that control (3.7) does not fit for this purpose. Indeed at the desired state (all x_i are equal to zero) the control is equal to zero and unable to keep the plant at the desired equilibrium point at presence of disturbances. We demonstrate how the disturbance rejection problem can be solved using dynamic actuators with variable structure. Let plant and actuator be integrators in the second-order system (Fig. 12). An external disturbance $f(t)$ is not accessible for measurement.

The control is designed as a piece-wise linear function not only of the output $x = x_1$ to be reduced to zero but also of actuator output y:

$$\begin{aligned} \dot{x}_1 &= y + f(t) \\ \dot{y} &= u, \\ u &= -kx_1 - k_y y. \end{aligned} \qquad \text{or} \qquad \begin{aligned} \dot{x}_1 &= x_2 \\ \dot{x}_2 &= -kx_1 - k_y x_2 - k_y f + \dot{f}. \end{aligned}$$

The state semi-planes $x_1 > 0$ and $x_1 < 0$ for the variable structure system with

$$k = \begin{cases} k_0 & \text{if} \quad x_1 s > 0 \\ -k_0 & \text{if} \quad x_1 s < 0 \end{cases}, \qquad k_y = \begin{cases} k_{y0} & \text{if} \quad ys > 0 \\ -k_{y0} & \text{if} \quad ys < 0 \end{cases},$$

$$s = cx_1 + x_2, \quad k_0, k_{yo}, c \text{ are constant}$$

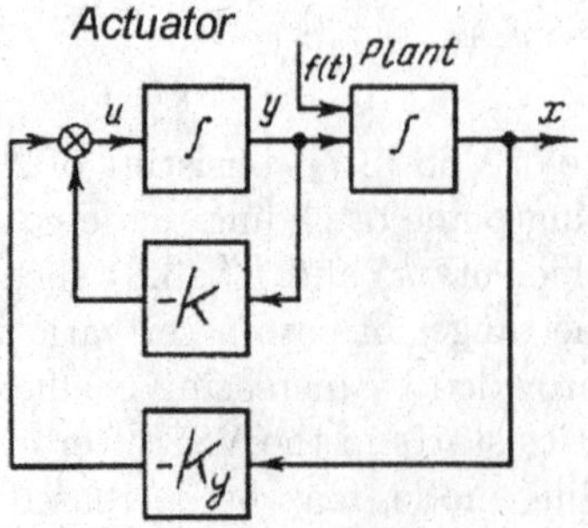

Fig. 12.

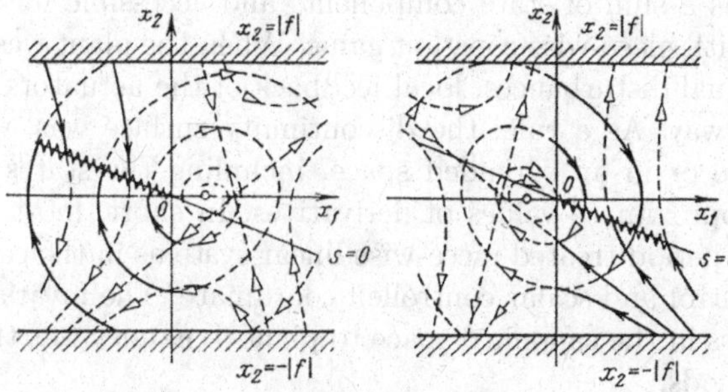

Fig. 13.

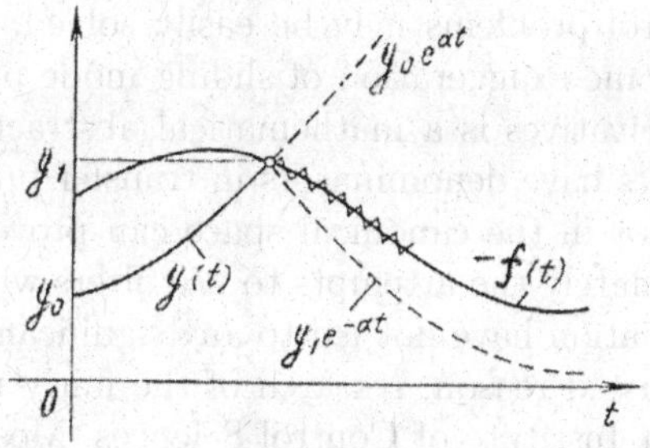

Fig. 14.

are shown in Fig. 13. For domain $x_2 < |f(t)|$ the singular points for each semi-plane are located in the opposite one and as a result state trajectories are oriented toward switching line $s = 0$. It means that sliding mode occurs in this line with motion equation $\dot{x} + cx = 0$ and solution tending to zero. The effect of disturbance rejection may be explained easily in structural language. Since the sign of actuator feedback is varied, the actuator output may be either diverging or converging exponential function (Fig. 14). In sliding mode due to high frequency switching, an average value of the output is equal to the disturbance with an opposite sign. It is clear that the disturbance, which can be rejected, should be between the diverging and converging exponential functions at any time. Similar approach stands behind the disturbance rejection method in VSS of an arbitrary order.

3.3 Comments for VSS in Canonical Space

In the sixties VSS studies were mainly focused on linear (time-invariant and time-varying) systems of an arbitrary order with scalar control and scalar variable to be controlled. These first studies utilized the space of an error coordinate and its time derivatives, or canonical space, while the control was designed as a sum of state components and accessible for measurement disturbances with piece-wise constant gains. When the plant was subjected to unknown external disturbances, local feedback of the actuators was designed in the similar way. As a rule, the discontinuity surface was a plane in the canonical space or in an extended space, including the states of the filters for deriving approximate values of derivatives. In short, most works of this period on VSS period treated piece-wise linear systems in the canonical space with scalar control and scalar controlled coordinate. The invariance property of sliding modes in the canonical space to plant dynamics was the key idea of all design methods.

The first attempts to apply the results of VSS theory demonstrated that the invariance property of sliding modes in canonical spaces had not been beneficial only. In a sense it decelerated development of the theory. The illusion that any control problems may be easily solved, should sliding mode be enforced, led to some exaggeration of sliding mode potential. The fact is, that the space of derivatives is a mathematical abstraction, and in practice any real differentiators have denominators in transfer functions; so the study of the system behavior in the canonical space can prove to be unacceptable idealization. Unfortunately the attempts to use filters with variable structure for multiple differentiation have not led to any significant success.

For the just discussed reason, research of the fairly narrow class of VSS, mainly carried out at Institute of Control Sciences, Moscow and by a group of mathematicians headed by E. Barbashin at Institute of Mathematics and Mechanics, Sverdlovsk, did not result in wide applications and did not produced significant echo in the scientific press. The results of this first stage of VSS development, i.e. analysis and design of VSS in the canonical space, were summarized in [5, 6]. In view of the limited field of practical applications in the frame of this approach (VSS with differentiating circuits), another extreme appeared, reflecting certain pessimism about implementation of any VSS with sliding modes.

The second stage of development of VSS theory began roughly in the late sixties and early seventies, when design procedures in the canonical space ceased to be looked as obligatory and studies were focused on systems of general form with arbitrary state components and with vector control and output values. The first attempts to revise VSS methodology in this new environment demonstrated that the pessimism about sliding mode control was unjustified and refusal to utilize the potential of sliding mode was an unreasonable extravagance [10, 11].

3.4 Preliminary Mathematical Remark

The basic design idea of the first stage of VSS theory was to enforce sliding mode in a plane of the canonical space. An equation of a sliding plane depending on a system output and its derivatives was interpreted as a sliding mode equation. Formally this interpretation is not legitimate. “To solve differential equation" means to find a time function such that substitution of the solution into the equation makes its right- and left-hand sides equal identically.

In our second-order example with control (2.3) and $0 < c < c*$, sliding mode existed on the switching line $s = 0$ and (2.4) was taken as the sliding mode equation. Its solution Ae^{-ct} being substituted into function s makes it equal to zero. Control (2.3) is not defined for $s = 0$ and respectively the right-hand side of the system equation (2.1) is not defined as well. Therefore we cannot answer the question whether the solution to (2.2) is the solution to the original system.

One of the founders of control theory in the USSR academician A.A. Andronov indicated that ambiguity in the system behavior is eliminated if minor non-idealities such as time delay, hysteresis, small time constants are recognized in the system model which results in so-called real sliding mode in a small neighborhood of the discontinuity surface. Ideal sliding motion is regarded as a result of limiting procedure with all non-idealities tending to zero [9]. The examples of such limiting procedure were also given. The relay second-order system was considered with motion equations in the canonical space and a straight line (2.2) as a set of discontinuity points for control. The behavior of the system was studied under the assumption that a time delay was inherent in the switching device and, consequently, the discontinuity points were isolated in time. It was shown that the solution of the second-order equation in sliding mode always tended to the solution of the first order equation (2.2) depending on c only irrespective of the plant parameters and disturbances, if the delay tended to zero. The validity of (3.6) as the model of sliding mode in the canonical space of an arbitrary-order system may be substantiated in the similar way.

4 Sliding Modes in Arbitrary State Spaces: Problem Statements

Now we demonstrate sliding modes in non-linear affine systems of general form

$$\dot{x} = f(x,t) + B(x,t)u \tag{4.1}$$

$$u_i = \begin{cases} u_i^+(x,t) & \text{if} \quad s_i(x) > 0 \\ u_i^-(x,t) & \text{if} \quad s_i(x) < 0 \end{cases} \tag{4.2}$$

where $x \in R^n$ is a state vector, $u \in R^m$ is a control vector, $u_i^+(x,t)$, $u_i^-(x,t)$ and $s_i(x)$ are continuous functions of their arguments, $u_i^+(x,t)$ $u_i^-(x,t)$. The control is designed as a discontinuous function of the state such that each component undergoes discontinuities in some surface in the system state space.

Similar to the above example, state velocity vectors may be directed towards one of the surfaces and sliding mode arises along it (arcs ab and cb in Fig. 15). It may arise also along the intersection of two of surfaces (arc bd).

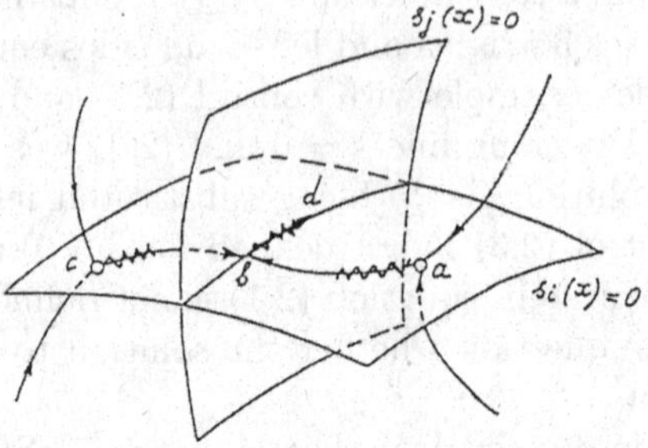

Fig. 15.

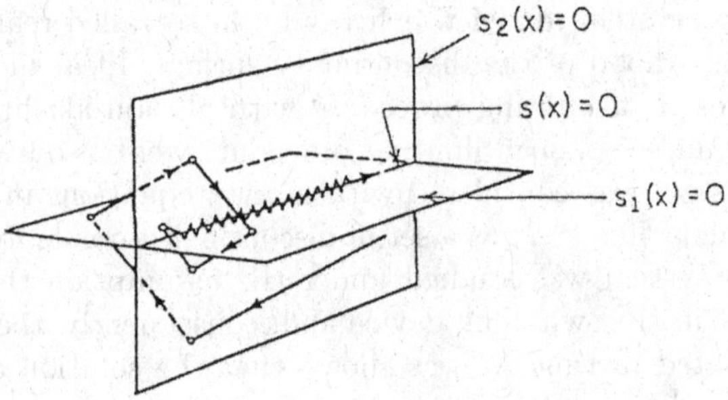

Fig. 16.

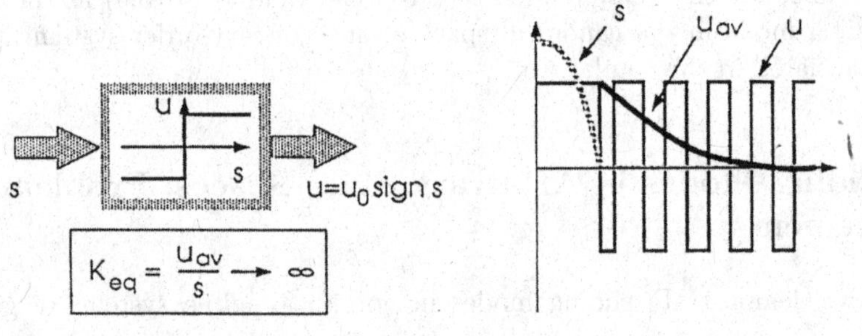

Fig. 17.

Figure 16 illustrates the sliding mode in the intersection even if it does not exist at each of the surfaces taken separately.

For the general case (4.1) sliding mode may exist in the intersection of all discontinuity surfaces $s_i = 0$, or in the manifold

$$s(x) = 0, \quad s^T(x) = [s_1(x), ..., s_m(x)] \tag{4.3}$$

of dimension $n - m$. Let us discuss the benefits of sliding modes, if it would be enforced in the control system. First, in sliding mode the input s of the element implementing discontinuous control is close to zero, while its output (exactly speaking its average value u_{av}) takes finite values (Fig. 17).

Hence, the element implements high (theoretically infinite) gain, that is the conventional tool to reject disturbance and other uncertainties in the system behavior. Unlike to systems with continuous controls, this property called invariance is attained using finite control actions. Second, since sliding mode trajectories belong to a manifold of a dimension lower than that of the original system, the order of the system is reduced as well. This enables a designer to simplify and decouple the design procedure. Both order reduction and invariance are transparent for the above two second-order systems.

In order to justify the above arguments in favor of using sliding modes in control systems, we, first, need mathematical methods for deriving equations of sliding modes in the intersection of discontinuity surfaces and, second, the conditions for the sliding mode to exist should be obtained. Only having these mathematical tools the design methods of sliding mode control for wide range of control problems may be developed.

5 Sliding Mode Equations: Equivalent Control Method

5.1 Problem Statement

The first mathematical problem concerns differential equations of sliding mode. For our second-order examples the equation was obtained using heuristic approach: the equation of the switching line $\dot{x} + cx = 0$ was interpreted as the motion equation. But even for an arbitrary time invariant second-order relay system

$$\begin{aligned} \dot{x}_1 &= a_{11}x_1 + a_{12}x_2 + b_1 u \\ \dot{x}_2 &= a_{21}x_1 + a_{22}x_2 + b_2 u, \\ u &= -M sign(s), \quad s = cx_1 + x_2; \quad M, a_{ij}, b_i, c \quad \text{are constant} \end{aligned} \tag{5.1}$$

the problem of mathematical description of sliding mode is quite a challenge and requires the design of special techniques. It arises due to discontinuities in control, since the relevant motion equations with discontinuous right-hand sides do not satisfy the conventional theorems on existence-uniqueness of solutions (a Lipshitz constant does not exists for discontinuous functions).

Discontinuous systems are not a subject of the conventional theory of differential equations dealing with continuous state functions.[1] The conventional theory does not answer even the fundamental questions, whether the solution exists and whether the solution is unique. Formally, even for our simple examples of second order systems in canonical form (2.1) our method of deriving the

[1] Strictly speaking, the conventional method require the right-hand sides of a differential equation to consist of functions $f(x)$ satisfying the Lipschitz condition $\|f(x_1) - f(x_2)\| < L\|x_1 - x_2\|$ with some positive number L, referred to as the Lipschitz constant, for any x_1 and x_2. The condition implies that the function does not grow faster than some linear one which is not the case for discontinuous functions if x_1 and x_2 are close to a discontinuity point.

sliding mode equations was not legitimate. The solution $x(t) = x(t_1)e^{-c(t-t_1)}$ should satisfy the original differential equation (2.1) rather than heuristically written equation (2.2). Direct substitution of $x(t)$ into (2.1) leads to $s(t) = 0$ and

$$(1 - a_2 + a_1)x(t_1)e^{-c(t-t_1)} \stackrel{?}{=} -M \ \text{sign}\ (0) + f(t).$$

Since the function sign $(\cdot)$ is not defined at zero point, we can not check whether the solution $x(t)$ is correct.

Uncertainty in behavior of discontinuous systems on the switching surfaces gives freedom on choosing an adequate mathematical model and gave birth to a number of lively discussions. For example, description of sliding mode in system (5.1) by a linear first-order differential equation, a common practice today, seemed unusual at first glance; this model was offered by Kornilov [12] and Popovski [13] and the approach was generalized for linear systems of an arbitrary order by Yu. Dolgolenko [14] in fifties.

The approach by Filippov [15], now recognized as classical, was not accepted by all experts in the area: at the first IFAC congress in 1960 Yu. Neimark (author of one more mathematical model of sliding mode based on convolution equation) offered an example of a system with two relay elements with a solution different from that of Filippov's method [16]. The discussion at the congress proved to be fruitful from the point of stating a new problem in the theory of sliding modes. The discussion led to the conclusion that two dynamic systems with identical equations outside a discontinuity surface may have different sliding mode equations. Most probably the problem of unambiguous description of sliding modes in discontinuous systems was first brought to light.

5.2 Regularization

In situations where conventional methods are not applicable, the common approach is to employ different methods of regularization or replacing the original problem by a closely similar one for which familiar methods are applicable. For systems with discontinuous controls, regularization approach has a simple physical interpretation. Uncertainty of system behavior at the discontinuity surfaces appears because the motion equations (4.1) and (4.2) are an ideal system model. Non-ideal factors such as small imperfections of switching devices (delay, hysteresis, small time constants), unmodeled dynamics of sensors and actuators etc. are neglected in the ideal model. Incorporating them into the system model makes discontinuity point isolated in time and eliminates ambiguity in the system behavior. Next, small parameters characterizing all these factors are assumed to tend to zero. If the limit of the solutions exists with the small parameters tending to zero, then they are taken as the solutions to the equations describing the ideal sliding mode. Such a limit procedure is the *regularization* method for deriving sliding mode equations in the dynamic systems with discontinuous control.

As was discussed in Chap. 3, such regularization with time delay was employed by Andronov for substantiation of sliding mode equation (2.2). Similar approach for nonlinear relay system with imperfections of time delay and hysteresis type was developed by Andre and Seibert [17]; it is interesting that their sliding mode equations coincided with those of Filippov's method [15] and the result may serve as substantiation of Filippov's method. At the same time the question may be asked, whether the method is applicable for the systems with different types of non-idealities. Generally speaking the answer is negative: a continuous approximation of a discontinuous function leads to motion equations different from those of Filippov's method [10]. So we should admit that for systems with a right-hand side as a nonlinear function of discontinuous control sliding mode equations cannot be derived unambiguously even for the case of scalar control. The "point-to-point" technique used in the cited papers here for scalar case is not applicable for the system with vector control and sliding modes in the intersection of a set of discontinuity surfaces (4.3).

To illustrate the regularization method, we consider a linear time-invariant system with one control input, being a scalar relay function of a linear combination of the state components:

$$\dot{x} = Ax + bu, \qquad x \in \Re^n, \tag{5.2}$$

A and b are $n \times n$ and $n \times 1$ constant matrices, $u = M$ sign (s), M is a scalar positive constant value, $s = cx$, $c = (c_1, c_2, ..., c_n) =$ const.

As in our examples of the systems in canonical space, the state trajectories may be oriented in a direction towards the switching plane $s(x) = 0$ in the state space $x^T = (x_1, x_2, \ldots, x_n)$. Hence the sliding mode occurs in the plane (Fig. 18) and the motion equation should be found. A similar problem was left unanswered for system 5.1.

Following the regularization procedure, small imperfections of a switching device should be taken into account. If a relay device is implemented with a hysteresis loop with the width 2Δ (Fig. 19), then the state trajectories oscillate in a Δ-vicinity of the switching plane (Fig. 20). The value of Δ is assumed

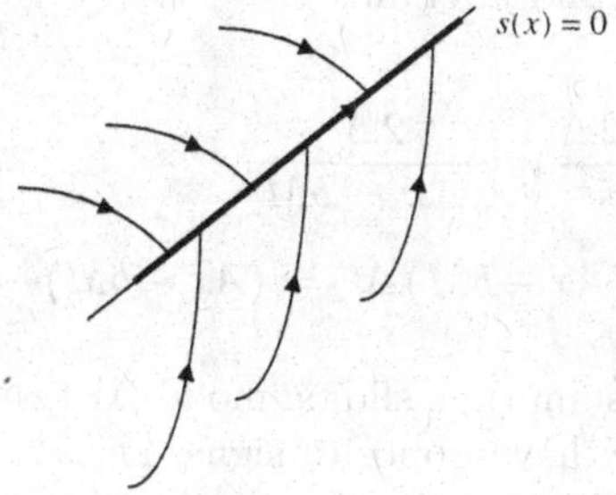

Fig. 18.

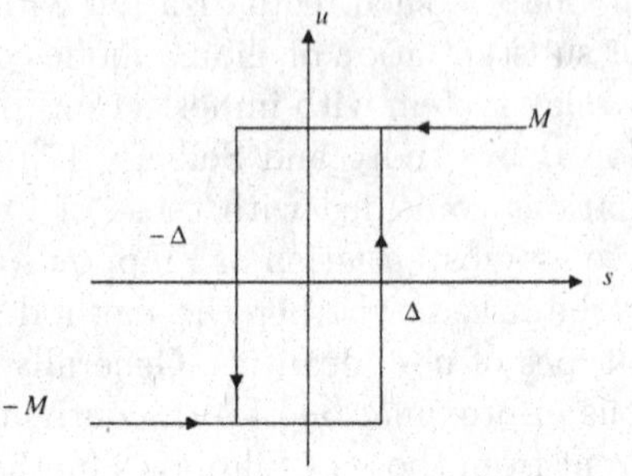

Fig. 19.

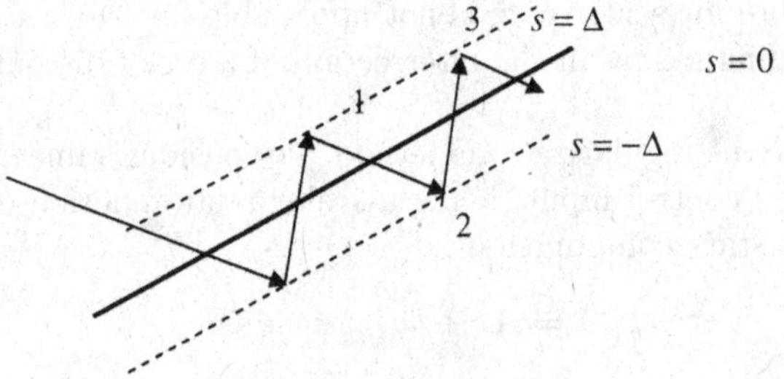

Fig. 20.

to be small such that the state trajectories may be approximated by straight lines with constant state velocity vectors $Ax+bM$ and $Ax-bM$ in the vicinity of some point x on the plane $s(x)=0$ (Fig. 20). Calculate times Δt_1 and Δt_2 intervals and increments Δx_1 and Δx_2 in the state vector for transitions from point 1 to point 2 and from point 2 to point 3, respectively:

$$\Delta t_1 = \frac{2\Delta}{\dot{s}^+} = \frac{-2\Delta}{cAx+cbM},$$
$$\Delta x_1 = (Ax+bM)\Delta t_1 = (Ax+bM)\frac{-2\Delta}{cAx+cbM}.$$

Similarly for the second interval

$$\Delta t_2 = \frac{2\Delta}{\dot{s}^-} = \frac{2\Delta}{cAx-cbM},$$
$$\Delta x_2 = (Ax-bM)\Delta t_2 = (Ax-bM)\frac{2\Delta}{cAx-cbM}.$$

Note that by our assumption, sliding mode exists in the ideal system therefore the values s and $\dot{s}$ have opposite signs, i.e. $\dot{s}^+ = cAx+cbM < 0$ and $\dot{s}^- = cAx-cbM > 0$. This implies that both time intervals Δt_1 and Δt_2 are positive. Note that the inequalities may hold if $cb < 0$.

The average state velocity within the time interval $\Delta t = \Delta t_1 + \Delta t_2$ may be found as

$$\dot{x}_{av} = \frac{\Delta x_1 + \Delta x_2}{\Delta t} = Ax - (cb)^{-1}bcAx.$$

The next step of the regularization procedure implies that the width of the hysteresis loop Δ should tend to zero. However we do not need to calculate $\lim_{\Delta \to 0}(\dot{x}_{av})$: the limit procedure was performed implicitly when we assumed that state trajectories are straight lines and the state velocities are constant. This is the reason why $\dot{x}_{av}$ does not depend on Δ. As it follows from the more accurate model the sliding mode in the plane $s(x) = 0$ is governed by

$$\dot{x} = (I_n - (cb)^{-1}bc)Ax \tag{5.3}$$

with initial state $s[x(0)] = 0$ and I_n being an identity matrix. It follows from (5.2) and (5.3) that

$$\dot{s} = c(I_n - (cb)^{-1}bc)Ax \equiv 0$$

hence the state trajectories of the sliding mode are oriented along the switching plane. The condition $s[x(0)] = 0$ enables one to reduce the system order by one. To obtain the sliding mode equation of $(n-1)$th order, one of the components of the state vector, let it be x_n, may be found as a function of the other $n-1$ components and substituted into the system (5.3). Finally the last equation for x_n can be disregarded.

Applying the above procedure to the second order system (5.1) results in a first order sliding mode equation along the switching line $s = c_1x_1 + c_2x_2 = 0$:

$$\dot{x}_1 = [a_{11} - a_{12}c_2^{-1}c_1 - (cb)^{-1}b_1(ca^1 - ca^2c_2^{-1}c_1)]x_1 + [d_1 - b_1(cb)^{-1}(cd)]f,$$

where $c = (c_1, c_2)$, $b^T = (b_1, b_2)$, $(a^1)^T = (a_{11}, a_{21})$, $(a^2)^T = (a_{12}, a_{22})$, $d^T = (d_1, d_2)$, and cb and c_2 are assumed to be different from zero. As we can see for this general case of a linear second order system, the sliding mode equation is of reduced order and depends on the plant parameters, disturbances and coefficients of switching line equations, but does not depend on control.

For the systems in canonical form (2.1) and (3.1), the above regularization method may serve as validation that the reduced order sliding mode equations (2.2) and (3.6) depend neither on plant parameters nor disturbances.

Exactly the same equations for our examples result from regularization based on an imperfection of "delay" type [17]. It is interesting to note that nonlinear systems of an arbitrary order with one discontinuity surface were studied in this paper and the motion equations proved to be the same for both types of imperfections – hysteresis and delay. This result may be easily interpreted in terms of relative time intervals for control input to take each of two extreme values. For a system of an arbitrary order with scalar control

$$\dot{x} = f(x, u), \qquad x, f \in \Re^n, \qquad u(x) \in \Re,$$

$$u(x) = \begin{cases} u^+(x) & \text{if} \quad s(x) > 0 \\ u^-(x) & \text{if} \quad s(x) > 0 \end{cases},$$

the components of vector f, scalar functions $u^+(x), u^-(x)$ and $s(x)$ are continuous and smooth, and $u^+(x) \neq u^-(x)$. We assume that sliding mode occurs on the surface $s(x) = 0$ and try to derive the motion equations using the regularization method. Again, let the discontinuous control be implemented with some imperfections of unspecified nature, control is known to take one of the two extreme values, $u^+(x)$ or $u^-(x)$, and the discontinuity points are isolated in time. As a result, the solution exists in the conventional sense and it does not matter whether we deal with small hysteresis, time delay or time constants neglected in the ideal model.

Like for the system (5.2) with hysteresis imperfection, the state velocity vectors $f^+ = f(x, u^+)$ and $f^- = f(x, u^-)$ are assumed to be constant for some point x on the surface $s(x) = 0$ within a short time interval$[t, t + \Delta t]$. Let the time interval Δt consists of two sets of intervals Δt_1 and Δt_2 such that$\Delta t = \Delta t_1 + \Delta t_2, u = u^+$ for the time from the set Δt_1 and $u = u^-$ for the time from the set Δt_2. Then the increment of the state vector after time interval Δt is found as

$$\Delta x = f^+ \Delta t_1 + f^- \Delta t_2$$

and the average state velocity as

$$\dot{x}_{av} = \frac{\Delta x}{\Delta t} = \mu f^+ + (1 - \mu) f^-,$$

where $\mu = \frac{\Delta t_1}{\Delta t}$ is relative time for control to take value u^+ and $(1 - \mu)$ – to take value u^-, $0 \leq \mu \leq 1$. To get the vector $\dot{x}$ the time Δt should be tended to zero. However we do not need to perform this limit procedure, it is hidden in our assumption that the state velocity vectors are constant within time interval Δt, therefore the equation

$$\dot{x} = \mu f^+ + (1 - \mu) f^- \tag{5.4}$$

represents the motion during sliding mode. Since the state trajectories during sliding mode are in the surface $s(x) = 0$, the parameter μ should be selected such that the state velocity vector of the system (5.4) is in the tangential plane to this surface, or

$$\begin{aligned} \dot{s} = \text{ grad } [s(x)] \cdot \dot{x} = \text{ grad } [s(x)][\mu f^+ + (1 - \mu) f^-] = 0, \quad \text{with} \\ \text{grad } [s(x)] = [\tfrac{\partial s}{\partial x_1} \ldots \tfrac{\partial s}{\partial x_n}]. \end{aligned} \tag{5.5}$$

The solution to (5.5) is given by

$$\mu = \frac{\text{grad }(s) \cdot f^-}{\text{grad }(s) \cdot (f^- - f^+)} \tag{5.6}$$

Substitution of (5.6) into (5.4) results in the sliding mode equation

$$\dot{x} = f_{sm}, \qquad f_{sm} = \frac{(\text{grad } s) \cdot f^-}{(\text{grad } s) \cdot (f^- - f^+)} f^+ - \frac{(\text{grad } s) \cdot f^+}{(\text{grad } s) \cdot (f^- - f^+)} f^-, \tag{5.7}$$

representing the motion in sliding mode with initial condition $s[x(0)] = 0$. Note that sliding mode occurs in the surface $s(x) = 0$, therefore the functions s and $\dot{s}$ have different signs in the vicinity of the surface (Fig. 18) and $\dot{s}^+ = (\text{grad}s) \cdot f^+ < 0, \dot{s}^- = (\text{grad } s) \cdot f^- > 0$. As follows from (5.6), the condition $0 \leq \mu \leq 1$ for parameter μ holds. It easy to check the condition $\dot{s} = (\text{grad } s) \cdot f_{sm} = 0$ for the trajectories of system (5.7) and to show that they are confined to the switching surface $s(x) = 0$. As it could be expected, direct substitution of grad $s = c$, $f^+ = Ax + bu^+$ and $f^- = Ax + bu^-$ into (5.7) results in the sliding mode equation (5.2) derived for the linear system (5.2) with the discontinuity plane $s(x) = cx = 0$ via hysteresis regularization.

5.3 Boundary Layer Regularization

The universal approach to regularization consists in introducing a boundary layer $\|s\| < \Delta, \Delta - const$ around manifold $s = 0$, where an ideal discontinuous control is replaced by a real one such that the state trajectories are not confined to this manifold but run arbitrarily inside the layer (Fig. 21). The nonidealities, resulting in motion in the boundary layer, are not specified. The only assumption for this motion is that the solution exists in the conventional sense. If, with the width of the boundary layer Δ tending to zero, the limit of the solution exists, it is taken as a solution to the system with ideal sliding mode. Otherwise we have to recognize that the equations beyond discontinuity surfaces do not derive unambiguously equations in their intersection, or equations of the sliding mode.

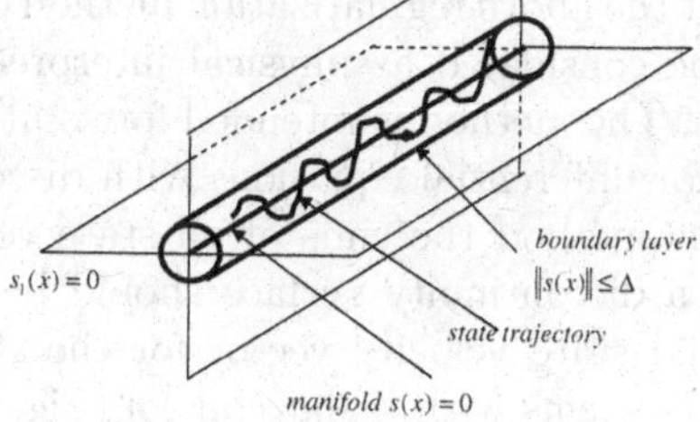

Fig. 21.

The boundary layer regularization enables substantiation of so-called *Equivalent Control Method* intended for deriving sliding mode equations in manifold $s = 0$ in affine systems

$$\dot{x} = f(x,t) + B(x,t)u \tag{5.8}$$

with $B(x,t)$ being $n \times m$ full rank matrix and $\det(GB) \neq 0$, $G(x) = \{\partial s/\partial x\}$. Following this method, the sliding mode equation with a unique solution may be derived.

First, the *equivalent control* should be found as the solution to the equation $\dot{s} = 0$ on the system trajectories (G and $(GB)^{-1}$ are assumed to exist):

$$\dot{s} = Gf + GBu = 0, \qquad u_{eq} = -\left(GB\right)^{-1} Gf.$$

Then the solution should be substituted into (5.8) for the control

$$\dot{x} = f - B\left(GB\right)^{-1} Gf \tag{5.9}$$

Equation (5.9) is the sliding mode equation with initial conditions $s(x(0), 0) = 0$.

Since $s(x) = 0$ in sliding mode, m components of the state vector may be found as a function of the rest $(n - m)$ ones: $x_2 = s_0(x_1)$, $x_2, s_0 \in \Re^m$; $x_1 \in \Re^{\text{n-m}}$ and, correspondingly, the order of sliding mode equation may be reduced by m:

$$\dot{x}_1 = f_1[x_1, t, s_0(x_1)], \quad f_1 \in \Re^{n-m}. \tag{5.10}$$

The idea of the equivalent control method may be easily explained with the help of geometric consideration. Sliding mode trajectories lie in the manifold $s = 0$ and the equivalent control u_{eq} being a solution to the equation $\dot{s} = 0$ implies replacing discontinuous control by such continuous one that the state velocity vector lies in the tangential manifold and as a result the state trajectories are in this manifold. It will be important for control design that sliding mode equation (5.10)

- Is of reduced order
- Does not depend on control
- Depends on the equation of switching surfaces

It is interesting that the above regularization method of deriving the sliding mode equation may be considered as physical interpretation of the famous Filippov's method [18]. The method is intended for solution continuation at a discontinuity surface for differential equations with discontinuous right-hand sides. According to the method the ends of all state velocity vectors in the vicinity of a point in a discontinuity surface should be complemented by a minimal convex set and state velocity vector for the sliding motion should belong to this set. For systems with scalar control (Fig. 22)

$$\dot{x} = f_0(x) + b(x)u, \qquad x, f, b \in \Re^n, \qquad u(x) \in \Re,$$

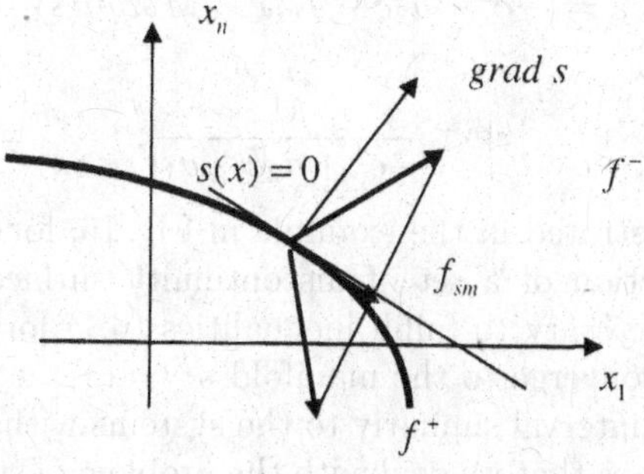

Fig. 22.

$$u(x) = \begin{cases} u^+(x) & \text{if} \quad s(x) > 0 \\ u^-(x) & \text{if} \quad s(x) > 0 \end{cases},$$

we have two points, ends of vectors $f^+ = f_o + bu^+, f^- = f_0 + bu^-$, and the minimal convex set is the straight line connecting the end. The intersection of the line with the tangential plane defines the state velocity vector f_{sm} in the sliding mode, or the right-hand side of the sliding mode equation.

$$\dot{x} = f_{sm}, \qquad f_{sm} = \frac{(grad \quad s) \cdot f^-}{(grad \quad s) \cdot (f^- - f^+)} f^+ - \frac{(grad \quad s) \cdot f^+}{(grad \quad s) \cdot (f^- - f^+)} f^-,$$

It is easy to check that the result of Filippov's method coincides with the equation derived by the equivalent control method.

6 Sliding Mode Existence Conditions

The second mathematical problem in the analysis of sliding mode as a phenomenon is deriving the conditions for sliding mode to exist. As to the second-order systems with scalar control studied in Sect. 2 the conditions may be obtained from geometrical considerations: the deviation from the switching surface s and its time derivative should have opposite signs in the vicinity of a discontinuity surface $s = 0$, or

$$\lim_{s \to +0} \dot{s} < 0, \qquad \text{and} \qquad \lim_{s \to -0} \dot{s} > 0. \tag{6.1}$$

Inequalities (6.1) are referred to as *reaching conditions* – the condition for the state to reach the surface $s = 0$ after a finite time for arbitrary initial conditions.

For the second-order relay system (2.1) the domain of sliding mode on $s = 0$ or for $\dot{x} = -cx$ (sector mn on the switching line, Fig. 3 may be found analytically from these conditions:

$$\dot{s} = (-c^2 + a_2c - a_1)x - Msign(s)$$

and

$$|x| < \frac{M}{|-c^2 + a_2c - a_1|}.$$

As it was demonstrated in the example in Fig. 16, for existence of sliding mode in an intersection of a set of discontinuity surfaces $s_i(x) = 0$, $(i = 1, ..., m)$ it is not necessary to fulfill inequalities (6.1) for each of them. The trajectories should converge to the manifold $s^T = (s_1, \ldots, s_m) = 0$ and reach it after a finite time interval similarly to the systems with scalar control. The term "converge" means that we deal with the problem of stability of the origin in m-dimensional subspace $(s_1, \ldots, s_m)$, therefore the existence conditions may be formulated in terms of the stability theory. The non-traditional condition: *finite time convergence* should take place. This last condition is important to distinguish the systems with sliding modes and the continuous system with state trajectories converging to some manifold asymptotically. For example the state trajectories of the system $\ddot{x} - x = 0$ converge to the manifold $s = \dot{x} - x = 0$ asymptotically since $\dot{s} = -s$, however it would hardly be reasonable to call the motion in $s = 0$ "sliding mode."

As to the systems with scalar control the conditions may be obtained from geometrical considerations: inequalities (6.1). However for vector cases it is not necessary to fulfill these inequalities for each discontinuity surfaces. As an illustration consider the motion projection on a two-dimensional subspace (s_1, s_2) governed by equations

$$\begin{aligned} \dot{s}_1 &= -sign \quad s_1 + 2sign \quad s_2 \\ \dot{s}_2 &= -2sign \quad s_1 - sign \quad s_2. \end{aligned}$$

The state trajectories are straight lines on the state plane (s_1, s_2) (Fig. 23). It is clear from the picture that for any point on $s_1 = 0$ or $s_2 = 0$ the state trajectories are not oriented towards the line therefore sliding mode does not exist at any of the switching lines taken separately. At the same time the

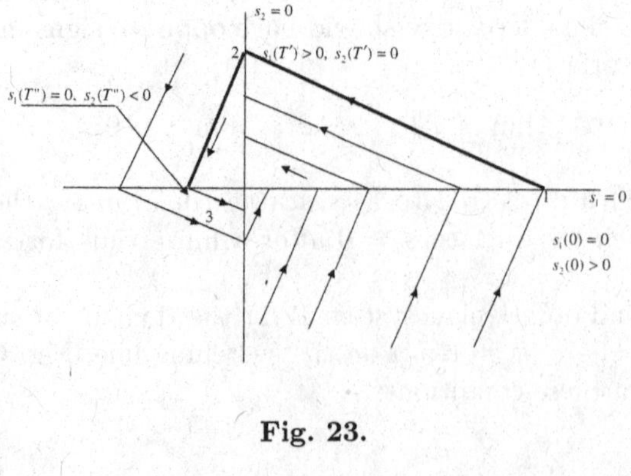

Fig. 23.

trajectories converge to the intersection of them – the origin in the subspace (s_1, s_2). Let us calculate the time needed for the state to reach it. For initial conditions

$$s_1(0) = 0, \quad s_2(0) > 0 \qquad \text{(point 1)}$$

$$\begin{matrix} \dot{s}_1 = 1 \\ \dot{s}_2 = -3 \end{matrix} \quad \text{for} \quad 0 < t < T'.$$

and

$$s_2(T') = 0, \quad T' = \frac{1}{3} s_2(0), \quad s_1(T') = \frac{1}{3} s_2(0) \qquad \text{(point 2)}.$$

For the further motion

$$\begin{matrix} \dot{s}_1 = -3 \\ \dot{s}_2 = -1 \end{matrix} \quad \text{for } T' < t < T' + T'',$$

and

$$s_1(T' + T'') = 0, \qquad T'' = \tfrac{1}{9} s_2(0), \quad s_2(T' + T'') = -\tfrac{1}{9} s_2(0) \qquad \text{or}$$

$$s_2(T_1) = -\tfrac{1}{9} s_2(0), \quad T_1 = T' + T'' = \tfrac{4}{9} s_2(0) \qquad \text{(point 3)}.$$

It means that

$$|s_2(T_i)| = (\tfrac{1}{9})^i \, |s_2(0)|, \qquad s_1(T_i) = 0,$$

$$\Delta T_i = T_i - T_{i-1} = \tfrac{4}{9} s_2(T_{i-1}) = \tfrac{4}{9} (\tfrac{1}{9})^{i-1} s_2(0), \quad i = 1, 2, ..., \quad T_0 = 0.$$

Since

$$\lim_{i \to \infty} [s_2(T_i)] = 0, \; s_1(T_i) = 0,$$

$$\lim_{i \to \infty} T_i = \lim_{i \to \infty} \textstyle\sum_{i=1}^{\infty} \Delta T_i = \tfrac{4}{9} \tfrac{1}{3} s_2(0) \tfrac{1}{1 - \frac{1}{9}} = \tfrac{1}{2} s_2(0)$$

the state will reach the manifold $(s_1, s_2) = 0$ after a finite time interval and then sliding mode will arise in this manifold as in all above systems with discontinuous scalar and vector controls. The example illustrates that the conditions for the two-dimensional sliding mode to exist can not be derived from analysis of scalar cases. Even more sliding mode may exist in the intersection of discontinuity surfaces although it does not exist on each of them taken separately.

Further we will deal with the conditions for sliding mode to exist for affine systems (5.8). To derive the existence conditions, the stability of the motion projection on subspace s

$$\dot{s} = Gf + GBu \qquad (6.2)$$

should be analyzed. The control (4.2) may be represented as

$$u(x,t) = u_0(x,t) + U(x,t)sign(s),$$

$$u_0(x) = \frac{u^+(x,t) + u^-(x,t)}{2},$$

diagonal matrix U with elements

$$U_i = \frac{u_i^+(x,t) - u_i^-(x,t)}{2}, \qquad i = 1, \dots, m$$

and

$$[sign(s)]^T = [sign(s_1), \dots, sign(s_m)]$$

Then the motion projection on subspace s is governed by

$$\dot{s} = d(x) - D(x)sign(s) \qquad \text{with} \qquad d = Gf + GBu_0, \qquad D = -GBU. \tag{6.3}$$

To find the stability conditions of the origin $s = 0$ for nonlinear system (6.3), or the conditions for sliding mode to exist, we will follow the standard methodology for stability analysis of nonlinear systems – try to find a Lyapunov function.

Definition 6.1 *The set $S(x)$ in the manifold $s(x) = 0$ is the domain of sliding mode if for the motion governed by (6.2) the origin in the subspace s is asymptotically stable with finite convergence time for each x from $S(x)$.*

Definition 6.2 *Manifold $s(x) = 0$ is referred to as sliding manifold if sliding mode exists at its each point, or $S(x) = \{x : s(x) = 0\}$.*

Theorem 6.3 *$S(x)$ is a sliding manifold for the system with motion projection on subspace s governed by $\dot{s} = -Dsign(s)$, if matrix $D + D^T > 0$ is positive definite.*

Theorem 6.4 *$S(x)$ is a sliding manifold for system (6.2) if*

$$D(x) + D^T(x) > 0, \qquad \lambda_0 > d_0\sqrt{m}, \qquad \lambda_{min}(x) > \lambda_0 > 0, \qquad \|d(x)\| < d_0,$$

$\lambda_{\min}$ is the minimal eigenvalue of matrix $\frac{D+D^T}{2}$, $\lambda_{\min} > 0$.

The statements of the both theorems may be proven using sum of absolute values of $s_i V = s^T sign(s) > 0$ as a Lyapunov function.

7 Design Principles

7.1 Decoupling and Invariance

The above mathematical results constitute the background of the design methods for sliding mode control involving two independent subproblems of lower

dimensions:
– Design of the desired dynamics for a system of the $(n-m)$th order by a proper choice of a sliding manifold $s = 0$.
– Enforcing sliding motion in this manifold which is equivalent to stability problem of the mth order system.

All the design methods will be discussed for affine systems (5.8). Affine systems are linear with respect to control but not necessary with the respect to state. Since the principle operating mode is in the vicinity of the discontinuity points, the effects inherent in the systems with infinite feedback gains may be obtained with finite control actions. As a result sliding mode control is an efficient tool to control dynamic high-order nonlinear plants operating under uncertainty conditions (e.g., unknown parameter variations and disturbances).

Formally the sliding mode is insensitive, or invariant to "uncertainties" in the systems satisfying the matching conditions $h(x,t) \in range(B)$ [18]. Vector $h(x,t)$ characterizes all factors in a motion equation

$$\dot{x} = f(x,t) + B(x,t)u + h(x,t),$$

whose influence on the control process should be rejected. The matching condition means that disturbance vector $h(x,t)$ may be represented as a linear combination of columns of matrix B.

In the sequel, we will deal with affine systems

$$\dot{x} = f(x,t) + B(x,t)u, \qquad x, f \in \Re^n, \qquad B(x) \in \Re^{n\times m}, \quad u(x) \in \Re^m, \quad (7.1)$$

$$u(x) = \begin{cases} u^+(x,t) & \text{if } s(x) > 0 \\ u^-(x,t) & \text{if } s(x) > 0 \end{cases} (component\text{−}wise),\ s(x)^T = [s_1(x) ... s_m(x)].$$

To obtain the equation of sliding mode in manifold $s(x) = 0$ under the assumption that matrix GB (matrix $G = \{\partial s/\partial x\}$ with rows as gradients of the components of vector s) is nonsingular, the equivalent control

$$u_{eq}(x,t) = -[G(x)B(x,t)]^{-1}G(x)f(x,,t)$$

should be substituted into (7.1) for the control $u(x)$ to yield

$$\begin{aligned} \dot{x} &= f_{sm}(x,t), \\ f_{sm}(x,t) &\doteq f(x,t) - B(x,t)[G(x)B(x,t)]^{-1}G(x)f(x,t). \end{aligned} \qquad (7.2)$$

Since $s(x) = 0$ in sliding mode, this system of m algebraic equations may be solved with respect to m component of the state vector constituting subvector x_2:

$$x_2 = s_0(x_1), \quad x_2 \in \Re^m, \quad x_1 \in \Re^{n-m}, \quad x^T = [\, x_1^T \;\; x_2^T \,]$$

and $s(x) = 0$.

Replacing x_2 by $s_0(x_1)$ in the first $n-m$ equations of (7.2) yields a reduced order sliding mode equation

$$\dot{x}_1 = f_{1sm}[x_1, s_0(x_1), t] \qquad \text{where}$$
$$f_{sm}^T(x,t) = f_{sm}^T(x_1, x_2, t) = [f_{1sm}^T(x_1, x_2, t) \quad f_{2sm}^T(x_1, x_2, t)]. \tag{7.3}$$

The motion equation (7.3) depends on function $s_0(x_1)$, i.e. on the equation of the discontinuity manifold. Function $s_0(x_1)$ may be handled as m-dimensional control for the reduced order system. It should be noted that the design problem is not a conventional one since right-hand sides in (7.2) and (7.3) depend not only on the discontinuity manifold equation but on the gradient matrix G as well. If a class of functions $s(x)$ is pre-selected, for instance linear functions or in the form of finite series, then both $s(x)$, G and, as a result, the right-hand sides in (7.3) depend on the set of parameters to be selected when designing the desired dynamics of sliding motion.

The second-order system (5.1) with a scalar control

$$\dot{x}_1 = a_{11}x_1 + a_{12}x_2 + b_1 u + d_1 f(t)$$
$$\dot{x}_2 = a_{12}x_1 + a_{22}x_2 + b_2 u + d_2 f(t)$$

$$u = -M \text{ sign } (s), \qquad s = c_1 x_1 + c_2 x_2.$$

may serve as an example. It was shown in Sect. 5 that sliding mode along the switching line

$$s = c_1 x_1 + c_2 x_2 = 0$$

is governed by the first-order equation

$$\dot{x}_1 = [a_{11} - a_{12}c_2^{-1}c_1 - (cb)^{-1}b_1(ca^1 - ca^2c_2^{-1}c_1)]x_1 + [d_1 - b_1(cb)^{-1}(cd)]f(t),$$

where $c = [c_1 \quad c_2]$, $b^T = [b_1 \quad b_2]$, $(a^1)^T = [a_{11} \quad a_{21}]$, $[a^2]^T = [a_{12} \quad a_{22}]$, $d^T = [d_1 \quad d_2]$ and cb and c_2 are assumed to be different from zero. The equation may be rewritten in the form

$$\dot{x}_1 = [a_{11} - a_{12}c_1^* - (c^*b)^{-1}b_1(c^*a^1 - c^*a^2c_1^*)]x_1 + [d_1 - b_1(c^*b)^{-1}(c^*d)]f(t)$$

with $c^* = [c_1^* \quad 1]$, and $c_1^* = c_2^{-1}c_1$. Hence only one parameter c_1^* should be selected to provide the desired motion of the first-order dynamics in our second order example.

7.2 Regular Form

The design procedure may be illustrated easily for the systems represented in the *Regular Form*

$$\begin{aligned} &\dot{x}_1 = f_1(x_1, x_2, t), \qquad x_1 \in R^{n-m} \\ &\dot{x}_2 = f_2(x_1, x_2, t) + B_2(x_1, x_2, t)u, \qquad x_2 \in R^m, \qquad \det(B_2) \neq 0. \end{aligned} \tag{7.4}$$

The state subvector x_2 is handled as a fictitious control in the first equation of (7.4) and selected as a function of x_1 to provide the desired dynamics in the first subsystem (the design problem in the system of the $(n-m)$th order with m-dimensional control): $x_2 = -s_0(x_1)$. Then the discontinuous control should be designed to enforce sliding mode in the manifold

$$s(x_1, x_2) = x_2 + s_0(x_1) = 0 \tag{7.5}$$

(the design problem of the mth order with m-dimensional control).

After a finite time interval sliding mode in the manifold (7.5) starts and the system will exhibit the desired behavior governed by

$$\dot{x}_1 = f_1[x_1, -s_0(x_1), t].$$

Note that the motion is of a reduced order and depends neither function $f_2(x_1, x_2, t)$ nor function $B_2(x_1, x_2, t)$ in the second equation of the original system (7.4).

Since the design procedures for the systems in Regular Form (7.4) are simpler than those for (7.1), it is of interest to find the class of systems (7.1) for which a nonlinear coordinate transformation exists such that it reduces the original system (7.1) to the form (7.4). By the assumption $rank(B) = m$, therefore (7.1) may be represented as

$$\begin{aligned} \dot{x}_1 &= f_1(x_1, x_2, t) + B_1(x_1, x_2, t)u, \qquad x_1 \in R^{n-m} \\ \dot{x}_2 &= f_2(x_1, x_2, t) + B_2(x_1, x_2, t)u, \qquad x_2 \in R^m, \qquad \det(B_2) \neq 0. \end{aligned}$$

For coordinate transformation

$$y_1 = \varphi(x, t), \qquad y_2 = x_2, \qquad y_1, \ \varphi \in \Re^{n-m},$$

the equation for y_1

$$\dot{y}_1 = \frac{\partial \varphi}{\partial x} f + \frac{\partial \varphi}{\partial x} Bu + \frac{\partial \varphi}{\partial t}$$

does not depend on control u, if vector function φ is a solution to matrix partial differential equation

$$\frac{\partial \varphi}{\partial x} B = 0. \tag{7.6}$$

Then the motion equations with respect to y_1 and y_2 are in the regular form. Necessary and sufficient conditions of solution existence and uniqueness for (7.6) may be found in terms of Pfaff 's forms theory and Frobenius theorem which constitute a well developed branch of mathematical analysis [19].

We illustrate now the method of reducing an affine system to the regular form for the case of scalar control

$$\dot{x} = f(x, t) + b(x, t)u, \qquad x \in \Re^n, \qquad u \in \Re, \tag{7.7}$$

$b(x, t)$ is an n-dimensional vector with components $b_i(x, t)$ and $b_n(x, t) \neq 0$.

Let a solution to an auxiliary system of $(n-1)$ order

$$dx_i/dx_n = b_i/b_n \quad i = 1, ..., n-1 \tag{7.8}$$

be set of functions

$$x_i = \tilde{\varphi}_i(x_n, t, c_1, c_2, ..., c_{n-1}), \quad i = 1, ..., n-1 \text{ with integration constants } c_i' s. \tag{7.9}$$

Solve these algebraic system of equations for c_i

$$c_i = \varphi_i(x),$$

and introduce the nonsingular coordinate transformation

$$y_i = \varphi_i(x), \qquad i = 1, ..., n-1 \tag{7.10}$$

Since $\frac{d\varphi_i}{dx_n} = (grad\varphi_i)^T b/b_n = 0$ or $(grad\varphi_i)^T b = 0$ the motion equation with respect to new state vector $(y_1, ..., y_{n-1}, x_n)$ is of form

$$\begin{aligned} \dot{y}_i &= (grad\varphi_i)^T f_i, \qquad i = 1, ..., n-1 \\ \dot{x}_n &= f_n + b_n u, \end{aligned}$$

Replacing x_i by the solution of (7.10) as functions of$(y_1, ..., y_{n-1}, x_n)$ leads to motion equations

$$\begin{aligned} \dot{y} &= f^*(y, x_n, t), \\ \dot{x}_n &= f_n^*(y, x_n) + b_n^*(y, x_n, t)u, \end{aligned} \tag{7.11}$$

y, f^* are $(n-1)$-dimensional vectors, f_n^*, b_n^* are scalar functions.

The system with respect to y and x_n is in the regular form (7.4) with $(n-1)$ and first order blocks.

Note the following characteristics for the design in the regular form:

1. In contrast to (7.2) and (7.3), the sliding mode equation does not depend on gradient matrix G, which makes the design problem at the first stage a conventional one-design of m-dimensional control x_2 in $(n-m)$-dimensional system with state vector x_1.
2. Calculation of the equivalent control to find the sliding mode equation is not needed.
3. The condition $\det(GB) = \det(B_2) \neq 0$ holds (Recall that this condition is needed to enforce sliding mode in the pre-selected manifold $s(x) = 0$).
4. Sliding mode is invariant with respect to functions f_2 and B_2 in the second block.

7.3 Block Control Principle

Due to the higher complexity of the modern plants and processes, the need of decomposition of the original system into independent subsystems of lower complexity is highly stressed and has been intensively studied. The objective of this section is to propose a sliding mode control design method based on block control principle [20, 21] reducing the design procedure to a set of simplest subproblems of low dimensions.

For a linear controllable system

$$\dot{x} = Ax + Bu \tag{7.12}$$

where $x \in R^n$, $u \in R^m$, the sliding mode control design principle may be demonstrated after reducing the system (7.12) to Regular Form

$$\begin{aligned} \dot{x}_1 &= A_{11}x_1 + A_{12}x_2 \\ \dot{x}_2 &= A_{21}x_1 + A_{22}x_2 + B_2 u \end{aligned} \tag{7.13}$$

where $x_1 \in R^{n-m}$, $x_2 \in R^m$, $u \in R^m$, $\det(B_2) \neq 0$, $rank(B_2) = \dim(u)$. The fictitious control of second subsystem, x_2, can be designed as

$$x_2 = A_{12}^+(-A_{11}x_1 + \Lambda)x_1, \tag{7.14}$$

where A_{12}^+ is the psudoinverse of A_{12} and Λ is the desired spectrum. To implement the control design (7.14), a sliding control can be designed as

$$u = -M \cdot sign(s). \tag{7.15}$$

where

$$s = x_2 - A_{12}^+(-A_{11}x_1 + \Lambda)x_1 \tag{7.16}$$

When sliding mode is enforced, s is equal to zero, condition (7.14) holds, and the system behavior depends only on the upper subsystem with the desired dynamics

$$\dot{x}_1 = \Lambda x_1. \tag{7.17}$$

A further development of the Regular Form is the so-called Block Control Form

$$\begin{aligned} \dot{x}_r &= A_r \; x_r + B_r x_{r-1} \\ \dot{x}_{r-1} &= A_{r-1} \begin{bmatrix} x_r \\ x_{r-1} \end{bmatrix} + B_{r-1}x_{r-2} \\ &\vdots \\ \dot{x}_2 &= A_2 \begin{bmatrix} x_r \\ \vdots \\ x_2 \end{bmatrix} + B_2 x_1 \\ \dot{x}_1 &= A_1 \begin{bmatrix} x_r \\ \vdots \\ x_1 \end{bmatrix} + B_1 u \end{aligned} \tag{7.18}$$

where $\dim(u) = rank(B_1)$, $rank(B_i) = \dim(x_i)$, $i = 2, \cdots, r$.

The fictitious control could be designed in following procedure such that every subsystem in the block control form has a desired spectrum. Let Λ_i,

$i = 2, \cdots, r$ be the desired spectra. The fictitious control x_{r-1} of the first subsystem should be designed as

$$x_{r-1} = -B_r^+ (A_r + \Lambda_r) x_r$$

to have the desired motion in the first block.

The fictitious control of the next block x_{r-2} can be selected to reduce

$$s_{r-1} = x_{r-1} + B_r^+ (A_r + \Lambda_r) x_r$$

to zero with the desired rate of convergence, determined by Λ_{r-1}. It can be found from the equation

$$\dot{s}_{r-1} = \tilde{A}_{r-1} \begin{bmatrix} x_r \\ x_{r-1} \end{bmatrix} + B_{r-1} x_{r-2}:$$

$$x_{r-2} = -B_{r-1}^+ (\tilde{A}_{r-1} \begin{bmatrix} x_r \\ x_{r-1} \end{bmatrix} - \Lambda_{r-1} s_{r-1})$$

($\tilde{A}_{r-1}$ depends on matrices of the first two blocks and Λ_r).

Follow the procedure step by step, the fictitious control, x_1 can be found

$$x_1 = -B_2^+ \left(\tilde{A}_2 - \Lambda_2 \right) \begin{bmatrix} x_r \\ \vdots \\ x_2 \end{bmatrix}.$$

The final step is to select discontinuous control $u = -M \cdot sign(s_1)$ enforcing sliding mode in the manifold

$$s_1 = x_1 + B_2^+ (A_2 + \Lambda_2) \begin{bmatrix} x_r \\ \vdots \\ x_2 \end{bmatrix} = 0.$$

After sliding mode is enforced, s_1 will be equal to zero which leads to convergence of all functions s_i to zero with the desired spectrum. It is important that the design task at each step is simple: it is of a reduced order with equal dimensions of the state and control.

7.4 Enforcing Sliding Modes

At the second stage of feedback design, discontinuous control should be selected such that sliding mode is enforced in manifold $s = 0$. As shown in Chap. 6, sliding mode will start at manifold $s = 0$ if the matrix $GB + (GB)^T$ is positive-definite and the control is of the form

$$u = -M(x,t) \quad sign(s) \qquad \text{(component-wise)}.$$

A positive scalar function $M(x,t)$ is chosen to satisfy the inequality

$$M(x,t) > \lambda^{-1} \|Gf\|,$$

where λ is the lower bound of the eigenvalues of the matrix $GB + (GB)^T$.

Now we demonstrate a method of enforcing sliding mode in manifold $s = 0$ for an arbitrary nonsingular matrix GB. The motion projection equation on subspace s can be written in the form

$$\dot{s} = GB(u - u_{eq}).$$

Remind that the equivalent control u_{eq} is the value of control such that time derivative of vector s is equal to zero.

Let $V = \frac{1}{2}s^T s$ be a Lyapunov function candidate and the control be of the form

$$u = -M(x,t)sign(s*), \qquad \text{s}* \; = (GB)^T s.$$

Then $\dot{V} = -M(x,t)[s*^T\, sign(s*) + s*^T \frac{u_{eq}}{M(x,t)}]$ and $\dot{V}$ is negative-definite for $M(x,t) > \|u_{eq}\|$. It means that sliding mode is enforced in manifold $s* = 0$ that is equivalent to its existence in manifold $s = 0$ selected on the first step of the design procedure. It is important that the conditions for sliding mode to exist are inequalities. Therefore upper estimate of the disturbances is needed rather than precise information on their values.

Example

To demonstrate the sliding mode control design methodology consider the conventional problem of linear control theory: *eigenvalue placement* in a linear time invariant multidimensional system

$$\dot{x} = Ax + Bu,$$

where x and u are $n-$ and m-dimensional state and control vectors, respectively, A and B are constant matrices, rank $(B) = m$. The system is assumed to be controllable.

For any controllable system there exists a linear feedback $u = Fx$ ($\boldsymbol{F}$ being a constant matrix) such that the eigenvalues of the feedback system, i.e. of matrix $A + BF$, take the desired values and, as a result, the system exhibits desired dynamic properties.

Now we will show that the eigenvalue placement task may be solved in the framework of the sliding mode control technique dealing with a reduced order system. The core idea is to utilize the methods of linear control theory for reduced order equations and to employ one of the methods of enforcing sliding modes with desired dynamics.

As it was demonstrated in this section, the design becomes simpler for systems represented in the regular form. Reducing system equations to the regular form will be performed as a preliminary step in the design procedures. Since rank $(B) = m$, matrix B may be partitioned (after reordering the state vector components) as

$$B = \begin{bmatrix} B_1 \\ B_2 \end{bmatrix}$$

where $B_1 \in \Re^{(n-m)\times m}$, $B_2 \in \Re^{m\times m}$ with $det\ B_2 \neq 0$. The nonsingular coordinate transformation

$$\begin{bmatrix} x_1 \\ x_2 \end{bmatrix} = Tx, \quad T = \begin{bmatrix} I_{n-m} & -B_1 B_2^{-1} \\ 0 & B_2^{-1} \end{bmatrix}$$

reduces the system equations to the regular form

$$\begin{aligned} \dot{x}_1 &= A_{11}x_1 + A_{12}x_2 \\ \dot{x}_2 &= A_{21}x_1 + A_{22}x_2 + u, \end{aligned}$$

where $x_1 \in \Re^{(n-m)}$, $x_2 \in \Re^m$ and A_{ij} are constant matrices for $i, j = 1, 2$.

It follows from controllability of (A, B) that the pair (A_{11}, A_{12}) is controllable as well. Otherwise an uncontrollable subspace exists in the first block which is an uncontrollable subspace in the original system as well. Handling x_2 as an m-dimensional fictitious control in the controllable $(n\text{-}m)$-dimensional first subsystem all $(n-m)$ eigenvalues may be assigned arbitrarily by a proper choice of matrix C in $x_2 = -Cx_1$. To provide the desired dependence between components x_1 and x_2 of the state vector, sliding mode should be enforced in the manifold $s = x_2 + Cx_1 = 0$, where $s^T = (s_1, ..., s_m)$ is the difference between the real values of x_2 and its desired value $-Cx_1$.

After sliding mode starts, the motion is governed by a reduced order system with the desired eigenvalues

$$\dot{x}_1 = (A_{11}x_1 - A_{12}C)x_1.$$

For a piece-wise linear discontinuous control

$$u = -(\alpha\,|x| + \delta)\ \text{sign}\ (s),$$

with

$$|x| = \sum_{i=1}^{n} |x_i| \qquad \text{sign}\ (s)^T = [\text{sign}\ (s_1) ... \ \text{sign}\ (s_m)];$$

α and δ being constant positive values,
calculate the time derivative of positive definite function

$$\begin{aligned} V &= \tfrac{1}{2}s^T s \\ \dot{V} &= s^T[(CA_{11} + A_{21})x_1 + (CA_{12} + A_{22})x_2] - (\alpha\,|x| + \delta)\,|s| \\ &\leq |s|\,|(CA_{11} + A_{21})x_1 + (CA_{12} + A_{22})x_2| - (\alpha\,|x| + \delta)\,|s|. \end{aligned}$$

It is evident that there exists such value of α that for any δ the time derivative $\dot{V}$ is negative which validates convergence of the state vector to manifold $s = 0$ and existence of sliding mode with the desired dynamics. The time interval preceding the sliding motion may be decreased by increasing parameters α and δ in the control.

7.5 Unit Control

The objective of this section is to demonstrate a design method for discontinuous control enforcing sliding mode in some manifold without individual selection of each component of control as a discontinuous state function. The approach implies design of control based on a Lyapunov function selected for a nominal (feedback or open loop) system. The control is to be found such that the time-derivative of the Lyapunov function is negative along the trajectories of the system with perturbations caused by uncertainties in the plant model and environment conditions.

The roots of the above approach may be found in papers by G. Leitmann and S. Gutman published in the 1970s [22, 23]. The design idea may be explained for an affine system

$$\dot{x} = f(t,x) + B(t,x)\,u + h(t,x) \tag{7.19}$$

with state and control vectors $x \in \Re^n$, $u \in \Re^m$, state-dependent vectors $f(t,x), h(t,x)$ and control input matrix $B(t,x)\,x \in \Re^{n\times m}$. The vector $h(t,x)$ represents the system uncertainties and its influence on the control process should be rejected.

The equation

$$\dot{x} = f(t,x)$$

represents an open loop nominal system which is assumed to be asymptotically stable with a known Lyapunov function

$$V(x) > 0,$$

$$W_o = dV/dt\,|_{h=0,u=0} = grad\,(V)^T f < 0,$$

$$(grad(V)^T = [\frac{\partial V}{\partial x_1} \cdots \frac{\partial V}{\partial x_n}].$$

The perturbation vector $h(t,x)$ is assumed to satisfy the matching conditions, hence there exists vector $\gamma(t,x) \in \Re^m$ such that

$$h(t,x) = B(t,x)\,\gamma(t,x)\,. \tag{7.20}$$

$\gamma(t,x)$ may be an unknown vector with known upper scalar estimate $\gamma_0(t,x)$

$$\|\gamma(t,x)\| < \gamma_0(t,x)\,.$$

Calulate the time derivative of Lyapunov function $V(x)$ along the trajectories of the perturbed system as

$$W = dV/dt = W_0 + grad\,(V)^T\,B\,(u+\gamma)$$

For control u depending on the upper estimate of the unknown disturbance, chosen as

$$u = -\rho\,(t,x)\,\frac{B^T grad\,(V)}{\|B^T grad\,(V)\|} \tag{7.21}$$

with a scalar function $\rho\,(t,x) > \gamma_0(x,t)$ and

$$\|B^T grad\,(V)\|^2 = \left[B^T grad\,(V)\right]\left[B^T grad\,(V)\right].$$

The time derivative of the Lyapunov function $V\,(x)$

$$\begin{aligned} W &= W_0 - \rho\,(t,x)\,\|B^T grad\,(V)\| + grad\,(V)^T\, B\gamma\,(t,x) \\ &< W_0 - \|B^T grad\,(V)\|\,[\rho\,(t,x) - \gamma_0\,(t,x)] < 0\,. \end{aligned}$$

is negative. This implies that the perturbed system with control (7.21) is asymptotically stable as well.

Two important features should be underlined for the system with control (7.21):

1. Control (7.21) undergoes discontinuities in $(n-m)$-dimensional manifold $s(x) = B^T grad\,(V) = 0$ and is a continuous state function beyond this manifold. This is the principle difference between control (7.1) and all the control inputs in the previous sections with individual switching functions for each control component.
2. The disturbance $h\,(t,x)$ is rejected *due to enforcing sliding mode* in the manifold $s\,(x) = 0$. Indeed, if the disturbance (7.20) is rejected, then control u should be equal to $-\gamma\,(t,x)$ which is not generally the case for the control (7.21) beyond the discontinuity manifold $s\,(x) = B^T grad\,(V) \neq 0$. It means that sliding mode occurs in the manifold $s = 0$ and the equivalent value of control u_{eq} is equal to $-\lambda\,(t,x)$.

Note that the norm of control (7.21) with the gain $\rho\,(t,x) = 1$

$$\left\|\frac{B^T grad\,(V)}{\|B^T grad\,(V)\|}\right\|$$

is equal to 1 for any value of the state vector. It explains the term "unit control" for the control (7.21).

Now we demonstrate how this approach can be applied to enforce sliding mode in manifold $s = 0$ for the system (7.19) with unknown disturbance vector $h(x,t)$ [24]. Function s is selected in compliance with some performance criterion at the first stage of feedback design. The control is designed as a discontinuous function of s

$$u = -\rho\,(t,x)\,\frac{D^T s\,(x)}{\|D^T s\,(x)\|} \tag{7.22}$$

with $D = (\partial s/\partial x)\,B$ and D being a nonsingular matrix.

The equation of a motion projection of the system (7.19) on subspace s is of the form

$$\dot{s} = (\partial s/\partial x)(f + h) + Du\,.$$

The conditions for the trajectories to converge to the manifold$s(x) = 0$ and sliding mode to exist in this manifold may be derived based on Lyapunov function

$$V = \frac{1}{2}s^T s > 0 \tag{7.23}$$

with time derivative

$$\begin{aligned} \dot{V} &= s^T (\partial s/\partial x)(f + h) - \rho(t,x)\left\|D^T s(x)\right\| \\ &< \left\|D^T s(x)\right\| \left[\left\|D^{-1}(\partial s/\partial x)(f + h)\right\| - \rho(t,x)\right] \end{aligned} \tag{7.24}$$

For $\rho(t,x) > \left\|D^{-1}(\partial s/\partial x)(f + h)\right\|$ the value of $\dot{V}$ is negative therefore the state will reach the manifold $s(x) = 0$ after a finite time interval for any initial conditions and then the sliding mode with the desired dynamics occurs. Finiteness of the interval preceding the sliding motion follows from inequality resulting from (7.23), (7.24)

$$\dot{V} < -\gamma V^{1/2} \qquad \gamma = const > 0$$

with the solution

$$V(t) < \left(-\frac{\gamma}{2}t + \sqrt{V_0}\right)^2, \qquad V_0 = V(0)\,.$$

Since the solution vanishes after some $t_s < \frac{2}{\gamma}\sqrt{V_0}$, the vector s vanishes as well and the sliding mode starts after a finite time interval.

It is of interest to note the principle difference in motions preceding the sliding mode in $s(x) = 0$ for the conventional component-wise control and unit control design methods. For the conventional method the control undergoes discontinuities, should any of the components of vector s changes sign, while the unit control is a continuous state function until the manifold $s(x) = 0$ is reached. Due to this reason unit control systems with sliding modes would hardly be recognized as VSS.

8 The Chattering Problem

The subject of this section is of great importance whenever we intend to establish the bridge between the recommendations of the theory and applications. Bearing in mind that the control has a high-frequency component, we should analyze the robustness or the problem of correspondence between an ideal sliding mode and real-life processes at the presence of unmodeled dynamics. Neglected small time constants (μ_1 and μ_2 in Fig. 24 with a linear plant) in plant models, sensors, and actuators lead to dynamics discrepancy (z_1 and

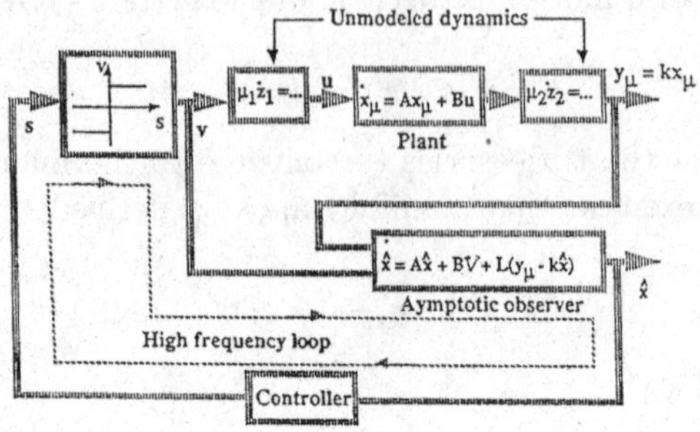

Fig. 24.

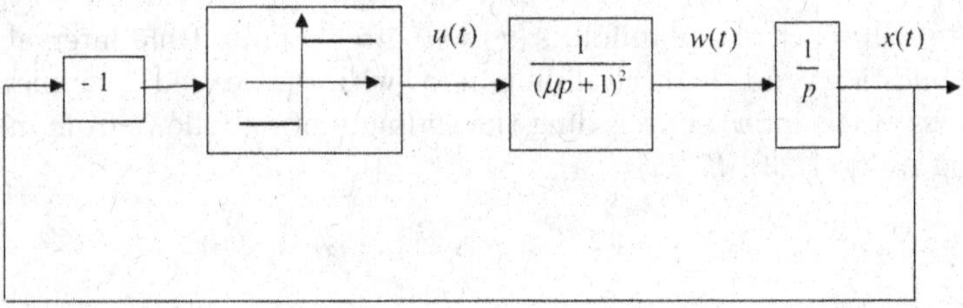

Fig. 25.

z_2 are the unmodeled-dynamics state vectors). In accordance with singular perturbation theory [25], in systems with continuous control a fast component of the motion decays rapidly and a slow one depends on the small time constants continuously. In discontinuous control systems the solution depends on the small parameters continuously as well. But unlike continuous systems, the switchings in control excite the unmodeled dynamics, which leads to oscillations in the state vector at a high frequency. The oscillations, usually referred to as *chattering*, are known to result in low control accuracy, high heat losses in electrical power circuits, and high wear of moving mechanical parts. These phenomena have been considered as a serious obstacle for applications of sliding mode control in many papers and discussions.

To qualitatively illustrate the influence of unmodeled dynamics on the system behavior, consider the simplest case shown in Fig. 25.

The motion equations may be written in the form

$$\begin{cases} \dot{x} = w \\ \dot{w} = v \\ \dot{v} = -\frac{2}{\mu}v - \frac{1}{\mu^2}w + \frac{1}{\mu^2}u. \end{cases} \tag{8.1}$$

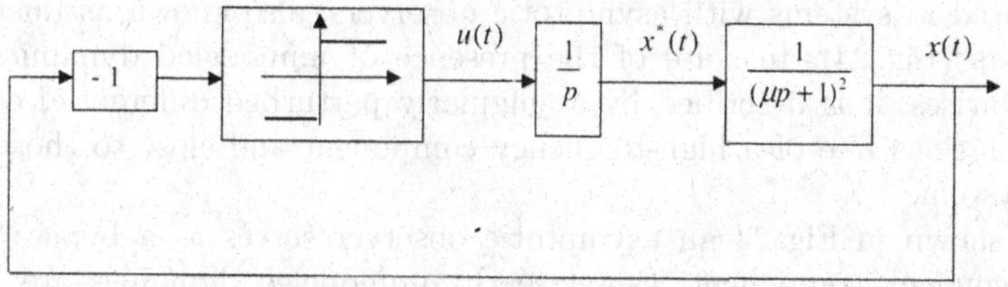

Fig. 26.

For the control $u = -M\,\mathrm{sign}(x)$, the sign-varying Lyapunov function

$$V = xv - 0.5w^2$$

has a negative time-derivative

$$\dot{V} = x\left(-\frac{2}{\mu}v - \frac{1}{\mu^2}w + \frac{1}{\mu^2}u\right) = -x\left(\frac{2}{\mu}v + \frac{1}{\mu^2}w\right) - \frac{1}{\mu^2}M\,|x|$$

for small magnitudes of v and w. This means that the motion is unstable in an $\epsilon(\mu)$-order vicinity of the manifold $s(x) = x = 0$. Alternatively to Fig. 25, the block-diagram of the system (8.1) may be represented in the form depicted in Fig. 26.

The motion equations may now be written as

$$\dot{x}^* = -M\,\mathrm{sign}(x)$$
$$\mu^2\ddot{x} + 2\mu\dot{x} + x = x^*$$

Sliding mode can not occur in the systems since the time derivative $\dot{x}$ is a continuous time function and can not have sign opposite to x in the vicinity of the point $x = 0$ where the control undergoes discontinuities.

The value of $\dot{x}^*$ is bounded and, as follows from the singular perturbation theory [25], the difference between x and x^* is of $\epsilon(\mu)$-order. The signs of x and x^* coincide beyond the $\epsilon(\mu)$-vicinity of $s(x) = x = 0$, hence the magnitudes of x^* and x decrease, i.e. the state trajectories converge to this vicinity and after a finite time interval t_1 the state remains in the vicinity. According to the analysis of (8.1), the motion in the vicinity $x = 0$ is unstable.

The fact of instability explains why chattering may appear in the systems with discontinuous controls at the presence of unmodeled dynamics. The high frequency oscillations in the discontinuous control system may be analyzed in time domain as well. The brief periods of divergence occur after switches of the control input variable $u(t)$ when the output $w(t)$ of the actuator is unable to follow the abrupt change of the control command.

The proposed solutions to the chattering problem thus focus on either avoiding control discontinuities in general or move the switching action to a controller loop without any unmodeled dynamics. A recent study and practical experience showed that the chattering caused by unmodeled dynamics may be eliminated in systems with asymptotic observers, also known as Luenberger observers (Fig. 24). In spite of the presence of unmodeled dynamics, ideal sliding arises, it is described by a singularly perturbed differential equation with solutions free of a high-frequency component and close to those of the ideal system.

As shown in Fig. 24 an asymptotic observer serves as a bypass for the high-frequency component, therefore the unmodeled dynamics are not excited. Preservation of sliding modes in systems with asymptotic observers predetermined successful application of the of sliding mode control.

Another way to reduce chattering implies replacing the discontinuous control by its continuous approximation in a boundary layer. This may result in chattering as well at the presence of unmodeled fast dynamics if the gain in the boundary layer is too high. Since the values of the time constants, neglected in the ideal model, are unknown, the designer should be oriented towards the worst case and reduce the gain such that the unmodeled dynamics are not excited. As a results the disturbance rejection properties of discontinuous (or high gain) control are not utilized to full extent.

9 Discrete-Time Systems

Most sliding mode approaches are based on finite-dimensional continuous-time models and lead to a discontinuous controller. Once such a dynamic system is "in sliding mode," its motion trajectory is confined to a manifold in the state space, i.e. to the sliding manifold. For continuous-time systems, this reduction of the system order may only be achieved by a discontinuous controller, switching at theoretically infinite frequency.

When challenged with the task of implementing a sliding mode controller in a practical system, the control engineer has two options:

- *Direct, analog implementation of discontinuous controller with a very fast switching device, e.g. with power transistors.*
- *Discrete implementation of the sliding mode controller, e.g. with a digital micro-controller.*

The first method is only suitable for systems with a voltage input allowing the use of analog switching devices. Most other systems are usually based on a discrete micro-controller based implementation. However, a discontinuous controller designed for a continuous-time system model would lead to chattering when implemented without modifications in discrete time with a finite sampling rate. This discretization chattering is different from the chattering problem caused by unmodeled dynamics as discussed in Chap. 8. Discretization chattering is due to the fact that the switching frequency is limited to the sampling rate, but correct implementation of a sliding mode controller requires infinite switching frequency. The following example will illustrate the difference between ideal continuous time sliding mode and direct discrete implementation with discretization chattering. The subsequent sections of this chapter are dedicated to the development of a discrete-time sliding mode concept to eliminate the chattering.

9.1 Discrete-Time Sliding Mode Concept

Before developing the concept of discrete-time sliding mode, let us revisit the principle of sliding mode in continuous time systems with an ideal discontinuous controller from an engineering point of view. Assume that in a general continuous-time system

$$\dot{x} = f(x, u, t) \tag{9.1}$$

with a discontinuous scalar controller

$$u = \begin{cases} u_0 & \text{if} \quad s(x) > 0 \\ -u_0 & \text{if} \quad s(x) < 0 \end{cases} \tag{9.2}$$

sliding mode exists in some manifold $s(x) = 0$ (Fig. 27).

Note the following observations characterizing the nature of sliding mode systems:

- The time interval between the initial point $t = 0$ and the reaching of the sliding manifold $s(x) = 0$ at t_{sm} is finite, in contrast to systems with a continuous controller, which exhibit asymptotic convergence.
- Once the system is "in sliding mode" for all $t \geq t_{sm}$, its trajectory motion is confined to the manifold $s(x) = 0$ and the order of the closed-loop system dynamics is less than the order of the original system.
- After sliding mode has started at t_{sm}, the system trajectory cannot be back-tracked beyond the manifold $s(x) = 0$ like in systems without discontinuities. In other words, at any point $t_0 \geq t_{sm}$, it is not possible to determine the time instance t_{sm} or to reverse calculate the trajectory for $t < t_{sm}$ based on information of the system state at t_0.

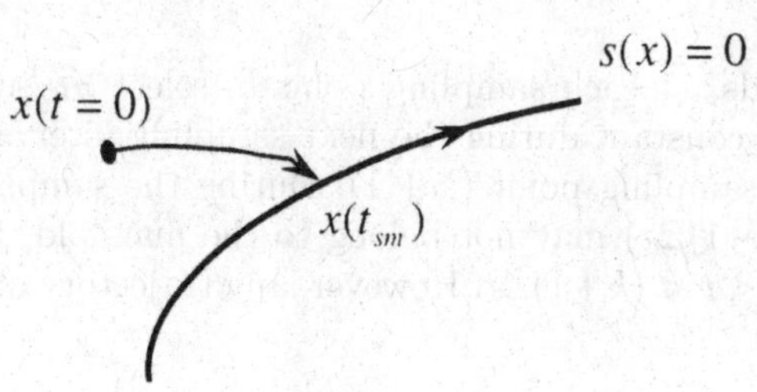

Fig. 27.

However, during both time intervals before and after reaching the sliding manifold, the state trajectories are continuous time functions and the relation between two values of the state at the ends of a finite time interval $t = [t_0,\ t_0 + \Delta t]$ may be found by solving (9.1) as

$$x(t_0 + \Delta t) = F\left(x(t_0)\right) \tag{9.3}$$

where $F\left(x(t)\right)$ is a continuous function as well. When derived for each sampling point $t_k = k\Delta t, k = 1, 2, ...,$ this equation is nothing but the discrete time representation of for the continuous time prototype (9.1), i.e.

$$x_{k+1} = F(x_k), \qquad x_k = x(k\Delta t).$$

Starting from time instance t_{sm}, the state trajectory belongs to the sliding manifold with

$$s\left(x(t)\right) = 0, \qquad \text{or for some} \qquad k_{sm} \geq \frac{t_{sm}}{\Delta t}, \qquad s(x_k) = 0, \qquad \forall k \geq k_{sm}.$$

It seems reasonable to call this motion “sliding mode in discrete time,” or “discrete-time sliding mode.” Note that the right-hand side of the motion equation of the system with discrete time sliding mode is a continuous state function.

So far, we have generated a discrete time description of a continuous-time sliding mode system. Next, we need to derive a discrete-time control law which may generate sliding mode in a discrete-time system. Let us return to the system (9.1) and suppose that for any constant control input u for any initial condition $x(0)$, the solution to (9.1) may be found in closed form, i.e.

$$x(t) = F\left(x(0), u\right).$$

Asume that control u may be chosen arbitrarily and follow the procedure below:

1. Select constant $u\left(x(t=0), \Delta t\right)$ for a given time interval Δt such that $s\left(x(t = \Delta t)\right) = 0$.
2. Next, find constant $u\left(x(t = \Delta t), \Delta t\right)$ such that $s\left(x(t = 2\Delta t)\right) = 0$.
3. In general, for each $k = 1, 2, \ldots,$ choose constant $u\left(x_k, \Delta t\right)$ such that $s\left(x_{k+1}\right) = 0$.

In other words, at each sampling point k, select u_k such that this control input, to be constant during the next sampling interval Δt, will achieve $s\left(x_{k+1}\right) = 0$ at sampling point $(k+1)$. During the sampling interval, state $x\left(k\Delta t < t < (k+1)\Delta t\right)$ may not belong to the manifold, i.e. $s\left(x(t)\right) \neq 0$ is possible for $k\Delta t < t < (k+1)\Delta t$. However, any trajectory of the discrete-time system

$$\begin{aligned} x_{k+1} &= F\left(x_k, u_k\right) \\ u_k &= u(x_k) \end{aligned}$$

hits the sliding manifold at each sampling point, i.e. $s(x_k) = 0 \; \forall k = 1, 2, ...,$ is fulfilled.

Since $F(x(0), u)$ for any constant u tends to $x(0)$ as $\Delta t \to 0$, the function $u(x(0), \Delta t)$ may exceed the available control resources u_0. As a result, the control steers state x_k to zero only after a finite number of steps k_{sm}. Thus the manifold is reached after a finite time interval $t_{sm} = k_{sm}\Delta t$ and thereafter the state x_k remains on the manifold. In analogy to continuous-time systems, this motion may be referred to as "discrete-time sliding mode." Note that sliding mode is generated in the discrete-time system with control $-u_0 \leq u \leq u_0$ as a continuous function of the state x_k and being piece-wise constant during the sampling interval.

The above example clarifies the definition of the term "discrete-time sliding mode" introduced [26] for an arbitrary finite-dimensional discrete-time system.

Definition 9.1 (Discrete-time sliding mode) *In the discrete-time dynamic system*

$$x_{k+1} = F(x_k, u_k), \quad x \in \Re^n, \;\; u \in \Re^m, \; m \leq n \tag{9.4}$$

discrete-time sliding mode takes place on a subset Σ of the manifold $\sigma = \{x : s(x) = 0\}, s \in \Re^m$, if there exists an open neighborhood $\aleph$ of this subset such that for each $x \in \aleph$ it follows that $s(F(x_{k+1})) \in \Sigma$.

In contrast to continuous-time systems, sliding mode may arise in discrete-time systems with a continuous function in the right-hand side of the closed loop system equation. Nevertheless, the aforementioned characteristics of continuous-time sliding mode have been transferred to discrete-time sliding mode. Practical issues will be discussed in the subsequent section using linear systems as an example.

9.2 Linear Discrete-Time Systems with Known Parameters

This section deals with discrete-time sliding mode controllers for linear time-invariant continuous-time plants. Let us assume that a sliding mode manifold is linear for an nth-order discrete-time system $x_{k+1} = F(x_k)$, i.e. $s_k = Cx_k$, $C \in \Re^{m \times n}$ with m control inputs. According to definition the sliding mode existence condition is of the form

$$s_{k+1} = C(F(x_k)) = 0 \tag{9.5}$$

to design a discrete-time sliding mode controller based on condition (9.5), consider the discrete-time representation of the linear time-invariant system

$$\dot{x}(t) = Ax(t) + Bu(t) + Dr(t) \tag{9.6}$$

with state vector $x(t) \in \Re^n$, control $u(t) \in \Re^m$, reference input $r(t)$, and constant system matrices A, B, and D. Transformation to discrete time with sampling interval Δt yields

$$x_{k+1} = A^* x_k + B^* u_k + D^* r_k, \tag{9.7}$$

where

$$A^* = e^{A\Delta t}, \quad B^* = \int_0^{\Delta t} e^{A(\Delta t - t)} B \, \mathrm{d}\tau, \quad D^* = \int_0^{\Delta t} e^{A(\Delta t - t)} D \, \mathrm{d}\tau \tag{9.8}$$

and the reference input $r(t)$ is assumed to be constant during the sampling interval Δt. In accordance with (9.5), discrete-time sliding mode exists if the matrix CB^* has an inverse and the control u_k is designed as the solution of

$$s_{k+1} = CA^* x_k + CD^* r_k + CB^* u_k = 0 \tag{9.9}$$

In other words, control u_k should be chosen as

$$u_k = -\left(CB^*\right)^{-1} \left(CA^* x_k + CD^* r_k\right). \tag{9.10}$$

By analogy with continuous-time systems, the control law (9.10) yielding motion in the manifold $s = 0$ will be referred to as "equivalent control." To reveal the structure of $u_{k_{eq}}$, let us represent it as the sum of two linear functions

$$u_{k_{eq}} = -\left(CB^*\right)^{-1} s_k - \left(CB^*\right)^{-1} \left(\left(CA^* - C\right) x_k + CD^* r_k\right) \tag{9.11}$$

and

$$s_{k+1} = s_k + \left(CA^* - C\right) x_k + CD^* r_k + CB^* u_k. \tag{9.12}$$

As it was mentioned in the previous section, $u_{k_{eq}}$ can exceed the available control resources with $\Delta t \to 0$ for initial $s_k \neq 0$, since $\left(CB^*\right)^{-1}\left(CA^* - C\right)$ and $\left(CB^*\right)^{-1} CD^*$ take finite values. Hence the real-life bounds for control u_k should be taken into account. Suppose that the control can vary within $\|u_k\| \leq u_0$ and the available control resources are such that

$$\left\|\left(CB^*\right)^{-1}\right\| \cdot \left\|\left(CA^* - C\right) x_k + CD^* r_k\right\| < u_0. \tag{9.13}$$

The controller

$$u_k = \begin{cases} u_{k_{eq}} & \text{for } \left\|u_{k_{eq}}\right\| \leq u_0 \\ u_0 \dfrac{u_{k_{eq}}}{\left\|u_{k_{eq}}\right\|} & \text{for } \left\|u_{k_{eq}}\right\| > u_0 \end{cases} \tag{9.14}$$

complies with the bounds of the control resources. As shown above, $u_k = u_{k_{eq}}$ for $\left\|u_{k_{eq}}\right\| \leq u_0$ yields motion in the sliding manifold $s = 0$. To proof convergence to this domain, consider the case $\left\|u_{k_{eq}}\right\| > u_0$, but in compliance with condition (9.13). From (9.11)-(9.14) it follows that

$$s_{k+1} = \left(s_k + \left(CA^* - C\right)x_k + CD^*r_k\right)\left(1 - \frac{u_0}{\left\|u_{k_{eq}}\right\|}\right) \qquad \textit{with } u_0 < \left\|u_{k_{eq}}\right\| \tag{9.15}$$

Thus

$$\begin{aligned} \|s_{k+1}\| &= \left\|\left(s_k + \left(CA^* - C\right)x_k + CD^*r_k\right)\right\| \left(1 - \frac{u_0}{\left\|u_{k_{eq}}\right\|}\right) \\ &\leq \|s_k\| + \left\|\left(CA^* - C\right)x_k + CD^*r_k\right\| - \frac{u_0}{\left\|(CB^*)^{-1}\right\|} \\ &< \|s_k\| \end{aligned} \tag{9.16}$$

Hence $\|s_k\|$ decreases monotonously and, after a finite number of steps, $\left\|u_{k_{eq}}\right\| < u_0$ is achieved. Discrete-time sliding mode will take place from the next sampling point onwards.

Controller (9.14) provides chattering-free motion in the manifold $s = 0$ in contrast to the direct implementation of a discontinuous controller resulting in discretization chattering in the vicinity of the sliding manifold. Similarly to the case of continuous-time systems, the equation $s = Cx = 0$ enables the reduction of system order, and the desired system dynamics "in sliding mode" can be designed by appropriate choice of matrix C.

9.3 Linear Discrete-Time Systems with Unknown Parameters

Complete information of the plant parameters is required for implementation of controller (9.14), which may not be available in practice. To extend the discrete-time sliding mode concept to systems with unknown parameters, suppose that system (9.7) operates under uncertainty conditions: the matrices A and D, and the reference input r_k are assumed unknown and may vary in some ranges. Similarly to (9.14) the controller

$$u_k = \begin{cases} -\left(CB^*\right)^{-1} s_k & \text{for } \left\|\left(CB^*\right)^{-1} s_k\right\| \leq u_0 \\ -u_0 \frac{(CB^*)^{-1} s_k}{\left\|(CB^*)^{-1} s_k\right\|} & \text{for } \left\|\left(CB^*\right)^{-1} s_k\right\| > u_0 \end{cases} \tag{9.17}$$

respects the bounds of the control resources. Furthermore, controller (9.17) does not depends on the plant parameters A and D and the reference input r_k. Substitution of (9.17) into the system equations of the previous section leads to

$$s_{k+1} = s_k\left(1 - \frac{u_0}{\left\|\left(CB^*\right)^{-1} s_k\right\|}\right) + \left(CA^* - C\right)x_k + CD^*r_k$$

for $u_0 < \left\| (CB^*)^{-1} s_k \right\|$ and similarly to (9.16)

$$
\begin{aligned}
\|s_{k+1}\| &\leq \|s_k\| \left(1 - \frac{u_0}{\|(CB^*)^{-1} s_k\|}\right) + \|(CA^* - C)\, x_k + CD^* r_k\| \\
&\leq \|s_k\| - \frac{u_0 \|s_k\|}{\|(CB^*)^{-1} s_k\|} + \|(CA^* - C)\, x_k + CD^* r_k\| \\
&\leq \|s_k\| - \frac{u_0}{\|(CB^*)^{-1}\|} + \|(CA^* - C)\, x_k + CD^* r_k\| \\
&< \|s_k\| .
\end{aligned}
$$

Hence, as for the case with complete knowledge of system parameters discussed in Sect. 9.2, the value of $\|s_k\|$ decreases monotonously and after a finite number of steps, control$\|u_k\| < u_0$ will be within the available resources. Substituting this value from (9.17) into (9.13) results in

$$
s_{k+1} = s_k + (CA^* - C)\, x_k + CD^* r_k .
$$

Since the matrices$(CA^* - C)$ and CD^* are of Δt-order, the system motion will be in a Δt-order vicinity of the sliding manifold $s = 0$. Convergence to the vicinity of the sliding manifold is achieved in finite time; thereafter, the motion trajectory does not follow the sliding manifold exactly, but rather remains within a Δt-order vicinity. This result should be expected from systems operating under uncertainty conditions, since we are dealing with an open-loop system during each sampling interval. In contrast to discrete-time systems with direct implementation of the discontinuous controller, this motion is free of discretization chattering as well.

10 Infinite-Dimensional Systems

All design methods discussed in the previous sections are oriented towards dynamic processes governed by ordinary differential equations or finite-dimensional discrete time equations. Many processes of modern technology are beyond these classes and their models should be treated as infinite-dimensional:

- Partial differential equations
- Integro-differential equations
- Systems with delays.

In particular the systems with state Q depending on time t and spatial variable x are governed by partial differential equation

$$
\partial Q(x,t)/\partial t = AQ(x)
$$

with an unbounded operator A (Examples: $\partial^2/\partial x^2$, $\partial^4/\partial x^4$).

Theoretical generalizations to infinite-dimensional cases involve principle difficulties since the both mathematical and design methods should be revised. The table illustrates why new developments are needed for sliding mode control applications for infinite-dimensional systems.

Finite dimensional systems:

$$\dot{x} = f(x) + B(x)u, \quad x \in \Re^n, \quad u \in \Re^m,$$
$$u = \begin{cases} u+(x) & \text{if} \quad s(x) > 0 \\ u^-(x) & \text{if} \quad s(x) < 0 \end{cases} \qquad \text{component-wise}$$
$$\dot{s} = \{\partial s/\partial x\} f + \{\partial s/\partial x\} Bu$$

Infinite-dimensional systems:

$$\partial Q(x,t)/\partial t = AQ(x) + bu(x),$$
$$\dot{s} = grad(s)AQ + grad(s)bu$$

Q and u are infinite-dimensional vectors, $s(Q) = cQ(x) = 0$ is a sliding manifold, b and c are bounded operators.

	Finite-dimensional	Infinite-dimensional
Motion equations	*Lipshitz condition beyond* $s(x) = 0$	*A is unbounded*
Existence Conditions	*"u should suppress* $grad(s)f(x)$*"*	*A is unbounded and "should suppress" does not work*
Control design	*Component-wise*	*No components*

The basic concepts of sliding mode control analysis and design methods can be found in [27, 28]. The two examples will be given in this section as illustration of sliding mode control applications for heat and mechanical systems.

10.1 Distributed Control of Heat Process

Consider control of one-dimensional heat process $Q(x,t)$ with heat isolated ends described by the second order nonlinear parabolic equation

$$\rho(x)\dot{Q} = (k(x,Q)Q')' - q(x,Q)Q + u(x,t) + f(x,t)$$
$$0 \le x \le 1, \quad t \ge 0, \quad Q(x,0) = Q_0(x), \quad Q'(0,t) = Q'(x,t) = 0$$

With distributed control $u(x,t)$ and bounded unknown disturbance $|f(x,t)| \le C$. ρ, k and q are unknown positive heat characteristics. The control task is to create temperature field $Q*(x,t)$. Without loss of generality one can confine oneself to the case $Q*(x,t) = 0$. The norm of deviation of the temperature field from the desired distribution will be used as the Lyapunov functional

$$v(t) = \frac{1}{2}\int_0^1 \rho(x)Q^2(x,t)dx.$$

Compute the time derivative of functional $v(t)$ on the system trajectories with the given boundary conditions

$$\dot{v}(t) = \int_0^1 Q(kQ')'dx - \int_0^1 qQ^2dx + \int_0^1 Q(u+f)dx$$

$$= -\int_0^1 k(Q')^2dx - \int_0^1 qQ^2dx + \int_0^1 Q(u+f)dx.$$

Taking $u = -M signQ(x,t)$, $M > C$, obtain $\dot{v}(t) < 0$ for $\int_0^1 Q^2(x,t)dx \neq 0$, therefore $v(t) \to 0$ with $t \to \infty$. Hence the discontinuous control $u = -M signQ(x,t)$ provides convergence of the heat field distribution to zero in metrics L_2. It is shown in [29], that point wise convergence stems from convergence in metrics L_2.

10.2 Flexible Mechanical System

This section discusses a flexible shaft as an example for a distributed system. Consider a flexible shaft with length l and inertial load J acting as a torsion bar as depicted in Fig. 28. Let $e(t)$ be the absolute coordinate of the left end of

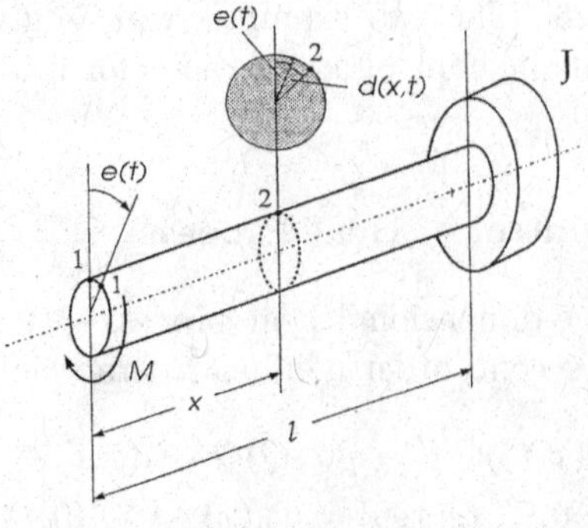

Fig. 28.

the bar with input torque M and let $d(x,t)$ be the relative deviation at location $0 \le x \le l$ at time t. Hence the absolute deviation of a point $0 < x < l$ at time t is described by $q(x,t) = e(t) + d(x,t)$ and governed by

$$\frac{\partial^2 q(x,t)}{\partial t^2} = a^2 \frac{\partial^2 q(x,t)}{\partial x^2}$$

where a is the flexibility constant depending on the geometry and the material of the bar. The boundary conditions corresponding to the input torque M and the load inertia J are

$$-M = a^2 \frac{\partial q(0,t)}{\partial x}, \qquad -J\frac{\partial^2 q(l,t)}{\partial t^2} = a^2 \frac{\partial^2 q(l,t)}{\partial x^2}.$$

Consider the input torque M as the control input $u(t)$ and the load position $q(l,t)$ as the system output. To find the transfer function $W(p)$ via Laplace transformation, assume zero initial conditions

$$q(x,0) = 0, \qquad \frac{\partial q(x,0)}{\partial t} = 0 \tag{10.1}$$

to yield

$$\begin{aligned} a^2 Q(0,p) &= -U(p), \\ p^2 Q(x,p) &= a^2 Q''(x,p), \\ a^2 Q'(l,p) &= -Jp^2 Q(l,p) \end{aligned} \tag{10.2}$$

where $Q(x,p)$ denotes the Laplace transform of $q(x,t)$ with spatial derivatives $Q'(x,p) = \frac{\partial Q(x,p)}{\partial x}$ and $Q''(x,p) = \frac{\partial^2 Q(x,p)}{\partial x^2}$, and $U(p)$ represents the Laplace transform of input variable $u(t)$. The solution to the boundary value problem (10.2) is given by

$$Q(x,p) = -\frac{(1-\frac{J}{a}p)e^{-\frac{l-x}{a}p} + (1+\frac{J}{a}p)e^{\frac{l-x}{a}p}}{ap\left(-(1-\frac{J}{a}p)e^{-\frac{l}{a}p} + (1+\frac{J}{a}p)e^{\frac{l}{a}p}\right)} U(p)$$

from which $W(p)$ may be found by setting $x = l$ to yield

$$W(p) = \frac{2e^{-\tau p}}{ap\,(1+\frac{J}{a}p) + (1-\frac{J}{a}p)\,e^{-2\tau p}},$$

where $\tau = \frac{l}{a}$ describes the "delay" between the right end and the left end of the bar. The corresponding differential-difference equation may be written in the form

$$J\ddot{q}(t) + J\ddot{q}(t-\tau) + a\dot{q}(t) - a\dot{q}(t-2\tau) = 2u(t-\tau).$$

Denoting $x_1(t) = q(t)$, $x_2(t) = \dot{q}(t)$, and $z(t) = J\ddot{q}(t) + a\dot{q}(t)$, we obtain the motion equations as

$$\begin{aligned}\dot{x}_1(t) &= x_2(t)\\ \dot{x}_2(t) &= -\left(ax_2(t) + z(t)\right)/J\end{aligned} \tag{10.3}$$

$$z(t) = 2ax_2(t-2\tau) - z(t-2\tau) + 2u(t-\tau) \tag{10.4}$$

which corresponds to the block-control structure in Chap. 7. In the first design step, assign a desired control $z_d(t)$ for the first block

$$z_d(t) = -M \operatorname{sign}\left(kx_1(t) + x_2(t)\right).$$

In order to achieve $z(t) = z_d(t)$ in the second design step, choose control input $u(t)$ in compliance with the methodology developed for discrete-time systems in Chap. 9

$$u(t) = u_{eq}(t) = \frac{1}{2}\left(-2ax_2(t-\tau) + z(t-\tau) - M \operatorname{sign}\left(kx_1(t+\tau)+x_2(t+\tau)\right)\right).$$

The manifold $s(t) = z_d(t) - z(t) = 0$ is reached within finite time $t < \tau$ and sliding mode exists thereafter.

If control $u(t)$ is bounded by $|u(t)| \le u_0$, choose

$$u(t) = \begin{cases} u_{eq}(t) & \text{for} \quad |u_{eq}(t)| \le u_0 \\ u_0 \operatorname{sign} u_{eq}(t) & \text{for} \quad |u_{eq}(t)| > u_0 \end{cases}$$

and there exists an open domain containing the origin of the state space of system ((10.3)–(10.4)) such that for all initial conditions in this domain, sliding mode occurs along the manifold $s(t) = 0$. The values of $x_1(t+\tau)$ and $x_2(t+\tau)$ have to be calculated as the solution of (10.3) with known input $z(t)$ from (10.4)

$$x(t+\tau) = e^{At}x(t) + \int_0^\tau e^{At}Bu(t-\xi)\,\mathrm{d}\,\xi$$

with

$$A = \begin{pmatrix} 0 & 1 \\ 0 & -a_2 \end{pmatrix}, \qquad B = \begin{pmatrix} 0 \\ -\frac{1}{J} \end{pmatrix}.$$

If only output $y(t) = x_1(t)$ can be measured, but not its time derivative $\dot{y}(t) = x_2(t)$, an asymptotic observer should be used to estimate the state $x_2(t)$.

11 Control of Induction Motor

Control of electric drives is one of the most challenging applications of sliding modes due to wide use of motors and low efficiency of the conventional linear control methodology for such essentially non-linear high-order plants as a.c. motors. Implementation of sliding mode control by means of the most common electric converters has turned out to be simple enough since "on-off" is the only operation mode for them and discontinuities in control actions are dictated by the very nature of power converters.

The most simple, reliable and economic of all motors, maintenance-free induction motors supersede DC motors in today's technology, although, in terms of controllability, induction motors seem the most complicated. Application of sliding mode control methodology to induction motors is demonstrated in this section [30].

The behavior of an induction motor is described by a nonlinear high-order system of differential equations:

$$\frac{dn}{dt} = \frac{1}{J}(M - M_L), \qquad M = \frac{x_H}{x_R}(i_\alpha \phi_\beta - i_\beta \phi_\alpha)$$

$$\frac{d\phi_\alpha}{dt} = -\frac{r_R}{x_R}\phi_\alpha - n\phi_\beta + r_R\frac{x_H}{x_R}i_\alpha \tag{11.1}$$

$$\frac{d\phi_\beta}{dt} = -\frac{r_R}{x_R}\phi_\beta + n\phi_\alpha + r_R\frac{x_H}{x_R}i_\beta$$

$$\frac{di_\alpha}{dt} = \frac{x_R}{x_s x_R - x_H^2}\left(-\frac{x_H}{x_R}\frac{d\phi_\alpha}{dt} - r_s i_\alpha + u_\alpha\right)$$

$$\frac{di_\beta}{dt} = \frac{x_R}{x_s x_R - x_H^2}\left(-\frac{x_H}{x_R}\frac{d\phi_\beta}{dt} - r_s i_\beta + u_\beta\right)$$

$$\begin{bmatrix} u_\alpha \\ u_\beta \end{bmatrix} = \frac{2}{3}\begin{bmatrix} e_{R\epsilon} & e_{S\alpha} & e_{T\alpha} \\ e_{R\alpha} & e_{S\beta} & e_{T\beta} \end{bmatrix} \times \begin{bmatrix} u_R \\ u_S \\ u_T \end{bmatrix} \tag{11.2}$$

where n is a rotor angle velocity, and two-dimensional vectors $\phi^T = (\phi_\alpha, \phi_\beta)$; $i^T = (i_\alpha, i_\beta)$, $u^T = (u_\alpha, u_\beta)$ are rotor flux, stator current, and voltage in the fixed coordinate system (α, β), respectively; M and M_L are a torque developed by a motor and a load torque, for an induction motor controlled by modern power converters u_R, u_S, u_T are phase voltages, which may take only two values either u_0 or $-u_0$; e_R, e_S, e_T are unit vectors of phase windings, R, S, T; and J, x_H, x_S, x_R r_R, r_S are motor parameters.

The control goal is to make one of the mechanical coordinates, for example, an angle speed $n(t)$, be equal to a reference input $n_0(t)$ and the magnitude of the rotor flux$\|\phi(t)\|$ be equal to its scalar reference input $\phi_0(t)$. The deviations from the desired motions are described by the functions

$$\begin{aligned} s_1 &= c_1[n_0 - n(t)] + \tfrac{d}{dt}[n_0 - n(t)] \\ s_2 &= c_2[\phi_0 - \|\phi(t)\|] + \tfrac{d}{dt}[\phi_0 - \|\phi(t)\|] \end{aligned} \tag{11.3}$$

and c_1, c_2 are const positive values.

The static inverter forms three independent controls u_R, u_S, u_T so one degree of freedom can be used to satisfy some additional criterion. Let the voltages u_R, u_S, u_T constitute a three-phase balanced system, which means that the equality

$$s_3 = \int_0^t (u_R + u_S + u_T)\, dt = 0 \tag{11.4}$$

should hold for any t.

If all three functions s_1, s_2, s_3 are equal to zero then, in addition to balanced system condition, the speed and flux mismatches decay exponentially since $s_1 = 0$, $s_2 = 0$ with c_1, $c_2 > 0$ are first-order differential equations. This means that the design task is reduced to enforcing the sliding mode in the manifold $s = 0$, $s^T = (s_1, s_2, s_3)$ in the system (11.1) with control $u^T = (u_R, u_S, u_T)$. Projection of the system motion on subspace s can be written as

$$\frac{ds}{dt} = F + Du$$

where vector $F^T = (f_1, f_2, 0)$ and matrix D are calculated on the trajectories of ((11.1)–(11.4)). They do not depend on control and are continuous functions of motor state and inputs. Matrix D is of the form

$$D = \begin{bmatrix} D_1 \\ d \end{bmatrix},$$

$$D_1 = \frac{kx_H}{x_S x_R - x_H^2} \begin{bmatrix} \frac{1}{J} & 0 \\ 0 & \frac{r_R}{\|\phi\|} \end{bmatrix} \times \begin{bmatrix} \phi_\beta & -\phi_\alpha \\ \phi_\alpha & \phi_\beta \end{bmatrix} \times \begin{bmatrix} e_{R\epsilon} & e_{S\alpha} & e_{T\alpha} \\ e_{R\alpha} & e_{S\beta} & e_{T\beta} \end{bmatrix}$$

and

$$d = (1,1,1)\,, \quad k - const, \qquad \det D \neq 0.$$

Discontinuous control will be designed using the Lyapunov function

$$V = \frac{1}{2} s^T s \geq 0.$$

Find its time derivative on the system state trajectories:

$$\frac{dV}{dt} = s^T (F + Du)$$

Substitution of control

$$u = -u_0 \operatorname{sgn} s^* \qquad s^* = D^T s \qquad (s^*)^T = (s_1^*, s_2^*, s_3^*)$$

yields

$$\frac{dV}{dt} = (s^*)^T F^* - u_0 |s^*|$$

where

$$F^{*T} = \left(D^{-1}F\right)^T = (f_1^*, f_2^*, f_3^*)$$
$$|s^*| = |s_1^*| + |s_2^*| + |s_3^*|.$$

The conditions

$$u_0 > |f_i^*|, \qquad i = 1, 2, 3$$

provide negativeness of dV/dt and hence the origin in the space s^* (and by virtue of $\det D \neq 0$ in the space s as well) is asymptotically stable. Hence sliding mode arises in the manifold $s^* = 0$, which enables one to steer the variables under control to the desired values. Note that the existence condition are inequalities, therefore the only range of parameter variations and a load torque should be known for the choice of necessary values of phase voltages.

The equations of discontinuity surface $s^* = 0$ depend on an angle acceleration, rotor flux, and its time derivative. These values may be found using asymptotic observers under the assumption that an angle speed n and stator currents i_R, i_S, i_T are measured directly. Bearing in mind that

$$\begin{bmatrix} i_\alpha \\ i_\beta \end{bmatrix} = \frac{2}{3} \begin{bmatrix} e_{R\epsilon} \; e_{S\alpha} \; e_{T\alpha} \\ e_{R\alpha} \; e_{S\beta} \; e_{T\beta} \end{bmatrix} \begin{bmatrix} i_R \\ i_S \\ i_T \end{bmatrix}$$

design an observer with the state vector$\left(\hat{\phi}_\alpha, \hat{\phi}_\beta\right)$ as an estimate of rotor flux components ϕ_α and ϕ_β and with i_α and i_β as inputs:

$$\frac{d\hat{\phi}_\alpha}{dt} = \frac{r_R}{x_R}\hat{\phi}_\alpha - n\hat{\phi}_\beta + r_R\frac{x_H}{x_R}i_\alpha$$

$$\frac{d\hat{\phi}_\beta}{dt} = -\frac{r_R}{x_R}\hat{\phi}_\beta + n\hat{\phi}_\alpha + r_R\frac{x_H}{x_R}i_\beta.$$

To obtain the equations for the estimation error$\bar{\phi}_\alpha = \hat{\phi}_\alpha - \phi_\alpha$, $\bar{\phi}_\beta = \hat{\phi}_\beta - \phi_\beta$ the equations for ϕ_α and ϕ_β in (11.1) should be subtracted from the observer equations:

$$\frac{d\bar{\phi}_\alpha}{dt} = -\frac{r_R}{x_R}\bar{\phi}_\alpha - n\bar{\phi}_\beta$$

$$\frac{d\bar{\phi}_\beta}{dt} = -\frac{r_R}{x_R}\bar{\phi}_\beta + n\bar{\phi}_\alpha$$

The time derivative of Lyapunov function

$$V = \frac{1}{2}\left(\bar{\phi}_\alpha^2 + \bar{\phi}_\beta^2\right) > 0$$

on the solutions of the estimate equation

$$\frac{dV}{dt} = -\frac{r_R}{x_R}\left(\bar{\phi}_\alpha^2 + \bar{\phi}_\beta^2\right) = -2\frac{r_R}{x_R}V < 0$$

is negative, which testifies to exponential convergence of V to zero and the estimates $\hat{\phi}_\alpha$ and $\hat{\phi}_\beta$ to the real values of ϕ_α and ϕ_α. The known values ϕ_α, ϕ_β, i_α and i_β enable one to find the time derivatives $d\phi_\alpha/dt$ and $d\phi_\beta/dt$ from the estimator equations and then $d\left\|\phi\right\|/dt$ needed for designing the discontinuity surface $s_2 = 0$.

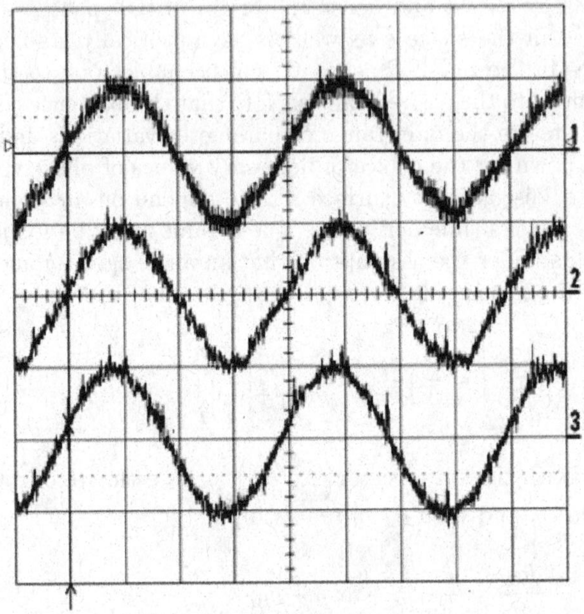

Curve 1 - Real speed
Curve 2 - Speed estimate
Curve 3 - Speed reference input 150sin[(π/2)t] r/min

Fig. 29.

The equation of the discontinuity surface s_1 depends on acceleration dn/dt. Since the values of ϕ_α and ϕ_β have been found and the currents i_α and i_β are measured directly, the motor torque may be calculated:

$$M = \left(\frac{x_H}{x_R}\right)(i_\alpha \phi_\beta - i_\beta \phi_\alpha).$$

Under the assumption that the load torque M_L varies slowly the value of dn/dt may be found using the conventional linear state observer.

In the further studies the sliding mode observer was developed for so-called sensorless systems to estimate the motor flux and speed simultaneously. Figure 29 shows the results of experiments for Westinghaus 5-hp 220V Y-connected four-pole induction motor. Real speed is measured by an optical encoder for speed estimation verification only and it is not used for the feedback control.

References

1. S.V. Emel'yanov, Method of designing complex control algorithms using an error and its first time-derivative only, *Automation and Remote Control*, v.18, No.10,

1957 (In Russian).
2. A.M. Letov, Conditionally stable control systems (on a class of optimal system), *Automation and Remote Control*, v.18, No.7, 1957 (In Russian).
3. S.V. Emel'yanov, Burovoi I.A. and et. al., Mathematical models of processes in technology and development of variable structure control systems, *Metallurgy*, No.21, Moscow, 1964 (In Russian).
4. S.V. Emel'yanov, V.I. Utkin and et.al., Theory of Variable Structure Systems. Nauka, Moscow, 1970 (In Russian).
5. S.V. Emel'yanov, V.A. Taran, On a class of variable structure control systems, Proc. of USSR Academy of Sciences, Energy and Automation, No.3, 1962 (In Russian).
6. E.A. Barbashin, Introduction to stability theory, Nauka, Moscow, 1967 (In Russian).
7. T.A. Bezvodinskaya and E.F. Sabayev, Stability conditions in large for variable structure systems, *Automation and Remote Control*, v.35, No.10 (P.1), 1974.
8. N.E. Kostyleva, Variable structure systems for plants with zeros in a transfer function, Theory and Application of Control Systems, Nauka, Moscow, 1964 (In Russian).
9. A.A. Andronov, A.A. Vitt and S.E. Khaikin, Theory of oscillations, Fizmatgiz, Moscow, 1959 (In Russian).
10. V. Utkin, Sliding Modes and their Applications in Variable Structure Systems. Nauka, Moscow, 1974 - in Russian (English translation by Mir Publ., Moscow, 1978).
11. V. Utkin, Sliding Modes in Control and Optimization, Springer Verlag, 1992. (Revised version of the book on Russian published by Nauka, 1981).
12. Yu.G. Kornilov, On effect of controller insensitivity on dynamics of indirect control, *Automation and Remote Control*, v.11, No.1, 1950 (In Russian).
13. A.M. Popovski, Linearization of sliding operation mode for a constant speed controller, *Automation and Remote Control*, v.11, No.3, 1950 (In Russian).
14. Yu.V. Dolgolenko, Sliding modes in relay indirect control systems, *Proceeding of 2nd All-Union Conference on Control*, v.1, Moscow, 1955 (in Russian).
15. A.F. Filippov, Application of the theory of differential equations with discontinuous right-hand sides to non-linear problems of automatic control, *Proceedings of 1st IFAC Congress II*, Butterworths, London, 1961.
16. Yu.I. Neimark, Note on A. Filippov's paper, Proceedings *of 1st IFAC Congress II*, Butterworths, London, 1961.
17. J. Andre and P. Seibert, Uber stuckweise lineare Differential-gluichungen bei Regelungsproblem auftreten, I, II, *Arch. Der Math.*, vol. 7, Nos. 2 und 3.
18. B. Drazenovic, The invariance conditions in variable structure systems, *Automatica*, v.5, No.3, Pergamon Press, 1969.
19. P.K. Rashevski, Geometric approach to partial differential equations, Gostechizdat, Moscow, 1947 (in Russian).
20. V.Utkin and et al., A hierarchical principle of the control decomposition based on motion separation, *Preprints of 9th World Congress of IFAC*, Budaprst, Hungary, Pergamon Press, vol. 5, Colloquim 09.4, pp. 134–139, 1984.
21. S. Drakunov and et al., Block control principle I, *Automation and Remote Control*, vol. 51, pp. 601–609 and vol. 52, pp. 737–746, 1990.
22. S. Gutman and G. Leitmann, Stabilizing Feedback Control for Dynamic Systems with Bounded Uncertainties. *Proceedings of IEEE Conference on Decision and Control*, pp. 94–99, 1976.

23. S. Gutman, Uncertain Dynamic Systems - a Lyapunov Min-Max Approach. *IEEE Trans.*, AC-24, No.3, pp. 437–449, 1979.
24. E.P. Ryan and M. Corless, Ultimate Boundness and Asymptotic Stability of a Class of Uncertain Systems via Continuous and Discontinuous Feedback Control, *IMA J. Math. Cont. and Inf.*, No.1, pp. 222–242, 1984.
25. P.V. Kokotovic, R.B. O'Malley, and P. Sannuti, Singular Perturbations and Order Reduction in Control Theory, *Automatica*, vol. 12, pp. 123-132, 1976.
26. S.V. Drakunov and V.I. Utkin, "Sliding Mode in Dynamic Systems", *Int. Journal of Control*, vol. 55, pp. 1029–1037, 1990.
27. V. Utkin and Yu. Orlov, Sliding Mode Control in Infinite-Dimensional Systems, *Automatica*, vol. 23, No. 6, pp. 753–757, 1987.
28. V. Utkin and Yu. Orlov, Sliding Mode Control of Infinite-Dimensional Systems. Nauka. Moscow, 1990 (in Russian).
29. A. Breger and et al., Sliding Mode Control of Distributed - Parameter Entities Subjected to a Mobile Multicycle Signal. *Automation and Remote Control*, vol. 41, No.3 (P.1), pp. 346–355, 1980.
30. V. Utkin, J. Guldner, and J. Shi, Sliding Mode Control in Electro-Mechanical Systems, Taylor&Francis, 1999.

Books (in English)

1. V. Utkin, *Sliding Modes and their Applications in Variable Structure Systems*, Mir Publ., Moscow (Translation of the book published by Nauka, Moscow, 1974 in Russian), 1978 [This book presents the first results on MIMO sliding mode control systems].
2. U. Itkis, *Control Systems of Variable Structure*, Wiley, New York, 1976 [This book deals with SISO control systems in canonical form].
3. A. Filippov, *Differential Equations with Discontinuous Right-Hand Sides*, Kluver, 1988 [Mathematical aspects of discontinuous dynamic systems are studied in this book].
4. *Deterministic Non-Linear Control*, A.S. Zinober, Ed., Peter Peregrinus Limited, UK, 1990 [The book is written by the author team and covers different control methods, including sliding mode control for non-linear dynamic plants operating under uncertainty conditions].
5. *Variable Structure Control for Robotics and Aerospace Application*, K-K.D. Young, Ed., Elsevier Science Publishers B.V., Amsterdam, 1993 [Applications of sliding mode control for electric drives, pulse-with modulators, manipulator, flexible mechanical structures, magnetic suspension, flight dynamics are studied in the book].
6. *Variable Structure and Lyapunov Control*, A.S. Zinober, Ed., Springer Verlag, London, 1993 [The book is prepared based on the results of VSS' 1992 workshop held in Sheffield, UK].
7. V. Utkin, *Sliding Modes in Control and Optimization*, Springer Verlag, Berlin, 1992 [The book includes mathematical, design and application aspects of sliding mode control].
8. *Variable Structure Systems, Sliding Mode and Nonlinear Control*, K.D. Young and U. Ozguner (Eds), Springer Verlag, 1999 [The book is prepared based on the results of VSS'1998 workshop held in Longboat Key, Florida, USA].
9. C. Edwards and S. Spurgeon, *Sliding Mode Control: Theory and Applications*, Taylor and Francis, London, 1999 [The conventional theory is presented along with new methods of control and observation and new application areas].

10. V. Utkin, J. Guldner, and J. Shi, *Sliding Mode Control in Electro-Mechanical Systems*, Taylor and Francis, London, 1999 [Theoretical design methods are presented in the context of applications for control of electric drives, alternators, vehicle motion, manipulators and mobile robots].
11. J.-J.E. Slotine and W. Li, *Applied Nonlinear Control*, Prentice Hall Englewood Cliffs, New Jersey, 1991 [Text book with a chapter on sliding mode control].
12. V. Utkin. *Sliding Mode Control*, Section 6.43.21.14, Chap. 6.43 "Control, Robotics and Automation", Encyclopedia, EOLSS Publisher Co Ltd, 2003 [Survey of Sliding Mode Control Analysis and Design Methods].

List of Participants

1. Agrachev Andrei A.
SISSA-Trieste, Italy
agrachev@ma.sissa.it
(lecturer)
2. Allegretto Walter
University of Alberta, Canada
wallegre@math.ualberta.ca
3. Angeli David
University di Firenze, Italy
angeli@dsi.unifi.it
4. Bacciotti Andrea
Politecnico di Torino, Italy
bacciotti@polito.it
5. Bianchini Rosa Maria
University di Firenze, Italy
bianchin@math.unifi.it
6. Burlion Laurent
CNRS LSS-Supélec, France
laurent.burlion@lss.supelec.fr
7. Cellina Arrigo
University la Bicocca Milano, Italy
cellina@matapp.unimib.it
8. Chaillet Antoine
LSS-CNRS, France
Antoine.Chaillet@lss.supelec.fr
9. Cinili Fabio
University di Roma Tor Vergata, Italy
cinili@ing.uniroma2.it
10. Demyanov Alexei
St Petersburg State University, Russia
alex@ad9503.spb.edu
11. Fabbri Roberta
University di Firenze, Italy
fabbri@dsi.unifi.it
12. Falugi Paola
University di Firenze, Italy
paola@control.dsi.unifi.it
13. Farina Marcello
Politecnico di Milano, Italy
farina@elet.polimi.it
14. Frosali Giovanni
University di Firenze, Italy
frosali@dma.unifi.it
15. Gessuti Daniel
University degli Studi di Padova, Italy
danielgessuti@inwind.it
16. Ghilardi Chiara
University di Firenze, Italy
ghilardi@control.dsi.unifi.it
17. Girejko Ewa
Bialystok Technical University, Poland
ewagir@wp.pl

18. Koeppl Heinz
Graz University of Technology, Austria
heinz.koeppl@tugraz.at
19. Kouteeva Galina
St Petersburg State University, Russia
star@gk1662.spb.edu
20. Kule Memet
Yildiz Technical University, Turkey
memetkule@mynet.com
21. Levaggi Laura
University di Genova, Italy
levaggi@dima.unige.it
22. Liuzzo Stefano
University di Roma Tor Vergata, Italy
liuzzo@ing.uniroma2.it
23. Lyakhova L.Sofya
University of Bristol, UK
s.lyakhova@bristol.ac.uk
24. Makarenkov Y.Oleg
Voronezh State University, Russia
omakarenkov@kma.vsu.ru
25. Malgorzata Wyrwas
Bialystok Technical University, Poland
wyrwas@pb.bialystok.pl
26. Marchini Elsa Maria
University di Milano, Italy
elsa@matapp.unimib.it
27. Marigonda Antonio
University di Padova, Italy
amarigo@math.unipd.it
28. Martynova S. Irina
Voronezh State University, Russia
i_martynova@inbox.ru
29. Mason Paolo
SISSA-Trieste, Italy
mason@sissa.it
30. Matthias Kawski
Arizona State University, USA
kawski@asu.edu
31. Morse A. Stephen
Yale University
morse@sysc.eng.yale.edu
(**lecturer**)
32. Nistri Paolo
University di Siena, Italy
pnistri@dii.unisi.it
(**editor**)
33. Notarstefano Giuseppe
University di Padova, Italy
notarste@dei.unipd.it
34. Panasenko Elena
Tambov State University, Russia
panlena_t@mail.ru
35. Papini Duccio
University di Siena, Italy
papini@dii.unisi.it
36. Pesetskaya Tatiana
Institute of Mathematics, Belorus
tanya_pesetskaya@yahoo.com
37. Pirozzi Salvatore
Seconda University di Napoli, Italy
salvatore.pirozzi@unina2.it
38. Poggiolini Laura
University di Firenze, Italy
poggiolini@dma.unifi.it
39. Saccon Alessandro
University di Padova, Italy
asaccon@dei.unipd.it
40. Sarychev Andrey
University di Firenze, Italy
asarychev@unifi.it
41. Serres Ulysse
SISSA-Trieste, Italy
serres@sissa.it
42. Shcherbakova Nataliya
SISSA-Trieste, Italy
chtch@sissa.it
43. Skiba Robert Adam
Nicolaus Copernicus University, Poland
robo@mat.uni.torun.pl

44. Sontag Eduardo
Rutgers University, USA
sontag@control.rutgers.edu
(**lecturer**)
45. Spadini Marco
University di Firenze, Italy
spadini@dma.unifi.it
46. Spindler Karlheinz
Fachhochschule Wiesbaden, Germany
spindler@r5.mnd.fh-wiesbaden.de
47. Stefani Gianna
University di Firenze, Italy
stefani@dma.unifi.it
(**editor**)
48. Sussmann Héctor
Rutgers University, USA
sussmann@math.rutgers.edu
(**lecturer**)
49. Tosques Mario
University di Parma, Italy
mario.tosques@unipr.it
50. Turbin Mikhail
Voronezh State University, Russia
mrmike@math.vsu.ru
51. Utkin I. Vadim
Ohio State University
utkin@ee.eng.ohio-state.edu
(**lecturer**)
52. Verrelli Cristiano Maria
University di Roma Tor Vergata, Italy
verrelli@ing.uniroma2.it
53. Zecca Pietro
CIME, University di Firenze
cime@math.unifi.it
54. Zezza PierLuigi
University di Firenze, Italy
pzezza@unifi.it
55. Zolezzi Tullio
University di Genova, Italy
zolezzi@dima.unige.it

LIST OF C.I.M.E. SEMINARS

Published by C.I.M.E

Published by Ed. Cremonese, Firenze

1966 39. Calculus of variations
40. Economia matematica
41. Classi caratteristiche e questioni connesse
42. Some aspects of diffusion theory

1967 43. Modern questions of celestial mechanics
44. Numerical analysis of partial differential equations
45. Geometry of homogeneous bounded domains

1968 46. Controllability and observability
47. Pseudo-differential operators
48. Aspects of mathematical logic

1969 49. Potential theory
50. Non-linear continuum theories in mechanics and physics and their applications
51. Questions of algebraic varieties

1970 52. Relativistic fluid dynamics
53. Theory of group representations and Fourier analysis
54. Functional equations and inequalities
55. Problems in non-linear analysis

1971 56. Stereodynamics
57. Constructive aspects of functional analysis (2 vol.)
58. Categories and commutative algebra

1972 59. Non-linear mechanics
60. Finite geometric structures and their applications
61. Geometric measure theory and minimal surfaces

1973 62. Complex analysis
63. New variational techniques in mathematical physics
64. Spectral analysis

1974 65. Stability problems
66. Singularities of analytic spaces
67. Eigenvalues of non linear problems

1975 68. Theoretical computer sciences
69. Model theory and applications
70. Differential operators and manifolds

Published by Ed. Liguori, Napoli

1976 71. Statistical Mechanics
72. Hyperbolicity
73. Differential topology

1977 74. Materials with memory
75. Pseudodifferential operators with applications
76. Algebraic surfaces

Published by Ed. Liguori, Napoli & Birkhäuser

1978 77. Stochastic differential equations
78. Dynamical systems

1979 79. Recursion theory and computational complexity
80. Mathematics of biology

1980	81. Wave propagation	
	82. Harmonic analysis and group representations	
	83. Matroid theory and its applications	

Published by Springer-Verlag

1981	84. Kinetic Theories and the Boltzmann Equation	(LNM 1048)
	85. Algebraic Threefolds	(LNM 947)
	86. Nonlinear Filtering and Stochastic Control	(LNM 972)
1982	87. Invariant Theory	(LNM 996)
	88. Thermodynamics and Constitutive Equations	(LNP 228)
	89. Fluid Dynamics	(LNM 1047)
1983	90. Complete Intersections	(LNM 1092)
	91. Bifurcation Theory and Applications	(LNM 1057)
	92. Numerical Methods in Fluid Dynamics	(LNM 1127)
1984	93. Harmonic Mappings and Minimal Immersions	(LNM 1161)
	94. Schrödinger Operators	(LNM 1159)
	95. Buildings and the Geometry of Diagrams	(LNM 1181)
1985	96. Probability and Analysis	(LNM 1206)
	97. Some Problems in Nonlinear Diffusion	(LNM 1224)
	98. Theory of Moduli	(LNM 1337)
1986	99. Inverse Problems	(LNM 1225)
	100. Mathematical Economics	(LNM 1330)
	101. Combinatorial Optimization	(LNM 1403)
1987	102. Relativistic Fluid Dynamics	(LNM 1385)
	103. Topics in Calculus of Variations	(LNM 1365)
1988	104. Logic and Computer Science	(LNM 1429)
	105. Global Geometry and Mathematical Physics	(LNM 1451)
1989	106. Methods of nonconvex analysis	(LNM 1446)
	107. Microlocal Analysis and Applications	(LNM 1495)
1990	108. Geometric Topology: Recent Developments	(LNM 1504)
	109. H_∞ Control Theory	(LNM 1496)
	110. Mathematical Modelling of Industrial Processes	(LNM 1521)
1991	111. Topological Methods for Ordinary Differential Equations	(LNM 1537)
	112. Arithmetic Algebraic Geometry	(LNM 1553)
	113. Transition to Chaos in Classical and Quantum Mechanics	(LNM 1589)
1992	114. Dirichlet Forms	(LNM 1563)
	115. D-Modules, Representation Theory, and Quantum Groups	(LNM 1565)
	116. Nonequilibrium Problems in Many-Particle Systems	(LNM 1551)
1993	117. Integrable Systems and Quantum Groups	(LNM 1620)
	118. Algebraic Cycles and Hodge Theory	(LNM 1594)
	119. Phase Transitions and Hysteresis	(LNM 1584)
1994	120. Recent Mathematical Methods in Nonlinear Wave Propagation	(LNM 1640)
	121. Dynamical Systems	(LNM 1609)
	122. Transcendental Methods in Algebraic Geometry	(LNM 1646)
1995	123. Probabilistic Models for Nonlinear PDE's	(LNM 1627)
	124. Viscosity Solutions and Applications	(LNM 1660)
	125. Vector Bundles on Curves. New Directions	(LNM 1649)

1996	126. Integral Geometry, Radon Transforms and Complex Analysis	(LNM 1684)
	127. Calculus of Variations and Geometric Evolution Problems	(LNM 1713)
	128. Financial Mathematics	(LNM 1656)
1997	129. Mathematics Inspired by Biology	(LNM 1714)
	130. Advanced Numerical Approximation of Nonlinear Hyperbolic Equations	(LNM 1697)
	131. Arithmetic Theory of Elliptic Curves	(LNM 1716)
	132. Quantum Cohomology	(LNM 1776)
1998	133. Optimal Shape Design	(LNM 1740)
	134. Dynamical Systems and Small Divisors	(LNM 1784)
	135. Mathematical Problems in Semiconductor Physics	(LNM 1823)
	136. Stochastic PDE's and Kolmogorov Equations in Infinite Dimension	(LNM 1715)
	137. Filtration in Porous Media and Industrial Applications	(LNM 1734)
1999	138. Computational Mathematics driven by Industrial Applications	(LNM 1739)
	139. Iwahori-Hecke Algebras and Representation Theory	(LNM 1804)
	140. Hamiltonian Dynamics - Theory and Applications	(LNM 1861)
	141. Global Theory of Minimal Surfaces in Flat Spaces	(LNM 1775)
	142. Direct and Inverse Methods in Solving Nonlinear Evolution Equations	(LNP 632)
2000	143. Dynamical Systems	(LNM 1822)
	144. Diophantine Approximation	(LNM 1819)
	145. Mathematical Aspects of Evolving Interfaces	(LNM 1812)
	146. Mathematical Methods for Protein Structure	(LNCS 2666)
	147. Noncommutative Geometry	(LNM 1831)
2001	148. Topological Fluid Mechanics	to appear
	149. Spatial Stochastic Processes	(LNM 1802)
	150. Optimal Transportation and Applications	(LNM 1813)
	151. Multiscale Problems and Methods in Numerical Simulations	(LNM 1825)
2002	152. Real Methods in Complex and CR Geometry	(LNM 1848)
	153. Analytic Number Theory	(LNM 1891)
	154. Imaging	to appear
2003	155. Stochastic Methods in Finance	(LNM 1856)
	156. Hyperbolic Systems of Balance Laws	(LNM 1911)
	157. Symplectic 4-Manifolds and Algebraic Surfaces	to appear
	158. Mathematical Foundation of Turbulent Viscous Flows	(LNM 1871)
2004	159. Representation Theory and Complex Analysis	(LNM 1931)
	160. Nonlinear and Optimal Control Theory	(LNM 1932)
	161. Stochastic Geometry	(LNM 1892)
2005	162. Enumerative Invariants in Algebraic Geometry and String Theory	to appear
	163. Calculus of Variations and Non-linear Partial Differential Equations	(LNM 1927)
	164. SPDE in Hydrodynamics: Recent Progress and Prospects	to appear
2006	165. Pseudo-Differential Operators, Quantization and Signals	to appear
	166. Mixed Finite Elements, Compatibility Conditions, and Applications	to appear
	167. From a Microscopic to a Macroscopic Description of Complex Systems	to appear
	168. Quantum Transport: Modelling, Analysis and Asymptotics	to appear
2007	169. Geometric Analysis and Partial Differential Equations	to appear
	170. Nonlinear Optimization	to appear
	171. Arithmetic Geometry	to appear

Lecture Notes in Mathematics

For information about earlier volumes
please contact your bookseller or Springer
LNM Online archive: springerlink.com

Vol. 1741: E. Lombardi, Oscillatory Integrals and Phenomena Beyond all Algebraic Orders (2000)
Vol. 1742: A. Unterberger, Quantization and Non-holomorphic Modular Forms (2000)
Vol. 1743: L. Habermann, Riemannian Metrics of Constant Mass and Moduli Spaces of Conformal Structures (2000)
Vol. 1744: M. Kunze, Non-Smooth Dynamical Systems (2000)
Vol. 1745: V. D. Milman, G. Schechtman (Eds.), Geometric Aspects of Functional Analysis. Israel Seminar 1999-2000 (2000)
Vol. 1746: A. Degtyarev, I. Itenberg, V. Kharlamov, Real Enriques Surfaces (2000)
Vol. 1747: L. W. Christensen, Gorenstein Dimensions (2000)
Vol. 1748: M. Ruzicka, Electrorheological Fluids: Modeling and Mathematical Theory (2001)
Vol. 1749: M. Fuchs, G. Seregin, Variational Methods for Problems from Plasticity Theory and for Generalized Newtonian Fluids (2001)
Vol. 1750: B. Conrad, Grothendieck Duality and Base Change (2001)
Vol. 1751: N. J. Cutland, Loeb Measures in Practice: Recent Advances (2001)
Vol. 1752: Y. V. Nesterenko, P. Philippon, Introduction to Algebraic Independence Theory (2001)
Vol. 1753: A. I. Bobenko, U. Eitner, Painlevé Equations in the Differential Geometry of Surfaces (2001)
Vol. 1754: W. Bertram, The Geometry of Jordan and Lie Structures (2001)
Vol. 1755: J. Azéma, M. Émery, M. Ledoux, M. Yor (Eds.), Séminaire de Probabilités XXXV (2001)
Vol. 1756: P. E. Zhidkov, Korteweg de Vries and Nonlinear Schrödinger Equations: Qualitative Theory (2001)
Vol. 1757: R. R. Phelps, Lectures on Choquet's Theorem (2001)
Vol. 1758: N. Monod, Continuous Bounded Cohomology of Locally Compact Groups (2001)
Vol. 1759: Y. Abe, K. Kopfermann, Toroidal Groups (2001)
Vol. 1760: D. Filipović, Consistency Problems for Heath-Jarrow-Morton Interest Rate Models (2001)
Vol. 1761: C. Adelmann, The Decomposition of Primes in Torsion Point Fields (2001)
Vol. 1762: S. Cerrai, Second Order PDE's in Finite and Infinite Dimension (2001)
Vol. 1763: J.-L. Loday, A. Frabetti, F. Chapoton, F. Goichot, Dialgebras and Related Operads (2001)
Vol. 1764: A. Cannas da Silva, Lectures on Symplectic Geometry (2001)
Vol. 1765: T. Kerler, V. V. Lyubashenko, Non-Semisimple Topological Quantum Field Theories for 3-Manifolds with Corners (2001)
Vol. 1766: H. Hennion, L. Hervé, Limit Theorems for Markov Chains and Stochastic Properties of Dynamical Systems by Quasi-Compactness (2001)
Vol. 1767: J. Xiao, Holomorphic Q Classes (2001)
Vol. 1768: M. J. Pflaum, Analytic and Geometric Study of Stratified Spaces (2001)
Vol. 1769: M. Alberich-Carramiñana, Geometry of the Plane Cremona Maps (2002)
Vol. 1770: H. Gluesing-Luerssen, Linear Delay-Differential Systems with Commensurate Delays: An Algebraic Approach (2002)
Vol. 1771: M. Émery, M. Yor (Eds.), Séminaire de Probabilités 1967-1980. A Selection in Martingale Theory (2002)
Vol. 1772: F. Burstall, D. Ferus, K. Leschke, F. Pedit, U. Pinkall, Conformal Geometry of Surfaces in S^4 (2002)
Vol. 1773: Z. Arad, M. Muzychuk, Standard Integral Table Algebras Generated by a Non-real Element of Small Degree (2002)
Vol. 1774: V. Runde, Lectures on Amenability (2002)
Vol. 1775: W. H. Meeks, A. Ros, H. Rosenberg, The Global Theory of Minimal Surfaces in Flat Spaces. Martina Franca 1999. Editor: G. P. Pirola (2002)
Vol. 1776: K. Behrend, C. Gomez, V. Tarasov, G. Tian, Quantum Comohology. Cetraro 1997. Editors: P. de Bartolomeis, B. Dubrovin, C. Reina (2002)
Vol. 1777: E. García-Río, D. N. Kupeli, R. Vázquez-Lorenzo, Osserman Manifolds in Semi-Riemannian Geometry (2002)
Vol. 1778: H. Kiechle, Theory of K-Loops (2002)
Vol. 1779: I. Chueshov, Monotone Random Systems (2002)
Vol. 1780: J. H. Bruinier, Borcherds Products on O(2,1) and Chern Classes of Heegner Divisors (2002)
Vol. 1781: E. Bolthausen, E. Perkins, A. van der Vaart, Lectures on Probability Theory and Statistics. Ecole d' Eté de Probabilités de Saint-Flour XXIX-1999. Editor: P. Bernard (2002)
Vol. 1782: C.-H. Chu, A. T.-M. Lau, Harmonic Functions on Groups and Fourier Algebras (2002)
Vol. 1783: L. Grüne, Asymptotic Behavior of Dynamical and Control Systems under Perturbation and Discretization (2002)
Vol. 1784: L. H. Eliasson, S. B. Kuksin, S. Marmi, J.-C. Yoccoz, Dynamical Systems and Small Divisors. Cetraro, Italy 1998. Editors: S. Marmi, J.-C. Yoccoz (2002)
Vol. 1785: J. Arias de Reyna, Pointwise Convergence of Fourier Series (2002)
Vol. 1786: S. D. Cutkosky, Monomialization of Morphisms from 3-Folds to Surfaces (2002)
Vol. 1787: S. Caenepeel, G. Militaru, S. Zhu, Frobenius and Separable Functors for Generalized Module Categories and Nonlinear Equations (2002)
Vol. 1788: A. Vasil'ev, Moduli of Families of Curves for Conformal and Quasiconformal Mappings (2002)

Vol. 1789: Y. Sommerhäuser, Yetter-Drinfel'd Hopf algebras over groups of prime order (2002)
Vol. 1790: X. Zhan, Matrix Inequalities (2002)
Vol. 1791: M. Knebusch, D. Zhang, Manis Valuations and Prüfer Extensions I: A new Chapter in Commutative Algebra (2002)
Vol. 1792: D. D. Ang, R. Gorenflo, V. K. Le, D. D. Trong, Moment Theory and Some Inverse Problems in Potential Theory and Heat Conduction (2002)
Vol. 1793: J. Cortés Monforte, Geometric, Control and Numerical Aspects of Nonholonomic Systems (2002)
Vol. 1794: N. Pytheas Fogg, Substitution in Dynamics, Arithmetics and Combinatorics. Editors: V. Berthé, S. Ferenczi, C. Mauduit, A. Siegel (2002)
Vol. 1795: H. Li, Filtered-Graded Transfer in Using Noncommutative Gröbner Bases (2002)
Vol. 1796: J.M. Melenk, hp-Finite Element Methods for Singular Perturbations (2002)
Vol. 1797: B. Schmidt, Characters and Cyclotomic Fields in Finite Geometry (2002)
Vol. 1798: W.M. Oliva, Geometric Mechanics (2002)
Vol. 1799: H. Pajot, Analytic Capacity, Rectifiability, Menger Curvature and the Cauchy Integral (2002)
Vol. 1800: O. Gabber, L. Ramero, Almost Ring Theory (2003)
Vol. 1801: J. Azéma, M. Émery, M. Ledoux, M. Yor (Eds.), Séminaire de Probabilités XXXVI (2003)
Vol. 1802: V. Capasso, E. Merzbach, B. G. Ivanoff, M. Dozzi, R. Dalang, T. Mountford, Topics in Spatial Stochastic Processes. Martina Franca, Italy 2001. Editor: E. Merzbach (2003)
Vol. 1803: G. Dolzmann, Variational Methods for Crystalline Microstructure – Analysis and Computation (2003)
Vol. 1804: I. Cherednik, Ya. Markov, R. Howe, G. Lusztig, Iwahori-Hecke Algebras and their Representation Theory. Martina Franca, Italy 1999. Editors: V. Baldoni, D. Barbasch (2003)
Vol. 1805: F. Cao, Geometric Curve Evolution and Image Processing (2003)
Vol. 1806: H. Broer, I. Hoveijn. G. Lunther, G. Vegter, Bifurcations in Hamiltonian Systems. Computing Singularities by Gröbner Bases (2003)
Vol. 1807: V. D. Milman, G. Schechtman (Eds.), Geometric Aspects of Functional Analysis. Israel Seminar 2000-2002 (2003)
Vol. 1808: W. Schindler, Measures with Symmetry Properties (2003)
Vol. 1809: O. Steinbach, Stability Estimates for Hybrid Coupled Domain Decomposition Methods (2003)
Vol. 1810: J. Wengenroth, Derived Functors in Functional Analysis (2003)
Vol. 1811: J. Stevens, Deformations of Singularities (2003)
Vol. 1812: L. Ambrosio, K. Deckelnick, G. Dziuk, M. Mimura, V. A. Solonnikov, H. M. Soner, Mathematical Aspects of Evolving Interfaces. Madeira, Funchal, Portugal 2000. Editors: P. Colli, J. F. Rodrigues (2003)
Vol. 1813: L. Ambrosio, L. A. Caffarelli, Y. Brenier, G. Buttazzo, C. Villani, Optimal Transportation and its Applications. Martina Franca, Italy 2001. Editors: L. A. Caffarelli, S. Salsa (2003)
Vol. 1814: P. Bank, F. Baudoin, H. Föllmer, L.C.G. Rogers, M. Soner, N. Touzi, Paris-Princeton Lectures on Mathematical Finance 2002 (2003)
Vol. 1815: A. M. Vershik (Ed.), Asymptotic Combinatorics with Applications to Mathematical Physics. St. Petersburg, Russia 2001 (2003)
Vol. 1816: S. Albeverio, W. Schachermayer, M. Talagrand, Lectures on Probability Theory and Statistics. Ecole d'Eté de Probabilités de Saint-Flour XXX-2000. Editor: P. Bernard (2003)
Vol. 1817: E. Koelink, W. Van Assche (Eds.), Orthogonal Polynomials and Special Functions. Leuven 2002 (2003)
Vol. 1818: M. Bildhauer, Convex Variational Problems with Linear, nearly Linear and/or Anisotropic Growth Conditions (2003)
Vol. 1819: D. Masser, Yu. V. Nesterenko, H. P. Schlickewei, W. M. Schmidt, M. Waldschmidt, Diophantine Approximation. Cetraro, Italy 2000. Editors: F. Amoroso, U. Zannier (2003)
Vol. 1820: F. Hiai, H. Kosaki, Means of Hilbert Space Operators (2003)
Vol. 1821: S. Teufel, Adiabatic Perturbation Theory in Quantum Dynamics (2003)
Vol. 1822: S.-N. Chow, R. Conti, R. Johnson, J. Mallet-Paret, R. Nussbaum, Dynamical Systems. Cetraro, Italy 2000. Editors: J. W. Macki, P. Zecca (2003)
Vol. 1823: A. M. Anile, W. Allegretto, C. Ringhofer, Mathematical Problems in Semiconductor Physics. Cetraro, Italy 1998. Editor: A. M. Anile (2003)
Vol. 1824: J. A. Navarro González, J. B. Sancho de Salas, $\mathscr{C}^\infty$ – Differentiable Spaces (2003)
Vol. 1825: J. H. Bramble, A. Cohen, W. Dahmen, Multiscale Problems and Methods in Numerical Simulations, Martina Franca, Italy 2001. Editor: C. Canuto (2003)
Vol. 1826: K. Dohmen, Improved Bonferroni Inequalities via Abstract Tubes. Inequalities and Identities of Inclusion-Exclusion Type. VIII, 113 p, 2003.
Vol. 1827: K. M. Pilgrim, Combinations of Complex Dynamical Systems. IX, 118 p, 2003.
Vol. 1828: D. J. Green, Gröbner Bases and the Computation of Group Cohomology. XII, 138 p, 2003.
Vol. 1829: E. Altman, B. Gaujal, A. Hordijk, Discrete-Event Control of Stochastic Networks: Multimodularity and Regularity. XIV, 313 p, 2003.
Vol. 1830: M. I. Gil', Operator Functions and Localization of Spectra. XIV, 256 p, 2003.
Vol. 1831: A. Connes, J. Cuntz, E. Guentner, N. Higson, J. E. Kaminker, Noncommutative Geometry, Martina Franca, Italy 2002. Editors: S. Doplicher, L. Longo (2004)
Vol. 1832: J. Azéma, M. Émery, M. Ledoux, M. Yor (Eds.), Séminaire de Probabilités XXXVII (2003)
Vol. 1833: D.-Q. Jiang, M. Qian, M.-P. Qian, Mathematical Theory of Nonequilibrium Steady States. On the Frontier of Probability and Dynamical Systems. IX, 280 p, 2004.
Vol. 1834: Yo. Yomdin, G. Comte, Tame Geometry with Application in Smooth Analysis. VIII, 186 p, 2004.
Vol. 1835: O.T. Izhboldin, B. Kahn, N.A. Karpenko, A. Vishik, Geometric Methods in the Algebraic Theory of Quadratic Forms. Summer School, Lens, 2000. Editor: J.-P. Tignol (2004)
Vol. 1836: C. Năstăsescu, F. Van Oystaeyen, Methods of Graded Rings. XIII, 304 p, 2004.
Vol. 1837: S. Tavaré, O. Zeitouni, Lectures on Probability Theory and Statistics. Ecole d'Eté de Probabilités de Saint-Flour XXXI-2001. Editor: J. Picard (2004)
Vol. 1838: A.J. Ganesh, N.W. O'Connell, D.J. Wischik, Big Queues. XII, 254 p, 2004.
Vol. 1839: R. Gohm, Noncommutative Stationary Processes. VIII, 170 p, 2004.
Vol. 1840: B. Tsirelson, W. Werner, Lectures on Probability Theory and Statistics. Ecole d'Eté de Probabilités de Saint-Flour XXXII-2002. Editor: J. Picard (2004)

Vol. 1841: W. Reichel, Uniqueness Theorems for Variational Problems by the Method of Transformation Groups (2004)
Vol. 1842: T. Johnsen, A. L. Knutsen, K3 Projective Models in Scrolls (2004)
Vol. 1843: B. Jefferies, Spectral Properties of Noncommuting Operators (2004)
Vol. 1844: K.F. Siburg, The Principle of Least Action in Geometry and Dynamics (2004)
Vol. 1845: Min Ho Lee, Mixed Automorphic Forms, Torus Bundles, and Jacobi Forms (2004)
Vol. 1846: H. Ammari, H. Kang, Reconstruction of Small Inhomogeneities from Boundary Measurements (2004)
Vol. 1847: T.R. Bielecki, T. Björk, M. Jeanblanc, M. Rutkowski, J.A. Scheinkman, W. Xiong, Paris-Princeton Lectures on Mathematical Finance 2003 (2004)
Vol. 1848: M. Abate, J. E. Fornaess, X. Huang, J. P. Rosay, A. Tumanov, Real Methods in Complex and CR Geometry, Martina Franca, Italy 2002. Editors: D. Zaitsev, G. Zampieri (2004)
Vol. 1849: Martin L. Brown, Heegner Modules and Elliptic Curves (2004)
Vol. 1850: V. D. Milman, G. Schechtman (Eds.), Geometric Aspects of Functional Analysis. Israel Seminar 2002-2003 (2004)
Vol. 1851: O. Catoni, Statistical Learning Theory and Stochastic Optimization (2004)
Vol. 1852: A.S. Kechris, B.D. Miller, Topics in Orbit Equivalence (2004)
Vol. 1853: Ch. Favre, M. Jonsson, The Valuative Tree (2004)
Vol. 1854: O. Saeki, Topology of Singular Fibers of Differential Maps (2004)
Vol. 1855: G. Da Prato, P.C. Kunstmann, I. Lasiecka, A. Lunardi, R. Schnaubelt, L. Weis, Functional Analytic Methods for Evolution Equations. Editors: M. Iannelli, R. Nagel, S. Piazzera (2004)
Vol. 1856: K. Back, T.R. Bielecki, C. Hipp, S. Peng, W. Schachermayer, Stochastic Methods in Finance, Bressanone/Brixen, Italy, 2003. Editors: M. Fritelli, W. Runggaldier (2004)
Vol. 1857: M. Émery, M. Ledoux, M. Yor (Eds.), Séminaire de Probabilités XXXVIII (2005)
Vol. 1858: A.S. Cherny, H.-J. Engelbert, Singular Stochastic Differential Equations (2005)
Vol. 1859: E. Letellier, Fourier Transforms of Invariant Functions on Finite Reductive Lie Algebras (2005)
Vol. 1860: A. Borisyuk, G.B. Ermentrout, A. Friedman, D. Terman, Tutorials in Mathematical Biosciences I. Mathematical Neurosciences (2005)
Vol. 1861: G. Benettin, J. Henrard, S. Kuksin, Hamiltonian Dynamics – Theory and Applications, Cetraro, Italy, 1999. Editor: A. Giorgilli (2005)
Vol. 1862: B. Helffer, F. Nier, Hypoelliptic Estimates and Spectral Theory for Fokker-Planck Operators and Witten Laplacians (2005)
Vol. 1863: H. Führ, Abstract Harmonic Analysis of Continuous Wavelet Transforms (2005)
Vol. 1864: K. Efstathiou, Metamorphoses of Hamiltonian Systems with Symmetries (2005)
Vol. 1865: D. Applebaum, B.V. R. Bhat, J. Kustermans, J. M. Lindsay, Quantum Independent Increment Processes I. From Classical Probability to Quantum Stochastic Calculus. Editors: M. Schürmann, U. Franz (2005)
Vol. 1866: O.E. Barndorff-Nielsen, U. Franz, R. Gohm, B. Kümmerer, S. Thorbjønsen, Quantum Independent Increment Processes II. Structure of Quantum Lévy Processes, Classical Probability, and Physics. Editors: M. Schürmann, U. Franz, (2005)
Vol. 1867: J. Sneyd (Ed.), Tutorials in Mathematical Biosciences II. Mathematical Modeling of Calcium Dynamics and Signal Transduction. (2005)
Vol. 1868: J. Jorgenson, S. Lang, $Pos_n(R)$ and Eisenstein Series. (2005)
Vol. 1869: A. Dembo, T. Funaki, Lectures on Probability Theory and Statistics. Ecole d'Eté de Probabilités de Saint-Flour XXXIII-2003. Editor: J. Picard (2005)
Vol. 1870: V.I. Gurariy, W. Lusky, Geometry of Müntz Spaces and Related Questions. (2005)
Vol. 1871: P. Constantin, G. Gallavotti, A.V. Kazhikhov, Y. Meyer, S. Ukai, Mathematical Foundation of Turbulent Viscous Flows, Martina Franca, Italy, 2003. Editors: M. Cannone, T. Miyakawa (2006)
Vol. 1872: A. Friedman (Ed.), Tutorials in Mathematical Biosciences III. Cell Cycle, Proliferation, and Cancer (2006)
Vol. 1873: R. Mansuy, M. Yor, Random Times and Enlargements of Filtrations in a Brownian Setting (2006)
Vol. 1874: M. Yor, M. Émery (Eds.), In Memoriam Paul-André Meyer - Séminaire de Probabilités XXXIX (2006)
Vol. 1875: J. Pitman, Combinatorial Stochastic Processes. Ecole d'Eté de Probabilités de Saint-Flour XXXII-2002. Editor: J. Picard (2006)
Vol. 1876: H. Herrlich, Axiom of Choice (2006)
Vol. 1877: J. Steuding, Value Distributions of L-Functions (2007)
Vol. 1878: R. Cerf, The Wulff Crystal in Ising and Percolation Models, Ecole d'Eté de Probabilités de Saint-Flour XXXIV-2004. Editor: Jean Picard (2006)
Vol. 1879: G. Slade, The Lace Expansion and its Applications, Ecole d'Eté de Probabilités de Saint-Flour XXXIV-2004. Editor: Jean Picard (2006)
Vol. 1880: S. Attal, A. Joye, C.-A. Pillet, Open Quantum Systems I, The Hamiltonian Approach (2006)
Vol. 1881: S. Attal, A. Joye, C.-A. Pillet, Open Quantum Systems II, The Markovian Approach (2006)
Vol. 1882: S. Attal, A. Joye, C.-A. Pillet, Open Quantum Systems III, Recent Developments (2006)
Vol. 1883: W. Van Assche, F. Marcellàn (Eds.), Orthogonal Polynomials and Special Functions, Computation and Application (2006)
Vol. 1884: N. Hayashi, E.I. Kaikina, P.I. Naumkin, I.A. Shishmarev, Asymptotics for Dissipative Nonlinear Equations (2006)
Vol. 1885: A. Telcs, The Art of Random Walks (2006)
Vol. 1886: S. Takamura, Splitting Deformations of Degenerations of Complex Curves (2006)
Vol. 1887: K. Habermann, L. Habermann, Introduction to Symplectic Dirac Operators (2006)
Vol. 1888: J. van der Hoeven, Transseries and Real Differential Algebra (2006)
Vol. 1889: G. Osipenko, Dynamical Systems, Graphs, and Algorithms (2006)
Vol. 1890: M. Bunge, J. Funk, Singular Coverings of Toposes (2006)
Vol. 1891: J.B. Friedlander, D.R. Heath-Brown, H. Iwaniec, J. Kaczorowski, Analytic Number Theory, Cetraro, Italy, 2002. Editors: A. Perelli, C. Viola (2006)
Vol. 1892: A. Baddeley, I. Bárány, R. Schneider, W. Weil, Stochastic Geometry, Martina Franca, Italy, 2004. Editor: W. Weil (2007)
Vol. 1893: H. Hanßmann, Local and Semi-Local Bifurcations in Hamiltonian Dynamical Systems, Results and Examples (2007)

Vol. 1894: C.W. Groetsch, Stable Approximate Evaluation of Unbounded Operators (2007)
Vol. 1895: L. Molnár, Selected Preserver Problems on Algebraic Structures of Linear Operators and on Function Spaces (2007)
Vol. 1896: P. Massart, Concentration Inequalities and Model Selection, Ecole d'Été de Probabilités de Saint-Flour XXXIII-2003. Editor: J. Picard (2007)
Vol. 1897: R. Doney, Fluctuation Theory for Lévy Processes, Ecole d'Été de Probabilités de Saint-Flour XXXV-2005. Editor: J. Picard (2007)
Vol. 1898: H.R. Beyer, Beyond Partial Differential Equations, On linear and Quasi-Linear Abstract Hyperbolic Evolution Equations (2007)
Vol. 1899: Séminaire de Probabilités XL. Editors: C. Donati-Martin, M. Émery, A. Rouault, C. Stricker (2007)
Vol. 1900: E. Bolthausen, A. Bovier (Eds.), Spin Glasses (2007)
Vol. 1901: O. Wittenberg, Intersections de deux quadriques et pinceaux de courbes de genre 1, Intersections of Two Quadrics and Pencils of Curves of Genus 1 (2007)
Vol. 1902: A. Isaev, Lectures on the Automorphism Groups of Kobayashi-Hyperbolic Manifolds (2007)
Vol. 1903: G. Kresin, V. Maz'ya, Sharp Real-Part Theorems (2007)
Vol. 1904: P. Giesl, Construction of Global Lyapunov Functions Using Radial Basis Functions (2007)
Vol. 1905: C. Prévôt, M. Röckner, A Concise Course on Stochastic Partial Differential Equations (2007)
Vol. 1906: T. Schuster, The Method of Approximate Inverse: Theory and Applications (2007)
Vol. 1907: M. Rasmussen, Attractivity and Bifurcation for Nonautonomous Dynamical Systems (2007)
Vol. 1908: T.J. Lyons, M. Caruana, T. Lévy, Differential Equations Driven by Rough Paths, Ecole d'Été de Probabilités de Saint-Flour XXXIV-2004 (2007)
Vol. 1909: H. Akiyoshi, M. Sakuma, M. Wada, Y. Yamashita, Punctured Torus Groups and 2-Bridge Knot Groups (I) (2007)
Vol. 1910: V.D. Milman, G. Schechtman (Eds.), Geometric Aspects of Functional Analysis. Israel Seminar 2004-2005 (2007)
Vol. 1911: A. Bressan, D. Serre, M. Williams, K. Zumbrun, Hyperbolic Systems of Balance Laws. Cetraro, Italy 2003. Editor: P. Marcati (2007)
Vol. 1912: V. Berinde, Iterative Approximation of Fixed Points (2007)
Vol. 1913: J.E. Marsden, G. Misiołek, J.-P. Ortega, M. Perlmutter, T.S. Ratiu, Hamiltonian Reduction by Stages (2007)
Vol. 1914: G. Kutyniok, Affine Density in Wavelet Analysis (2007)
Vol. 1915: T. Bıyıkoğlu, J. Leydold, P.F. Stadler, Laplacian Eigenvectors of Graphs. Perron-Frobenius and Faber-Krahn Type Theorems (2007)
Vol. 1916: C. Villani, F. Rezakhanlou, Entropy Methods for the Boltzmann Equation. Editors: F. Golse, S. Olla (2008)
Vol. 1917: I. Veselić, Existence and Regularity Properties of the Integrated Density of States of Random Schrödinger (2008)
Vol. 1918: B. Roberts, R. Schmidt, Local Newforms for GSp(4) (2007)
Vol. 1919: R.A. Carmona, I. Ekeland, A. Kohatsu-Higa, J.-M. Lasry, P.-L. Lions, H. Pham, E. Taflin, Paris-Princeton Lectures on Mathematical Finance 2004. Editors: R.A. Carmona, E. Çinlar, I. Ekeland, E. Jouini, J.A. Scheinkman, N. Touzi (2007)
Vol. 1920: S.N. Evans, Probability and Real Trees. Ecole d'Été de Probabilités de Saint-Flour XXXV-2005 (2008)
Vol. 1921: J.P. Tian, Evolution Algebras and their Applications (2008)
Vol. 1922: A. Friedman (Ed.), Tutorials in Mathematical BioSciences IV. Evolution and Ecology (2008)
Vol. 1923: J.P.N. Bishwal, Parameter Estimation in Stochastic Differential Equations (2008)
Vol. 1924: M. Wilson, Littlewood-Paley Theory and Exponential-Square Integrability (2008)
Vol. 1925: M. du Sautoy, L. Woodward, Zeta Functions of Groups and Rings (2008)
Vol. 1926: L. Barreira, V. Claudia, Stability of Nonautonomous Differential Equations (2008)
Vol. 1927: L. Ambrosio, L. Caffarelli, M.G. Crandall, L.C. Evans, N. Fusco, Calculus of Variations and Non-Linear Partial Differential Equations. Lectures given at the C.I.M.E. Summer School held in Cetraro, Italy, June 27–July 2, 2005. Editors: B. Dacorogna, P. Marcellini (2008)
Vol. 1928: J. Jonsson, Simplicial Complexes of Graphs (2008)
Vol. 1929: Y. Mishura, Stochastic Calculus for Fractional Brownian Motion and Related Processes (2008)
Vol. 1930: J.M. Urbano, The Method of Intrinsic Scaling. A Systematic Approach to Regularity for Degenerate and Singular PDEs (2008)
Vol. 1931: M. Cowling, E. Frenkel, M. Kashiwara, A. Valette, D.A. Vogan, Jr., N.R. Wallach, Representation Theory and Complex Analysis. Venice, Italy 2004. Editors: E.C. Tarabusi, A. D'Agnolo, M. Picardello (2008)
Vol. 1932: A.A. Agrachev, A.S. Morse, E.D. Sontag, H.J. Sussmann, V.I. Utkin, Nonlinear and Optimal Control Theory. Cetraro, Italy 2004. Editors: P. Nistri, G. Stefani (2008)

Recent Reprints and New Editions

Vol. 1702: J. Ma, J. Yong, Forward-Backward Stochastic Differential Equations and their Applications. 1999 – Corr. 3rd printing (2007)
Vol. 830: J.A. Green, Polynomial Representations of GL_n, with an Appendix on Schensted Correspondence and Littelmann Paths by K. Erdmann, J.A. Green and M. Schoker 1980 – 2nd corr. and augmented edition (2007)
Vol. 1693: S. Simons, From Hahn-Banach to Monotonicity (Minimax and Monotonicity 1998) – 2nd exp. edition (2008)
Vol. 470: R.E. Bowen, Equilibrium States and the Ergodic Theory of Anosov Diffeomorphisms. With a preface by D. Ruelle. Edited by J.-R. Chazottes. 1975 – 2nd rev. edition (2008)
Vol. 523: S.A. Albeverio, R.J. Høegh-Krohn, S. Mazzucchi, Mathematical Theory of Feynman Path Integral. 1976 – 2nd corr. and enlarged edition (2008)